AF353115

Prospects for Electric Mobility: Systemic, Economic and Environmental Issues

Prospects for Electric Mobility: Systemic, Economic and Environmental Issues

Editor

Amela Ajanovic

MDPI • Basel • Beijing • Wuhan • Barcelona • Belgrade • Manchester • Tokyo • Cluj • Tianjin

Editor
Amela Ajanovic
Energy Economics Group
TU WIEN
Vienna
Austria

Editorial Office
MDPI
St. Alban-Anlage 66
4052 Basel, Switzerland

This is a reprint of articles from the Special Issue published online in the open access journal *Energies* (ISSN 1996-1073) (available at: www.mdpi.com/journal/energies/special_issues/electric_mobility).

For citation purposes, cite each article independently as indicated on the article page online and as indicated below:

LastName, A.A.; LastName, B.B.; LastName, C.C. Article Title. *Journal Name* **Year**, *Volume Number*, Page Range.

ISBN 978-3-0365-1420-8 (Hbk)
ISBN 978-3-0365-1419-2 (PDF)

Contents

About the Editor . . . . **vii**

Preface to "Prospects for Electric Mobility: Systemic, Economic and Environmental Issues" . **ix**

Amela Ajanovic, Reinhard Haas and Manfred Schrödl
On the Historical Development and Future Prospects of Various Types of Electric Mobility
Reprinted from: *Energies* **2021**, *14*, 1070, doi:10.3390/en14041070 . . . . **1**

Gürkan Kumbaroğlu, Cansu Canaz, Jonathan Deason and Ekundayo Shittu
Profitable Decarbonization through E-Mobility
Reprinted from: *Energies* **2020**, *13*, 4042, doi:10.3390/en13164042 . . . . **27**

Marina Siebenhofer, Amela Ajanovic and Reinhard Haas
How Policies Affect the Dissemination of Electric Passenger Cars Worldwide
Reprinted from: *Energies* **2021**, *14*, 2093, doi:10.3390/en14082093 . . . . **51**

Amela Ajanovic, Marina Siebenhofer and Reinhard Haas
Electric Mobility in Cities: The Case of Vienna
Reprinted from: *Energies* **2021**, *14*, 217, doi:10.3390/en14010217 . . . . **75**

Felix Guthoff, Nikolai Klempp and Kai Hufendiek
Quantification of the Flexibility Potential through Smart Charging of Battery Electric Vehicles
and the Effects on the Future Electricity Supply System in Germany
Reprinted from: *Energies* **2021**, *14*, 2383, doi:10.3390/en14092383 . . . . **93**

Maximilian Schulz and Kai Hufendiek
Discussing the Actual Impact of Optimizing Cost and GHG Emission Minimal Charging of
Electric Vehicles in Distributed Energy Systems
Reprinted from: *Energies* **2021**, *14*, 786, doi:10.3390/en14030786 . . . . **113**

Lukáš Dvořáček, Martin Horák, Michaela Valentová and Jaroslav Knápek
Optimization of Electric Vehicle Charging Points Based on Efficient Use of Chargers and
Providing Private Charging Spaces
Reprinted from: *Energies* **2020**, *13*, 6750, doi:10.3390/en13246750 . . . . **133**

**Mona Kabus, Lars Nolting, Benedict J. Mortimer, Jan C. Koj, Wilhelm Kuckshinrichs, Rik W.
De Doncker and Aaron Praktiknjo**
Environmental Impacts of Charging Concepts for Battery Electric Vehicles: A Comparison of
On-Board and Off-Board Charging Systems Based on a Life Cycle Assessment
Reprinted from: *Energies* **2020**, *13*, 6508, doi:10.3390/en13246508 . . . . **161**

Carlo Corinaldesi, Georg Lettner, Daniel Schwabeneder, Amela Ajanovic and Hans Auer
Impact of Different Charging Strategies for Electric Vehicles in an Austrian Office Site
Reprinted from: *Energies* **2020**, *13*, 5858, doi:10.3390/en13225858 . . . . **193**

Jasmine Ramsebner, Albert Hiesl and Reinhard Haas
Efficient Load Management for BEV Charging Infrastructure in Multi-Apartment Buildings
Reprinted from: *Energies* **2020**, *13*, 5927, doi:10.3390/en13225927 . . . . **211**

Timo Kern, Patrick Dossow and Serafin von Roon
Integrating Bidirectionally Chargeable Electric Vehicles into the Electricity Markets
Reprinted from: *Energies* **2020**, *13*, 5812, doi:10.3390/en13215812 . . . . **235**

Enea Mele, Anastasios Natsis, Aphrodite Ktena, Christos Manasis and Nicholas Assimakis
Electromobility and Flexibility Management on a Non-Interconnected Island
Reprinted from: *Energies* **2021**, *14*, 1337, doi:10.3390/en14051337 . . . . **265**

Albert Hiesl, Jasmine Ramsebner and Reinhard Haas
Modelling Stochastic Electricity Demand of Electric Vehicles Based on Traffic Surveys—The
Case of Austria
Reprinted from: *Energies* **2021**, *14*, 1577, doi:10.3390/en14061577 . . . . **285**

Katarzyna Turoń, Andrzej Kubik and Feng Chen
Electric Shared Mobility Services during the Pandemic: Modeling Aspects of Transportation
Reprinted from: *Energies* **2021**, *14*, 2622, doi:10.3390/en14092622 . . . . **305**

Paula Brezovec and Nina Hampl
Electric Vehicles Ready for Breakthrough in MaaS? Consumer Adoption of E-Car Sharing and
E-Scooter Sharing as a Part of Mobility-as-a-Service (MaaS)
Reprinted from: *Energies* **2021**, *14*, 1088, doi:10.3390/en14041088 . . . . **325**

Christian Wankmüller, Maximilian Kunovjanek, Robert Gennaro Sposato and Gerald Reiner
Selecting E-Mobility Transport Solutions for Mountain Rescue Operations
Reprinted from: *Energies* **2020**, *13*, 6613, doi:10.3390/en13246613 . . . . **351**

Lukáš Janota, Tomáš Králík and Jaroslav Knápek
Second Life Batteries Used in Energy Storage for Frequency Containment Reserve Service
Reprinted from: *Energies* **2020**, *13*, 6396, doi:10.3390/en13236396 . . . . **371**

About the Editor

Amela Ajanovic

Amela AJANOVIC is an assistant professor at the Energy Economics Group at Vienna University of Technology. She holds a master's degree in electrical engineering and a PhD in energy economics at TU Wien. Her main research interests are renewable fuels and alternative automotive technologies, energy and transport policies, as well as the transition to a sustainable energy system and long-term energy scenarios. She has been involved in several national and international research projects related to these topics. She has authored more than 50 papers and journal articles and she is in the editorial boards of five academic journals.

Preface to "Prospects for Electric Mobility: Systemic, Economic and Environmental Issues"

The transport sector is experiencing huge pressure due to need for a significant reduction in greenhouse gas emissions and air pollution. Due to the intensified policy support, the electrification of road transport had progressed over the last decade. However, despite some good examples such as Norway, in most countries, mobility is still far away from sustainable low-carbon transformation. With the increasing number of electric vehicles over the last decade, especially in China, Europe and the USA, some new challenges and opportunities for the future have been noticed. The Special Issue "Prospects for Electric Mobility: Systemic, Economic and Environmental Issues"contributes to the better understanding of the current situation as well as of the future prospects and impediments related to the use of electric vehicles. The seventeen papers published in this Special Issue range from the review of historical developments of electricity use in different transport modes over the analysis of policy impacts on disseminations of electric vehicles, optimization and development of charging infrastructure, up to the future prospects of electro mobility. They analyse current challenges, possible solutions and new discussion points relevant to the sustainable transformation of the transport sector.

Amela Ajanovic
Editor

Review

On the Historical Development and Future Prospects of Various Types of Electric Mobility

Amela Ajanovic [1],*, Reinhard Haas [1] and Manfred Schrödl [2]

[1] Energy Economics Group, Vienna University of Technology (TU WIEN), 1040 Vienna, Austria; haas@eeg.tuwien.ac.at
[2] Institute of Energy Systems and Electrical Drives, Vienna University of Technology (TU WIEN), 1040 Vienna, Austria; manfred.schroedl@tuwien.ac.at
* Correspondence: ajanovic@eeg.tuuwien.ac.at; Tel.: +43-1-58801-370364

Abstract: Environmental problems such as air pollution and greenhouse gas emissions are caused by almost all transport modes. A potential solution to these problems could be electric mobility. Currently, efforts to increase the use of various types of electric vehicles are under way virtually worldwide. While in recent years a major focus was put on the electrification of passenger cars, electricity has already, for more than hundred years, been successfully used in some public transport modes such as tramways and metros. The core objective of this paper is to analyze the historical developments and the prospects of electric mobility in different transport modes and their potential contribution to the solution of the current environmental problems. With respect to the latter, we analyze the effect of the electricity generation mix on the environmental performance of electric vehicles. In addition, we document major policies implemented to promote various types of e-mobility. Our major conclusions are: (i) The policies implemented will have a major impact on the future development of electric mobility; (ii) The environmental benignity of electric vehicles depends to a large extent on the electricity generation mix.

Keywords: electric mobility; environment; policies; public transport; passenger cars

Citation: Ajanovic, A.; Haas, R.; Schrödl, M. On the Historical Development and Future Prospects of Various Types of Electric Mobility. *Energies* **2021**, *14*, 1070. https://doi.org/10.3390/en14041070

Academic Editor: Hugo Morais

Received: 12 January 2021
Accepted: 7 February 2021
Published: 18 February 2021

Publisher's Note: MDPI stays neutral with regard to jurisdictional claims in published maps and institutional affiliations.

1. Introduction

The transformation and electrification of the transport system has become a major strategy in the fight against environmental problems and climate change. In the EU, the transport sector is responsible for a quarter of total greenhouse gas (GHG) emissions. The largest amount of these emissions is caused by road transport, especially passenger cars. In opposition to all other sectors, in which GHG emissions have been decreasing over the last few decades, transport has had by far the worst dynamic performance (see Figure 1). Therefore, electrification of the transportation sector would significantly reduce local air pollution that causes respiratory illnesses and cancer. Moreover, depending on the electricity used, electrification of transport could significantly reduce the amount of greenhouse gases which contribute to global warming. A major challenge today is to identify the proper solutions for the transport sector which can support the transition towards an overall more sustainable energy system. One of the mostly discussed strategies is the "Avoid-Shift-Improve" (The "Avoid" refers to the need to avoid unnecessary travel and reduce trip distances, the "Shift" refers to the need to improve trip efficiency using more sustainable transport modes, and the "Improve" refers to the need to improve transport practices and technologies.) strategy. Although a major focus is currently placed on the electrification of passenger cars, in the future, with the increasing shift to public and shared mobility, it will be important to ensure a high electrification level of almost all transport modes.

1

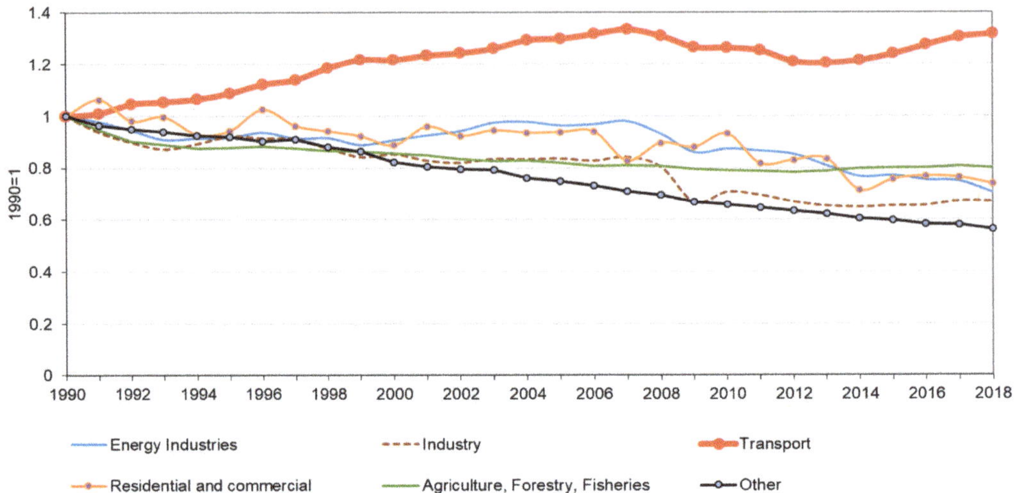

Figure 1. Development of GHG-emissions in the EU-27 in the different sectors (and transport modes) from 1990–2018 (1990 = 1) (Data source: [1]).

Currently, efforts to increase the use of electricity in the transport sector are underway worldwide. Since many transport modes, such as railways, tramways and metros, are already very well electrified, in the past few years a major focus has been placed on the electrification of road transport, especially passenger cars and buses. Figure 2 shows the example of Austria that the public transport (e.g., railways, underground, tramways and e-buses) contribute to more than 95% of the total kilometers driven with electric vehicles.

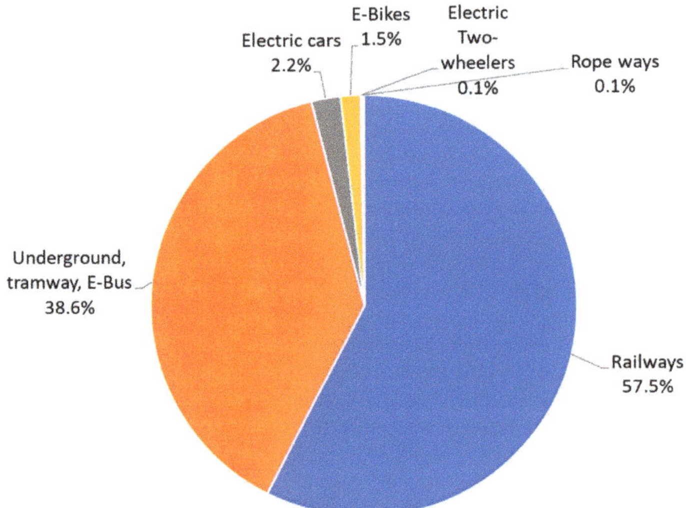

Figure 2. Shares of different modes and technologies of electric mobility by the number of kilometers driven in Austria in the year 2018 (Data source: [2]).

Although both private and public electric mobility have a long history, they are still seen as a major option for the reduction of the GHG-emissions in the future. Especially in urban areas electric mobility is considered as an important means to solve the local environmental problems. Moreover, the shift from private cars to public transport in combination with electrification is often seen as a major strategy for heading towards sustainable transport systems.

However, surprisingly, in the literature there are only very few contributions analyzing e-mobility in all dimensions and possible applications. Many papers have focused on battery electric vehicles, despite their current minor relevance in comparison to public transport (e.g., trains, railways, underground, etc.), whereas there is a very low number of studies dealing with the electrification of other transport modes.

The core objective of this paper is to document the historical development of electricity use in the transport sector, as well as to analyze the current situation with respect to electric mobility considering all relevant individual and public transport modes. Moreover, the impact of electrification on emission reduction is evaluated in view of different electricity generation portfolios.

To enable beneficial electrification, it is important that the costs of e-mobility are acceptable, that they are competitive with corresponding conventional transport modes, that e-mobility significantly contributes to emission reduction, and that, with increasing use of renewable energy sources (RES), electric vehicles can contribute to better grid management [3].

An example of how different electric transport modes are placed in the electricity system is shown in a stylized description for the case of Austria in Figure 3. The grid issue is of special interest for the electrification of mobility. On the transmission (TM) level, an almost autonomous parallel TM grid exists for Austrian railways (AR). They have their own power plants, mainly from hydro sources. Most other e-mobility applications, such as underground, buses and electric vehicles (EVs), are powered at different levels of the distribution grid. With respect to EVs charged privately, it is important to state that enforcements of network connections as well as of the distribution lines may be required for several applications.

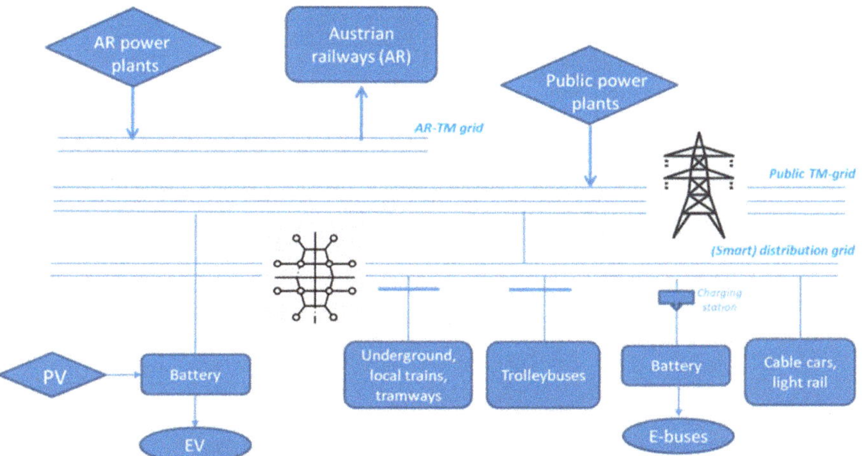

Figure 3. A stylized figure of how different modes and technologies of electric mobility are placed in the electricity system in the example of Austria.

This review paper builds on publicly available data sources and all used materials and data are cited in the corresponding sections. They are described with sufficient details to allow others to replicate and build on these results.

It builds on a comprehensive literature review, data collection and documentation by transport modes for private and public mobility, and the environmental assessments are built on methods presented in our previous publications [4,5].

This paper starts with a comprehensive documentation of the historical development of e-mobility in major private and public transport modes. In Section 3, the state of the art in view of e-mobility is presented, indicating the major current challenges. In the following section, a supporting policy framework is discussed. The environmental benefits of e-mobility in relation to the electricity generation mixes are analyzed in Section 5. The major conclusions are derived at the end of the paper.

2. History of E-Mobility

Although e-mobility had been very frequently discussed over the last decade, it is important to stress that electricity use for mobility has a long history. Electricity has been used successfully for years in different transport modes such as trams, underground, trolleybuses, and railways.

In this section, we will briefly document the history of electricity use in the transport sector and consider the major private and public transport modes.

2.1. Private Cars

The history of battery electric vehicles (BEVs) is very well documented in different papers and reports (e.g., [6–11]). The fact is that BEVs are not a new automotive technology. They have a long history of about 200 years. About 100 years ago they already played a significant role in passenger car mobility. The history of electric vehicles can be divided into four major segments: (i) 1830s, entering the markets; (ii) from about 1890 to 1920, increasing popularity of EVs; (iii) after 1920, their declining popularity; and (vi) starting in the 1970s, increasing attention [12].

As documented by Høyer (2015) [13], the early history of EVs started with the first tested lightweight electric vehicles constructed in the USA, the United Kingdom and the Netherlands in the 1830s. However, the "first golden age" of EVs was undoubtedly between 1880–1920 in the USA. In 1900, in competition with steam-powered vehicles and gasoline internal combustion engine (ICE) cars, the top-selling cars in the USA were BEVs [14]. Electric vehicles were dominant in the large and developed urban areas such as New York, Boston and Chicago. In these cities the ratio was two electric vehicles to one gasoline ICE vehicle [15].

At that time, there was no clear preference for one of the available automotive technologies, but each technology had some advantages and disadvantages [6]. For example, steam-powered vehicles were cheaper and faster but required a long time to start and frequent water filling stops. The gasoline cars were more dirty, difficult to start and slightly more expensive. However, they were able to manage long travel distances without interruption. The major advantage of BEVs was that they did not have the same disadvantages associated with gasoline cars such as noise, vibrations and smell [16], but they were slow and expensive [6]. Since cars were used mostly used in urban areas until the end of the 19th century, due to the low number of the good roads outside cities, the limited driving range of BEVs was not a problem. However, already at the beginning of the 20th century, the need for travel activity increased and car manufactures tried to find ways to make BEVs more suitable for travelling longer distances in the countryside. For example, (i) fast changeable batteries were developed to enable longer driving distances; (ii) regenerative braking systems were introduced, utilizing the capacity of the motor to act as a generator re-charging the battery in the case of driving downhill; (iii) the hybrid vehicle was invented [8,10,13,17]. Although hybrid electric vehicles were able to provide silent mobility with a longer driving range, they were never seriously considered before the early 1970s, mostly due to cost issues [8].

As shown in Figure 4, the first historical peak in EV's global stock was reached in 1912 with about 30,000 cars on the streets [14,18]. However, this peak was followed by a fast decline, which started in the 1920s, mostly due to the beginning of the mass-production of ICE vehicles [16,19], as well as the significant decrease in gasoline prices due to the discovery of new oil fields in Texas.

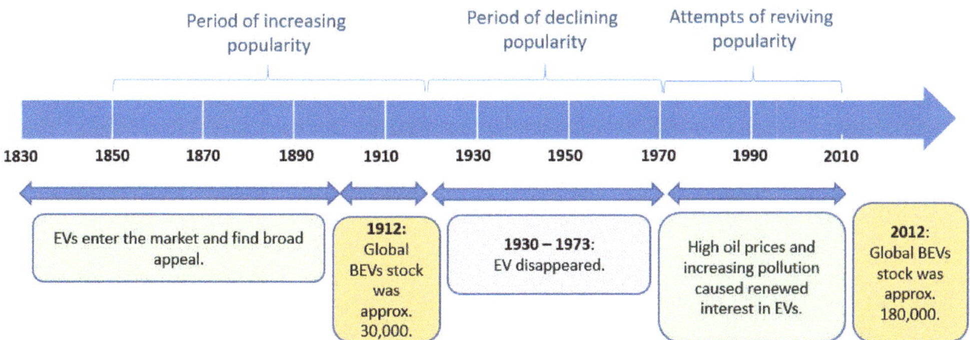

Figure 4. Major milestones in the history of battery electric vehicles.

Such developments, in combination with increasing electricity prices [13,16,19], were the major reasons why BEVs were no longer present in the car markets at the beginning of the 1930s. The renewed interest in EVs started with the first oil crisis in the 1970s, and was intensified with the increasing environmental problems caused by the use of fossil fuels in the transport sector. However, just at the beginning of the 21st century, due to significant technological improvements, electric cars looked more promising than ever before.

Yet, the first attempts to increase the sale of BEVs were not successful, mostly for four reasons: (i) high battery costs; (ii) rather short driving distance; (iii) limited infrastructure; and (iv) a long time to charge the batteries. The current take-up of BEVs is mainly due to supportive policies (see Section 4).

2.2. Public Transport

Although a major focus today is placed on the electrification of private cars, the use of electricity in public transport has a long and continuous history. In this section, the historical development of the major public transport modes based on electricity are presented.

2.2.1. Electric Railways

The earliest battery electric locomotive powered by galvanic cells was constructed in 1837, Scotland. A few years later a larger electric locomotive named Galvani was constructed, which was first tried out on the Edinburgh and Glasgow Railway in September 1842. However, its limited battery power prevented its general use. This first electric locomotive was demolished by railway workers, because it was seen as a threat to their job security [20,21]. The first practical AC electric locomotive was designed and demonstrated in 1891 [22].

The first electric train for passengers was invented by Werner von Siemens in Berlin in 1879. Currently, the oldest electric railway in the world, which is still in operation, is the Volk's Electric Railway, which opened in Brighton in 1883 [23].

The earliest electrification projects in Europe were focused on mountainous regions to enable easier electricity supply in the regions where hydro power was available. Electric locomotives have been a suitable option on steeper lines.

The first electric mainline railroads were built in the first two decades of the 20th century, in North America, Japan and many European countries. However, since these were often associated with high financial expenses, which was a challenge for the private rail companies, electrification progressed very slowly. Likewise, the accompanying construction of power plants to generate the required electricity was associated with high capital costs. These mostly isolated electrification efforts slowed abruptly or came to a complete stop with the beginning of World War I, especially in Europe. At the beginning of the 20th century, however, the railroad companies started replacing the steam locomotives on some lines due to the many advantages of electric traction.

Figure 5 shows a rough picture of the development of the electrification of railways in selected countries in the period from 1931 to 2019. It can be noticed that in 1930s the level of electrification was very low all over the world. Only Switzerland managed significant progress in this period, and had electrified almost 50% of its network already by 1930.

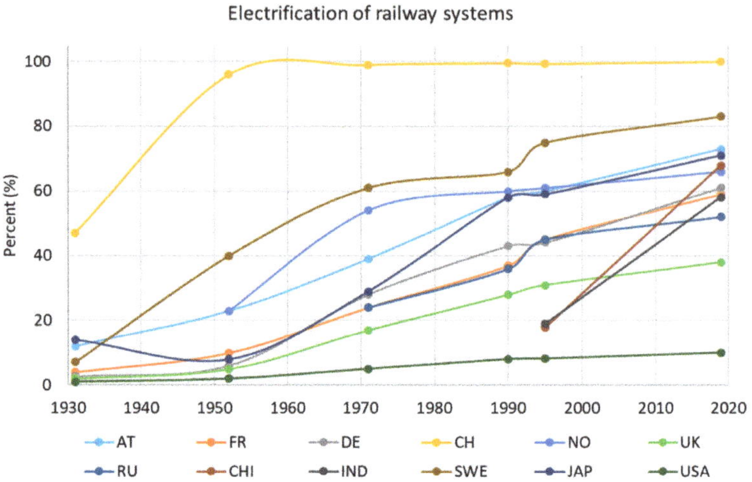

Figure 5. The electrification of railways in several countries, 1930 to 2019.

At the beginning of the Second World War, the expansion of electric railways slowed down again, and priority was given to the production of war materials. Full railway electrification did not take place on a large scale until after the war. As Figure 5 illustrates, countries such as Sweden or Norway, which were exempted from more intensive fighting and bombing, were able to significantly increase the electrification levels of their railway lines. Destruction of infrastructure and lack of capital during the first post-war years resulted in a slow growth and even a decrease of electric lines, especially in Germany and Japan. From the 1950s onwards, an intensive electrification of railway lines took place throughout Europe, as well as in Japan. This development can be clearly seen in Figure 5. However, the increasing degree of individual motorization of the population had a sometimes drastic effect on the railway operators. The consequence was the abandonment of several thousand unprofitable line kilometers. For example, the United Kingdom and France reduced their routes on a large scale. By 1990, the share of electric lines had increased further, partly due to the general conversion to electric operation.

While in Europe, a moderate but continuous growth in electrified lines over the last three decades took place, the growth of Chinese high-speed lines took on an enormous scale from 2008 onwards, mostly due to very high investment [24], and the rapid expansion of its high-speed rail network. In Japan, the electrification of the railways was almost completed before the turn of the millennium.

Of specific interest is the development in the USA. Due to almost private ownership, no costly networks were constructed and, hence, electrification remained very low. In contrast with Europe, in the U.S., railways are mainly used for freight transport. Hence, the corresponding emission of pollutants did not play an important role.

A specific category of electric railways is high-speed trains, with the speeds of at least 250 km/h. The development of high-speed train network lengths over the last three decades in the most important regions worldwide is illustrated in Figure 6.

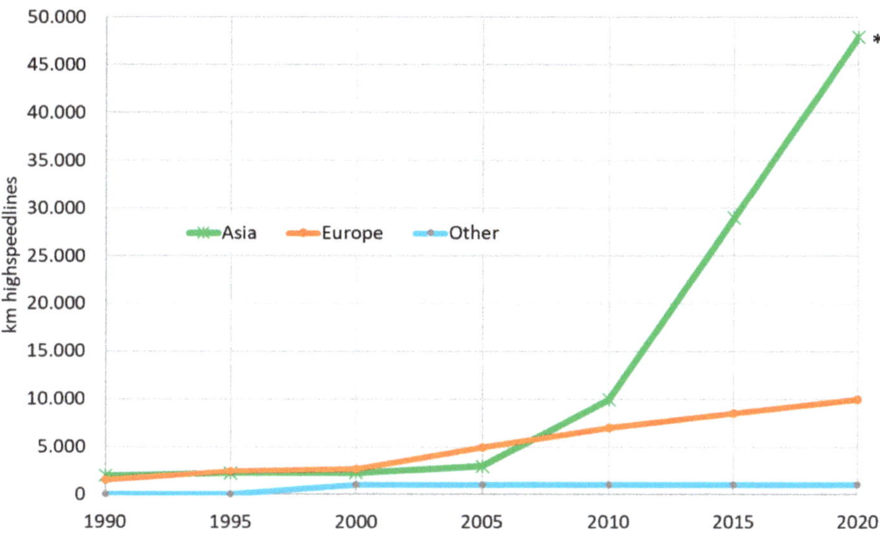

*) Numbers for 2020 preliminary assuming the same growth from 2018 to 2020 as between 2015 and 2018

Figure 6. The development of high-speed train network lengths in the period 1990 to 2020 (own analysis, [24]).

After the great success of the TGV in France, many countries, especially in Europe, tried to build their own High-Speed Rail (HSR) lines. In 1992 this was done in Spain and in 1997 in Belgium and Germany, and in the 2000s in Great Britain, Austria, South Korea, Taiwan, China, the Netherlands and other Western European countries. Turkey, Morocco and Saudi Arabia followed in 2017 [24].

2.2.2. Tram

Trams are transportation vehicles based on railways. They were originally derived from railway networks towards public urban passenger mobility services. Trams emerged in the first years of the 19th century in South Wales, UK, where a small part of Swansea and Mumbles Railway situated in urban areas was remodeled to be used for trams. However, that very first horse-driven model of the tram does not have many common features with modern tramways [25].

The first electric tram worldwide was invented in 1875 by Fyodor Pirotsky, and the first public electric tramway was put in service in St. Petersburg, but only in September 1880. The first public electric tramway operated in permanent service was opened in Lichterfelde, near Berlinin, in 1881. This was the first successful electric tram operated commercially, and it achieved a speed of 24 mile/h and transported 12,000 passengers in its first three months alone [26]. It initially drew its current from the rails, from an overhead wire. The basic principle of that first electric tramway system is still in operation today.

After the first successful demonstrations, electric tramways spread very quickly across European cities at the end of 19th century (see Table 1).

After the successful experiments and implementation of electric tramways in various European cities, they became a common transport mode all around the world.

Electric trams systems in Toronto, Canada were introduced in 1892. In the US, the first commercial installation of an electric tram was in 1884 in Cleveland, Ohio [34].

In Australia the first electric tramway was a Sprague system, firstly shown at the Melbourne Centennial Exhibition in 1888. It was followed by the commercial use of electric tram systems starting from 1889 in many Australian cities (e.g., Adelaide, Sydney, Brisbane, Hobart, Ballarat, Bendigo, Fremantle, Geelong, Kalgoorlie, Launceston, Leonora,

Newcastle, and Perth). However, by the 1970s, only Melbourne remained operating a tram system in Australia [35].

The electric railroad system in Kyoto was the first tram system in Japan, starting the transport of passengers in 1895 [36]. However, by the 1960s, the tram had generally died out in Japan [37,38].

Table 1. Some examples of the first electric tramways in Europe [27–33].

Year	Country	City
1887	Hungary	Budapest
1891	Czech Republic	Prague
1892	Ukraine	Kiev
1893	Germany	Dresden
	France	Lyon
	Italy	Milan
	Italy	Genoa
1894	Italy	Rome
	Sweden	Oslo
	Germany	Plauen
	Serbia	Belgrade
1895	United Kingdom	Bristol
	Bosnia and Herzegovina	Sarajevo
1896	Spain	Bilbao
1897	Denmark	Copenhagen
	Austria	Vienna
1898	Italy	Florence
	Italy	Turin
1899	Finland	Helsinki
	Spain	Madrid
	Spain	Barcelona

2.2.3. Trolleybus

The trolleybus is an electric vehicle powered by electricity using overhead electrification that first appeared at the beginning of the 20th century in Europe. However, its technological development can be traced back in the early 1880s.

In 1882 Siemens demonstrated the new concept of an EVs powered from a fixed source but was capable of being steered like other road vehicles. This vehicle, called an Electromote, was a light wagonette running without rails, and it was the first trolleybus [39]. The first trolleybus was developed in Germany and put in service in 1901 in Konigstein-Bad. The first commercial trolleybus operation in the US was in 1910 in Hollywood. Starting from 1911 the trolleybus was also in use in the United Kingdom [40]. In the early 1920s, there were efforts to establish trolleybus services in many cities. For example, Toronto used four trolleybuses for three years, starting in 1922. On New York's Staten Inlands, 23 vehicles were in operation. However, most of these buses were in operation for just a few years [40]. These early vehicles were not particularly elegant. In contrast to the tram, which was very fast established as the major urban public transport mode, the trolleybus remained in a primitive state until the mid-1920s. Just throughout the 1930s, with improved technology, trolleybuses gained attention for reliability, speed and comfort. For example, by 1939, there were 35 trolleybus systems in operation in London involving 3429 vehicles [39]. Furthermore, in the USA, trolleybuses were widely accepted as important city transportation services. For example, in 1927, 12 trolleybuses were running in New York and in 1929 26 vehicles were in service in Salt Lake City. The largest trolleybus system with 300 trolleybuses was installed in Seattle before the Second World War.

During the war, independence of imported fossil fuels was a huge advantage for electric vehicles, including tram and trolleybuses; however, in this period their infrastructure was gravely damaged in most European urban areas. After the war, due to the development of large diesel buses, as well as increasing growth of the urban population, more flexible diesel buses appeared to be a better option for mobility in many cities. For example, in 1954, London declared its plan to replace all its trolleybuses with diesel buses, and the last trolleybus system in the UK was closed in 1972. A similar pattern was also seen in other countries. However, especially in South-Eastern Europe, e.g., in Sarajevo, trolleybuses still provide a significant contribution to public transport.

Although trolleybuses have certain advantages, such as their quietness, their vibration-free operation, their long lifetimes and low maintenance requirements, as well as their absence of local pollution, they have been ignored in many countries which have discontinued trolleybus operation due to their disadvantages, such as their operational inflexibility and costs. The trolleybus and overhead lines were expensive and energy costs were also relatively high. The cost of operating trolleybuses was becoming significantly greater than that of motor buses in the same service. The availability of mass-produced diesel buses led to a gradual decline and, in some cases, total abandonment of trolleybuses. For example, in England all trolleybus systems were abandoned by 1972. In Germany, which at one time had 50 operating trolleybuses, only six remained by 1977 [40]. Mostly Eastern European countries continued with trolleybus operations.

However, just few years after trolleybuses were banned in most cities, oil crisis and increasing petrol prices made them attractive again for some countries. The renewed interest in trolleybuses started in the early 1970s as a result of changes in the costs, as well as increased concern for the environment, and this continues to the present. In Europe, interest in the electrification of mobility, including different demonstration programs, started at the beginning of 1970s, mostly due to supporting measures provided by governments and public authorities. At that time, commercial vehicles were considered to be more suitable for the early diffusion of electric vehicles.

Some countries re-equipped existing trolleybus systems (e.g., France, The Netherland, the USA), and some opened a new system (e.g., Belgium, South America). However, this revival of the trolleybus was moderate due to the expansion of light rapid transit and underground systems worldwide. Yet, trams, underground systems, light rapid transit and trolleybuses have a lot of joint characteristics. They are quiet, powered by electricity and are locally pollution-free.

The major milestones in the development of trolleybuses are depicted in Figure 7.

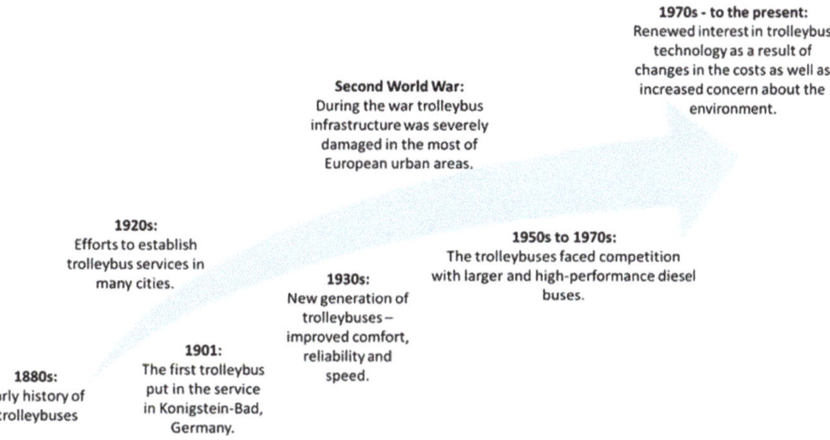

Figure 7. The major milestones in the development of trolleybuses.

2.2.4. Underground

The first underground railway system worldwide was constructed in London. It was originally opened in 1863 for steam-powered locomotive trains, and in 1890 it was the world's first electrified underground network [41,42]. Three years later an electric railway in Liverpool was opened. On the continent, Budapest opened the first electrified underground line in 1896. The first line of the Paris Metro opened in 1900. The Berlin "U-Bahn" opened in 1902. Since many sections of the line were elevated, it was also called "Hochbahn" (high railway). Germany's next U-Bahn was opened in Hamburg in 1904. In Vienna, and old two-line Metropolitan Railway, which was in operation since 1898, was transformed to a modern underground railway system in 1978.

New York City built its first rapid transit line in 1868, and the first section of 14.5 km of the New York subway opened in 1904 [42].

In 1913, in Buenos Aires, the first subway in the Southern Hemisphere opened as an underground tramway [41].

In Japan the first subway line opened in 1927 in Tokyo. During a time in which steam engine locomotives were the widest-used type of locomotives, the Tokyo metro started the development of the 1000 Series electric train for the use underground [43].

The construction of the Moscow metro started in 1931 and already in 1935, the first stations were opened to the citizens. This first metro line had a length of 11 km [44,45].

The first discussions about the Beijing metro system started in the early 1950s when the Chinese capital had about 5000 vehicles and a population of about three million. The then-premier, Zhou Enlai, said, "Beijing is building the subway purely for defense reasons. If it was for transport, purchasing 200 buses would solve the problem." [45,46]. Yet, Beijing's subway, which opened in 1971, is currently one of the most frequented in the world, transporting approximately 10 million passengers per day. In 2002 the system began its rapid expansion. However, the existing grid is still not able to adequately meet Beijing's mass transit demand.

3. E-Mobility: State of the Art

In the past, the use of electricity in the transport sector was mostly focused on public transport. However, increasing emissions, especially from the road transport, have changed the priorities over the last few decades. Currently, major effort has been put into the electrification of road transport, especially passenger cars. With the decrease in battery prices, as well as significant technical improvements, interest in the electrification of almost all transport modes is rapidly increasing.

3.1. Road Transport

Road transport is currently on the frontline of electrification, especially passenger cars and city buses.

3.1.1. Passenger Cars

Road transport, especially light duty vehicles, cause the largest amount of transport emissions. The electrification of road transport is seen as an essential strategy for meeting the European emission reduction goals. Electrification of transport combines the advantages of more energy-efficient automotive technologies with an increasing replacement of fossil fuels by renewable energy sources. However, electrification will not just change the powertrains used, but also the conditions of its use, leading to new user behavior and preferences.

Because of this complexity, the number of BEVs on the road is still very low but is continuously increasing, mostly due to the different kinds of supporting policy measures implemented as well as the increasing installation of public charging infrastructure.

In 2010, just about 17,000 electric cars were driven worldwide. However, there were just a few countries with more than 1000 electric vehicles: China, Japan, Norway, UK and USA. The majority of electric cars were used in the scope of different pilot and demonstration projects, and they were largely supported by governments through various incentive schemes and tax waivers. In 2010, the ratio of consumer spending on EVs and government spending was about 60% to 40%. In the meantime, acceptance of EVs has significantly increased, so that the ratio of consumer spending on EVs is more than 85%. The drop in government spending is mostly due to changes in incentive schemes in the US and China [47].

In spite of this, there were about 7.2 million electric cars in 2019 worldwide, and the leading countries/regions in the electrification of passenger vehicles were China, Europe and the United States. Almost half (47%) of the world's EVs stock was in China, 25% in Europe and 20% in the United States.

Figure 8 shows the development of the global stock of rechargeable electric vehicles. Battery electric vehicles accounted for the majority (67%) of the world's EVs in 2019.

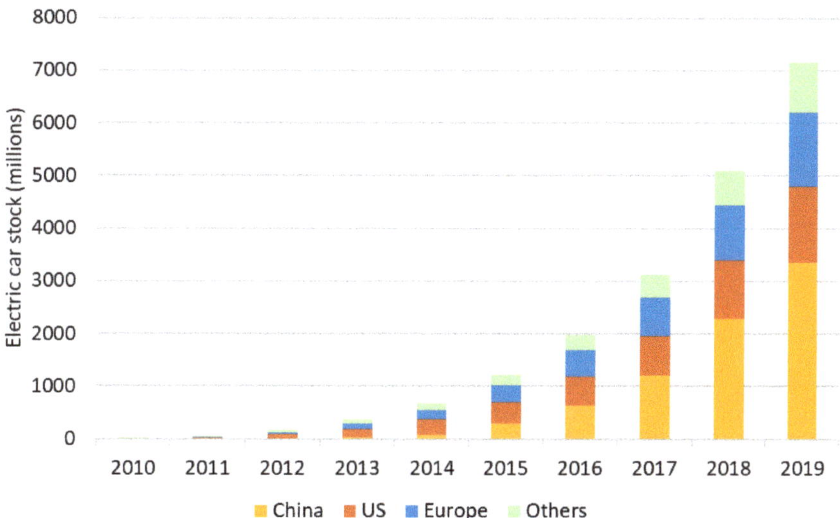

Figure 8. The global stock of rechargeable EVs, 2010–2019 (Data source: [47]).

However, based on the share of EVs in the total vehicle stock, Norway is the worldwide leader. For example, the electric car market share in Norway was 56% in 2019, followed by The Netherlands (15%) and Sweden (12%). In most other countries, the electric car market share is well below 5% [47].

Although 25% of the global stock of EVs is in Europe, the penetration of BEVs on the EU market is relatively slow. In spite of the low numbers and market shares (just about 2 % of new registered passenger cars), new BEVs registrations in the EU have been continuously rising in recent years [47].

3.1.2. Electro Micro-Mobility

Interest in electro micro-mobility (e.g., e-scouters, e-bikes) has been rapidly increasing since their emergence in 2017 [47]. They are especially of interest in urban areas and for shared mobility. Since in most of the countries/regions (e.g., China, the EU, USA) about a half of passenger-kilometers are short trips of under 8 km, there is huge potential for electro micro-mobility [48]. Such mobility can lower local air pollution and noise in cities. Yet, their full environmental impact is dependent on their total life-cycle emissions, which are determined mostly by the carbon intensity of electricity used, as well as embodied emissions. The impact of embodied emissions is very dependent on the lifetime of the electro micro-mobility. Moreover, there is an important question: which modes are replaced by micro-mobility, e.g., public transport or private cars. For example, in most cases, e-bikes are just replacing normal bicycles.

However, besides personal mobility, some of the electric micro-mobility vehicles can play an important role in last-mile delivery in urban areas. E-cargo bikes are already being used in several European cities for different delivery and courier services.

3.1.3. Two-Wheeled Electric Vehicles

Electrification of the two-wheeled vehicles has been intensified over the last few years, especially in Asia. The low weight and energetic need of two-wheelers make them suitable for electrification. They are of special interest for use in urban areas for short distances.

Currently, China is the leader in the two-wheeled electric vehicle market, with about 300 million units on the roads [47]. Major reasons for high deployment of two-wheeled vehicles in China are regulations and modest prices. For example, electric two-wheelers are exempted from registration taxes and are allowed to be used in bicycle lanes. Moreover, several cities have banned fossil fuel powered two-wheelers from downtown areas [49].

However, in other Asian countries, which use nearly 900 million two-wheeled vehicles, electrification is still marginal. Although about 20% of CO_2 emissions and 30% of particulate emissions in India are caused by motorized two-wheelers [50], India has just about 0.6 million electric two-wheelers [51]. The sales of electric two-wheelers increased from 54,800 units in 2018 to 126,000 units in 2019; however, this is still very low number compared to the total two-wheeler sale which reached a record of 21 million units in 2019 [47,52].

In Europe and the United States, electric two-wheelers are in competition with electric bikes, which do not require a driver's license or insurance [53]. Still, an increased use of shared electric two-wheelers can be noticed in Europe.

3.1.4. Electro Buses

Urban buses are the first road transport mode where electrification over the last few years is already having a significant impact [54].

The number of electric buses is growing all over the world. However, roughly 98% of electric buses are currently deployed in Chinese urban areas. Already in 2016, China registered on average 340 electric city buses per day. At the same time, in Europe about 70 new buses were put on the road, including all bus categories (e.g., urban, intercity, coaches) and all fuel types. Europe and United States, as well as other countries, still have a minor role to play in the adoption of electric buses. The new electric bus registrations in China and other countries/regions are shown in Figure 9.

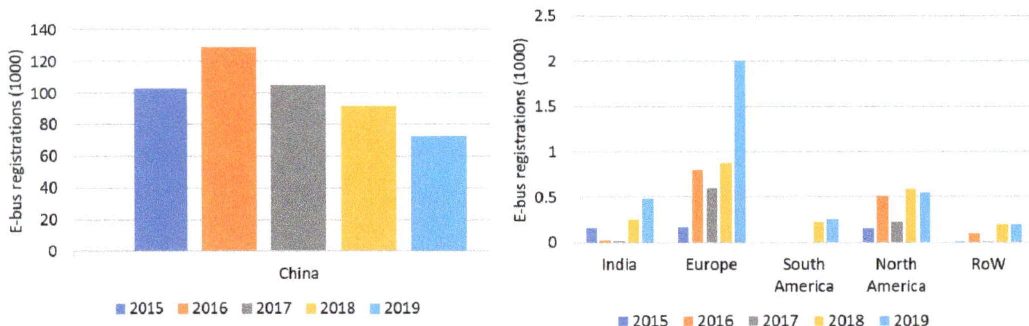

Figure 9. New e-bus registrations by country/region, 2015–2019 (Data source: [47]).

Although the number of E-bus registrations in China has been decreasing over the last few years, mostly due to the reduction of subsidies, the numbers in all other countries are negligible in comparison to China.

Worldwide, there were about 513,000 electric buses in 2019, see Figure 10. The majority of them, about 95%, were made and sold in China [47]. In China the share of e-buses in the total municipal bus fleet is about 18% [55]. The Chinese electrification plans for public transport are pretty ambitious. A good example is the city of Shenzhen in which all ICE buses have been replaced with electro buses (about 16,500 buses) [55].

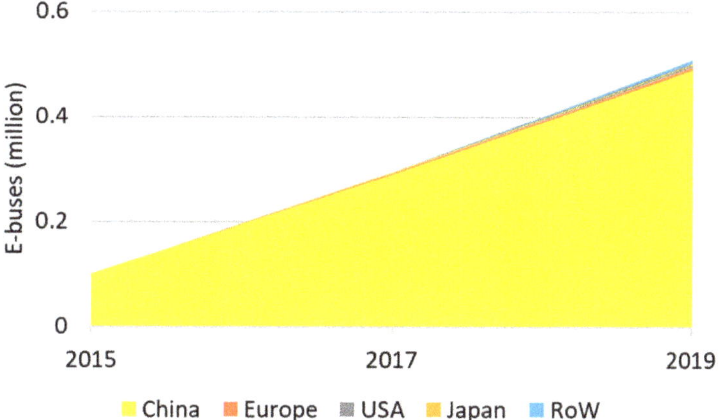

Figure 10. Development of the global stock of E-buses (Data source: [56]).

Over the last few years in Europe the number of e-buses has remarkably increased. In 2019, Europe registered 1900 electric buses [47]. The percentage of e-buses in overall city bus sales was about 10%. According to ACEA, 4% of new bus registrations in 2019 were e-buses [57]. The majority of electric buses in Europe is used in four countries: the UK (800), the Netherlands (800), France (600) and Germany (450) [47]. The Netherlands and the UK are leading European countries in electric bus adoption.

In 2019, the North America's electric bus fleet consisted of 2255 e-buses, of which about 500 were new registrations [58]. furthermore, India and South America are markets with big potential in bus electrification [47,59] (see Figure 9).

Despite the high efficiency of e-buses and their low maintenance costs, their high purchase price is a significant obstacle for faster market penetration. The share of the purchase costs in the total costs of ownership (TCO) of e-buses is about two times higher than in the case of diesel buses (see Figure 11).

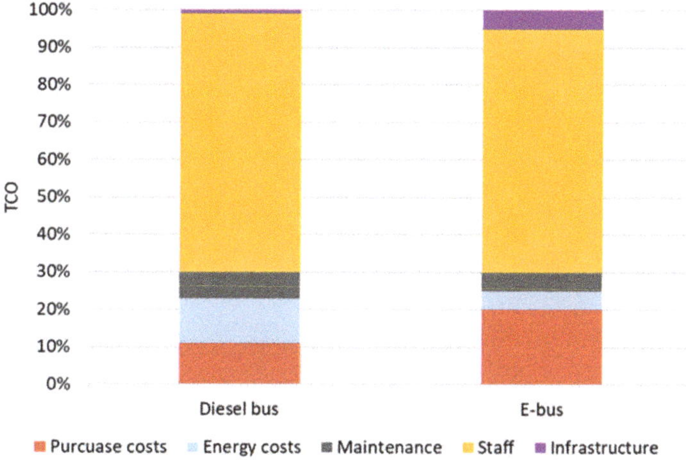

Figure 11. TCO for diesel and e-buses (Data source: [55], own analysis).

However, total energy consumption can increase by 50% due to the use of e-bus climate systems, leading to a substantial reduction in the driving range.

In any case, use of e-buses in urban areas can significantly reduce local air pollution, but full environmental benefits are very dependent on emissions related to electricity generation mix, as well as manufacturing and recycling of batteries.

3.1.5. Electric Trucks

Currently, electric trucks are mostly used in niche markets and in the scope of different pilot and demonstration projects. The battery pack with its very low volumetric energy density is the major impediment of a wider dissemination of electric trucks. The energy density limits driving range and load capacity. However, when driving on an uncongested highway, e-trucks can reach powertrain-to-wheel efficiencies of about 85%, while a conventional truck can achieve efficiencies of no higher than 30%. Currently, most e-trucks operate in urban areas in the scope of different municipality operations, such as delivery or garbage collection [60].

The deployment of e-vans and e-tracks is shown in Figure 12. It is obvious that China has predominance in this vehicle segment too.

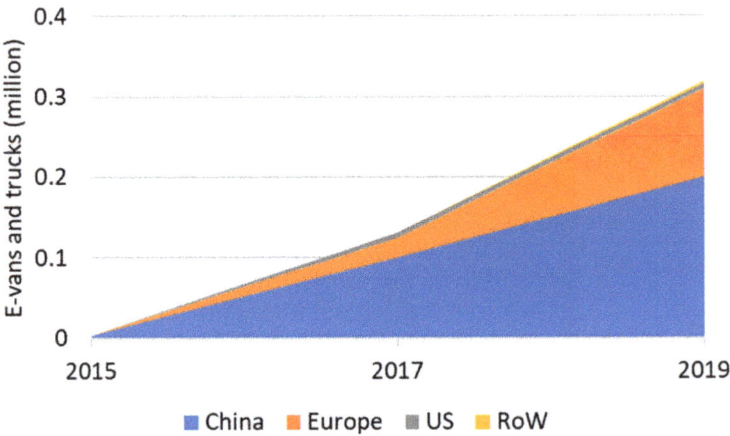

Figure 12. Development of electric vans and trucks, 2015–2019 (Data source: [56]).

In 2019 more than 6000 medium- and heavy-duty electric trucks were sold in China, mostly due to government subsidies but also due to improvements in battery performance, their cost reductions, as well as the increasing number of truck models. Currently, the most-used type of battery in trucks is lithium-ion chemistries [61]. However, in Europe and the USA, medium- and heavy-duty electric trucks are still largely used just in demonstration projects. Figure 13 shows the numbers of medium- and heavy-duty e-trucks sold worldwide in the period 2010–2019.

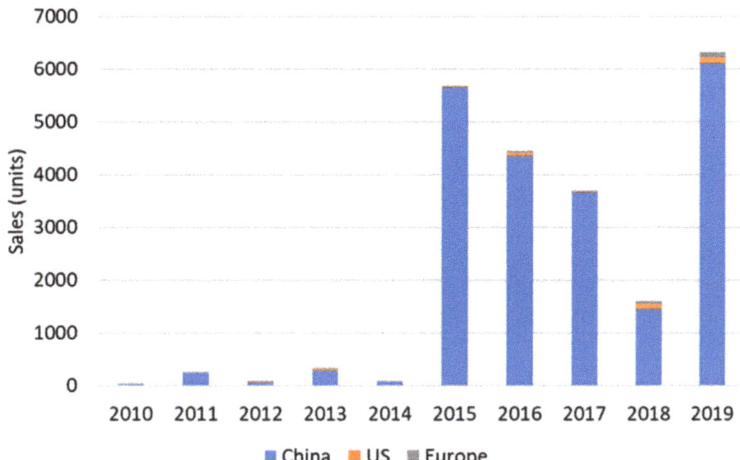

Figure 13. Global sales of e-trucks (Data source: [47]).

3.1.6. Ropeway Transport System

The ropeways have so far mainly gained attention in mountain areas for tourist purposes. The first ropeways were constructed about 100 years ago in the Alps, in Switzerland and Austria. However, in recent years they are becoming a novel option in urban public transport. Ropeways are spreading practically all over the world. Residents of Medellín in Colombia, for example, have been using cable cars to get to work since 2004. In the Turkish capital Ankara, the largest urban ropeway project on the Eurasian continent was realized in 2014. At a height of 60 m, residents of the suburbs use it to float into the city center. The largest ropeway grid in an urban area has been in the Bolivian capital La Paz since 2014. It stretches 33 km, linking the city center to the densely built-up poor neighborhoods. It comprises ten ropeway lines with hundreds of gondolas transporting around 300,000 passengers a day [62].

In regions with a hilly topography, electric ropeways could be a good alternative to buses and trains. Since electric ropeways are generally considered an environmentally benign technology with a small ecological footprint, their popularity is increasing worldwide, particularly in developing and emerging countries.

However, as with every other technology, they have some advantages and disadvantages. Besides their low environmental footprint, they have a high capacity of up to 5000 people per hour and direction. To achieve such transport performance on the road, double-articulated buses would have to run every two minutes. Ropeways are also technically mature systems and are statistically among the safest means of transport in the world [63]. Unlike buses, they also overcome steep gradients and do not get in the way of traffic lights or other vehicles, and thus avoid traffic jams. Among the biggest advantages are their low cost and short construction times, ideally only a few months. Moreover, ropeways have a high energy efficiency. According to surveys commissioned by the German Federal Environment Agency, ropeways consume 5.8 kWh per 100 passenger kilometers—only half of the already quite efficient underground, and they are clearly favorable in comparison to electric buses or individual BEVs (see Figure 14).

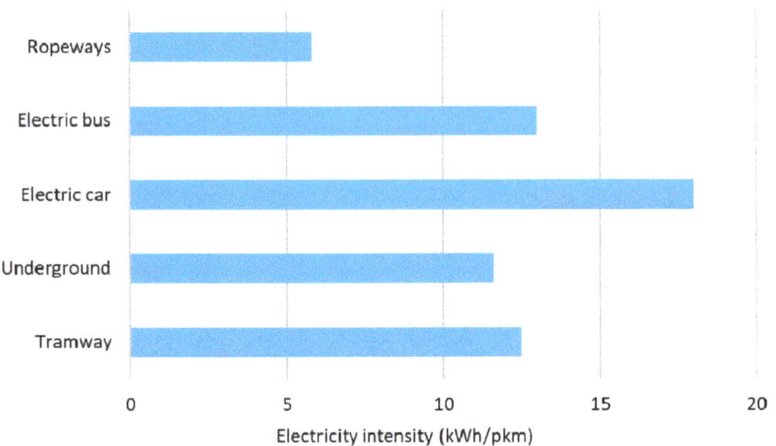

Figure 14. Electricity intensity of various modes of e-mobility in cities (Data source: [64], own analysis).

Nevertheless, there are some technical limits. The rope drive is only suitable for stretches of about five kilometers in length. With several sections, distances of any length could be covered, but then another disadvantage would come into play: ropeways can hardly manage more than 25 km per hour in circulating operation [65]. At higher speeds, boarding and alighting would no longer be manageable.

3.2. Other Transport Modes

Currently major focus is placed on the electrification of road mobility, especially passenger cars. However, there are many activities also in the electrification of other transport modes.

Although not very frequently discussed, rail transport is the major electrified transport mode today. In Europe, about 60% of the railway grid is already electrified and about 80% of rail transport is running on these lines. Furthermore, almost no technical obstacles for further electrification exist. However, further electrification of the rail transport is dependent on a cost–benefits ratio. For example, there is no interest in replacing diesel trains with electric ones on the low-density lines. Considering the costs for electrification of rail infrastructure and the expected emission savings, the best solution is the electrification of busy lines.

In contrast to the problem-free electrification of rail transport, electrification of shipping and aviation is still a big challenge.

After a number of pilot projects of small battery electric aircraft flights over rather short distances, the first commercial passenger aircraft flight of a full electrified airplane took place in December 2019; however, this was just for 15 min. The major problem for the electrification of aviation is the relatively low battery energy density. In spite of the fact that a growing number of aviation companies is developing small electric planes, mostly for test and demonstration purposes, the electrification of the aviation sector is still in its very early stages [47].

Electric ship propulsion actually has a long history, reaching back more than 100 years; however, this is in very limited numbers [66,67]. What are usually considered the first generation electric propulsion ships are those built in the 1920s, although there are also earlier examples of diesel-electric propulsion ships, e.g., the river tanker Vandal launched in 1903 [68]. An example from 1935 is the passenger liner "S/S Normandie" with 4×29 MW synchronous electric motors, one on each of the four propeller shafts. However, until the 1980s electric propulsion was not used very often, but with the rapid development of high-power semiconductor switching devices, interest in electric ship propulsion rose again.

Currently, the electrification of shipping is making progress, yet it is still very limited due to the required ranges and battery performances. At present electricity is used just in some ferries and short-distance vessels. Norway is a worldwide leader in electrification of mobility, with about 20 electric ferries in use. However, also other countries, such as China, Finland, Denmark, the Netherlands and Sweden are starting with the use of electric ships. For river navigation and short distance maritime transport, electrification could bring significant benefits in improvement of air quality. However, the major challenge is cost competitiveness.

4. Policies

As in many other sectors of an economy also in transport private initiatives do not always bring about the optimal solution for society, e.g., due to market distortions or neglection of environmental externalities, such as local air pollution or global GHG emissions. In such cases, local or federal governments usually interfere to correct for these failures and support the change in the "right" direction. The major instruments to do so are regulations, taxes, subsidies and standards [69–71]. In the following, we discuss the major policy interferences regarding the progress of electrification of public and private mobility.

In the beginning of the 20th century in most USA- and European cities it was full competitiveness in the private and public passenger transport [6,13]. No subsidies for any specific vehicle type (e.g., steam, electric, petrol, etc.) were provided, no specific taxes were charged and no standards were implemented.

After this first phase, private electrification efforts virtually died out. What followed were the initiatives for electrifying national railway systems by federal governments, as well as electrification of the public transport (e.g., trams, underground trains and trolleybuses) by the municipalities of cities [72]. However, these efforts were quite different from country to country and policy framework played an important role. One of the major issues in this context is the ownership structure. Indeed, as with the provision of infrastructure in the electricity system itself, but also with respect to public transport, the infrastructure for electricity, e.g., the overhead lines, which are a major cost factor, depended on the ownership structure. For example, in the USA, where the railway companies are privately owned, up to today only a very moderate electrification has taken place [73].

Over the last decade, electrification of mobility was driven by ambitious targets and policies set with the goal to reduce fossil fuel use in the transport sector, as well as to reduce local and global environmental problems [74–76]. Worldwide, governments have introduced a broad portfolio of policies which should enable a more sustainable development of the transport sector including all transport modes. Mostly, used policy instruments are national GHG reduction targets, fuel efficiency targets, CO_2 emissions standards, vehicle sale targets/mandated, and different kinds of supporting monetary measures (e.g., incentives, subsidies, tax exemptions or reduction, etc.) [69,70] Moreover, there is a broad portfolio of non-monetary measures such as free parking areas for EVs, possibility for EV drivers to drive on bus lanes, avoidance for EVs to drive in city centers as well as zero emission zones [77].

Besides these direct measures implemented to accelerate the uptake of EVs, diesel-emission scandals and announced bans of ICE vehicles are indirectly supporting the electrification of mobility. For example, it is announced that all new cars and vans sold in Norway should be zero-emission vehicles starting from 2025. The same goal has also been announced by other countries, starting, however, from 2030 or 2040 [47]. Moreover, European mandatory CO_2 emission target for the new passenger cars—95 gCO_2/km in 202—should initiate the production and dissemination of more green automotive powertrains with, in the ideal case, nearly zero emissions. Currently, very ambitious targets and more severe emission testing procedures are set for the years 2025 and 2030 [78]. For the achievement of these targets, the increasing use of more environmentally benign vehicles such as BEVs is essential.

There is a broad portfolio of the policy instruments used for the promotion and support of EVs. However, as discussed in the literature [79–81], measures and policies related to the purchase of EVs (e.g., tax exemptions, subsidies) are considered to be the financial instruments with the highest effectiveness, especially in countries with high registration tax rates for conventional vehicles, like Norway or the Netherlands. As discussed by Ajanovic et al. [79], consumers usually do not consider total costs of car ownership, so that benefits during the operation of the cars, such as lower or zero annual circulation taxes, often provide only a small price signal and finally, have less of an impact on ordering an electric vehicle [82].

Currently, the range of subsidies for the purchase of a BEV is between 4000 and 6000 EUR [47]. Since in most of countries these subsidies are very important and their use is increasing over time, China, as a leader in the electrification of road transport, is already in a position to reduce direct subsidies. The policy framework for EVs in China is in transition from direct to more indirect supporting measures, including the development of the charging infrastructure.

Although the USA has a long history in promotion of energy efficient vehicles, starting with the Corporate Average Fuel Economy Standards in the 1970s, they have only a 20% share in the global stock of the rechargeable EVs. With the proposed vehicle fuel-efficiency standards in 2020, the annual improvement in fuel-economy standards should be reduced from 4.7% to 1.5% for model years 2021 through to 2026. Moreover, it was decided to not extend the federal tax credit for the purchase of electric vehicles [47]. However, California is a leader in the adoption of ambitious policies for the promotion of zero-emission medium- and heavy-duty vehicles.

In the EU, many policy goals are focused on the reduction of GHG emissions and an increase of RES in electricity generation, as well as in the transport sector. Almost all policies implemented and targets set support the use of EVs, directly or indirectly, for example, EU CO_2 emissions regulations (no. 333/2014) [83], the European climate and energy package [75,76], the White Paper on Transport [84], the European Green Deal [85]. Besides policies and goals at the EU level, almost all EU countries have a wide range of supporting policies for the promotion of EVs implemented on the national and local level.

5. Environmental Issues of Electricity

One of the most important reasons to increase the use of e-mobility is to reduce the local and global emissions. However, whether this effect will be reached in practice and to what extent depends almost solely on the primary energy mix used in local or national power plants for electricity generation. The environmental benignity of electric vehicles in relation to the electricity mix used is comprehensively analyzed in the literature [86–89]. Nordelöf et al. (2014) [86] provide a review article investigating the usefulness of various types of lifecycle assessment (LCA) studies of electrified vehicles. Van Mierlo et al. (2017) [87] analyze the impact of the electricity production on the overall LCA performance of EVs and how the energy source mix for electricity generation influences the impact. Moro/Helmers (2017) [88] conduct an analysis on the advantages and drawbacks of the Well-to-Wheel (WTW) methodology when compared with the life cycle approach based on the EU electricity generation mix. In the recent publication, Ajanovic et al. (2021) [89] provide assessment of CO_2 emissions of various transport modes for the case study of Vienna.

In practice, a broad range of primary energy sources is available for electricity generation and there is a significant difference in electricity generation mixes across the countries. Figure 15 shows the mix of inputs for electricity production in selected countries/regions. While in China the electricity mix is dominated by coal, about 70%, in Japan and the US the share of coal is much lower, and the largest share has natural gas leading to a lower specific emission. In Europe, due to the continuously increasing use of RES, total share of fossil energy in the electricity generation mix is about 40%. However, there are significant differences across European countries. An exceptional country is Norway, with very high uses of RES, almost solely hydro, for electricity generation.

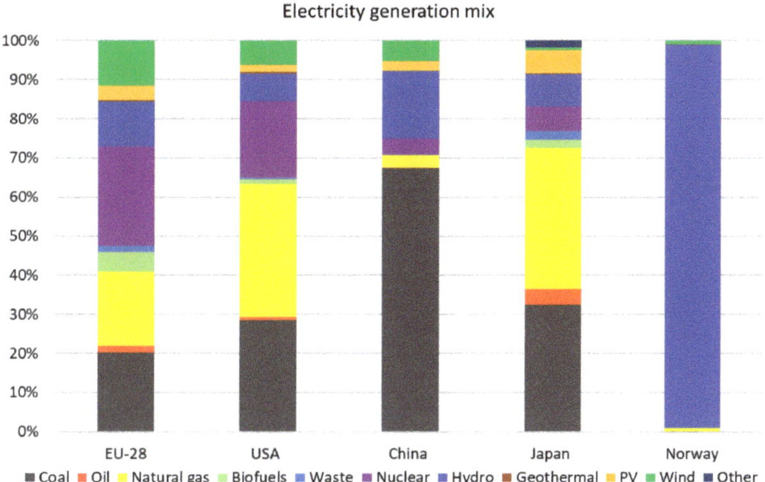

Figure 15. Electricity generation mix in selected countries 2018 (Data source: [90], own analysis).

The portfolio of energy inputs for electricity generation is changeable over time, as well as the corresponding carbon intensity of the electricity mix, see Figure 16. This figure depicts the development of specific CO_2 emissions of electricity generated in various countries from 2000 to 2018. As can be seen, the specific emissions have decreased in all regions shown since 2000 by almost the same percentage. The IEA expects almost continuous further decreases up to 2040 [91]. An exemption is Norway, where almost 100% of electricity is generated in hydro power plants and a further decrease of CO_2 emissions of electricity is virtually not possible.

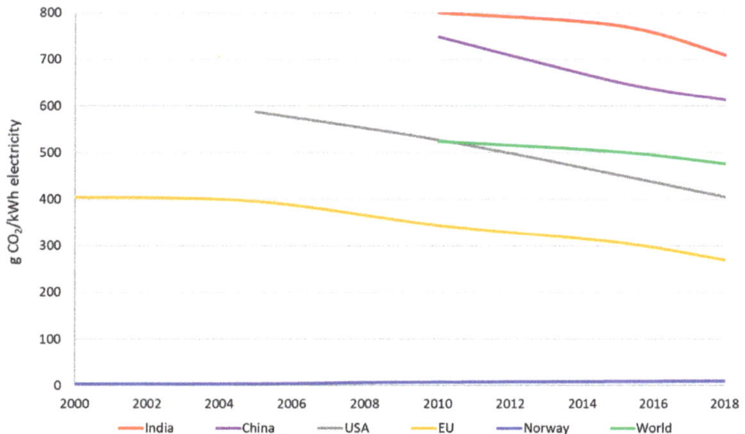

Figure 16. Development of specific CO_2 emissions of electricity generated in various countries and world-wide 2000–2018 [91,92].

A comparison of specific CO_2 emissions of electricity generated in various countries compared to gasoline and diesel is illustrated in Figure 17. This figure shows that related to kWhs, CO_2 emissions are higher for electricity.

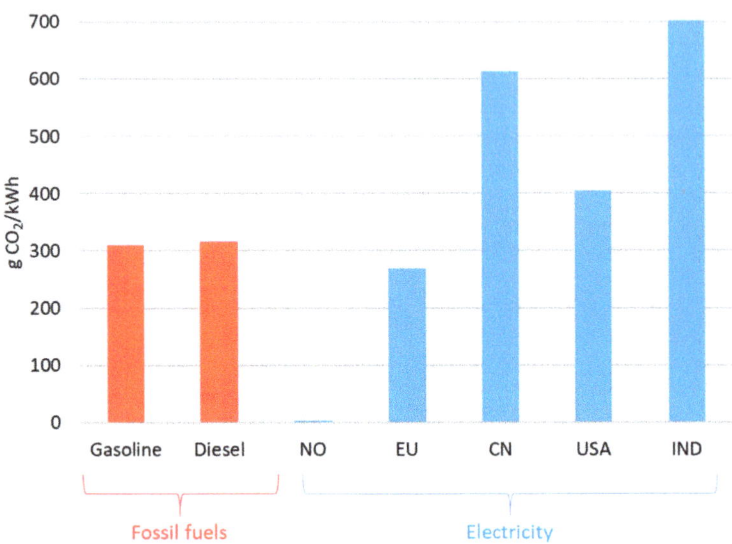

Figure 17. Specific CO$_2$ emissions of electricity generated in various countries compared to gasoline and diesel.

As can be seen, in some of the countries with a high use of coal in electricity generation (e.g., China and India), a specific carbon intensity of electricity mix is much higher than those of gasoline and diesel.

However, because of different efficiency of the end-use conversion systems, mainly ICEs vs. electric motors, the environmental comparison has to be conducted related to km driven. Since some of the transport modes, such as underground and tramway systems, are almost completely electrified worldwide, there is no need to compare them with corresponding fossil systems. Currently, the major focus is placed on emission reduction from the passenger car transport. In the following section we discuss the environmental benefits of electrification, taking BEV as an example.

As shown is Figure 18, whereas for 100% RES the CO$_2$ emissions per km driven are almost zero, they are around 170 gCO$_2$ per km driven for coal power plants, respectively about 90 gCO$_2$/km for pure natural gas plants. For different mixes, of course, the corresponding figures are in between. The figure shows the reductions compared to gasoline cars for a share of 40% RES in a natural gas-dominated country and for 60% RES in a coal-dominated one. This graph illustrates undoubtedly that CO$_2$ reduction is higher the bigger the share of renewables in the electricity production portfolio.

Using the same type of BEVs in different countries could lead to different total emissions, depending on the electricity used. The total CO$_2$ emissions of BEVs in various countries, compared to gasoline and diesel cars, are illustrated in Figure 19. These emissions are calculated for different countries/regions assuming an average driving range of 15,000 km driven per year. The total emissions are split up into Well-to-Tank (WTT), Tank-to-Wheel (TTW) and lifecycle car emissions, as depicted in Figure 19. The embedded emissions of car materials and manufacturing are included in the lifecycle car emissions.

As it can be seen in Figure 19, only in Norway does e-mobility really lead to a remarkable reduction in CO$_2$ emissions. In all other countries, CO$_2$ emissions from e-mobility are lower, e.g., in the EU for about 50%, in the USA for about 30%, but in India, total emissions of BEVs are almost at the same level as those of conventional cars.

However, with the increasing share of RES in electricity generation, which is set as a goal in many countries, the environmental benefits of electric vehicles will be continuously improving.

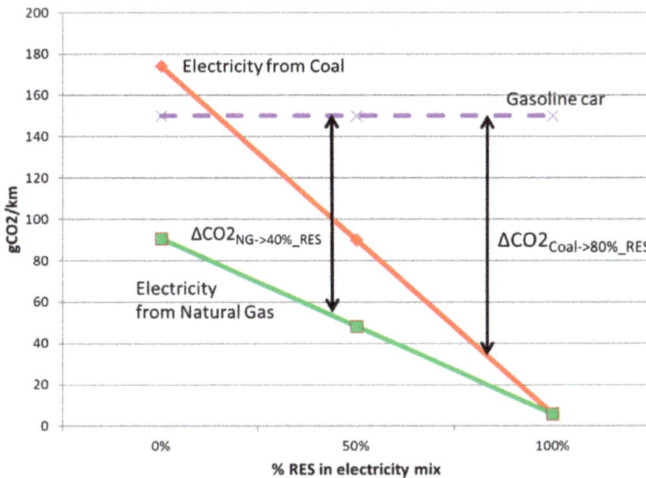

Figure 18. Specific WTW-CO_2-emissions of fuels (excl. LCA emissions of the vehicle) in dependence of the share of RES in the electricity generation portfolio (based on [93]).

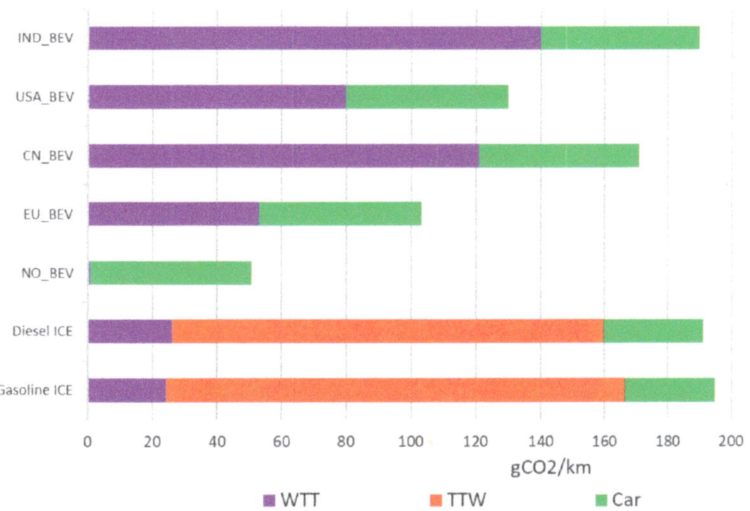

Figure 19. Total CO_2 emissions of BEVs in various countries compared to gasoline and diesel cars (based on the electricity generation mix of 2018).

6. Conclusions

Currently, there is a broad range of electricity use in the transport sector from individual private mobility such as electric cars, scooters and e-bikes, over different kinds of urban public mobility (e.g., underground, trolleybuses, cable cars, etc.) up to trucks and railways. Although the electrification of shipping and aviation is very limited, it is also progressing. That is to say, virtually every transport mode can be electrified.

However, a broader deployment of e-mobility in most applications will not be possible without more or less severe political interferences. The major policy measures currently used are different kinds of monetary and non-monetary incentives, which could have a direct as well as an indirect impact on the dissemination of e-mobility. For the faster deployment of e-mobility, it is important to have a combination of different instruments, such as

(i) subsidies and tax reliefs; (ii) sometimes even more important are indirect measures, such as diesel ban in cities or the introduction of emission-free zones; (iii) implementation of CO_2-taxes; (iv) tighter emission standards for the whole fleet; and (v) legislation with a "right to charge" in the garages of urban apartment buildings.

A specific challenge for the faster dissemination of e-mobility is further development of batteries and a reduction of their costs. Currently, BEVs are still more expensive than petrol cars. However, fuel costs are already lower and, due to technological learning, it is expected that by 2030 the overall costs per km driven will even out. The introduction of CO_2 taxes would accelerate this development.

The most important issue for public applications will be affordability for the public, e.g., the municipality of a city. Of course, as the past has shown, this is not a problem in the rich cities of the Western world. But it is a severe one in emerging and even more in developing countries, where underground transport can hardly be financed and hence, cheaper solutions such as light rail systems or ropeways will be the more proper solutions.

Along with all types of e-mobility goes the issue of infrastructure development. The construction of the necessary crucial infrastructure, such as overhead lines or other networks for electricity, as well as fast charging stations, is of very high relevance. However, the deployment of the infrastructure is depends on regulations and policy frameworks, which means the involvement of different stakeholders and policymakers.

Besides financial and policy issues, topography also has an impact on the deployment of different kinds of e-mobility. For example, in regions with a hilly topography, electric ropeways could be a better solution than e-buses or e-trains. Their major advantages are the lower energy demands per person per km and lower investment costs in comparison with the underground. In cities with a very high population density, a light rail system above the roads could be a good solution, e.g., in Bangkok. In any case economics from society's point of view will play a crucial and predominant role.

Finally, it has to be stressed once more that the major reason for promoting and implementing any type of e-mobility is to cope with the pressing environmental situation, local pollution, as well as global GHG emissions. In this context, for the environmental performance, it is of great relevance how the electricity used for e-mobility will be generated. Only if it is ensured by highly credible sources that the electricity is generated from RES, will e-mobility definitively contribute to a more environmentally benign and sustainable transport system. After all, the two absolutely crucial issues for the future deployment of all types of e-mobility are (i) political interferences; and (ii) electricity generation mix.

Author Contributions: Conceptualization, A.A. and R.H.; methodology, A.A.; validation, A.A., M.S. and R.H.; formal analysis, A.A., R.H.; investigation, A.A., R.H.; resources, A.A.; data curation, A.A.; writing—original draft preparation, A.A.; writing—review and editing, A.A., R.H., M.S.; visualization, A.A.; project administration, A.A.; funding acquisition, A.A. All authors have read and agreed to the published version of the manuscript.

Funding: The present work was funded by the Vienna Science and Technology Fund (WWTF) through the TransLoC project ESR17-067.

Institutional Review Board Statement: Not applicable.

Informed Consent Statement: Not applicable.

Data Availability Statement: Not applicable.

Conflicts of Interest: The authors declare no conflict of interest.

References

1. European Commission. Statistical Pocketbook 2020. EU Transport in Figures. Available online: https://ec.europa.eu/transport/facts-fundings/statistics/pocketbook-2020_en (accessed on 1 October 2020).
2. VCÖ. Available online: https://www.vcoe.at/publikationen/infografiken/oeffentlicher-verkehr (accessed on 15 November 2020).

3. Farnsworth, D.; Shipley, J.; Sliger, J.; Lazar, J. Beneficial Electrification of Transportation. Montpelier, VT: Regulatory Assistance Project. January 2019. Available online: https://www.raponline.org/wp-content/uploads/2019/01/rap-farnsworth-shipley-sliger-lazar-beneficial-electrification-transportation-2019-january-final.pdf (accessed on 14 December 2020).

4. Ajanovic, A.; Haas, R. On the Environmental Benignity of Electric Vehicles. *J. Sustain. Dev. Energy Water Environ. Syst.* **2019**, *7*, 416–431. [CrossRef]

5. Ajanovic, A.; Haas, R. Economic and Environmental Prospects for Battery Electric- and Fuel Cell Vehicles: A Review. *Fuel Cells* **2019**, *19*, 515–529. [CrossRef]

6. Ajanovic, A. The future of electric vehicles: Prospects and impediments. *WIREs Energy Environ.* **2015**, *4*, 521–536. [CrossRef]

7. Anderson, J.; Anderson, C.D. *Electric and Hybrid Cars: A History*; McFarland & Co.: London, UK, 2005.

8. Wakefeld, E.H. *History of the Electric Automobile: Battery-Only Powered Cars*; Society of Automotive Engineers Inc.: Warrendale, PA, USA, 1994.

9. Wakefeld, E.H. *History of the Electric Automobile: Hybrid Electric Vehicles*; Society of Automotive Engineers Inc.: Warrendale, PA, USA, 1998.

10. Chan, C.C. The state of the art of electric, hybrid, and fuel cell vehicles. *Proc. IEEE* **2007**, *95*, 704–718. [CrossRef]

11. Chan, C.C. The state of the art of electric and hybrid vehicles. *Proc. IEEE* **2002**, *90*, 247–275. [CrossRef]

12. PBS. Timeline: History of the Electric Car. Available online: https://www.pbs.org/now/shows/223/electric-car-timeline.html (accessed on 20 November 2020).

13. Høyer, K.G. The history of alternative fuels in transportation: The case of electric and hybrid cars. *Util. Policy* **2008**, *16*, 63–71.

14. Santini, D.J. Electric Vehicle Waves of History: Lessons Learned about Market Deployment of Electric Vehicles. 2011. Available online: https://www.intechopen.com/books/electric-vehicles-the-benefits-and-barriers/plug-in-electric-vehicles-a-century-later-historical-lessons-on-what-is-different-what-is-not- (accessed on 20 November 2020).

15. Reiner, R.; Cartalos, O.; Evrigenis, A.; Viljamaa, K. Challenger for a European Market for Electric Vehicles. IP/A/ITRE/NT/2010-004. June 2010. Available online: https://www.europarl.europa.eu/document/activities/cont/201106/20110629ATT22885/201 10629ATT22885EN.pdf (accessed on 14 December 2020).

16. Plug-in Hybrid Cars. Available online: http://eartheasy.com/move_plug-in_cars.html (accessed on 15 November 2020).

17. Propfe, B.; Kreyenberg, D.; Wind, J.; Schmid, S. Market penetration analysis of electric vehicles in the German passenger car market towards 2030. *Int. J. Hydrogen. Energy* **2013**, *38*, 5201–5208. [CrossRef]

18. The Forgotten Fleet: Looking Back on Early Electric Vehicles for a Better Future. Available online: https://newsroom.posco.com/en/looking-back-early-electric-vehicles/ (accessed on 20 November 2020).

19. Green Car Congress: Energy, Technologies and Policies for Sustainable Mobility, Electric (Battery). Available online: http://www.greencarcongress.com/electric_battery/index.html (accessed on 1 October 2020).

20. Day, L.; McNeil, I. *Biographical Dictionary of the History of Technology*; CRC Press: Boca Raton, FL, USA, 1998; ISBN 978-0-415-06042-4.

21. William, G. The Underground Electric. In *Our Home Railways*; Frederick Warne and Co.: London, UK, 1910.

22. Frey, S. *Railway Electrification: System & Engineering*; White Word Publications: Delhi, India, 2012; ISBN 978-81-323-4395-0.

23. Richmond Union Passenger Railway. IEEE History Center. Available online: https://ethw.org/Milestones:Richmond_Union_Passenger_Railway,_1888 (accessed on 20 November 2020).

24. Leboeuf, M. *High Speed Rail. Fast Track to Sustainable Mobility*; UIC Passenger Department: Paris, France, 2018; ISBN 978-2-7461-2700-5.

25. History and Different Types of Trams. Available online: http://www.trainhistory.net/railway-history/tram/ (accessed on 14 December 2020).

26. Highlights of Electrification. Available online: http://siemenselevator.org/history (accessed on 14 December 2020).

27. Guarnieri, M. Electric tramways of the 19th century. *IEEE Ind. Electron. Mag.* **2020**, *14*, 71–77. [CrossRef]

28. City of Sarajevo. Sarajevo through History. Available online: https://web.archive.org/web/20141023042858/http://www.sarajevo.ba/en/stream.php?kat=79 (accessed on 14 December 2020).

29. City of Belgrade. Important Years in City History. Available online: http://www.beograd.rs/index.php?lang=cir&kat=beoinfo&sub=201239%3f (accessed on 14 December 2020).

30. Trams of Hungary and Much More. Available online: http://hampage.hu/trams/e_index.html (accessed on 14 December 2020).

31. RATB —Regia Autonoma de Transport Bucureşti. Available online: https://web.archive.org/web/20150318064322/http://www.ratb.ro/index.php?page=meniu&id_rubrica_meniu=13 (accessed on 1 October 2020).

32. LPP. Historical Highlights. Available online: https://web.archive.org/web/20120304092909/http://www.jhl.si/en/lpp/?m=51&k=1605 (accessed on 1 October 2020).

33. Fasting, Kåre: Sporveier i Oslo gjennom 100 år. AS Oslo Sporveier, Oslo. 1975, pp. 49–50. Available online: https://en.wikipedia.org/wiki/Tram#cite_note-44 (accessed on 1 October 2020).

34. American Public Transportation Association. Milestones in U.S. Public Transportation History. Available online: https://web.archive.org/web/20090303212350/http://apta.com/research/stats/history/mileston.cfm (accessed on 1 October 2020).

35. ACT Light Rail. Available online: https://web.archive.org/web/20190402162522/https://www.actlightrail.info/p/routes-for-light-rail.html (accessed on 1 October 2020).

36. Kyoto City Official Website. Available online: https://www.city.kyoto.lg.jp/sogo/page/0000022084.html (accessed on 1 October 2020).

37. Freedman, A. *Tokyo in Transit: Japanese Culture on the Rails and Road*; Stanford University Press: Palo Alto, CA, USA, 2011.

38. JFS. The Rebirth of Trams: The Promise of Light Railway Transit (LRT). JFS Newsletter No.64. December 2007. Available online: https://www.japanfs.org/en/news/archives/news_id027840.html (accessed on 1 October 2020).
39. Brunton, L.A. The Trolleybus Story. Available online: https://ieeexplore.ieee.org/stamp/stamp.jsp?arnumber=141004 (accessed on 14 December 2020).
40. Vumbaco, B.J. Special Report 200: The Trolley Bus: Where it is and Where it's Going. 1982. Available online: http://onlinepubs.trb.org/Onlinepubs/sr/sr200/200.pdf (accessed on 1 December 2020).
41. Post, R.C. *Urban Mass Transit: The Life Story of a Technology*; Greenwood Technographies: Westport, CT, USA, 2006.
42. Railway Technology. World's Oldest Metro Systems. Available online: https://www.railway-technology.com/features/worlds-oldest-metro-systems/ (accessed on 14 December 2020).
43. JSA. An Overview of Japanese Subway Systems. Available online: http://www.jametro.or.jp/en/japan/ (accessed on 1 October 2020).
44. A Brief History of the Moscow Metro. Available online: https://theculturetrip.com/europe/russia/articles/a-brief-history-of-the-moscow-metro/ (accessed on 1 October 2020).
45. A Short History of World Metro Systems —In Pictures. Available online: https://www.theguardian.com/cities/gallery/2014/sep/10/-sp-history-metro-pictures-london-underground-new-york-beijing-seoul (accessed on 14 December 2020).
46. Li, S.; Liu, Y.; Purevjav, A.O.; Yang, L. Does subway expansion improve air quality? *J. Environ. Econ. Manag.* **2019**, *96*, 213–235. [CrossRef]
47. IEA. *Global EV Outlook 2020 Entering the Decade of Electric Drive?* International Energy Agency: Paris, France, 2020.
48. McKinsey & Company. Micromobility's 15,000-mile Checkup. 2019. Available online: www.mckinsey.com/industries/automotive-and-assembly/our-insights/micromobilitys-15000-mile-checkup (accessed on 14 December 2020).
49. Cherry, C. Electric Two-Wheelers in China: Promise, Progress and Potential. 2010. Available online: https://www.accessmagazine.org/wp-content/uploads/sites/7/2016/01/access37_electric_cycles_China.pdf (accessed on 14 December 2020).
50. Viswanathan, V.; Sripad, S. The Key to an Electric Scooter Revolution in India is Getting the Battery Right. 2019. Available online: https://qz.com/india/1737200/an-electric-scooter-revolution-in-india-needs-betterbatteries/ (accessed on 14 December 2020).
51. IEA. Global EV Outlook 2019: Scaling-up the Transition to Electric Mobility. 2019. Available online: https://webstore.iea.org/global-ev-outlook-2019 (accessed on 14 December 2020).
52. Statista. Two-Wheeler Sales in India from Financial year 2011 to 2020. Available online: https://www.statista.com/statistics/318023/two-wheeler-sales-in-india/ (accessed on 30 January 2021).
53. Ostiari, E. The Electrification of Transportation A Solution for the Ecological Transition. Mirova. December 2019. Available online: https://www.mirova.com/sites/default/files/2019-12/ElectrificationTransports2019.pdf (accessed on 14 December 2020).
54. Transport & Environment. Electric Buses Arrive on Time. November 2018. Available online: https://www.transportenvironment.org/sites/te/files/publications/Electric%20buses%20arrive%20on%20time.pdf (accessed on 15 December 2020).
55. Malkov, A.; Kilefors, P.; Ishchenko, R.; Lindstrom, L.; Ovanesov, A.; Guzman, R.; Song, N.; Arsenyeva, Y. Electric Buses. Viewpoint. Arthur D. Little Luxembourg S.A. 2020. Available online: https://www.adlittle.at/en/insights/viewpoints/electric-buses (accessed on 15 December 2020).
56. Bloomberg, N.E.F. Electric Vehicle Outlook. 2020. Available online: https://bnef.turtl.co/story/evo-2020/page/4/1 (accessed on 15 December 2020).
57. ACEA. Fuel Types of New Buses. Available online: https://www.acea.be/press-releases/article/fuel-types-of-new-buses-diesel-85-hybrid-4.8-electric-4-alternative-fuels-6 (accessed on 15 December 2020).
58. EV-Volumes, EV Data Center. Available online: www.ev-volumes.com/datacenter/ (accessed on 15 December 2020).
59. Yutong. 63 Yutong Dual-Source Trolleybuses Enter Mexico across the Ocean! 2019. Available online: https://en.yutong.com/pressmedia/yutong-news/2019/2019JNOhRqN4yq.html (accessed on 15 December 2020).
60. IEA. Trucks and Buses. 2020. Available online: https://www.iea.org/reports/trucks-and-buses (accessed on 1 January 2021).
61. IEA. Global EV Outlook 2017: Two Million and Counting. 2017. Available online: www.iea.org/publications/freepublications/publication/GlobalEVOutlook2017.pdf (accessed on 15 December 2020).
62. Wissen.de. Seilbahnen in Städten: Das Verkehrsmittel der Zukunft? 2020. Available online: https://www.wissen.de/seilbahnen-staedten-das-verkehrsmittel-der-zukunft (accessed on 15 December 2020).
63. Reichenbach, M.; Puhe, M. Flying high in urban ropeways? A socio-technical analysis of drivers and obstacles for urban ropeway systems in Germany. *Transp. Res. Part D Transp. Environ.* **2018**, *61*, 339–355. [CrossRef]
64. FIS. Daten und Fakten zum Energieverbrauch des Schienenverkehrs. Available online: https://www.forschungsinformationssystem.de/servlet/is/342234/ (accessed on 15 December 2020).
65. Hoffmann, K.; Liehl, R. *Cable-Drawn Urban Transport Systems*; WTT Press: Southampton, UK, 2005; Volume 77, ISSN 1743-3509.
66. ETO. History of Electric Propulsion Technology. Available online: https://electrotechnical-officer.com/history-of-electric-propulsion-technology/ (accessed on 17 December 2020).
67. Babb, C.E. The Curious History of Electric Ship Propulsion. 2015. Available online: https://futureforce.navylive.dodlive.mil/2015/08/electric-ship-propulsion/ (accessed on 17 December 2020).
68. Hubbell, D. The Vandal Was the First Diesel-Electric Vessel. 2019. Available online: https://www.waterwaysjournal.net/2019/11/19/the-vandal-was-the-first-diesel-electric-vessel/ (accessed on 17 December 2020).
69. Ajanovic, A.R. Haas: The impact of energy policies in scenarios on GHG emission reduction in passenger car mobility in the EU-15. *Renew. Sustain. Energy Rev.* **2017**, *68*, 1088–1096. [CrossRef]

70. Cansino, J.M.; Sánchez-Braza, A.; Sanz-Díaz, T. Policy Instruments to Promote Electro-Mobility in the EU28: A Comprehensive Review. *Sustainability* **2018**, *10*, 2507. [CrossRef]
71. Ajanovic, A.R.; Haas, F. Wirl: Reducing CO2 emissions of cars in the EU: Analyzing the underlying mechanisms of standards, registration taxes and fuel taxes. *Energy Effic.* **2016**, *9*, 925–937. [CrossRef]
72. Glotz-Richter, M.; Koch, H. Electrification of Public Transport in Cities (Horizon 2020 ELIPTIC Project). *Transp. Res. Procedia.* **2016**, *14*, 2614–2619. [CrossRef]
73. EESI. Electrification of U.S. Railways: Pie in the Sky, or Realistic Goal? Available online: https://www.eesi.org/articles/view/electrification-of-u.s.-railways-pie-in-the-sky-or-realistic-goal (accessed on 30 January 2021).
74. Regulation (EU) 2019/631 of the European Parliament and of the Council of 17 April 2019 Setting CO2 Emission Performance Standards for New Passenger Cars and for New Light Commercial Vehicles, and Repealing Regulations (EC) No 443/2009 and (EU) No 510/2011). Available online: https://eur-lex.europa.eu/legal-content/EN/TXT/PDF/?uri=CELEX:32019R0631&from=EN (accessed on 30 January 2021).
75. European Commission. 2020 Climate & Energy Package. 2018. Available online: https://ec.europa.eu/clima/policies/strategies/2020_en (accessed on 10 May 2020).
76. European Commission. 2030 Climate & Energy Framework. 2018. Available online: https://ec.europa.eu/clima/policies/strategies/2030_en (accessed on 10 May 2020).
77. Ajanovic, A.; Haas, R. Dissemination of electric vehicles in urban areas: Major factors for success. *Energy* **2016**, *115*, 1451–1458. [CrossRef]
78. European Commission. Proposal for a Regulation of the European Parliament and of the Council Setting Emission Performance Standards for New Passenger Cars and for New Light Commercial Vehicles as Part of the Union's Integrated Approach to Reduce CO2 Emissions from Light-Duty Vehicles and Amending Regulation (EC) No. 715/2007 (Recast). 2018. Available online: https://eurlex.europa.eu/legal-content/EN/TXT/?uri=CELEX:52017PC0676R(01) (accessed on 17 August 2020).
79. Ajanovic, A.; Haas, R. On the economics and the future prospects of battery electric vehicles. *Greenhouse Gas Sci.Technol.* **2020**, *10*, 1151–1164. [CrossRef]
80. Velten, E.K.; Stoll, T.; Meinecke, L. Measures for the Promotion of Electric Vehicles. Ecologic Institute, Berlin, Commissioned by Greenpeace e.V. 2019. Available online: https://www.ecologic.eu/sites/files/publication/2019/3564-foeerderung_von_e-autos-engl-langfassung.pdf (accessed on 17 December 2020).
81. Haugneland, P.; Lorentzen, E.; Bu, C.; Hauge, E. Put a Price on Carbon to Fund EV Incentives—Norwegian EV Policy Success. EVS30 Symposium, Stuttgart. 2017. Available online: https://elbil.no/wp-content/uploads/2016/08/EVS30-Norwegian-EV-policy-paper.pdf (accessed on 17 December 2020).
82. Runkel, M.; Mahler, A.; Ludewig, D.; Rückes, A.L. Loss of Revenues in Passenger Car Taxation due to Incorrect CO2 Values in 11 EU States. 2018. Available online: http://www.foes.de/pdf/2018-03-10_FOES_Taxation_loss_due_incorrect_CO2_values.pdf (accessed on 10 May 2020).
83. EUR-Lex, Regulation (EU) No. 333/2014 of the European Parliament and of the Council of 11 March 2014 amending Regulation (EC) No. 443/2009 to Define the Modalities for Reaching the 2020 Target to Reduce CO2 Emissions from New Passenger Car. 2014. Available online: https://eur-lex.europa.eu/legal-content/EN/TXT/PDF/?uri=CELEX:32014R0333&from=EN (accessed on 10 May 2020).
84. European Commission. Roadmap to a Single European Transport Area: Towards a Competitive and Resource Efficient Transport System. 2011. Available online: https://doi.org/10.2832/30955 (accessed on 10 May 2020).
85. European Commission, Communication from the Commission to the European Parliament, the European Council, the Council, the European Economic and Social Committee and the Committee of the Regions. The European Green Deal, Brussels, 11.12.2019 COM (2019) 640 Final. Available online: https://eur-lex.europa.eu/resource.html?uri=cellar:b828d165-1c22-11ea-8c1f-01aa75ed71a1.0002.02/DOC_1&format=PDF (accessed on 10 May 2020).
86. Nordelöf, A.; Messagie, M.; Tillman, A.M.; Söderman, M.L.; Van Mierlo, J. Environmental impacts of hybrid, plug-in hybrid, and battery electric vehicles—what can we learn from life cycle assessment? *Int. J. Life Cycle Assess* **2014**, *19*, 1866–1890. [CrossRef]
87. Van Mierlo, J.; Messagie, M.; Rangaraju, S. Comparative environmental assessment of alternative fueled vehicles using a life cycle assessment. *Transp. Res. Procedia* **2017**, *25C*, 3439–3449. [CrossRef]
88. Alberto, M.; Helmers, E. A new hybrid method for reducing the gap between WTW and LCA in the carbon footprint assessment of electric vehicles. *Int. J. Life Cycle Assess* **2017**, *22*, 4–14.
89. Ajanovic, A.; Siebenhofer, M.; Haas, R. Electric Mobility in Cities: The Case of Vienna. *Energies* **2021**, *14*, 217. [CrossRef]
90. IEA. Data and Statistics. Available online: https://www.iea.org/data-and-statistics?country=JAPAN&fuel=Energy%20supply&indicator=ElecGenByFuel (accessed on 15 December 2020).
91. IEA. Carbon Intensity of Electricity Generation in Selected Regions in the Sustainable Development Scenario, 2000–2040. Available online: https://www.iea.org/data-and-statistics/charts/carbon-intensity-of-electricity-generation-in-selected-regions-in-the-sustainable-development-scenario-2000-2040 (accessed on 15 December 2020).
92. European Environmental Agency. Available online: https://www.eea.europa.eu/data-and-maps/ (accessed on 10 May 2020).
93. Ajanovic, A.; Haas, R. Driving with the sun: Why environmentally benign electric vehicles must plug in at renewables. *Solar Energy* **2015**, *121*, 169–180. [CrossRef]

Article

Profitable Decarbonization through E-Mobility

Gürkan Kumbaroğlu [1],*[ID]**, Cansu Canaz** [1]**, Jonathan Deason** [2] **and Ekundayo Shittu** [2][ID]

[1] Department of Industrial Engineering, Boğaziçi University, 34342 Istanbul, Turkey; cansu.canaz@gmail.com
[2] Department of Engineering Management and Systems Engineering, George Washington University, Washington, DC 20052, USA; jdeason@gwu.edu (J.D.); eshittu@gwu.edu (E.S.)
* Correspondence: gurkank@boun.edu.tr

Received: 13 May 2020; Accepted: 23 July 2020; Published: 5 August 2020

Abstract: This paper focuses on the interdependent relationship of power generation, transportation and CO_2 emissions to evaluate the impact of electric vehicle deployment on power generation and CO_2 emissions. The value of this evaluation is in the employment of a large-scale, bottom-up, national energy modeling system that encompasses the complex relationships of producing, transforming, transmitting and supplying energy to meet the useful demand characteristics with great technological detail. One of such models employed in this analysis is the BUEMS model. The BUEMS model provides evidence of win-win policy options that lead to profitable decarbonization using Turkey's data in BUEMS. Specifically, the result shows that a ban on diesel fueled vehicles reduces lifetime emissions as well as lifetime costs. Furthermore, model results highlight the cost-effective emission reduction potential of e-buses in urban transportation. More insights from the results indicate that the marginal cost of emission reduction through e-bus transportation is much lower than that through other policy measures such as carbon taxation in transport. This paper highlights the crucial role the electricity sector plays in the sustainability of e-mobility and the value of related policy prescriptions.

Keywords: electricity; transportation; e-mobility; CO_2 emissions; decarbonization; marginal cost; sustainability

---

1. Introduction

Globally, the energy sector is the main contributor to CO_2 emissions followed by transportation. For example, the International Energy Agency [1] indicates that global CO_2 emissions from fuel combustion in 2017 were 32.8 Gt, growing from 32.3 Gt in 2016. The report highlights that electricity/heat generation and transport account for two-thirds of total CO_2 emissions and were equally responsible for almost the entire global growth in emissions since 2010. Emissions from the transport sector account for approximately a quarter of energy-related CO_2 emissions worldwide, while road vehicles account for nearly three-quarters of transport CO_2 emissions. In the Fifth Assessment Report (AR5), the Intergovernmental Panel on Climate Change (IPCC) highlights that emissions from the transport sector have been increasing at a faster rate than any other energy end-use sector [2,3]. Emissions from transportation are on the rise, essentially due to greater overall volume of travel that outweighs improvements in the energy efficiency of vehicles.

Advancements in e-mobility technology with significant reductions in battery prices and increase in efficiencies expedite the diffusion prospects of plug-in electric vehicles. Similar advancements in hydrogen vehicles and vehicle to grid (V2G) technology brighten the future of e-mobility as a sustainable solution for the transport sector with zero emissions on the road. However, a systemic analysis is needed for understanding the overall implications on energy demand and CO_2 emissions of electrification in transport. In particular, it becomes important how the additional electricity required is supplied. Switching from internal combustion vehicles to electric vehicles may not be sustainable if the increased electricity demand is generated in coal-fired power plants as the net effect on CO_2 emissions would be a rise. While it is indeed true that plug-in electric vehicles do not directly emit greenhouse gas emissions, there are indirect emissions due to the electricity generated to recharge the vehicle batteries. The optimal configuration of the sources of electricity often result from the dynamics of the competition [4], state of technological progress [5] or responses to uncertainties in regulations [6] or reliability [7] of the technology portfolio. Some best practices for future grid emissions indicate that plug-in electric vehicles can reduce transportation emissions [8].

The composition of CO_2 emission sources in Turkey resembles the global picture. The energy sector is responsible for 41% of total CO_2 emissions while the transport sector accounts for 23% according to the latest National Inventory Submission [9]. Emissions from oil products, mainly used in the transport sector, account for 27%. Road transportation is responsible for 93% of CO_2 emissions from the transport sector, containing 90% of all passenger transport and 89% of all freight transport. Considering newly registered vehicles, the share of diesel cars has strongly increased to 63% over the past decade. Road transportation contains 90% of all passenger transport and 89% of all freight; it dominates Turkey's transport sector and is responsible for 93% of CO_2 emissions from the transport sector. While rail transportation accounts for 4.4% of freight and 1% of passenger transport, maritime supplies 6% of freight transport. Considering newly registered vehicles in 2017, the share of diesel cars has strongly increased by 63% over the past decade.

Based upon this interdependent relationship of power generation, transportation and CO_2 emissions, this study evaluates the impact of electric vehicle deployment on power generation and CO_2 emissions in Turkey using the Bottom-Up Energy Modeling System (BUEMS), which is designed as a large-scale linear optimization model. The objective of this paper is to answer an integrated set of questions related to how the transportation sector, despite the advances in vehicular fuel efficiencies and through plug-in electric vehicles, influences overall energy demand, cost and CO_2 emissions. This paper sheds light on the influence of electrifying transportation on the carbon footprints of the increased demand for electricity. The paper tests the central hypothesis that a transition to electric vehicles increases carbon emissions the higher the dependence of the aggregate electricity demand on fossil technologies. A contribution of the paper highlights the unintended and indirect consequences of increased transport electrification without considerations for grid decarbonization. While the extant literature on the topic has argued for reduced emissions, this paper deviates by underscoring that emissions may be higher depending on the carbon intensity of power generation and the underlying scenarios. More significantly, the implication for policy shows that the marginal cost of emission reduction through e-buses is much lower than through other policy measures like carbon taxation.

The rest of this paper proceeds with the description of the energy modeling applications in transportation. Section 2 presents an abridged summary of related literature within the context of energy modeling applications in transportation. Section 3 is focused on the underlying mathematical formulations that drive the BUEMS modeling structure. Section 4 presents an elaborate description of the reference energy model. Section 5 proceeds with the calibration of the model while Section 6 presents the results of the analysis. To understand the influence of specific differences in model parameters, Section 7 conducts a sensitivity analysis of the results from Section 6. Section 8 concludes.

2. Literature Review on Energy Modeling Applications in Transport

A recent study compares the transport sector in China with that of the USA from a decarbonization perspective by using the Integrated MARKAL-EFOM System (TIMES) model [10]. In the study, decarbonization characteristics and alternatives of transport in China and the USA for future years are modeled by carrying out carbon tax scenarios and analyses. Based on the scenarios and their results, biofuel will reduce the carbon emissions in the near-term, and substitute at least 20% of oil products in China and the USA by 2050. On the other hand, it is expected that electrification will help the decarbonization of the transport sector in the long-term, as long as less carbon-intensive power generation technologies become more economical.

In a model of light-duty plug-in electric vehicles for national energy and transportation planning in the USA, the switch from conventional gasoline vehicles (CVs) to hybrid electric vehicles (HEVs) and consequently to plug-in electric vehicles (PEVs) shows an increase in the capital cost and a decrease in the fuel cost [11]. The ideal composition of light-duty vehicles (LDVs) changes among groups due to a dependence of reduction in fuel cost on the travel pattern. With a stepped but aggressive introduction of PEVs combined with investments in renewable energy, cost from energy and transportation systems can be reduced by 5%, emissions from electricity generation and LDV tailpipes can be reduced by 10%, in a 40-year interval. When renewable sources are used for electricity generation instead of fossil fuels, cumulative GHG emissions can be decreased with a marginal cost increase. For instance, yearly GHG emissions can reach 50% of year-1 levels in year 40, cumulative GHG emissions can be decreased by 18.3% with only increasing the total system cost by 0.03%, checked against the case where the LDV fleet is provided electricity without any emission concern.

In evaluating the decarbonization of the Greek road transport sector using alternative technologies and fuels, [12] uses the Long-range Energy Alternatives Planning (LEAP) model. In the base scenario, targets for the development of renewable energies and the reduction of GHG emissions are also included. Hybrid vehicles, electric vehicles, fuel cell vehicles, biofuels and gas engine vehicles are defined under various scenarios. Results show that CO_2 emissions can be reduced by 38% while energy consumption is decreased by 26% compared to the base scenario. The share of gasoline and diesel fuel (internal combustion engines and hybrids) decreases from 76% to 59% for LDV (light duty vehicle) while energy consumption reduces from 91% to 66% for HDV (heavy duty vehicle) compared to the base scenario by the year 2050.

In a recent study, five modeling teams converged to understand the impact of road transport in energy demand and emissions by developing baseline projections for India's transportation sector [13]. The observations highlight key distinctions in the base-year data and future projections particularly on energy consumption by transport mode, and service demand for passenger and freight transport. The differences in regional locations are exacerbated by the economic composition of specific regions. For example, a study that examined transportation carbon emissions in 341 cities concludes that transport emissions were significantly lower in central and western cities than in eastern cities and the emissions per GDP contributed to policies aimed at reducing transport emissions [14].

These modeling initiatives underline not only the regional disparities, but also the commonalities in the interdependent behavior between electricity generation, transportation and CO_2 emissions. A recent study employed the Global Change Assessment Model (GCAM) to understand the scenario of paths to meet the Paris Agreement [15] and between energy sources and technology options [16]. The variations observed inform the urgency to examine this interdependency in Turkey particularly with the lens of a data-rich calibrated model. This examination is predicated on a modeling platform that offers an adequate representation of the multiple cross-sectional dependencies across sectors of the Turkish economy.

3. The BUEMS Modeling Framework

In this study, the bottom-up energy modeling system BUEMS [17,18] has been calibrated using the most recent Turkish energy and transport sector data for Turkey. BUEMS is a bottom-up model that represents the energy sector in a technologically detailed way. All the complex relationships of producing, transforming, transmitting and supplying energy to meet the useful demand characteristics are represented with great technological detail. The objective is the minimization of total energy system cost. The levels and prices of the various energy sources are at equilibrium in each period, which guarantees that the net total cost of supplying all levels of energy services is minimized, while satisfying a number of constraints as detailed below.

The objective function minimizes the discounted total costs ($TOTCOST$) of the country's energy system, summed throughout the planning horizon (over all periods), i.e.,

$$TOTCOST = \sum_t (1+d)^{p(1-t)} \times ANNCOST(t) \times \left(1 + (1+d)^{-1} + ... + (1+d)^{1-p}\right) \tag{1}$$

$$ANNCOST(t) = \sum_k [AnnInvCost(t,k) \times INV(t,k) + FixOM(t,k) \times CAP(t,k) +$$
$$VarOM(t,k) \times ACT(t,k)] + \sum_{c,l} MiningCost(t,l,c) \times Mining(t,l,c) +$$
$$ImportPrice(t,c) \times Import(t,c) - ExportPrice(t,c) \times Export(t,c) +$$
$$CO_2Tax(t) \times CO_2(t) \tag{2}$$

where p is the period length (in years) between two successive time periods, d is the discount rate, $ANNCOST(t)$ is the annual total energy cost at time t, $AnnInvCost(t,k)$ is the annualized investment cost per unit capacity of technology k at time t. $FixOM(t,k)$ and $VarOM(t,k)$ are the fixed (per unit capacity) and variable (per unit activity) operational and maintenance costs of technology k at time t. INV, CAP, and ACT represent respectively the investments in technology k, the capacity of technology k, and the activity level of technology k at time t. $MiningCost(t,l,c)$ and $ImportPrice(t,c)$ are the unit supply prices of resource c according to its source of supply (if it is a mining source, $MiningCost(t,l,c)$; if it is an import source, $ImportPrice(t,c)$). The index l stands for the type of mining facility. $ExportPrice(t,c)$ is the unit cost of a resource that is exported. $CO_2Tax(t)$ is the tax amount for unit CO_2 emissions if a CO_2 tax applies. As can be acknowledged from these explanations, the objective function is designed to cover just about all costs in the system while featuring the assumption of linearity (i.e., all costs increase proportionally with the related amounts of investment, activity, capacity, etc.).

The objective function is optimized over a number of constraints including the balance of energy demands, capacities and activities and accounting for emissions, i.e.,

$$\sum_k CAP(t,k) \geq D(t,d) \quad \forall\, d,t \tag{3}$$

$$ACT(t,k) \leq af(t,k) \times CAP(t,k) \quad \forall\, t \tag{4}$$

$$CAP(t,k) = RESID(t,k) + \sum_{t',t-t'<life\,of\,tech\,k}^{t} INV(t,k) \quad \forall\, k,t \tag{5}$$

$$\sum_{k,c\,\epsilon\,producing\,resource} Output(t,k,c) \times ACT(t,k) + \sum_l Mining(t,l,c) + Import(t,C) \geq$$
$$\sum_{k,c\,\epsilon\,processing\,resource} Input(t,k,c) \times ACT(t,k) + \sum_l Export(t,c) \quad \forall\, c \tag{6}$$

$$CO_2(t) = \sum_k ef(t,k) \times ACT(t,k) \tag{7}$$

where $D(t,d)$ is the demand for category d at time t, $af(t,k)$ is the annual availability factor of technology k at time t, $RESID(t,k)$ is the residual capacity of technology k. The residual capacities are considered already in place in the base year, $Output(t,k,c)$ is the level of energy resource/commodity c produced by a unit activity of technology k at time t, $Input(t,k,c)$ is the level of resource/commodity c processed by a unit activity of technology k at $time t$, $ef(t,k)$ is the CO_2 emission factor, i.e., it defines emissions produced by unit activity of technology k. Both $output(t,k,c)$ and $input(t,k,c)$ change over time reflecting efficiency improvements. The reduction in $output(t,k,c)$ over time reflects efficiency improvements in demand technologies while a reduction in $input(t,k,c)$ is the result of an efficiency improvement in conversion technologies. Efficiency values for all technologies and improvements over time can be found in [18] These constraints ensure that:

- Demand is always satisfied by available capacity, total activity levels of energy utilization sectors as end-use technologies are always equal to or greater than the demand requirements;
- In each period, summation of investments (realized at current periods and at past periods but still last because of its lifetime) and capacities (installed in prior time and still have additional lifespan) equals to the available capacity by the way of capacity transfer constraints;
- The activity level of an energy conversion technology is always less than or equal to its available capacity by force of activity-capacity relation constraints;
- Cumulative supply limit ensures that total supply level of a supply (resource) technology is always less than or equal to its cumulative supply level;
- Total generation of multiple output technologies for each output equals to a fraction of total activity level for each output.

The linkages of some of the technologies can be followed from Figure 1 where each node represents a technology or demand and each link represents an energy source. The figure contains a small part of a single energy demand, road transportation. This demand can be satisfied by different end-use technologies using gasoline, diesel fuel, electricity, hydrogen and compressed natural gas. These energy sources are produced by other technologies like power plants and refineries, which receive fuel supply from imports and mining. Supply technologies provide energy sources to the system with a unit supply cost. On the other hand, process technologies like refineries and power plants generate secondary energy forms by using other primary energy sources. End-use demand technologies like space-heating technologies satisfy useful energy demand by using primary and/or secondary energy sources. All technologies have defined capacities to be activated, which are selected in a model run under the objective of overall cost minimization. Input requirements and efficiency of technologies also have significant contribution to technology penetration. It should be noted that Figure 1 is a highly simplified fictitious example of sample relationships defined in the model. Technology and energy source options are numerous in the application. For example, on the supply side, there are 10 natural gas import options and five different coal extraction methods. On the demand side, on the other hand, there are seven different electricity space heating options each having different cost characteristics. In addition, parameters like efficiency, life time, residual capacity, bounds on different system variables, and many others have significant impact on the optimization process as they affect levelized costs.

As can be acknowledged from these explanations, the constraints of the model are designed to cover just about all energy related flows, interactions, limitations and requirements in the system, while featuring the critical, yet realistic linearity assumption.

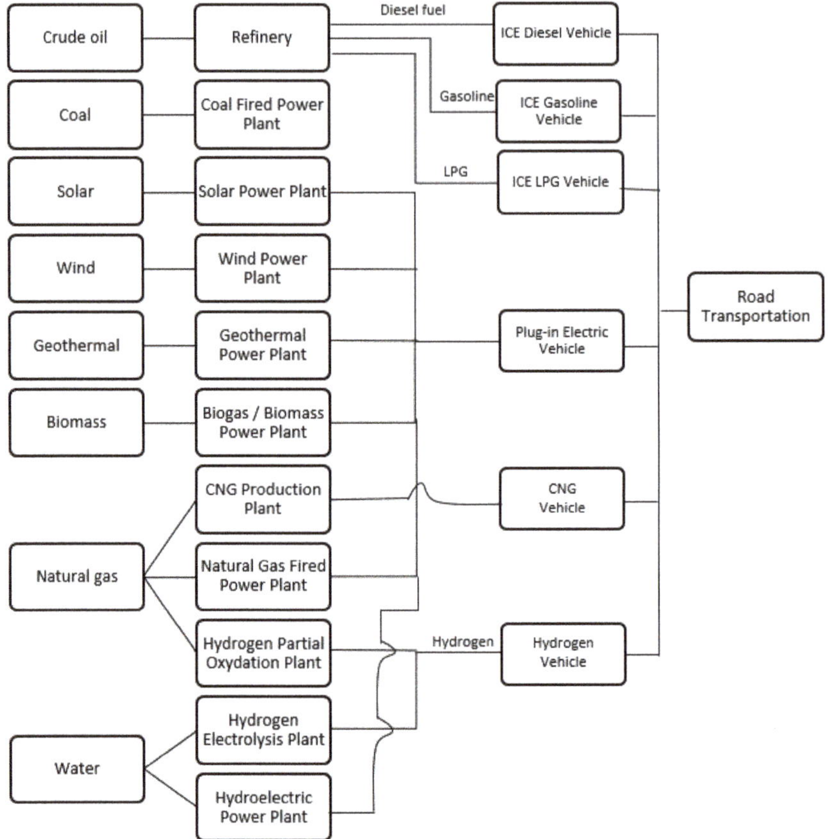

Figure 1. Partial view of Bottom-Up Energy Modeling System's (BUEMS's) Reference Energy System.

4. Development of the Database and Reference Energy System

The study started with a scientific research project supported by the Scientific and Technical Research Council of Turkey [19] supported by the Ministry of Energy and Natural Resources. Several graduate theses were conducted as part of the project [18,20–22], through which an extensive database was developed. Data needs associated with the current situation (sub-sectoral energy demands, resource availabilities, technology and cost parameters, including capacities, vintages, economic lifetimes, efficiencies and utilization rates, investment costs, fixed/variable operational costs), as well as resource availability projections up to the end of the planning horizon, were compiled through publicly available data sources, questionnaires and expert interviews.

A compilation of demand projections at the sub-sectoral level, future prices of energy carriers and technology and cost parameters of future technologies (up to the year 2050) were problematic as even experts could not provide reliable estimates. Therefore, regarding sub-sectoral demands for future periods, it was decided to project each sub-sectoral energy demand into the future period by multiplying its current value by the estimate of the associated sectoral gross domestic product increase rate up to that period, obtained from the Turkish Statistical Institute. Official forecasts, as published under various publicly available reports, were analyzed to ensure the consistency of sectoral projections with economic and natural resource abilities and growth targets of Turkey.

The consistency of sub-sectoral assumptions with overall economic growth patterns was verified by related public experts.

Regarding future prices of energy carriers, there exists a significant debate over how energy prices should be modeled. Despite a large body of empirical work, there is still much uncertainty regarding the true dynamics of energy price changes. In countries like Turkey where a major part of energy demand is met by imports, prices principally result from the interaction of world demand and supply. In 2018, 74.5% of Turkey's primary energy supply was imported. Thereby, 88% of the country's crude oil demand, 99.1% of its natural gas demand and 97.3% of its hard coal demand was brought into the country. In this study, price forecasts of the imported fossil fuels are based on IEA assumptions reported in the World Energy Outlook [23] whereas domestic fuels' prices are based on the guess of experts. IEA trajectories were used to reflect expert elicitation together with the inputs from the Ministry of Energy and Natural Resources. A long-term leveling off of natural gas price growth is assumed in line with expectations for increased domestic production (shale gas from the Trace region and off-shore production from the Mediterranean) putting downward pressure on natural gas prices. Technology costs and technical characteristics that are used to compute levelized costs and a supply curve (which is stepwise because it is built up by discrete technologies). All technology parameters (efficiencies, lifetimes, investment and operating costs) are taken from the publicly available EPA (United States Environmental Protection Agency) database. The supply-side technologies are matched with energy service demands to select from each of the energy sources, carriers, and transformation technologies to produce the least-cost solution. These demand levels are modeled in this application as price insensitive, i.e., demand remains unchanged if the supply prices change and vice-versa. It was a necessary assumption for the current study as there was neither availability of information on price elasticities nor an explicit representation of the macroeconomy integrated in the employed modeling framework. This is a common approach for these kinds of bottom-up energy sector models, although presenting a weakness in terms of micro-economic foundations.

Figure 2 shows the simplified Reference Energy System. This overview provides an indication of the sectoral breakdown, together with energy flows from primary sources to final demand categories, thereby passing through various conversion, process and demand technologies. It is, however, a simplified representation as all the detail cannot be included due to size limitations: The model is comprised of 209 energy carriers, 99 demand categories and 1545 technologies.

In BUEMS, the transport sector is grouped into four major modes: air, maritime, rail and road transportation. These major modes are also divided into sub-modes, as shown in Figure 3.

For all twenty-five sub-modes, BUEMS requires total billion vehicle kilometers traveled per year as the sectoral demand data of the transport sector. This data is the most critical data for the transport sector when different policy scenarios are applied and scenario results are compared. Demand data, total billion vehicle kilometers traveled, is calculated as total distance independently of passenger number or freight amount carried on the vehicle.

Fuel efficiencies, fixed and variable operation and maintenance costs, lifetimes, residual capacities, capacity bounds, activity levels, availability factors, investment prices, investment bounds are other input requirements for transport sector technologies in BUEMS. All the data required by the model for the transport sector is compiled for the base year 2012 and year 2017 according to the gathered data. Then, data between 2017 and 2052 is estimated based on a strategic plan of MoTI (Ministry of Transport and Infrastructure) and expert consultation.

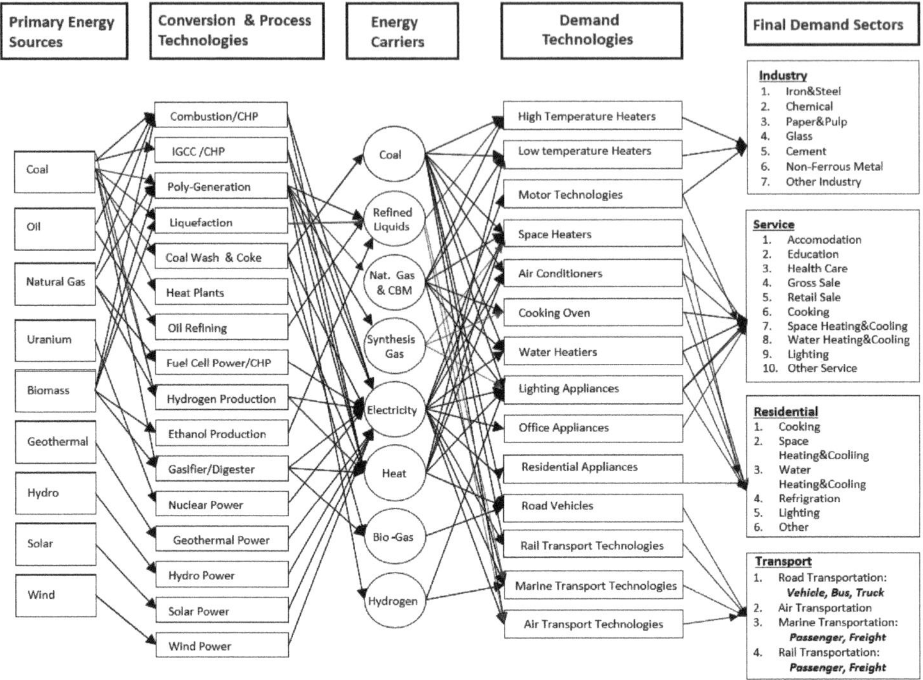

Figure 2. The BUEMS simplified reference energy system.

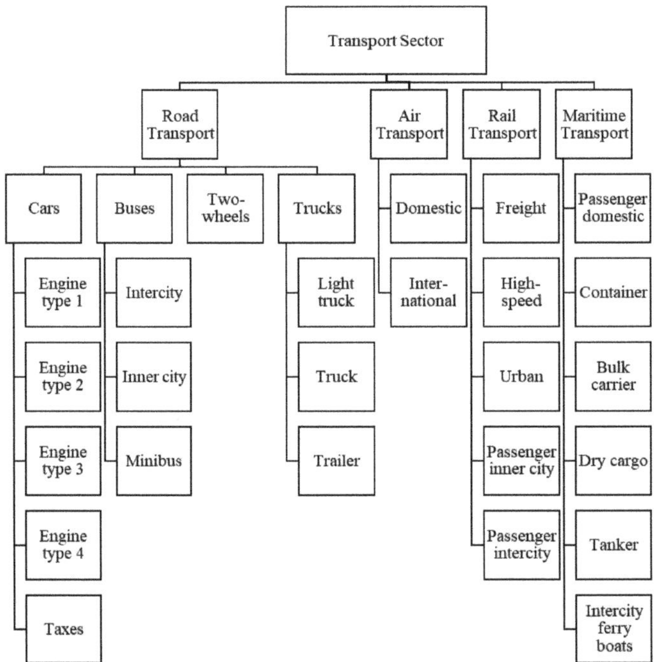

Figure 3. Sub-modes of the transport sector in BUEMS.

5. Model Calibration

BUEMS is calibrated according to the most recent publicly available energy and transport data for Turkey. Reference year and validation year are taken as 2012 and 2017, respectively, with model results in five-year time intervals from 2022 until 2052. The base scenario reflects the current energy flows in Turkey with its business-as-usual assumptions, together with a modest prediction for electric vehicles. The transport sector refinement, data update and calibration are based on a graduate study by [24] where further detail on model calibration can be found. Various electric vehicle diffusion and policy scenarios are carried out and results of these alternative scenarios are compared with the base scenario.

5.1. Data and Assumptions

This section describes demand data sources, their compilation and associated assumptions for the transport sector.

- Road Transport: Demand data, billion vehicle kilometers, for years 2012 and 2017 are compiled by using reports from the General Directorate of Highways (KGM) and statistics from the Statistical Institute of Turkey (TUIK) while demand data for future years are predicted based on expert consultation for all transport sub-modes in the transport sector.
- Air Transport: International flights with departures from Turkey are included in air transport demand data together with all domestic flights. Turkish Airlines Annual Reports are used to calculate total demand. In the calculation of international take-offs, only data for airlines based in Turkey were available.
- Rail Transport: Train kilometers except for rail urban are compiled from TUIK statistics and Turkish Railways Annual Statistics published by the Turkish State Railways. However, there exist no statistics related to rail urban sub-mode includes metro and tramway trains in the cities. Therefore, this information is gathered from the municipalities of the cities with metro and/or tramway infrastructure. Demand data for future years is estimated according to expert guess.
- Maritime Transport: Only domestic marine navigation is taken into consideration. Demand data for each sub-mode is calculated according to data obtained from the General Directorate of Maritime Affairs and TUIK. Demand data, billion vehicle kilometers, for future years is estimated according to expert guess.

A summary of the demand data by each of the transport modes for 2022 through to 2052, in five-year intervals, are presented in Table 1 below with details of their breakdowns in [24].

Table 1. Demand data by mode of transportation, 2022–2052.

Mode	2022	2027	2032	2037	2042	2047	2052
Road	155.30	182.10	215.86	258.17	311.07	377.10	459.48
Rail	0.11	0.12	0.12	0.13	0.14	0.15	0.16
Maritime	0.16	0.17	0.18	0.19	0.20	0.21	0.22
Air	1.34	1.50	1.65	1.78	1.89	1.96	2.02

5.2. Electric Vehicles Consumption Data

Upper and lower bounds are set according to the prediction of consumption values for electric vehicles to obligate electric vehicles deployment in BUEMS. The behavior and effects on the energy system and emissions of electric vehicle deployment are observed under different diffusion and energy policy scenarios. Electric vehicles are differentiated into three categories: electric cars, electric buses and electric rail vehicles.

For the diffusion of electric cars, only lower bounds are used. Bounds for consumption levels of electric cars are based on expert survey results estimating 140,000 electric vehicles (EVs) to be on the road in Turkey by the year 2022—see Appendix A for the survey instrument. While this estimate is also expressed by the government, it does not appear to be realistic given the limited increase in EV charging stations and low EV sales. Also, the national electric vehicle manufacturing project has been progressing slower than anticipated. Hence, for the base scenario, the official estimate is reduced by 50%. In other words, it is assumed that there will be 70,000 EVs on the road in Turkey by the year 2022. Electricity consumption level predictions of electric cars are finalized according to the below assumptions:

- An increase of 19.1% per year in the number of electric cars.
- A consumption of 25 kWh of electricity per 100 km [25].
- An annual mileage of 15,000 km.

In consequence of the above assumptions, finalized numbers and electricity consumption levels of electric cars, which are included in BUEMS as lower bounds in the base case reference scenario (BASE), are provided together with the corresponding number of cars in Table 2 below.

Table 2. BASE scenario electric cars' lower electricity consumption bounds (PJ).

Year	2022	2027	2032	2037	2042	2047	2052
Electricity Consumption—low	0.95	2.27	5.44	13.06	31.35	75.25	180.59
Number of Electric Cars—low	*70,000*	*168,000*	*403,200*	*967,680*	*2,322,432*	*5,573,837*	*13,377,208*

On the part of electric buses, both upper and lower bounds are used. To determine current numbers of electric buses, the municipalities are contacted one by one. As a result, the below assumptions and electricity consumption bounds are used for electric buses.

Assumptions for lower bounds:

- An estimation of 79 electric buses in 2022, increasing by 8% per year.
- A consumption of 135 kWh of electricity per 100 km [25].
- Annual mileage of 50,000 km.

Assumptions for upper bounds:

- An estimation of 99 electric buses in 2022, increasing by 16% per year.
- A consumption of 150 kWh of electricity per 100 km [25].
- Annual mileage of 50,000 km.

Finalized numbers and electricity consumption levels of electric buses that are included into BUEMS as upper and lower bounds are represented in Table 3 below.

Table 3. BASE scenario electricity consumption bounds for Electric Buses (PJ).

Year	2022	2027	2032	2037	2042	2047	2052
Electricity Consumption—low	0.02	0.03	0.04	0.06	0.09	0.13	0.19
Number of Electric Buses—low	*79*	*116*	*171*	*251*	*369*	*542*	*797*
Electricity Consumption—up	0.03	0.06	0.12	0.25	0.52	1.09	2.29
Number of Electric Buses—up	*99*	*208*	*437*	*917*	*1925*	*4043*	*8491*

For electric rail vehicles, upper and lower bounds are also prepared. Consumption levels of electric rail vehicles except for rail urban sub-mode are based on TCDD data. On the side of rail urban part, electricity consumption values of urban rail systems in Ankara is found from the municipality website. Then, electricity consumption calculations for other cities are obtained, as shown in Table 4.

Table 4. BASE scenario electricity consumption bounds for Rail Vehicles (PJ).

Year	2022	2027	2032	2037	2042	2047	2052
Electricity Consumption—low	4.61	4.88	5.17	5.48	5.81	6.16	6.53
Electricity Consumption—up	4.84	5.14	5.44	5.77	6.12	6.48	6.87

5.3. Scenario Definitions

For the base scenario ("BASE") definition, the current energy system of Turkey is represented with its business-as-usual assumptions, together with a modest prediction on minimal diffusion of EVs, as shown in Tables 2–4. In addition to BASE, various alternative scenarios are defined to evaluate the impacts of alternative policies. These are summarized in Table 5 and further described in detail as follows:

Table 5. Scenario Summary.

Scenario	Description
BASE	Business-as-usual assumptions with modest EV diffusion. Lower bounds at 70,000 electric cars in 2022 and 2.4 times growth in each period.
MOREECAR	Same assumptions as BASE except for accelerated EV diffusion. Lower bounds at 140,000 electric cars in 2022 and 2.4 times growth in each period.
NONEWCOAL_E	Same assumptions as BASE except for coal-fired power generation: coal-based electricity is limited so as not increase after 2027.
ISTBUS	Same assumptions as BASE except for electric buses: all buses in Istanbul are assumed to be electric by year 2027.
NODIESEL_T	Same assumptions as BASE except for diesel vehicles: usage of diesel fueled vehicles is stopped by year 2027.
CO2TAX50_T	Same assumptions as BASE except for CO2 emission tax: \$50 per ton of CO_2 emission tax is applied to the transport sector only starting in 2032.
CO2TAX50_E	Same assumptions as BASE except for CO2 emission tax: \$50 per ton of CO_2 emission tax is placed on the electricity generation sector only starting in 2032.

MOREECAR: In addition to the BASE scenario, a more rapid distribution of EVs is assumed in the MOREECAR scenario, which—in line with official estimates—starts with 140 thousand electric cars in 2022. The assumption of 140 thousand e-cars in 2022 is based on the official declaration by the Minister of Industry and Technology [26]. In this scenario, the electricity consumption levels of electric cars, included in BUEMS as lower bounds, are as follows. All assumptions other than the lower bounds on electric cars, as depicted in Table 6, are the same as in the BASE scenario.

Table 6. MOREECAR scenario electric cars' lower electricity consumption bounds (PJ).

Year	2022	2027	2032	2037	2042	2047	2052
Electricity Consumption	1.89	4.54	10.89	26.13	62.71	150.49	361.19
Number of Electric Cars	*140,000*	*336,000*	*806,400*	*1,935,360*	*4,644,864*	*11,147,674*	*26,754,417*

NONEWCOAL_E: In this scenario coal based electricity generation is limited starting from the year 2027 as no new coal fired power plant is constructed after 2027. Thus it is made sure that the growing electricity demand due to the deployment of EVs is not coming from new coal-fired power plants. Accordingly, the electricity sector grows with renewables and natural gas fired power generation instead of coal. All assumptions other than the construction ban on coal fired power plants are the same as in the BASE scenario. The year 2027 has been chosen for a ban on new coal fired power generation as it is the first model year coming after 2023. The official power sector strategy is to utilize all domestic coal reserves by the year 2023 [27]. The utilization of indigenous coal reserves is part of the policy priority to reduce import dependence, and a ban on new coal-fired power generation is possible once domestic resources are exploited.

NODIESEL_T: A ban on new diesel cars is used as an alternative policy instrument to assure an accelerated diffusion of electric cars. The UK, for example, announced a ban on the sales of new diesel and gasoline powered cars in 2035. The impacts of such a policy on the diffusion of electric cars are explored under the NODIESEL scenario. The introduction of new diesel cars and buses is banned in this scenario starting from the year 2027, thereby restricting coal based electricity generation so as to direct the transport sector's incremental electricity demand due to the ban towards less carbon-intensive energy sources. The ban on diesel fueled vehicles expanding rapidly in Europe is expected to affect Turkey gradually with a stop in the manufacturing of diesel cars by 2030 [28]. The scenario bans the introduction of new diesel vehicles in 2027, which is the latest model year before the expected stop of manufacturing. All assumptions other than the diesel ban on new cars and restriction of coal-fired power generation, as depicted in Table 7, are the same as in the BASE scenario.

Table 7. Consumption bounds for coal-fired power generation in the NODIESEL_T scenario (PJ).

Year	2022	2027	2032	2037	2042	2047	2052
Coal-fired Power Generation—up	386.10	439.70	483.67	532.03	585.24	643.76	708.14

ISTBUS: This scenario is created on the consideration of what will happen if all the buses in Istanbul are electric vehicles by the year 2027. Therefore, according to the inner city bus transport data of Istanbul, the electric buses' bounds are revised. It should be noted that the annual growth rate of electricity consumption bounds from 2027 until 2052 averages 15.4%, as shown in Table 8. The rate of increase in the number of buses gets slightly higher due to efficiency increase over time, reaching between 190,000 and 622,787 e-buses by 2052 with a 10–17% average annual increase. Also, coal fired power plants are restricted in the ISTBUS scenario to direct the growing electricity demand of electric buses to renewable energy. The restrictions on coal-fired power generation are the same as in the NODIESEL_T scenario, as shown in Table 9. The Istanbul Municipality has announced plans to convert all inner city bus transport to electricity by the year 2030 [29]. The scenario assumes all bus transport in Istanbul to become electric by 2027, which is the latest model year before the planned transformation.

All assumptions other than the bounds on electric buses and coal-fired power generation as depicted in Tables 8 and 9 are the same as in the BASE scenario.

Table 8. Electricity consumption bounds for electric buses in the ISTBUS scenario (PJ).

Year	2022	2027	2032	2037	2042	2047	2052
Electricity Consumption—low	0.02	3.64	6.71	13.39	27.96	59.80	129.45
Electricity Consumption—up	0.04	4.64	8.60	17.20	35.96	76.95	166.57

Table 9. Consumption bounds for coal-fired power generation in the ISTBUS scenario (PJ).

Year	2022	2027	2032	2037	2042	2047	2052
Coal-fired Power Generation—up	386.10	439.70	483.67	532.03	585.24	643.76	708.14

It should be noted that when the lower bounds are binding, for example, Table 6, the presented results are the low bound results, while the results with the upper bounds significance, Tables 7 and 9, we present the upper bound fixed results as such. Table 8 is the exception with the upper and lower bounds results presented.

In the following section, scenario results will show that with the increase in electric cars, CO_2 emissions from electricity generation rise while transport sector emissions decrease. The rise in emissions is essentially due to increased coal-fired power generation. Therefore, in order to limit the growth of emissions, various scenarios restricting coal use were defined along with scenarios with CO_2 bounds and CO_2 taxes. The carbon tax (50\$/ton CO_2) is applied on the transport sector and electricity sector both individually and simultaneously under alternative scenarios.

CO2TAX50_T: In this scenario, an emission tax of \$50 per ton of CO_2 is introduced in transportation starting from the year 2032. The tax is applied on all modes of transportation, yet applied on the transport sector only. The tax amount of \$50 per ton of CO_2 is pretty much in line with assumptions and findings in the scientific literature. For example, a fixed tax trajectory was set such that the tax begins in 2020 at either \$25 or \$50 and rises at either 1% or 5% per year [30]. Starting at the \$50 level without further escalation is assumed in this scenario. All assumptions other than the emission tax are the same as in the BASE scenario.

CO2TAX50_E: In this scenario, an emission tax of \$50 per ton of CO_2 is introduced in power generation starting from the year 2032. The tax is applied in the electricity sector only. The tax amount of \$50 per ton of CO_2 is in line with assumptions and findings in the scientific literature as mentioned in the previous scenario description. All assumptions other than the emission tax are the same as in the BASE scenario.

6. Model Results

The total energy consumption of transportation includes electricity, petroleum, diesel, biodiesel, hydrogen, jet fuel, CNG, LPG, ethanol and methanol as fuel types, as shown in Figure 4. According to the results of the base scenario, diesel fuel is the dominant fuel from the beginning of the planning horizon, 56.4% in 2012 and 53.4% in 2017. However, its share in total transport energy consumption decreases over time, reducing to 31.5% in 2047 and 22.5% in 2052 being essentially substituted by gasoline fueled vehicles and EVs. The increase in the use of petroleum use is a result of the substitution of gasoline for diesel fueled vehicles and the increase in the demand of residual fuel oil used in ships, together with the rapid increase in overall transport demand. Its rising share indicates the long-run price competitiveness of petroleum. Electricity consumption in transportation gradually increases because of electric car deployment with a share of 0.3% in 2012, 0.5% in 2017, 6.3% in 2047 and 13.8% in 2052. Hydrogen fuel, which has no incentive in the model like plug-in EVs, gets involved at the end of the planning horizon only with a consumption of 1.6 PJ in 2052. While jet fuel and petroleum consumption continue to grow depending on the increase of transport demand, methanol consumption decreases for each period. Lastly, LPG, which is being used because of economic reasons in Turkey as it is cheaper than other alternatives being subsidized, drops over time from a considerable share of 15.49% in 2012, becoming zero by the year 2047 due to a considerable increase in electric cars.

The electricity consumption of the transport sector is dominated by electric cars. While it is negligibly small in 2012 and only 0.1 PJ in 2017, it reaches 75.2 PJ in 2047 and 180.6 PJ in 2052, as shown in Figure 5.

CO_2 emissions are increasing in all scenarios; total emissions of the BASE scenario skyrocketed from 359,280 kton in 2017 to 1,408,500 kton in 2052, which is almost quadruple. Transport sector emissions are increasing from 65.6 kton in 2017 to 83.1 kton in 2052 implying a 26.7% increase. Most interestingly, the more EVs are deployed, the higher the total emissions, as can be observed by comparing the MOREECAR and BASE scenarios in Figure 6. This is because of increased coal usage for power generation; therefore it is essential to limit the expansion of coal. That is why all other scenarios have a limitation or policy in place to reduce coal usage. When the results of these scenarios are investigated it is seen that emissions decrease by 0.4%, 4%, 11%, 29% and 51% in 2052 for the CO2TAX50_E, ISTBUS, MOREECAR, NODIESEL_T and CO2TAX50_T scenarios, respectively, in comparison to the BASE scenario.

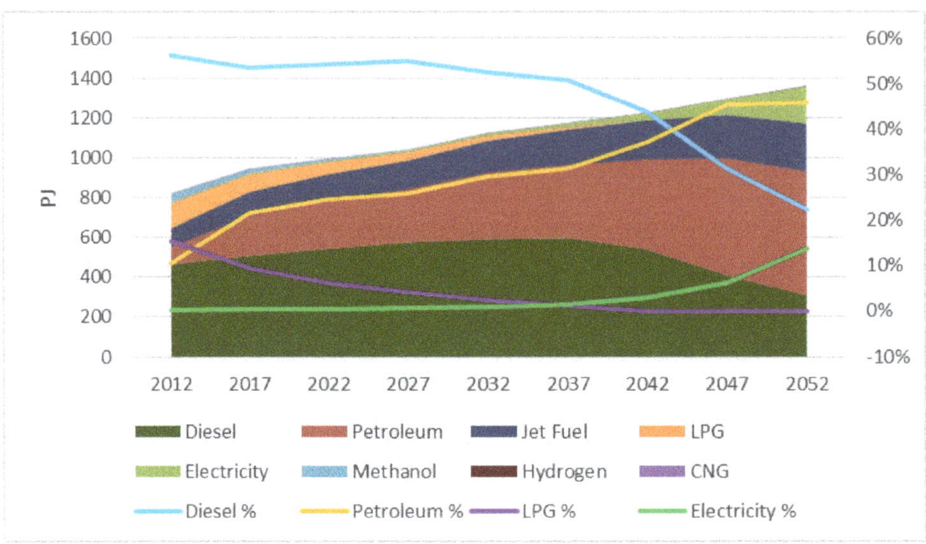

Figure 4. Energy consumption of the transport sector by fuel type: BASE scenario.

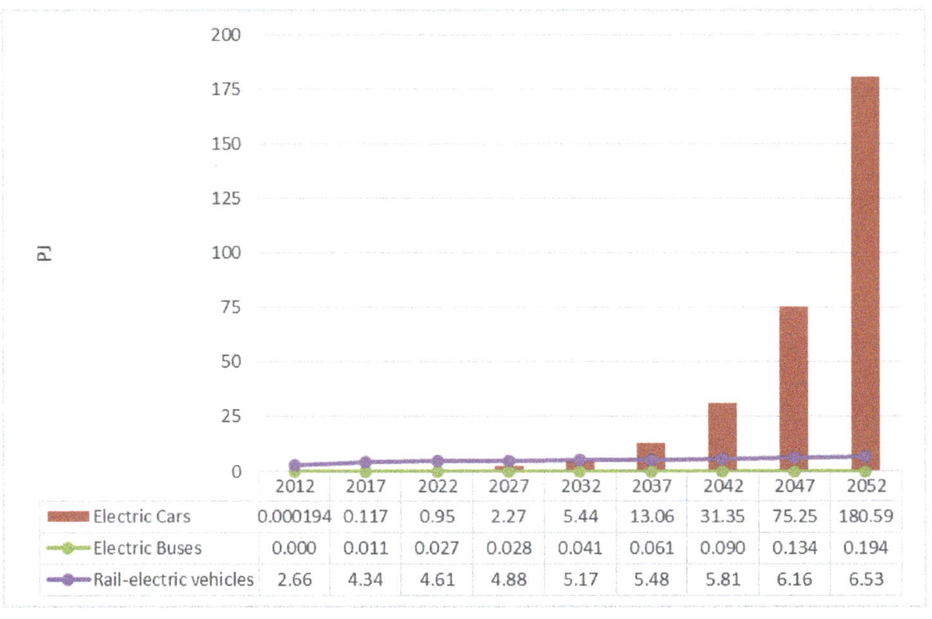

Figure 5. Electricity consumption of the transport sector by vehicle type: BASE scenario.

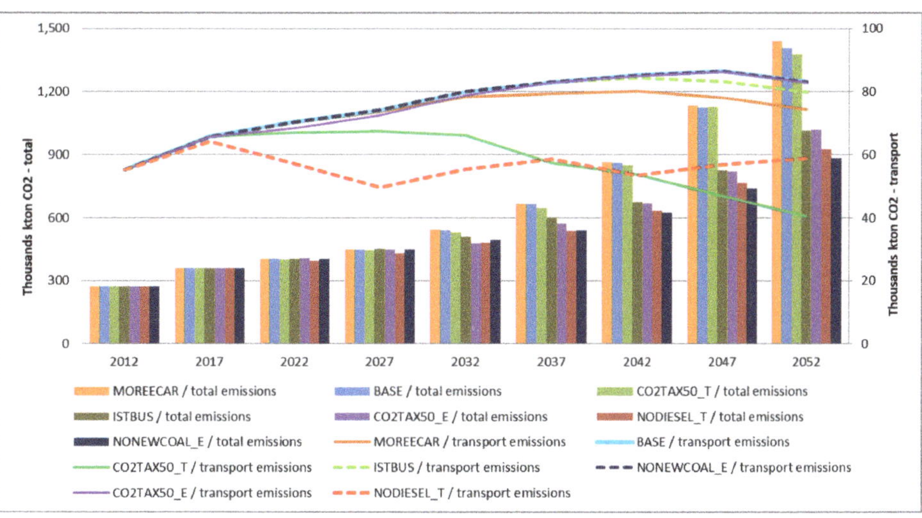

Figure 6. Total and transport sector emissions by scenario.

Turkey's skyrocketing total emissions are almost proportional to the change in coal-fired power generation, as can be observed from Figure 7. Total emissions increase by 1049 Mton (from 359 Mton in 2017 to 1408 Mton in 2052), of which 49.5% comes from increased coal-fired power generation. In 2052, coal-fired power generation amounts to 3721 PJ in the BASE and 3939 PJ in the MOREECAR scenario. It reduces to 1520 PJ, 708 PJ and 440 PJ in the CO2TAX50_E, ISTBUS and NONEWCOAL_E scenarios, respectively. Accordingly, in terms of emission reduction, the most effective policy options, ordered by reduction amount, are (i) banning new investment in coal-fired power plants, (ii) making bus transportation in Istanbul electric, and (iii) introducing a carbon tax on electricity generation. However, are these options cost-effective? The next section compares the system-wide cost implications of the scenarios.

Lifetime emissions and lifetime cost values, discounted to the year 2019, are depicted in Figure 8. It can be seen that all alternative scenarios have lower total emissions than the BASE scenario except for MOREECAR. Also, all of them have higher total system cost compared to the BASE scenario except for NODIESEL_T. In other words, the NODIESEL_T scenario has lower lifetime emissions and lower lifetime costs—a profitable decarbonization option pointed out by the model. It should be noted that the BASE scenario includes lower bounds for diesel vehicles so as to yield the most likely reference path under business-as-usual conditions. Obviously, this business-as-usual pathway with increasing diesel vehicles is not the cheapest one as a lower cost pathway emerges when there is a ban on new diesel fueled vehicles. The diesel consumption in the transport sector reduces drastically as a result of the ban on new vehicles, yet does not vanish until the end of the lifetime of the existing diesel-fueled vehicle stock. The diesel gap between the BASE scenario and NODIESEL_T reduces over time as the share of diesel fuel declines in the long term in the BASE scenario anyway. The diesel consumption of the transport sector in the year 2027 is 573.48 PJ in the BASE scenario versus 215.69 PJ in the NODIESEL_T scenario. Then, in year 2052, diesel consumption gets 306.05 PJ in scenario BASE and 113.17 PJ in the NODIESEL_T scenario.

While NODIESEL_T offers a win-win opportunity, the MOREECAR scenario indicates a lose-lose option: it has a higher total system cost and higher total emissions than the BASE scenario. This is because the growing electricity demand due to increased e-mobility is met by coal-fired electricity generation. Therefore, in the absence of decarbonization in electricity generation, the diffusion of e-mobility is not sustainable. It helps only to reduce emissions on the road while that reduction is offset by increased emissions from electricity generation.

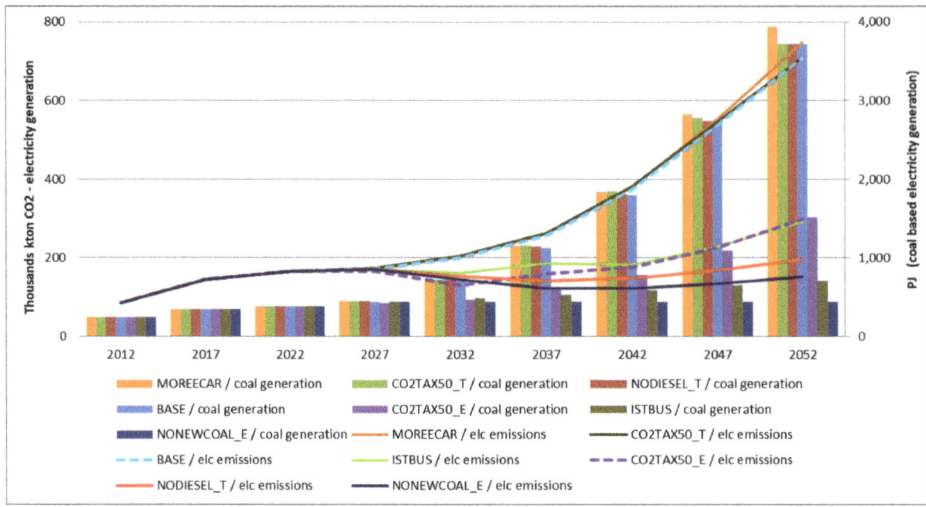

Figure 7. Emissions from electricity generation and coal-based electricity generation levels by scenario.

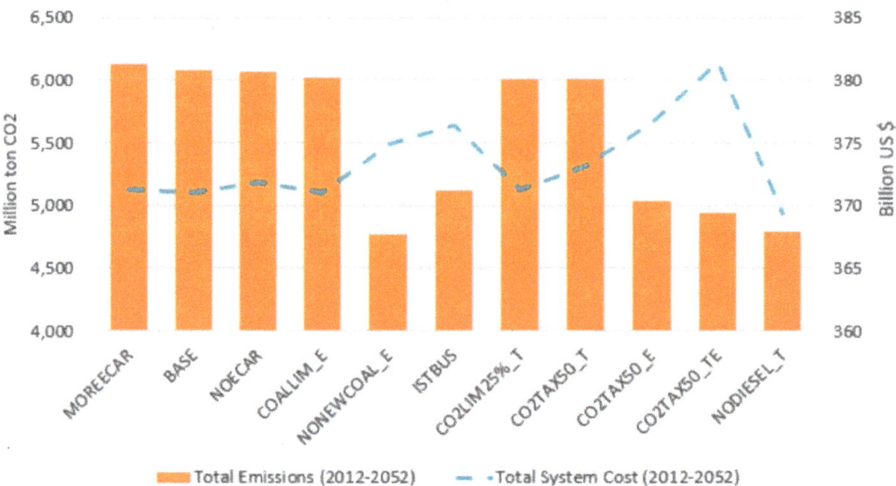

Figure 8. Total emissions and total system cost by scenario throughout the planning horizon.

The ratio of lifetime cost changes to lifetime CO_2 changes, referred to as marginal abatement cost, is shown in Table 10. The win-win NODIESEL_T scenario indicates a profit of 1.29\$ per ton CO_2 reduced. The lose-lose MOREECAR scenario on the other hand has no marginal abatement cost because its emissions are higher than in the BASE scenario, hence no abatement.

When scenarios are ranked according to lifetime marginal abatement cost, the lowest cost options are associated with the electricity sector (NONEWCOAL_E and CO2TAX50_E), followed by urban e-bus (ISTBUS). Carbon taxation in transportation (CO2TAX50_T) appears to be the least cost-effective policy option.

Table 10. Lifetime Marginal Abatement Costs (cost increase/emission savings).

Parameter	CO$_2$ Savings (Million ton)	Cost Increase (Million US\$)	Marginal Abatement Cost (US\$ ton CO$_2$)
BASE	Ref. point	Ref. point	Ref. point
NODIESEL_T	1286	−1656	−1.29
NONEWCOAL_E	1314	3932	2.99
CO2TAX50_E	1043	5702	5.47
ISTBUS	963	5369	5.58
CO2TAX50_T	78	2135	27.37
MOREECAR	−47	277	N/A

Table 11 presents the periodical breakdown of the abatement cost figures for each scenario. It can be observed that the abatement declines over time in all scenarios except the win-win scenario NODIESEL_T where profit declines. In the NODIESEL_T scenario there is an adaptation process for shifting away from diesel vehicles, which is highly profitable in the short run as explained previously, and diminishes in the longer term when the shift is completed. In the other scenarios, the initial investments into low-carbon technologies induce high up-front costs but lower operating and maintenance costs, essentially due to renewables, results in savings in the longer term. It can be seen that a CO$_2$ tax on electricity generation (CO2TAX50_E) is initially profitable as well, has an abatement cost of 30.66 \$/ton CO$_2$ in 2032 (the year when carbon tax is launched), which gradually decreases thereafter.

Table 11. Overall CO$_2$ abatement costs by scenario.

Scenarios	2022	2027	2032	2037	2042	2047	2052
NONEWCOAL_E	–	–	15.75	7.13	4.05	2.73	1.88
ISTBUS	–	–	9.70	15.99	6.87	4.68	3.47
CO2TAX50_T	9.75	78.14	132.55	44.49	58.65	49.24	1.18
CO2TAX50_E	–	−44.07	30.66	16.68	6.51	3.68	2.41
NODIESEL_T	−56.92	−33.08	−4.53	1.37	1.98	2.17	1.74
MOREECAR	–	–	–	–	–	–	–

The primary energy supply mix composition of energy technologies for the BASE and alternative scenarios are depicted in Figures 9 and 10, respectively.

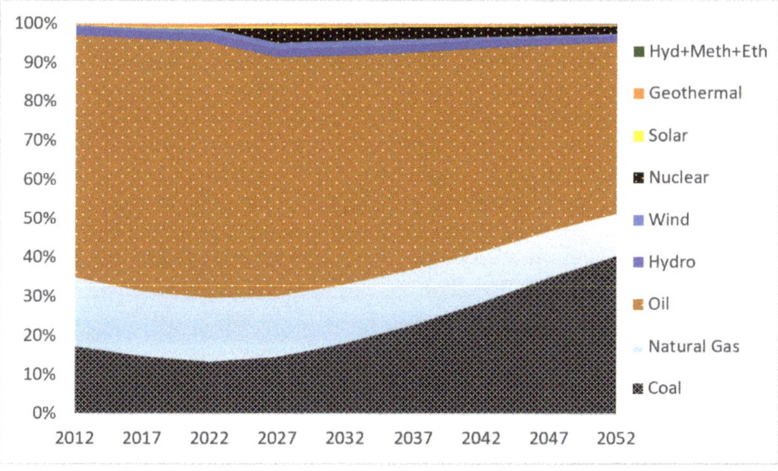

Figure 9. Primary energy supply mix in the BASE scenario.

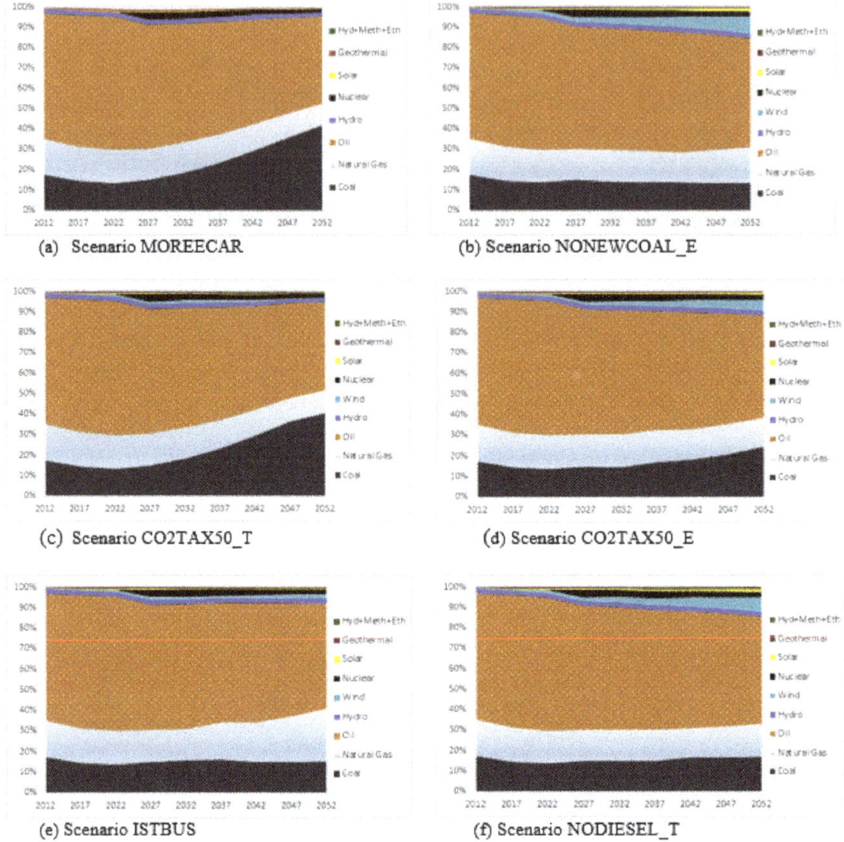

Figure 10. Primary energy supply mix in the alternative scenarios.

7. Sensitivity Analysis

Transport sector demand data and the cost of EVs are among the most important parameters affecting scenario results in this study. Therefore, the sensitivity analysis focuses on these two parameters, thereby elaborating on their impacts on CO_2 emissions and energy system costs. It is found that the results are much more sensitive to demand than to the cost of EVs. According to Figure 11, a 20% rise in transport sector demand causes a 0.7% increase in total emissions while 20% rise in EV cost causes less than 0.1% increase. On the other hand, a 20% fall in EV cost reduces emissions by 0.2% while a 20% fall in transport sector demand provides a 0.7% decrease.

As can be seen from Figure 12, a 20% rise in transport sector demand causes a 1.5% increase on total system cost while a 20% rise in EV cost causes a 0.13% increase. On the other hand, 20% fall in EV cost provides a 1.47% decrease in total system cost while a 20% fall in transport sector demand leads to a 0.17% decrease.

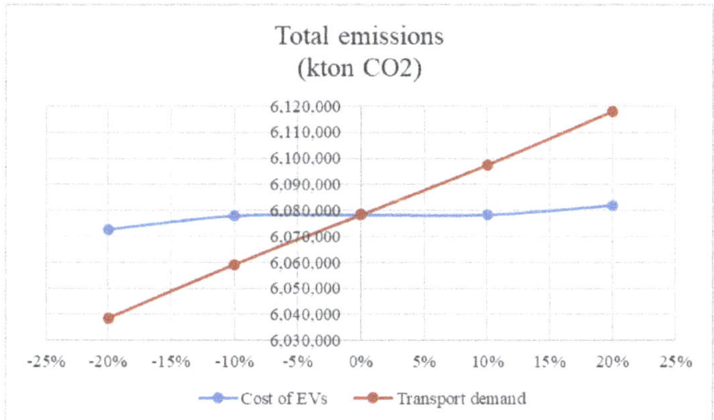

Figure 11. Total emissions change with different data for transport sector demand and cost of electric vehicles (EVs) in the base scenario.

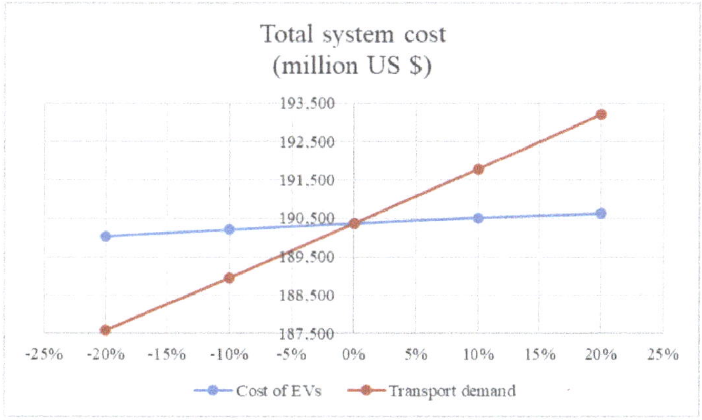

Figure 12. Total system cost change with different data for transport sector demand and cost of EVs in the base scenario.

8. Conclusions

Evaluation of electric vehicle deployment with respect to energy and environmental impacts is a multilateral task in need of a technologically detailed multisectoral analysis. To evaluate the long-term impacts of alternative policies, the bottom-up model BUEMS has been used and calibrated under various scenarios according to the most recent Turkish energy and transport sector data. All the complex relationships of producing, transforming, transmitting and/or supplying energy sources according to the useful demand characteristics are represented with great technological detail.

Model results highlight win-win policy options that lead to profitable decarbonization. In particular, it is found that a ban of diesel fueled vehicles reduces lifetime emissions as well as lifetime costs. Hence, the diesel ban scenario results in negative abatement costs over the planning horizon. This finding is in support of all the urban policies being implemented worldwide to get rid of diesel fueled vehicles in the cities, which has become very popular especially after the Dieselgate scandal in 2015. German cities started introducing bans on older diesel vehicles with the court several years ago. Major cities like Paris and London are now following this trend, which leads to profitable decarbonization in Turkey. Therefore, while there still is no such agenda, Turkish municipalities should consider introducing a ban on diesel fueled vehicles to exploit the profitable emission reduction potential. Besides banning diesel fueled vehicles, Turkish municipalities should also consider introducing electric buses for public transportation as pointed out by model results. The marginal cost of emission reduction through e-buses is much lower than through other policy measures like carbon taxation in transport.

It should be noted that the electricity sector plays a crucial role on the sustainability of e-mobility. It is found that the elimination of emissions on the road through e-mobility is more offset by emissions from power plants if electricity is generated by coal. This is a particularly important finding for a country like Turkey, which has both ambitious e-mobility diffusion plans as well as policy priority goals on the utilization of domestic coal reserves for electricity generation. From a sustainability point of view, these two do not go hand in hand as pointed out by the scenario results. Decarbonization of electricity generation is therefore of primary importance for electrification of transportation. It is found that a ban on new coal-fired power generation is a cost-effective way for reducing emissions in the electricity sector. Introducing an emission tax in the electricity sector is about five-fold more cost effective than introducing an emission tax in transportation (5.47 \$/ton CO_2 in electricity vs. 27.37 \$/ton CO_2 in transport for 50 \$/ton CO_2 emission tax).

For further research, model resolution can be increased so as to introduce a daily load duration curve and allow for optimization of instantaneous charging needs with network supply/demand balance. Also, the model can be developed into a price-elastic version to reflect the impact of price changes on the demand for electricity, other energy sources and technology choices. Furthermore, the storage potential of renewable power generation or their intermittent nature [31] can be explored in more detail using a stochastic model extension.

Author Contributions: Conceptualization, G.K. and C.C.; methodology, G.K. and E.S.; validation, J.D. and E.S.; resources, J.D. and E.S.; data curation, G.K. and C.C.; writing–original draft preparation, G.K. and C.C.; writing–review and editing, J.D. and E.S.; project administration, G.K., J.D. and E.S.; funding acquisition, G.K. and E.S. All authors have read and agreed to the published version of the manuscript.

Funding: This research was funded by the Scientific and Technical Research Council of Turkey under Scientific Research Project #104M348. This research was also partially funded by the United States National Science Foundation under Award No. 1847077.

Acknowledgments: The authors would like to acknowledge the support from Bogazici University under Scientific Research Project BAP12281, as well as the Fulbright Commission under Grant #FY-2019-TR-SS-17 for featuring the international collaboration.

Conflicts of Interest: The authors declare no conflict of interest. The funders had no role in the design of the study; in the collection, analyses, or interpretation of data; in the writing of the manuscript, or in the decision to publish the results.

Abbreviations

The following abbreviations are used in this manuscript:

CO_2	Carbon dioxide
USA	United States of America
EV	Electric Vehicle
BUEMS	Bottom-Up Energy Modeling System
IEA	International Energy Agency
IPCC	Intergovernmental Panel on Climate Change
V2G	Vehicle to Grid
CV	Conventional gasoline Vehicle
HEV	Hybrid Electric Vehicle
PEV	Plug-in Electric Vehicle
LDV	Light-Duty Vehicle
GCAM	Global Change Assessment Model
TIMES	The Integrated MARKAL-EFOM System
LEAP	Long-range Energy Alternatives Planning model

Appendix A. Survey Instrument

1- What are your future predictions about the number of electric vehicles in Turkey based on the fact that the number of electric vehicles reached 657 in total with 155 sales in 2018 according to TEHAD (Turkey Electric & Hybrid Vehicles Association) data?

	2018	2022	2032	2042	2052
Number of electric vehicles	657				

2- What are your forecasts about the number of charging stations in the future in regard to existence of 400 charging stations in Turkey that are publicly available and actively in use as of the end of 2017 according to EMRA?

	2017	2022	2032	2042	2052
Total number of charging stations (private & public)	1500				
Number of publicly available charging stations	400				

3- What are your future expectations for electricity consumption and charging efficiency of electric vehicles?

	2018	2022	2032	2042	2052
Electricity consumption (kWh per 100km)	15-20				
Charging efficiency (%)	85%-90%				

4- What are your predictions related to fixed and variable maintenance costs (TL or $ per km) of electric vehicles in the future?

	2018	2022	2032	2042	2052
Fixed O&M cost	1200 TL / 60k km				
Variable O&M cost)	5 TL / 100 km (fuel cost)				
Average battery cost	250 $ / kWh				

5- What is the most important milestone in transition to electric vehicle usage and what kind of incentives should be practiced for the electric vehicles deployment?

Figure A1. Electric vehicle deployment projection survey in Turkey.

References

1. IEA. CO$_2$ Emission from Fuel Combustion, Statistics Report. Technical Report. 2019. Available online: https://www.iea.org/reports/co2-emissions-from-fuel-combustion-2019 (accessed on 30 July 2020).
2. Pachauri, R.K.; Allen, M.R.; Barros, V.R.; Broome, J.; Cramer, W.; Christ, R.; Church, J.A.; Clarke, L.; Dahe, Q.; Dasgupta, P.; et al. *Climate Change 2014: Synthesis Report. Contribution of Working Groups I, II and III to the Fifth Assessment Report of the Intergovernmental Panel on Climate Change*; World Meteorological Organization (WMO): Geneva, Switzerland, 2014. Available online: https://epic.awi.de/id/eprint/37530/1/IPCC_AR5_SYR_Final.pdf (accessed on 30 July 2020).
3. IPCC. *Climate Change 2014, Mitigation of Climate Change. Contribution of IPCC AR5 WG3 2014: Working Group III to the Fifth Assessment Report of the Intergovernmental Panel on Climate Change*; Cambridge University Press: Cambridge, UK; New York, NY, USA, 2014.
4. Shittu, E.; Parker, G.; Jiang, X. Energy technology investments in competitive and regulatory environments. *Environ. Syst. Decis.* **2015**, *35*, 453–471. [CrossRef]
5. Shittu, E. Energy technological change and capacity under uncertainty in learning. *IEEE Trans. Eng. Manag.* **2013**, *61*, 406–418. [CrossRef]
6. Kamdem, B.G.; Shittu, E. Optimal commitment strategies for distributed generation systems under regulation and multiple uncertainties. *Renew. Sustain. Energy Rev.* **2017**, *80*, 1597–1612. [CrossRef]

7. Deluque, I.; Shittu, E.; Deason, J. Evaluating the reliability of efficient energy technology portfolios. *EURO J. Decis. Process.* **2018**, *6*, 115–138. [CrossRef]

8. Alexander, M.; Tonachel, L. Projected Greenhouse Gas Emissions for Plug-in Electric Vehicles. *World Electr. Veh. J.* **2016**, *8*, 987–995. [CrossRef]

9. UNFCCC. National Inventory Report Turkey. 2019. Available online: https://unfccc.int/documents/19481 (accessed on 4 August 2020).

10. Zhang, H.; Chen, W.; Huang, W. TIMES modelling of transport sector in China and USA: Comparisons from a decarbonization perspective. *Appl. Energy* **2016**, *162*, 1505–1514. [CrossRef]

11. Wu, D.; Aliprantis, D.C. Modeling light-duty plug-in electric vehicles for national energy and transportation planning. *Energy Policy* **2013**, *63*, 419–432. [CrossRef]

12. Tsita, K.G.; Pilavachi, P.A. Decarbonizing the Greek road transport sector using alternative technologies and fuels. *Therm. Sci. Eng. Prog.* **2017**, *1*, 15–24. [CrossRef]

13. Paladugula, A.L.; Kholod, N.; Chaturvedi, V.; Ghosh, P.P.; Pal, S.; Clarke, L.; Evans, M.; Kyle, P.; Koti, P.N.; Parikh, K.; et al. A multi-model assessment of energy and emissions for India's transportation sector through 2050. *Energy Policy* **2018**, *116*, 10–18. [CrossRef]

14. Li, F.; Cai, B.; Ye, Z.; Wang, Z.; Zhang, W.; Zhou, P.; Chen, J. Changing patterns and determinants of transportation carbon emissions in Chinese cities. *Energy* **2019**, *174*, 562–575. [CrossRef]

15. Lazarou, S.; Christodoulou, C.; Vita, V. Global Change Assessment Model (GCAM) considerations of the primary sources energy mix for an energetic scenario that could meet Paris agreement. In Proceedings of the 2019 54th International Universities Power Engineering Conference (UPEC), Bucharest, Romania, 3–6 September 2019; pp. 1–5.

16. Lazarou, S.; Markovic, F.; Dagoumas, A. Energy and climate policy with the GCAM model: Assessing energy sources and technology options. *Int. J. Renew. Energy Res. (IJRER)* **2018**, *8*, 2299–2309.

17. Karali, N. Design and Development of a Large-Scale Energy Model. Ph.D. Thesis, Bogazici University, Istanbul, Turkey, 2012.

18. Işık, M. Energy, Economy and Environment Integrated Large-Scale Modeling and Analysis of the Turkish Energy System. Ph.D. Thesis, Bogazici University, Istanbul, Turkey, 2016.

19. Tubitak. *Development of the Boğaziçi University Energy Modeling System and Evaluation of Greenhouse Gas Emission Restrictions on the Turkish Economy*; Technical Report; Scientific and Technical Research Council of Turkey: Istanbul, Turkey, 2018.

20. Tetik, E. Turkish Electricity Sector: A Bottom-Up Approach. Master's Thesis, Isik University, Istanbul, Turkey, 2017.

21. Günel, G. Analysis of CO_2 Emission Reduction, Energy and Economy Interactions in Turkey Using BUEMS-MACRO. Master's Thesis, Boğaziçi University, Istanbul, Turkey, 2018.

22. Yıldırım, I. Energy Policy Analysis with the Boğaziçi University Energy Modeling System BUEMS: Prospects for the Diffusion of Renewable Energy Tecnologies and Utilization of Domestic Coal Reserves in Turkey. Master's Thesis, Boğaziçi University, Istanbul, Turkey, 2019.

23. IEA. World Energy Outlook 2009. Available online: http://large.stanford.edu/courses/2013/ph241/robert s2/docs/WEO2009.pdf (accessed on 30 July 2020).

24. Canaz, C. Transport Sector Energy Use, Electric Vehicle Deployment and CO2 Emissions in Turkey: An Evaluation Using the Boğaziçi University Energy Modeling System. Master's Thesis, Boğaziçi University, Istanbul, Turkey, 2019.

25. IEA. Global EV Outlook 2018: Towards Cross-Modal Electrification. Technical Report. 2018. Available online: https://www.oecd.org/publications/global-ev-outlook-2018-9789264302365-en.htm (accessed on 30 July 2020).

26. Erdem, U. Varank'a Göre 2022'de 140 bin Elektrikli Araç Yollara çıkacak. 2019. Available online: https://www.hurriyet.com.tr/ekonomi/varanka-gore-2022de-140-bin-elektrikli-arac-yollara-ci kacak-41109629 (accessed on 30 July 2020).

27. MENR. Electric Energy Markets and Supply Security Strategy Document. 2009. Available online: https://www.enerji.gov.tr/File/?path=ROOT (accessed on 30 July, 2020).

28. Soczu. Avrupa'daki dizel Yasakları Türkiye'yi de Etkileyecek. 2020. Available online: https://www.so zcu.com.tr/2020/otomotiv/avrupadaki-dizel-yasaklari-turkiyeyi-de-etkileyecek-5566743/ (accessed on 30 July 2020).

29. Milliyet. Istanbul Otobusleri 'Elektrik'lenecek. 2016. Available online: https://www.milliyet.com.tr/gundem/istanbul-otobusleri-elektrik-lenecek-2355216 (accessed on 30 July 2020).
30. Barron, A.R.; Fawcett, A.A.; Hafstead, M.A.; McFarland, J.R.; Morris, A.C. Policy insights from the EMF 32 study on US carbon tax scenarios. *Clim. Chang. Econ.* **2018**, *9*, 1840003. [CrossRef] [PubMed]
31. Jiang, X.; Parker, G.; Shittu, E. Envelope modeling of renewable resource variability and capacity. *Comput. Oper. Res.* **2016**, *66*, 272–283. [CrossRef]

Article

How Policies Affect the Dissemination of Electric Passenger Cars Worldwide

Marina Siebenhofer *, **Amela Ajanovic and Reinhard Haas**

Energy Economics Group (EEG), Institute of Energy Systems and Electrical Drive, Vienna University of Technology (TU Wien), Gusshausstraße 25-29, E370, 1040 Vienna, Austria; ajanovic@eeg.tuwien.ac.at (A.A.); haas@eeg.tuwien.ac.at (R.H.)
* Correspondence: siebenhofer@eeg.tuwien.ac.at

Abstract: Road transportation is one of the largest emitters of greenhouse gas emissions. The EU set the target to reduce overall transport emissions by 60% by 2050 compared to 1990. Electric mobility is considered a proper means to achieve this goal. Battery electric vehicles (BEVs) are a mature technology. The high investment costs, limited driving range and a charging infrastructure that is not extensive yet are currently the main challenges. This work analyses how policies affect the dissemination of BEVs in selected countries with remarkable market shares of BEVs. The core objective is to investigate how policies affect BEV economics compared to conventional car economics. Financial policies and their effects on BEVs for the major markets of China, the USA and Europe were analysed. To do so, the total cost of ownership (TCO) was calculated for each country. The major conclusions were: (i) The investment cost of a car had the most significant impact on the TCO; (ii) Low TCO as an incentive was not enough to ensure successful BEV dissemination; (iii) Non-monetary incentives such as access to certain zones and the usage of bus lanes for BEVs combined with registration taxes, low electricity prices and high fuel prices were very favourable conditions.

Keywords: battery electric vehicles; emissions; electric mobility; policies; transport

Citation: Siebenhofer, M.; Ajanovic, A.; Haas, R. How Policies Affect the Dissemination of Electric Passenger Cars Worldwide. *Energies* **2021**, *14*, 2093. https://doi.org/10.3390/en14082093

Academic Editor: Miguel-Angel Tarancon

Received: 30 January 2021
Accepted: 6 April 2021
Published: 9 April 2021

Publisher's Note: MDPI stays neutral with regard to jurisdictional claims in published maps and institutional affiliations.

1. Introduction

The transport sector causes around 30% of the greenhouse gas (GHG) emissions in the EU; thereof, 72% is road transport. In contrast to other sectors (energy, industry, residential, agriculture, forestry and fishing), traffic emissions have been increasing rapidly since 1990 [1].

Figure 1 shows the development of CO_2 emissions in the transport sector (road, railway, aviation, other) for selected EU countries, over the period of 1990–2018, in relation to 1990. In most of the countries, emissions had been continuously increasing until 2007. After a financial crisis, emissions decreased in most of the countries. However, this decrease was followed by an emission-increasing trend starting from about 2013. Significant differences among countries could also be seen due to country-specific circumstances such as gross domestic product (GDP) and national policy framework.

If the number of conventional vehicles with internal combustion engines (ICEs) could be sharply reduced, the emissions from the transport sector would decrease. In contrast to ICEs, electric vehicles (EVs), especially battery electric vehicles (BEVs) and fuel cell vehicles (FCVs), have zero emissions at the point of use and could significantly contribute to the reduction of GHG emissions in combination with electricity produced from renewable energy sources (RES). In the best case, EVs could cause 75% to 90% lower GHG emissions than ICEs [2]. Furthermore, in combination with electricity from RES, battery electric vehicles (BEVs) could save about 65% to 70% in GHG emissions compared to plug-in hybrid electric vehicles (PHEVs) [3].

51

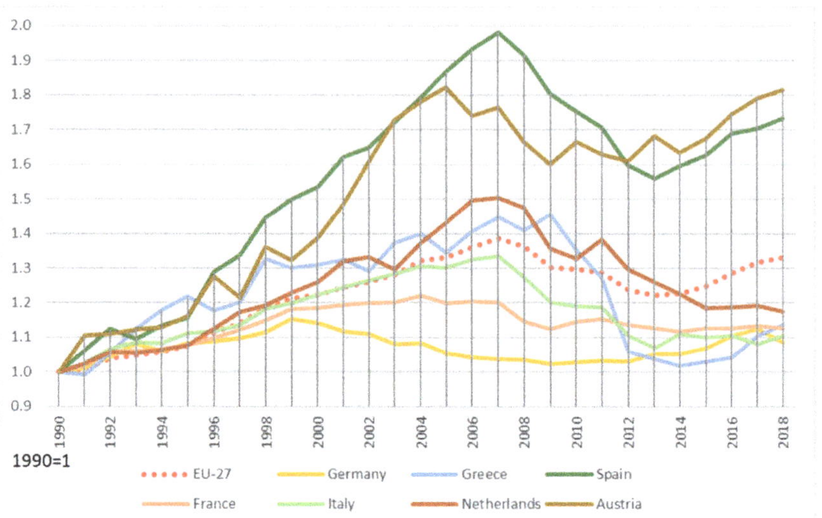

Figure 1. Development of CO_2 emissions from transport in selected EU countries [4].

Far-reaching arguments against switching to EVs are high investment costs compared to ICEs and technical limitations (e.g., limited driving range and charging infrastructure availability) [5]. The investment costs of an EV are about one-third higher in comparison to an ICE. Furthermore, the driving range of an EV is quite low, mostly about 130–250 km, compared to the 700 km range of an average ICE [6]. A requirement for the successful distribution of EVs is the availability of charging infrastructure. If this is not the case, range anxiety occurs [7].

Politicians have at least two options to address these problems. On the one hand, BEVs can be promoted directly by implementing monetary policies such as subsidies or with non-monetary ones such as the use of bus lanes for BEVs and free parking. On the other hand, BEVs can be pushed indirectly, e.g., through a CO_2 tax on all fuels and higher registration taxes on conventional vehicles than on BEVs. Another legal option is driving bans for diesel and gasoline cars in cities or emission-free zones.

The core objectives of this paper are to analyse the current economic state of BEVs in comparison to conventional cars and to identify proper policies to overcome the major current barrier of high investment costs. Regarding economics, we investigated the total cost of ownership (TCO), including existing national policy instruments such as fuel and registration taxes as well as subsidies. The TCO was calculated for selected major countries worldwide. The selection of countries was based on their relevance so far regarding market penetration of BEVs such as the major BEV markets of China, California in the USA and the most important European countries (Austria, Germany, The Netherlands and Norway). For these countries, detailed economic analyses were conducted for two cases of cars, small and large ones.

Regarding proper policies to overcome major current barriers, measures in different dimensions (subsidies, standards, taxes and legal frameworks) and applicable monetary and non-monetary incentives to promote BEVs (and reduce the number of ICEs) are elaborated upon. Furthermore, fuel and electricity prices as well as BEV distribution in selected countries (China, Japan, the USA, UK and some European countries (Austria, France, Germany, Sweden and Norway)) are examined. Moreover, amounts of subsidies, exemptions from taxes and non-monetary measures for the selected countries are elaborated upon.

The major new contribution of this paper is that a comprehensive and up-to-date survey, discussion and assessment on all policies existing in major countries regarding market deployment of BEVs—direct and indirect measures, monetary and non-monetary policies, technical and behavioural—is conducted and completed by a sound and detailed economic analysis. Hence, it also closes the existing research gap regarding state-of-the-art economics and the impact of promotion policies regarding BEVs. To the best of our knowledge, no such comprehensive analysis exists. Of course, policies to support E-Mobility have been investigated in several previous studies.

Rietman et al. [8] dealt with political measures and how they promoted E-Mobility. The study investigated the effectiveness of measures in 20 countries. The authors stated that cooperation between public and private sectors was essential to promote EVs. Moreover, monetary measures, in combination with measures for the charging infrastructure, were highly effective. Government measures suggested that government policies reflected the preferences of consumers. Furthermore, countries with higher purchasing power also had higher EV penetration. Cansino et al. [9] provided an overview of the most important measures to promote E-Mobility in the EU-28. The authors concluded that in addition to financial incentives for purchasing and supporting R&D projects, tax and infrastructure measures were the most effective ways to promote EVs. In addition, they found that in countries where CO_2-based taxation had been introduced, penetration rates were higher. Dijk et al. [10] examined the socio-technical development of E-Mobility. They summarised that E-Mobility was suffering from high oil prices, carbon limits, car-sharing and intermodality. Moreover, the EV market was mainly dependent on the progress of batteries, measures to reduce CO_2 emissions, new value propositions by companies and the image of EVs. Held et al. [11] analysed mobility policies to increase adoption rates of e-vehicles in 15 EU cities. They found that policies that had a more substantial impact on the TCO of EVs, combined with incentives for installing charging infrastructure and a public power grid combined with push factors that make the use of conventional cars unattractive, led to beneficial effects. Incentives should always have been linked to de-incentives; isolated measures were less effective. Gass et al. [12] analysed alternative policy instruments to promote EVs in Austria in their 2011 study. They concluded that up-front price support worked better than taxation systems. Moreover, significant learning effects should reduce the cost of EVs in the future, which can be achieved primarily by promoting research. Wang et al. [13] investigated incentive measures and their influence on the market share of EVs. They examined which measures, other than highly effective subsidies, could drive EV penetration. Charging infrastructure, fuel price, and access to bus lanes for EVs were key factors in driving EV penetration. In summary, the most important factor was not direct subsidies but road priority (access to high-occupancy vehicle lanes (HOVs) and bus lanes). The study by Vilchez et al. [14] identified factors that influenced the EV market in Europe. They concluded that the purchase price of EVs was still a barrier. Consumers preferred government incentives. In addition, socio-cultural characteristics of consumers also influenced purchase intentions. The impact of energy policies on scenarios regarding GHG emission reduction in passenger car mobility in the EU-15 was analysed by Ajanovic/Haas. The core message was that policymakers must set clear and rigorous priorities to reduce CO_2 emissions. The most important goals were to improve energy efficiency and reduce energy consumption [5,15]. Cherchi [6] provided a work about the measurement of the effect of informational and normative conformity for EVs compared to ICEs. She summarised that social conformity effects such as advice from non-experts to potential buyers were highly effective for the dissemination of EVs. Such effects could compensate for low driving ranges of EVs or differences in the purchase price. Furthermore, a combination of free/reduced parking fees with reserved parking spots was highly effective in spreading EVs. Palmer et al. [16] showed the impact of ownership costs on market share by calculating a TCO over a 16-year period. Among other findings, they found that long-term government support was essential to promote the dissemination of EVs.

In this article, an examination of how policies affect the distribution of BEVs and ICEs in selected countries is provided. The methods of approach are given in Section 2. A general overview of energy policy instruments in transport that affect the promotion of

BEVs is given in Section 3. Recent developments of BEVs are stated in Section 4. Incentives for BEVs in selected countries and a comparison of these is provided in Section 5. To see how current policies affect ICEs' and BEVs' costs from an individual perspective, a TCO is calculated for each of the selected countries; the results can be found in Section 6. Finally, the conclusions and future perspectives for further developing BEVs are given in Section 7.

2. Method of Approach

One of the arguments against switching to BEVs is high investment costs [6]. To compare the economic performance of an ICE with a BEV, a TCO was calculated. With the calculation of a TCO, direct and indirect mobility costs could be shown from an individual perspective. In this way, the effects of monetary measures and sales could be identified [17]. The aim was to identify the difference between ICEs' and BEVs' costs, considering the capital costs, operating and maintenance costs and electricity/fuel costs.

To analyse the economics of the BEVs, the TCO was calculated in € per year. For the calculation, 12,500 km per year was assumed [18]. The TCO in € per year was calculated as follows:

$$\mathrm{TCO} = C_{Cap} + C_{O\&M} + C_E \ [\text{€/year}] \tag{1}$$

C_{Cap} is the cost of capital for the vehicle including purchase subsidies, either ICE or BEV [€/year]; $C_{O\&M}$ is the cost of operation and maintenance [€/year], which includes the insurance, maintenance and repair costs; parking tolls and road charges are not considered; and C_E is the cost of energy, either fuel or electricity [€/year].

The capital cost per year depending on the initial investment cost IC_0 (including subsidies τ_{sub}) and the capital recovery factor (CRF) α is:

$$C_{Cap} = (IC_0 - \tau_{sub}) \times \alpha \ [\text{€/year}] \tag{2}$$

α describes the ratio of a constant annuity to the present value of this annuity's receipt for a given time. It is calculated using an interest rate z and a depreciation period n. With the annuity method, the depreciation costs and annual capital costs are determined with the same recovery factor. The depreciation period is assumed to be eight years, the same as the lifetime of vehicles. This period also covers the warranty period for a battery for driving distances between 80,000 and 24,000 km. An interest rate of 5% is assumed, as this is a standard interest rate from a bank loan [19].

$$\alpha = \frac{z \, (1+z)^n}{(1+z)^n - 1} \tag{3}$$

The cost of energy depends on the price for the fuel or electricity p_f [€/kWh or €/l], the km driven per year vkm and the fuel intensity FI [kWh/km].

$$C_E = p_f \times \text{vkm} \times FI \ [\text{€/year}] \tag{4}$$

The fuel price is the sum of the net price p_{net}; the value-added tax (VAT) τ_{VAT}, which, in the EU, was in the range from 17% to 27% in 2020 [20]; CO_2 based tax τ_{CO2} and an excise tax on fuels τ_{Excise}.

$$P_f = P_{net} + \tau_{VAT} + \tau_{CO2} + \tau_{Excise} \ [\text{€/kWh}] \tag{5}$$

In the following, we apply this method to calculate TCO for two specific cases (small and large cars), wherein we compare the TCO for selected countries with some relevance of BEVs as described in Section 4. This comparison was made to identify the current level of the economics of BEVs compared to conventional vehicles. Note that all currently applied policy instruments in our state of knowledge—taxes and subsidies—were considered and included in this investigation.

The first case included a comparison of TCO for small cars, the Volkswagen (VW) Golf and the VW E-Golf. We chose approximately the same type of vehicle to guarantee a valid comparison. Comparability was given due to similar engine power (VW Golf–110 kW,

VW E-Golf–100 kW) and similar weight (VW Golf—1211 kg, VW E-Golf—585 kg) of the compared cars; see Table 1. In the second case, the TCO was calculated for bigger cars, an Audi A5 and a Tesla Model 3. The comparability was also mad through similar engine power (Audi A5—195 kW, Tesla Model 3—211 kW) and similar weight (Audi A5—1600 kg, Tesla Model 3—1684 kg).

Table 1. Technical specifications for vehicles analysed.

Technical Specifications	Case 1		Case 2	
	VW Golf 1.0 TSI	**VW E-Golf**	**Audi A5**	**Tesla Model 3**
Energy Source	Gasoline	Electricity	Gasoline	Electricity
Engine Power (kW)	110	-	195	-
Electric Motor Power (kW)	-	100	-	211
Battery Capacity (kWh)	-	35.8	-	50
Fuel Tank Capacity (litres)	50	-	58.9	-
Performance				
Weight (kg)	1211	1585	1600	1684
Fuel Consumption (l/100 km)	4.3	-	5.9	-
Electricity Consumption (kWh/100 km)	-	12.7	-	15.3
Tailpipe CO_2 Emissions (g/km)	141	-	113	-
Driving Range (km)	1163	201	998	386
Purchase Price	18,640 €	16,540 €	34,220 €	29,870 €

In addition to the technical components, which show the comparability of the models, the price difference between the models should also be discussed at this point. The price difference between the two models in the first case (VW Golf: €18,640; E-Golf: €16,540) was approximately €2100. In the second case (Audi A5: 34,220 €; Tesla Model 3: 29,870 €) the difference was about 4350 €. Apart from that, the Tesla Model 3 was a more exclusive model than the Audi A5. These factors should be taken into account.

Furthermore, a sensitivity analysis was performed to determine the impact of a change in the interest rate or the depreciation time on the annual costs of a vehicle.

3. A Survey on Energy Policies in Transport

In this chapter, an overview of possible policies and measures to promote E-Mobility, focusing on BEVs, is provided. First of all, it is shown which policies exist and on which segment of the transport sector, e.g., mobility, they act. Figure 2 shows how mobility and policy measures in different dimensions interact. It depicts the complexity and the connections between policies and the service mobility itself. In the following examples, policies related to energy, infrastructure and vehicles are given.

1. Energy

To reduce emissions from the transport sector, it is essential that energy used in vehicles comes from low-carbon and renewable energy sources (such as sun, wind, biomass). If this is not ensured, then the vehicles, both ICEs and EVs, are powered by fossil fuels. With the promotion of renewable energy and renewable energy systems (RES), the vehicle fleet could become more sustainable. With, for example, taxes on fuel, mobility with conventional cars becomes more expensive. With standards and, for instance, a minimum of fuel intensity, mobility becomes again more environmentally friendly.

2. Infrastructure

The infrastructure can be directly influenced by allowing the use of bus lanes for BEVs. Other examples are free and/or reserved parking spaces for BEVs or emission-free zones where BEVs are permitted to access. However, most important is to develop the necessary charging infrastructure.

3. Vehicle

For example, BEVs can be promoted directly with subsidies on the purchase price, on the insurance or indirectly with a scrapping premium on ICEs. Exemptions from various taxes like the registration tax, the import tax or the VAT are further promotion options to push down vehicle prices. Furthermore, the total fleet can be promoted through CO_2 regulations, for example, by setting a limit value in gCO_2/km for fleet consumption.

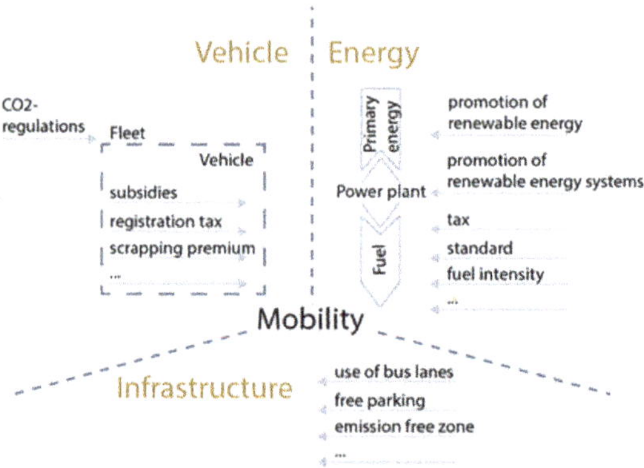

Figure 2. Mobility and policy measures in different dimensions.

Figure 3 shows four dimensions, which are fundamental for energy policy interventions in general. In principle, the promotion of BEVs can be affected in four different ways, either financial through subsidies and taxes or regulatory through standards or legal frameworks. Figure 3 also includes specific measures for BEVs in the four dimensions.

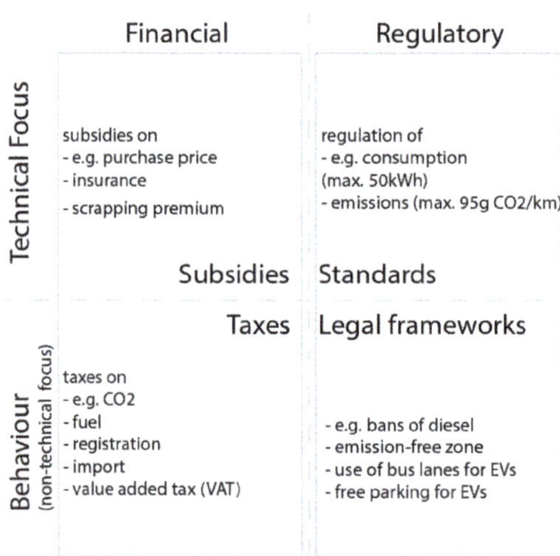

Figure 3. The four major dimensions of energy policies with a special focus on transport.

As shown in Table 2, policy instruments that can be implemented can be divided into monetary incentives and non-monetary incentives. Furthermore, the columns with the terms "direct policies" and "indirect policies" describe whether the incentives affect the BEVs directly or indirectly from an individual drivers' perspective. These possible incentives, in addition to specific measures implemented in countries, can be found in Section 5.

1. Monetary Incentives

Subsidies: Often, the government grants money off the purchase list price of BEVs. Other possibilities are discounts on car insurance or a so-called scrapping premium if the old ICE car is scrapped and in the best case, a BEV is purchased instead. Furthermore, the state can subsidise electricity prices, which affect the charging prices via the public loading infrastructure or mainly for the infrastructure at home.

Tax payments: In almost all countries, there are exemptions from taxes for BEVs (tax benefits). The proportion of tax to vehicle price/operation varies widely between countries. The exemption from car purchase taxes and registration taxes such as the value-added tax for BEVs is usual. Registration taxes are often based on the CO_2 emissions of a vehicle. Mostly, customers must pay a motor vehicle tax for car ownership; exemptions are also common.

In comparison, ICEs are exempt from taxes if they fall below a specific CO_2 or kW limit. In addition, there are electricity taxes to pay. Indirect policy instruments that affect BEVs monetarily are CO_2 taxes on petrol and diesel or high registration taxes on ICEs.

Other: BEVs can often park free of charge. Furthermore, often exemptions/reductions for tolls, congestion charging or low emission zone charging are given.

2. Non-monetary Incentives

Standards: Mandatory standards that affect BEVs' spread are stipulating minimum energy levels (energy efficiency standards) or maximum energy use levels (maximum energy consumption) on vehicles [21]. On the other side, indirect standards are CO_2 regulations (e.g., 95 gCO_2/km within the EU) or standards on diesel concerning the fuel intensity.

Legal: Furthermore, some laws/regulations allow EVs to use low/free-emission zones or bus lanes. These permits also have an indirect effect, as they exclude ICEs from using low/free-emission zones.

Other: Other non-monetary incentives are national EV sales targets. Because of the objectives, governments need to offer incentives for EV dissemination. Reserved parking spaces for EVs are common. Free parking for EVs is also indirect, as it excludes ICEs from using them. In Scandinavia, EVs are sometimes allowed to travel free of charge by ferry, and in some places, there are also free electric charging stations. Exemptions from road charges can also be given.

Which of the policies will be applied is very dependent on regional differences and country specific circumstances, such as GDP and national policy targets. Countries with higher GDP are able to provide higher subsidies and tax reductions. For countries with lower GDP, indirect monetary measures could be more attractive.

Table 2. Policy instruments to promote E-Mobility [19,22–35].

Policy Instruments		"Direct Policies"	"Indirect Policies"
Monetary Incentives	Subsidies	• E.g., vehicle purchase price • Vehicle insurance • Electricity prices • Private charging infrastructure	• E.g., reduction of subsidies for fossil fuels • Scrapping premium
	Tax Payment	Exemptions • E.g., car purchase and registration tax • Value-added tax (VAT) • Car ownership • Motor vehicle tax • Consumption of fuel/electricity • Electricity tax • Use of road infrastructure • Import tax Benefits • E.g., company tax	• E.g., CO_2 tax on petrol, diesel • Registration tax on ICEs • Fuel tax
	Other	• E.g., lower or no parking fees • Lower or no tolls • Lower no congestion charging • Lower or no low-emission zone charging	• E.g., higher parking fees for ICEs
Non-Monetary Incentives	Standards	• E.g., maximum energy consumption (negative) • Minimum energy efficiency/energy efficiency standards	• E.g., CO_2 regulations (e.g., 95 g/km) • Standards on diesel (fuel intensity)
	Legal	• E.g., use of bus lanes • Low/free-emission zones • Reserved parking spaces	• E.g., low/free-emission zones
	Other	• E.g., EV sales targets • Good charging infrastructure • Free parking • Free fairy crossings • Free charging • Exemption from road charges	• E.g., parking spaces only for EVs

4. Recent Developments of BEVs

In this chapter, the BEV market share in selected countries—the USA, China, Japan and some European countries—is highlighted. For the selected countries and the rest of the world, global electric car sales are also stated. Furthermore, a comparison between fuel and electricity prices is provided.

Figure 4 shows that Norway had the worldwide highest market share of BEVs. There had been a rapid share increase in Norway since 2013. In 2010, the market share was only 0.3%, but it was already 45.6% in 2020. The rate increased by a factor of 152 in 10 years. It is illustrated that The Netherlands, in second place, also had a rapid increase in BEV market share rates in recent years. While in 2014, the market share in The Netherlands amounted to only 0.7%, in 2019, it had increased to 15.2%. However, in 2020, the market share fell to 10.2%. As shown in Figure 5, Sweden also had a high market share, with only 0.8% in 2015, but the market share increased rapidly to 7.9% in 2020. Throughout the EU, the market share of BEVs was relatively low, at 4.2% in 2020. Similarly, China (3.9% in 2019), Japan

(0.5% in 2019) and the USA (1.5% in 2019) in 2019 had very low BEV shares. The decline in BEV market share in Japan from 0.6% in 2018 to 0.5% was mainly due to the high purchase prices of BEVs. In addition, there were very few fast chargers that were publicly available in 2019. Furthermore, there were very limited tax incentives for the purchase of BEVs in Japan. [36]. Figure 5 was added to illustrate the developments in the markets in countries with market shares below 6%.

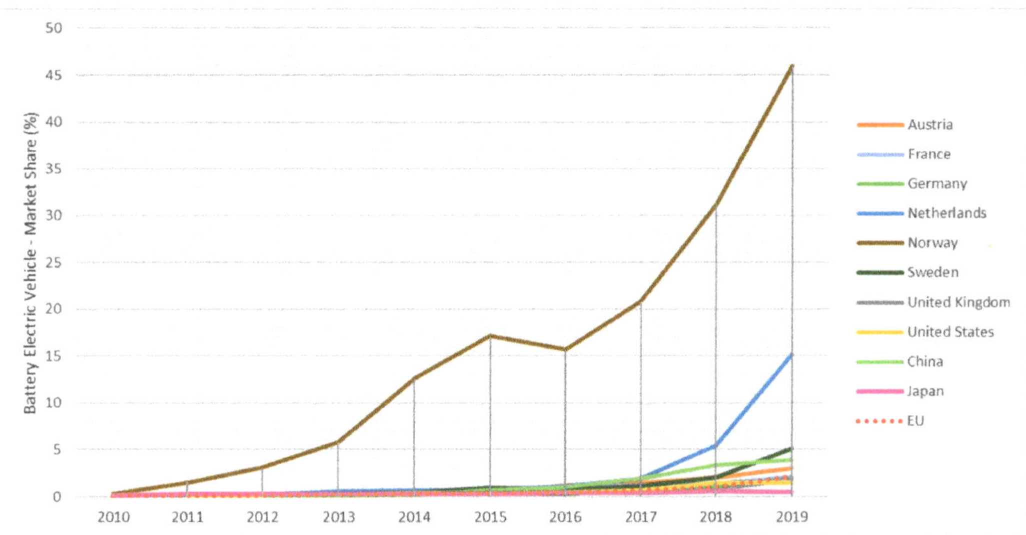

Figure 4. Battery electric vehicle market share in new registrations in major countries [36,37].

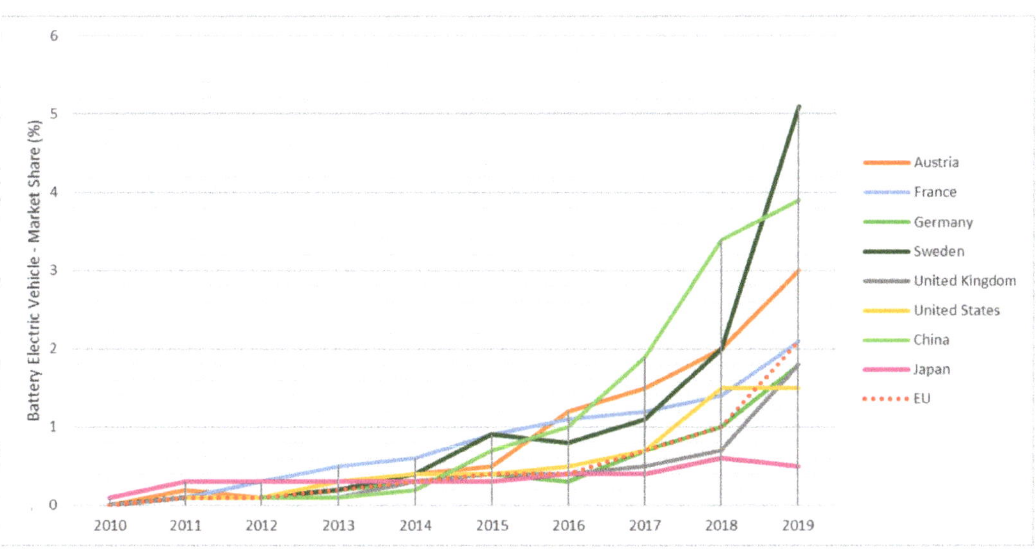

Figure 5. Battery electric vehicle market share in new registrations in major countries with a market share below 6% [36,37].

Figure 6 illustrates global electric car sales in major countries. As is shown in Figure 6, China had the highest amount of BEV sales. In 2011, the sales amounted to 0.01 million. Since then, the rate of EVs increased in 2018 to 1.18 million. In 2019, it was 1.10 million.

The EU followed China with an amount of 0.01 million EV sales in 2011, 0.40 million sales in 2018 and 0.59 million in 2020. The third key market was the USA, with 0.36 million EV sales in 2018 and 0.22 million in 2019. Throughout the world, the EV sales amounted to 2.3 million with a market share of 3.2% in 2019. Preliminary numbers for global BEV sale in 2020 are also depicted in Figure 6.

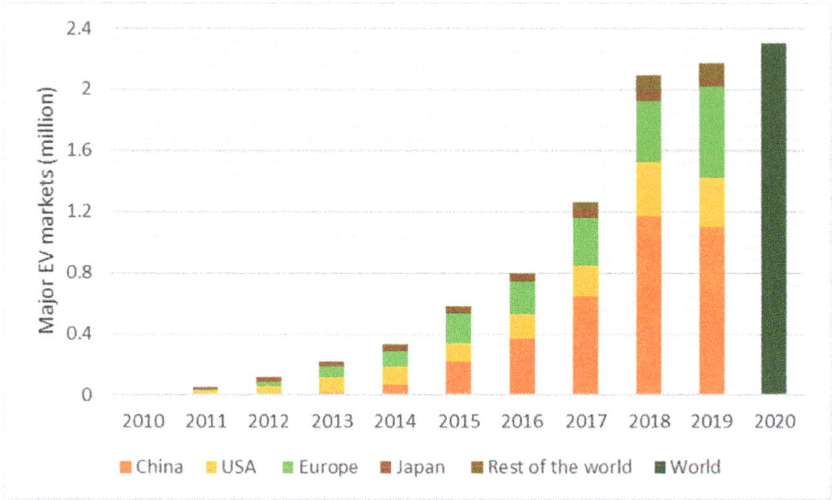

Figure 6. Global electric car sales in major countries [38].

In general, petrol prices are strongly related to the overall economic situation of a country. Prices vary from country to country, sometimes considerably. The higher the GDP, the higher the petrol and diesel prices. The USA is an exception to this with a high GDP but very low taxes on mineral oil products. The average gasoline price worldwide was 1.05 USD per litre in 2020 [39]. The attractiveness of BEVs depends strongly on fossil fuel prices and electricity prices. A good example is Norway, where electricity prices are very low in relation to fuel prices.

Figure 7 illustrates electricity prices in comparison to fossil fuel costs in passenger car transport in 2020. It can be said that regarding efficiency, one litre of gasoline was comparable with 8.94 kWh and one litre of diesel with 9.97 kWh. The lowest electricity price could be found in China, with 0.07 € per kWh, and in Norway, with 0.08 € per kWh. The largest difference between the electricity price and the gasoline price was in Norway. Germany had the highest electricity price, with 0.30 € per kWh, while the average price in the EU was 0.21 € per kWh. The fuel prices were only lower in a few countries than the electricity prices, e.g., in China and some European countries (both in Norway and diesel in The Netherlands). The lowest fuel prices were in the USA, where one kWh of gasoline cost 0.06 €. The highest costs for fuel could be found in the EU. In The Netherlands, a kWh of gasoline cost 0.18 € and in Norway, 0.17 €. The average gasoline price within the EU amounted to 1.14 € per kWh and the average diesel price to 1.12 € per litre. Usually, gasoline prices were higher than diesel prices due to higher taxes on gasoline.

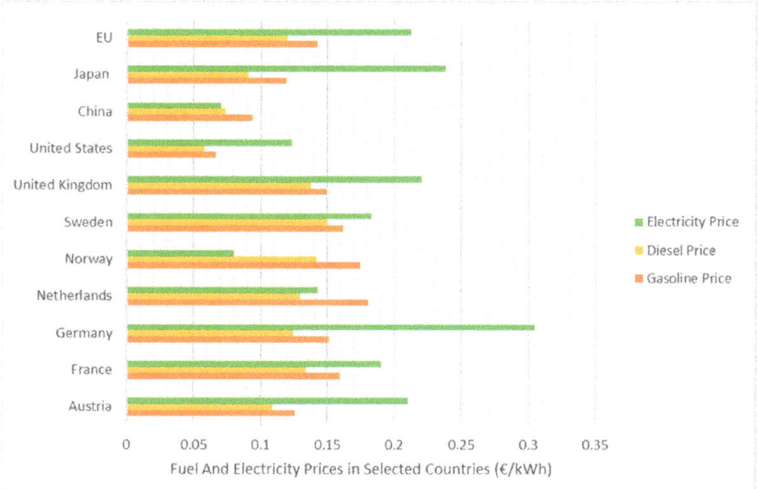

Figure 7. Fuel and electricity prices in €/kWh in the USA, China and selected European countries in 2020 [39,40].

5. Comparison of Incentives for BEVs in Selected Countries—EU, USA, China, Japan

In this chapter, a comparison of incentives for BEVs in selected countries—the USA, China, Japan and some European countries—for the year 2020 is provided. It is shown how different BEVs were promoted in selected countries. Note that especially all of the non-monetary measures were quite generic. Just because one country offered free parking for BEVs did not mean that BEVs could park for free at all parking spaces. Furthermore, incentives are changing all the time. Incentives that were valid for 2020 can be replaced by new ones or completely abolished in 2021.

As can be seen in Table 3, in all countries considered, except Norway, the customer got a purchase subsidy on a new BEV list price. Furthermore, the exemption of different taxes, like the VAT, registration tax (based on CO_2 emissions, bonus/malus), consumption tax and annual taxes (road and circulation taxes) was very common. Companies often received company tax benefits when purchasing BEVs. There were non-monetary incentives in all countries to make the purchase and the use of a BEV more attractive. Free parking and the usage of bus lanes were the most popular measures.

In Austria, a purchase subsidy of 5000 € on the list price of BEVs and an exemption from all car-related taxes was given [41]. Moreover, the promotion of 600 € for a home-installed wallbox was granted. [31]. In France, there was a very high subsidy, up to 7000 €, on a BEV purchase price. Moreover, the exemption from VAT and registration tax was guaranteed. The customer got tax credits for installing charging spots, and for changing from an ICE to a BEV, a scrapping premium was given [42]. In Germany, the purchase subsidy for a BEV amounted up to 6000 €, and there were the usual car-related tax exemptions [43]. In The Netherlands, a 4000 € subsidy on the list price of a BEV was guaranteed. The usual EU exemptions for BEVs were given [41]. Norway was unusual because there was no purchase subsidy on BEVs. However, BEVs were exempted from all car-related taxes. Furthermore, more non-monetary measures existed than in other countries: for example, no charges on toll roads or ferries, a maximum of 50% of the total amount on ferries and the usual measures like free parking and bus lane use [44]. Moreover, the customer got 815 € support for home charging infrastructure [45]. Sweden guaranteed up to 5700 € for a BEV. In addition to the usual exemptions from the taxes, a climate bonus of 5961 € on Zero-CO_2 emission cars was given [41].

In China, the purchase subsidy was between 2400 € and 3300 €. Furthermore, exemption from VAT and the vehicle and vessel tax were given. In China, BEVs were excluded from registration restrictions and driving bans and license plate quotas [46,47]. Japan granted a subsidy with up to 3154 € on the BEV's purchase price. However, BEVs were exempted from all car-related taxes such as the automobile tax, acquisition and road taxes [48]. In the United Kingdom, the subsidy on the purchase price amounted up to 3322 €. There were various tax benefits, and in London, BEVs got discounts/exemptions by entering the Congestion Zone and Ultra Low Emission Zone [49]. In the USA, there were federal subsidies of 2041 €; the local subsidies granted in the USA differed from state to state. For example, in California, there were additional rebates of 2450 €. The VAT of 8.25% was low; furthermore, the EV customer could get a federal tax credit between 2041 € and 6124 € [22].

Table 3. Incentives in selected countries in the year 2020 [12,21–30,43–45,50–52].

	Monetary Incentives		**Non-Monetary Incentives ***
	Subsidies	**Tax Exemptions, Benefits/Other**	
Austria	5000 €	Tax Exemptions, Benefits • VAT 20% • Up-front fuel consumption tax • Engine-related vehicle tax • Registration tax (all cars below 118 g/km (2020)) • Ownership tax • Company tax Other • 600 €/per wallbox, 1800 €/per wallbox for an apartment-building	• Free parking • Bus lane use
China	3300 € (driving range over 400 km), 2400 € (driving range up to 400 km)	Tax Exemptions, Benefits • VAT 16% • Vehicle and vessel tax	• Free parking, reserved parking • BEVs are not subject to registration restrictions or driving bans • Exemption from license plate quotes, new energy vehicle (NEV) license (drawing lots, competitive process) • Exemption from traffic restrictions
France	Up to 7000 € (list price under 45,000 €), 3000 € (price over 45,000 €),	Tax Exemptions, Benefits • VAT 20% • Registration tax (all cars below 138 g/km (2020)) • Company tax benefits Other • 300 € tax credit of purchase and installation costs of charging points • Scrapping premium	• Free parking

Table 3. *Cont.*

	Monetary Incentives		Non-Monetary Incentives *
	Subsidies	Tax Exemptions, Benefits/Other	
Germany	Up to 6000 €	**Tax Exemptions, Benefits** • VAT 19% • Registration fees (26.3 €) • Annual road tax • Annual circulation tax (10 years from the date of registration; after that, 50%) • Company tax benefits	• Free parking, reserved parking spots • Bus lane use
Japan	Up to 3154 €	**Tax Exemptions, Benefits** • VAT 10% • Road tax, toll • Exemption/reduction from certain car-related taxes (automobile weight tax, annual automobile tax, acquisition tax)	• Free parking • Priority/special lane use
Netherlands	4000 € (list price up to 45,000 €)	**Tax Exemptions, Benefits** • VAT 21% • Road tax • Registration tax • Extra benefits for leasing cars • Company tax benefits • Tax-deductible investments: zero-emission cars and charging points	• Free public charging points
Norway	none	**Tax Exemptions, Benefits** • VAT 25% (inclusive leasing) • Registration tax • Road taxes • Purchase tax • Import tax • Reduction of circulation tax on fuel • Company tax benefits **Other** • 815 € for home charging/infrastructure	• No charges on toll roads or ferries • Maximum of 50% of the total amount on ferries • Free parking • Bus lane use • Access only for BEVs
Sweden	Up to 5700 € (not more than 25% of the purchase price; linked to CO_2 emissions)	**Tax Exemptions, Benefits** • VAT 25% • Registration tax • Annual circulation tax (CO_2, weight) -> exemption first 5 years • Company tax benefits • Climate bonus for 0 g CO_2 car (5961 €)	• Free public charging points • Free parking

Table 3. *Cont.*

	Monetary Incentives		Non-Monetary Incentives *
	Subsidies	Tax Exemptions, Benefits/Other	
United Kingdom	Up to 3322 € (list price up to 45,144 €)	**Tax Exemptions, Benefits** • VAT 25% • Registration tax • Company tax benefits • Ownership tax (annual circulation tax)	• Free entry to London Congestion Zone and Ultra Low Emission Zone (ULEZ) (London) • Free parking
USA, California	2055 € for purchase or lease; California: additionally, 2466 € (Tesla Models and General Motors exempted)	**Tax Exemptions** • Excise tax • VAT 8.25% • Reduced vehicle registration fee of 30 € **Tax Benefits** • Company tax benefits • Federal tax credits: 2058 €–6175 € (Tesla Models 2020 exempted); per manufacturer 200,000 (Tesla and General Motors exempted)	• Free parking

All data apply to a vehicle purchased/used in 2020. * Quite generic, for example: Just because one country offered free parking for BEVs did not mean that BEVs could park for free at all parking spaces.

6. Results: Economic Assessment

In this chapter, the focus is on the interpretation of the results from the TCO. It is of particular interest to analyse the differences in the mobility costs between the selected countries (China, the USA (California), and some European countries (Austria, Germany, The Netherlands, Norway) for the year 2020. A look at the composition of an individual's expenses is given in detail. An examination of the differences between the cost composition of ICEs and BEVs is stated, and a general look at the contrast of the chosen cars is provided. In addition, a calculation of TCO of BEVs without subsidies on investment costs was done. Furthermore, a sensitivity analysis was performed to determine the impact of a change in the interest rate or the depreciation time on the annual costs of a vehicle.

First of all, the differences in the purchase price between ICEs and BEVs must be mentioned. In Case 1, the acquisition costs for a VW Golf, exclusive taxes, amounted to 18,640 € in the selected countries. The purchase price for a comparable E-Golf was slightly lower, at 16,540 €. Case 2 showed a larger difference between the purchase price for the Audi A5 (34,220 €) and the Tesla Model 3 (29,870 €). As already discussed in Section 2, the vehicles were comparable despite the price difference due to similar size and engine power.

Figure 8 (Case 1—Volkswagen Golf vs. Volkswagen E-Golf) and Figure 9 (Case 2—Audi A5 vs. Tesla Model 3) shows the TCO per year of an ICE and a BEV in selected countries. The costs include subsidies and taxes. A depreciation period of eight years and an interest rate of 5% were assumed.

In Case 1 (see Figure 8), total costs per year for a VW Golf ranged from 2520 € in the USA to 3960 € in Norway. For a VW E-Golf, the TCO per year ranged from 2650 € in the USA to 2980 € in Norway.

In Case 2 (see Figure 9), TCO per year for an Audi A5 ranged from 3990 € in the USA to 5960 € in Norway and Austria. For a Tesla Model 3, the costs were between 2760 € in China and 4460 € in Norway.

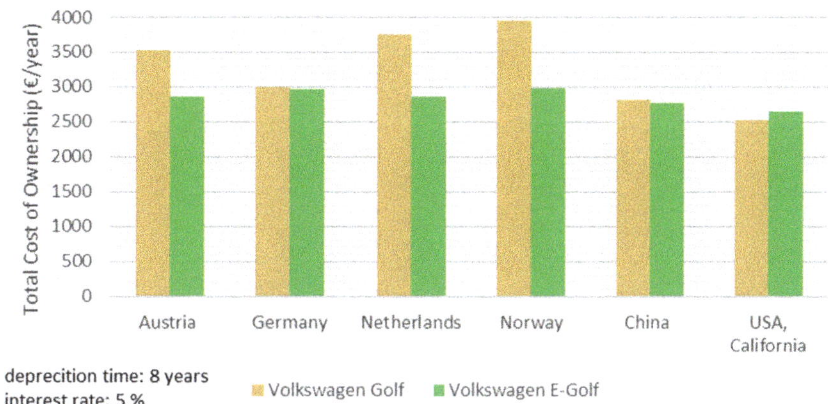

Figure 8. Total cost of ownership (TCO) (Case 1)—VW Golf vs. E-Golf.

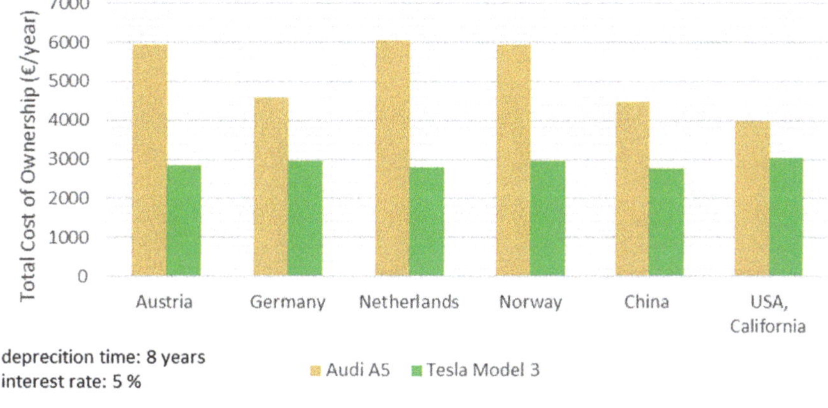

Figure 9. TCO (Case 2)—Audi A5 vs. Tesla Model 3.

In both cases, the lowest prices for the ICE were in the USA and the highest in Norway (Case 1) and The Netherlands (Case 2). The lowest costs for BEVs were in the USA (Case 1) and China (Case 2). The highest costs could be found in Norway, Germany and the USA. Case 2 shows the lowest price in the USA for the Tesla Model 3. However, the low costs of the Tesla Model 3 in the USA was only because Tesla was exempted from subsidies.

The differences in the costs were due to the divergencies in the costs of capital (C_{Cap}), operating & maintenance ($C_{O\&M}$) and energy (C_E).

Figure 10 (Case 1) and Figure 11 (Case 2) illustrate the composition of costs in detail. It also shows how high the costs would be without subsidies. Overall, the most significant price factor was the C_{Cap}, followed by the C_E. The $C_{O\&M}$ had the lowest impact on the total costs. In the following. detailed results for selected countries are presented.

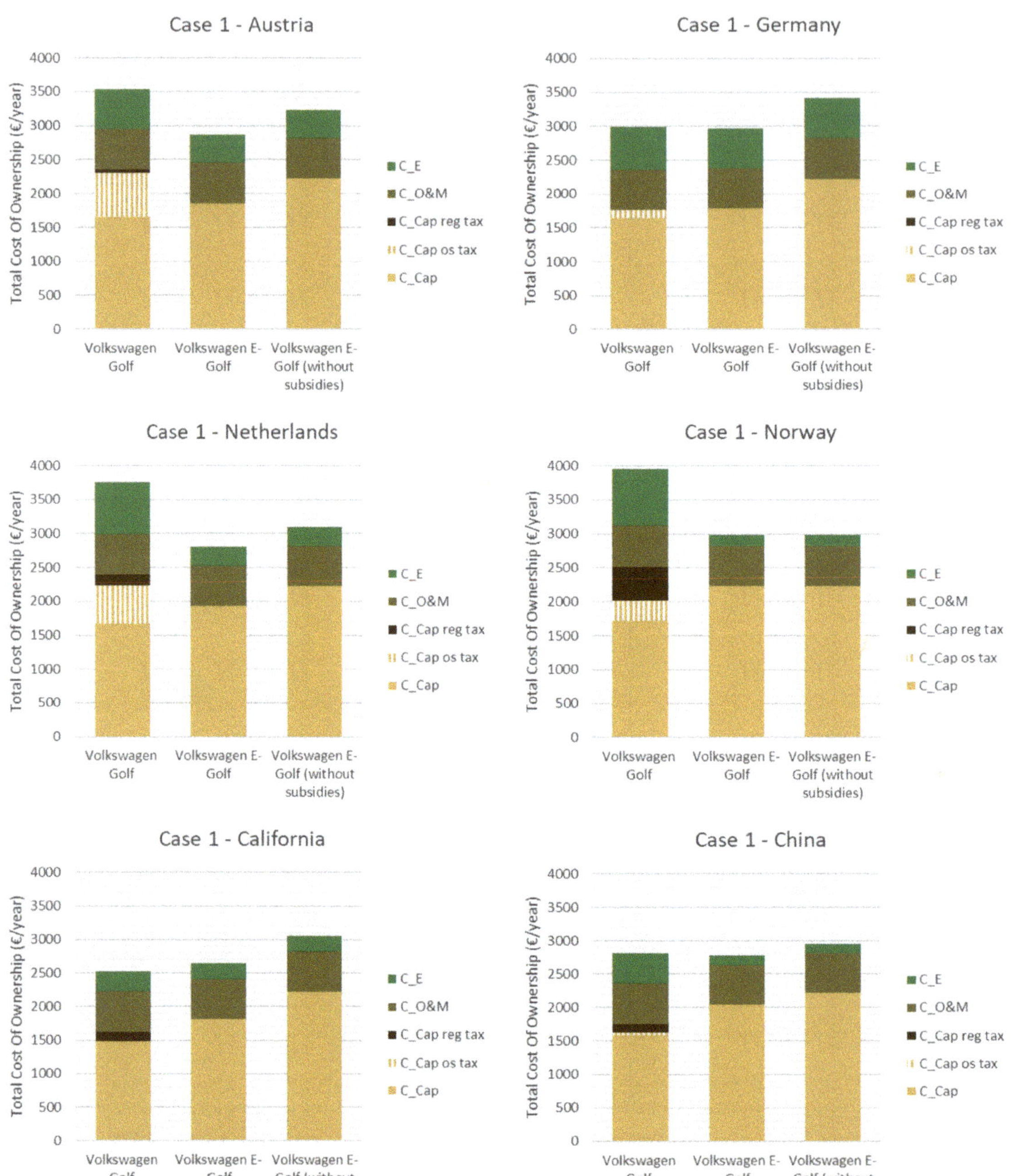

Figure 10. TCO: Case 1—Composition of the TCO in Case 1 for 2020.

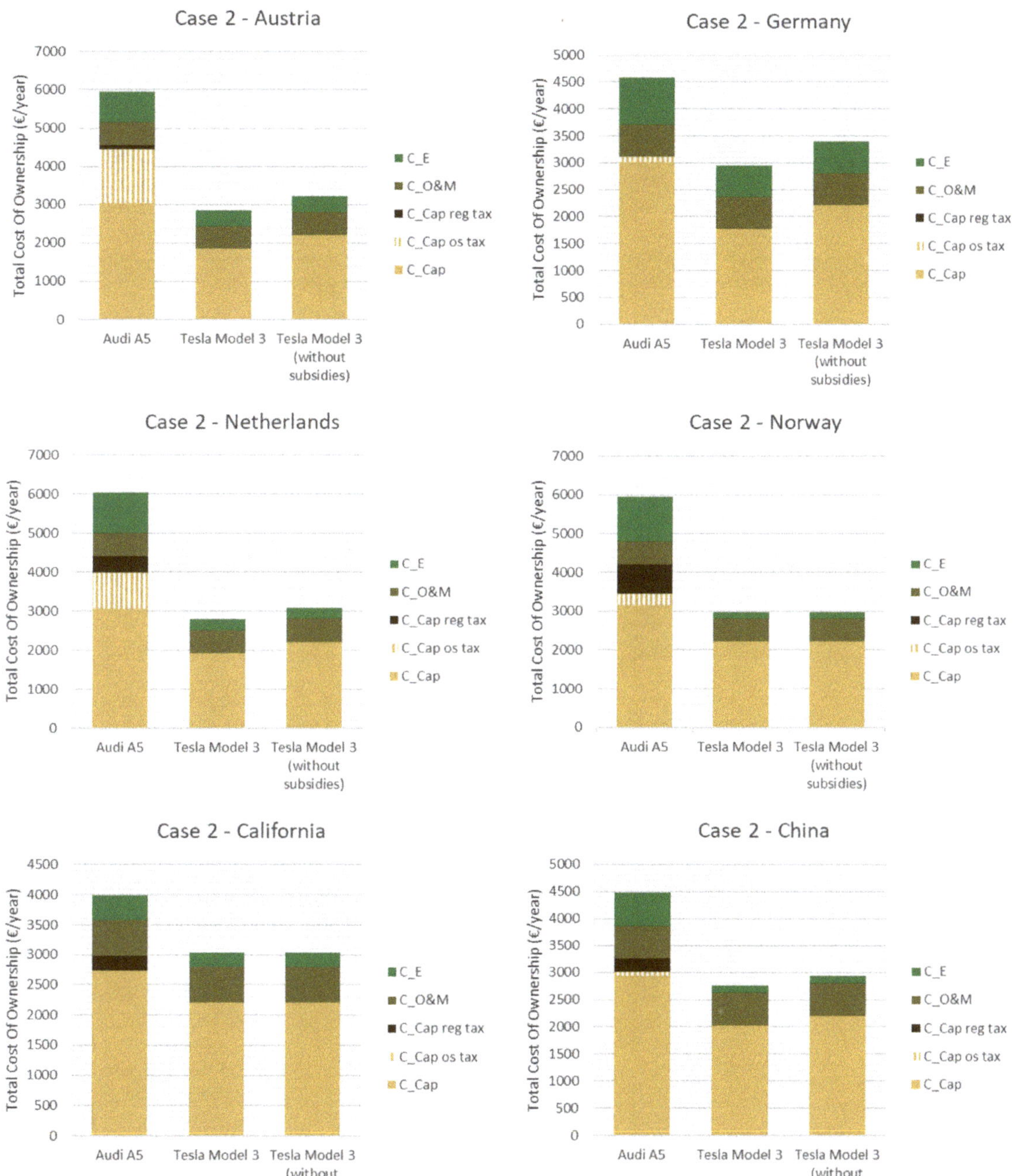

Figure 11. TCO: Case 2—Composition of the TCO in Case 2 for 2020.

The Netherlands, Austria

As shown in Figures 10 and 11, The Netherlands and Austria had very similar TCO. In The Netherlands, BEVs, per year, were slightly more expensive, and an ICE slightly less expensive than in Austria. The Netherlands had a VAT of 21%, and Austria, a VAT of 20%. A significant difference could be found in the amount of registration taxes. In The Netherlands, registration taxes were almost four times higher than in Austria. Among the countries considered, registration taxes were only higher in Norway. However, the ownership tax in Austria was higher than in any of the other countries considered. Gasoline prices in The Netherlands were the highest within the selected countries, but electricity prices were in the midfield and lower than in Austria. Austria granted a 5000 € subsidy, The Netherlands, a 6000 € subsidy for a BEV. The annual costs for the vehicles considered were very similar. The costs of a BEV in The Netherlands had more impact on the BEV market share than in Austria (in The Netherlands 10.2%, Austria 6% in 2020). The Netherlands showed that the high cost for a BEV could be compensated for through high registration taxes combined with subsidies and high gasoline prices.

Germany, China

In Germany and China, the annual costs of an ICE and BEV were approximately the same. Germany granted the highest subsidies, with 6000 € for a BEV, among the countries considered. In comparison, China granted 3000 € for a BEV. In China, the VAT, at 16%, is rather low; in Germany, the VAT is 19%. In Germany, there are no registration taxes on ICEs; the ownership tax is also very low. In addition, in China, the registration tax for an ICE is very low. However, there was a lottery procedure there, so whether an ICE could be registered at all was a random decision. In China, electricity prices were the lowest within the selected countries (0.07 €/kWh); fuel prices were also low. Only in the USA was the fuel price even lower than in China. In Germany, electricity prices were the highest within the compared countries (0.31 €/kWh). The BEV market share in China was 3.9% in 2019; in Germany, it was 1.8%. An average level of subsidies and low registration taxes in combination with low electricity prices thus seemed to have a higher impact than high subsidies in combination with low taxes. China also strongly promoted BEVs with the lottery process.

Norway

Norway is considered separately at this point. Norway had the largest BEV market share worldwide, with 45.6% in 2020. Nevertheless, the annual costs for a BEV were very high. On the other hand, ICEs were much more expensive than in the other countries considered. This was due to very high registration taxes and high gasoline prices. Only in The Netherlands was gasoline price higher. The costs of electricity, on the contrary, were very low (0.08 €/kWh). Lower electricity prices in the selected countries existed only in China (0.07 €/kWh). In Norway, there were no subsidies for BEVs granted. However, it balanced itself out again. Nowhere else were so many monetary measures granted to BEVs as in Norway. The VAT (25%) was, for the EU-average (21%), rather high. Norway showed that monetary measures, combined with high registration taxes on ICEs, had a high impact on disseminating BEVs.

USA (California)

The USA (California) had a unique position in terms of the cost of ICEs. The prices were much lower than in the other selected countries due to the very low VAT of 8.25%. Additionally, there were no ownership taxes on ICEs. Furthermore, the gasoline price (0.07 €/kWh) was the lowest within the selected countries, and the electricity price was in the midfield. For a BEV, there was a 5500 € subsidy granted; this was the second-highest value within the compared countries. However, Tesla and General Motors were exempted from the federal tax credit. In the USA, very few non-monetary measures were provided for BEVs.

Apart from the individual situation in selected countries, there were differences between the two cases that needed to be considered. As described briefly in the introduction to this Chapter, this was due to the cars' different purchase prices. The differences can be explained as follows:

In Case 1, the TCO for an ICE and BEV was approximately equal. The purchase price for a VW Golf in, for example, Austria amounted to 18,670 €, and for an E-Golf, 30,090 €.

However, there were subsidies for the E-Golf in the amount of 5000 €. In addition, the VW Golf price got higher because of the VAT (20%), registration tax and ownership tax on ICEs. Nevertheless, the slightly lower electricity costs could not compensate for the higher fuel costs. As a result, the E-Golf cost more than the VW Golf in Austria.

Case 2 showed large differences in the TCO of a BEV and an ICE. The net price of the Audi A5 in Austria for example, amounted to 34,280 €, and of the Tesla Model 3, to 29,920 €. The price difference between the cars was not as high as in Case 1. The Audi A5 cost more than the Tesla Model 3. Austria guaranteed the same amount of subsidies (5000 €) for the Tesla Model 3. On the other hand, the Tesla Model 3 consumed 0.16 kWh/km, which was similar to the VW E-Golf.

The price margins between ICEs and BEVs were reflected in the final results. This explained the large differences between costs of ICEs and BEVs when comparing the two cases.

In the following, we would like to show cost composition, without taking subsidies on the purchase price into account, using the examples of Austria, Norway and California. Cost allocations without subsidies for all countries considered are implemented in Figure 10 (Case 1) and Figure 11 (Case 2).

In Austria, in Case 1, the annual costs of 2860 € for a VW E-Golf would amount to 3230 € per year without subsidies. The E-Golf would be cheaper than the comparable VW Golf at 3530 € per year. Taking Case 2 into consideration, the annual cost of 2850 € for a Tesla Model 3 would be 3220 € without subsidies. A comparable Audi A5 would cost 5950 € per year. Therefore, in this case, the EV without subsidies would still be even much cheaper than ICE.

Norway was the only country considered where no subsidies were granted. Nevertheless, BEVs were much cheaper than comparable conventional cars. In no other country considered was the difference between ICEs and comparable BEVs higher than in Norway.

In California, in Case 1, the annual costs for a VW E-Golf without subsidies would amount to 3050 €. The VW E-Golf would be more expensive than the comparable VW Golf with yearly costs of 2520 €. Case 2 showed a different situation. Without subsidies, the Tesla Model 3 would cost 3040 € per year. Therefore, it was cheaper than the Audi A5, with an annual cost of 3990 €.

This demonstrated that subsidies had a significant impact on TCO.

Sensitivity analyses were done to see how depreciation time and the interest rate affected the costs. In the first step, the parameter for the depreciation time was changed. Thus, a period of 6 years and a period of 10 years were used for the calculation. The interest rate of 5% was not changed. In the second step, the interest rate parameters were changed to 4% and 6%. The depreciation time of 8 years was not changed.

The following effects of the interest rate were found: If the interest rate was set to 4%, the costs changed positively (the annual costs become lower). Conversely, an interest rate of 6% had a negative effect on costs (annual costs become higher).

A change in the depreciation time had the following effects: If a depreciation time of 6 years was assumed, the costs changed slightly positively (the annual costs became lower). Conversely, a depreciation time of 10 years had a slightly negative effect on costs (annual costs became higher).

A change in the interest rate parameter had a stronger effect on the annual costs than a change in the depreciation time.

In summary, the annual cost of an ICE was the highest in Norway, The Netherlands and Austria. In the USA (California), China and Germany, the annual costs for an ICE were the lowest. The high costs of conventional cars in Norway and The Netherlands were mainly due to the very high fossil fuel prices and registration taxes. In contrast, fuel prices in the USA (California) and China were very low.

In Norway, the high costs of BEVs were due to the lack of subsidies; in Germany, it was due to very high electricity prices. The very high BEV market share (52% in 2020) in Norway was explained by the fact that there were additional rebates for home charging infrastructure and more non-monetary measures than anywhere else as well as the high registration taxes on ICE.

In the USA (California), the annual cost of a BEV was the lowest. In Case 2, the high cost of the Tesla 3 was an exception because Tesla, in particular, was exempt from subsidies. Electricity prices were in the lower middle range; subsidies were quite high. Nevertheless, the market share of BEVs was only 1.5%. Low annual costs for a BEV were also found in China (market share 3.9% in 2019), which was due to very low electricity prices. This demonstrated that low TCO was not enough to ensure successful BEV dissemination.

7. Conclusions

Current policy strategies have a much higher impact on the dissemination of BEVs than pure private decisions [53]. It is essential to identify the most effective policies and provide recommendations for policymakers.

Today, almost no electric passenger cars purchased without policy interventions can be seen. The TCO analysis showed that investment costs of the vehicle itself had the most significant impact on the TCO and hence, on the unfavourable economics of BEVs. As demonstrated by the examples of Norway and The Netherlands, the number of BEVs deployed was highest in countries where proper policy measures are installed.

Norway had the highest BEV market share worldwide. Nevertheless, the annual costs for a BEV in Norway were very high. ICEs were also much more expensive than in the other selected countries considered. It was shown that measures such as the increase of prices for ICEs through high registration taxes and high fuel prices and non-monetary incentives such as access to certain zones and the usage of bus lanes for BEVs were highly effective. Furthermore, Norway's electricity prices were very low. The Netherlands also showed that the high cost of a BEV could be compensated for through high registration taxes and high gasoline prices. In addition, high subsidies were granted on BEVs in The Netherlands.

Low costs for BEVs alone did not ensure high EV dissemination. Nevertheless, the high price of BEVs was still a barrier [14]. Non-monetary incentives, combined with indirect policies such as high registration taxes and high fuel prices, are highly effective. In addition, it could be identified that incentives such as low electricity costs should always be linked to de-incentives such as high fuel prices [11]. All of these incentives are highly dependent on policy intervention (but also on spatial conditions).

The analysis of the TCO also demonstrated that a change in the interest parameter had a stronger effect on the annual costs of a vehicle than the depreciation time. It follows that a reduction in the depreciation time will probably not lead to a remarkable reduction in costs.

Other factors for the successful diffusion of BEVs that were not considered in this paper are the high impact of charging infrastructure, the GDP and CO_2-based taxes [8,9,11]. It is recommended that up-front price support works better than taxation systems [53]. In addition, essential are the technological progress of batteries and the image of BEVs [6,10].

Finally, the following recommendations for policy makers to reduce CO_2-emissions are summarised. These must be clearly and rigorously prioritized [15]. Forcing electric mobility in private passenger car transport requires policy interferences at least in the following dimensions:

(i) Pushing the purchases of BEVs by providing remarkable purchase subsidies at least for the next years to reduce the investment costs over time due to technological learning effects;

(ii) Implementation of high registration taxes to push up the prices of ICEs even more;

(iii) Legislative measures for the introduction of emission-free transport zones and free parking for BEVs would be helpful to disseminate BEVs, especially in urban areas;

(iv) Introducing CO_2 taxes would support the economic performance of BEVs indirectly by increasing the prices of petrol and diesel.

Further recommendations based on other studies:

(v)   Setting up the proper charging infrastructure: Along with all types of E-Mobility goes infrastructure development such as constructing charging infrastructure for BEVs [8,9,11]. Because this is a problem again concerning regulation, it means the involvement of policymakers.

(vi)   The relevance of the energy source for electricity generation: Regarding the environmental performance of any type of E-Mobility, the most crucial question is from which source the electricity used is generated and how this mix will develop in the future [19]. With the promotion of renewable energy and RES, electricity becomes as environmentally clean as possible, but again, this is largely a political decision.

Summing up, policy measures in different dimensions will be the major driving force for increasing BEVs' deployment in the future.

Author Contributions: Conceptualisation, M.S. and R.H.; methodology, M.S. and A.A.; formal analysis, M.S.; investigation, M.S. and A.A.; resources, M.S. and A.A.; data curation, M.S. and A.A.; writing—original draft preparation, M.S.; writing—review and editing, M.S., A.A. and R.H.; visualisation, M.S..; project administration, A.A.; funding acquisition, A.A. All authors have read and agreed to the published version of the manuscript.

Funding: This research was funded by the Vienna Science and Technology Fund (WWTF) through project ESR17-067.

Institutional Review Board Statement: Not applicable.

Informed Consent Statement: Not applicable.

Data Availability Statement: MDPI Research Data Policies.

Conflicts of Interest: The authors declare no conflict of interest.

References

1.   Europäisches Parlament. CO2-Emissionen von Autos: Zahlen und Fakten. Available online: www.europarl.europa.eu/news/de/headlines/society/20190313STO31218/co2-emissionen-von-autos-zahlen-und-fakten-infografik (accessed on 15 August 2020).

2.   Ajanovic, A. The future of electric vehicles: Prospects and impediments. *WIRE Energy Environ.* **2015**, *4*, 521–536. [CrossRef]

3.   Fritz, D.; Heinfellner, H.; Lichtblau, G.; Pölz, W.; Schodl, B. *Ökobilanz Alternativer Antriebe: Fokus Elektrofahrzeuge*; Umweltbundesamt GmbH: Wien, Austria, 2016; ISBN 9783990043851.

4.   European Environment Agency. European Environment Agency's Home Page. Available online: https://www.eea.europa.eu/ (accessed on 20 January 2021).

5.   Ajanovic, A.; Haas, R. Electric vehicles: Solution or new problem? *Environ. Dev. Sustain.* **2018**, *20*, 7–22. [CrossRef]

6.   Cherchi, E. A stated choice experiment to measure the effect of informational and normative conformity in the preference for electric vehicles. *Transp. Res. Part A Policy Pract.* **2017**, *100*, 88–104. [CrossRef]

7.   Illmann, U.; Kluge, J. Public charging infrastructure and the market diffusion of electric vehicles. *Transp. Res. Part D Transp. Environ.* **2020**, *86*, 102413. [CrossRef]

8.   Rietmann, N.; Lieven, T. How policy measures succeeded to promote electric mobility–Worldwide review and outlook. *J. Clean. Prod.* **2019**, *206*, 66–75. [CrossRef]

9.   Cansino, J.; Sánchez-Braza, A.; Sanz-Díaz, T. Policy instruments to promote electro-mobility in the EU28: A comprehensive review. *Sustainability* **2018**, *10*, 2507. [CrossRef]

10.   Dijk, M.; Orsato, R.J.; Kemp, R. The emergence of an electric mobility trajectory. *Energy Policy* **2013**, *52*, 135–145. [CrossRef]

11.   Held, T.; Gerrits, L. On the road to electrification–A qualitative comparative analysis of urban e-mobility policies in 15 European cities. *Transp. Policy* **2019**, *81*, 12–23. [CrossRef]

12.   Hauff, K.; Pfahl, S.; Degenkolb, R. Taxation of electric vehicles in Europe: A methodology for comparison. *WEVJ* **2018**, *9*, 30. [CrossRef]

13.   Wang, N.; Tang, L.; Pan, H. A global comparison and assessment of incentive policy on electric vehicle promotion. *Sustain. Cities Soc.* **2019**, *44*, 597–603. [CrossRef]

14.   Vilchez, J.J.G.; Smyth, A.; Kelleher, L.; Lu, H.; Rohr, C.; Harrison, G.; Thiel, C. Electric car purchase price as a factor determining consumers' choice and their views on incentives in Europe. *Sustainability* **2019**, *11*, 6357. [CrossRef]

15.   Alter Motive. *Deriving Effective Least-Cost Policy Strategies for Alternative Automotive Concepts and Alternative Fuels*; Vienna University of Technology: Wien, Austria, 2011.

16.   Palmer, K.; Tate, J.E.; Wadud, Z.; Nellthorp, J. Total cost of ownership and market share for hybrid and electric vehicles in the UK, US and Japan. *Appl. Energy* **2018**, *209*, 108–119. [CrossRef]

17. Lévay, P.Z.; Drossinos, Y.; Thiel, C. The effect of fiscal incentives on market penetration of electric vehicles: A pairwise comparison of total cost of ownership. *Energy Policy* **2017**, *105*, 524–533. [CrossRef]

18. Change in Distance Travelled by Car. Available online: https://www.odyssee-mure.eu/publications/efficiency-by-sector/transport/distance-travelled-by-car.html (accessed on 5 March 2021).

19. Ajanovic, A.; Haas, R. On the economics and the future prospects of battery electric vehicles. *Greenh. Gas Sci. Technol.* **2020**, *10*, 1151–1164. [CrossRef]

20. Asen, E. 2020 VAT Rates in Europe. Available online: https://taxfoundation.org/european-union-value-added-tax-2020/ (accessed on 3 March 2021).

21. Wiel, S.; McMahon, J.E. Governments should implement energy-efficiency standards and labels—cautiously. *Energy Policy* **2003**, *31*, 1403–1415. [CrossRef]

22. Electric Car Tax Credits: What's Available? Available online: https://www.energysage.com/electric-vehicles/costs-and-benefits-evs/ev-tax-credits/ (accessed on 26 November 2020).

23. Federal Tax Credits for Electric and Plug-in Hybrid Cars. Available online: https://www.fueleconomy.gov/feg/taxevb.shtml (accessed on 26 November 2020).

24. EAFO. Available online: https://www.eafo.eu/ (accessed on 26 November 2020).

25. How China's Electric Vehicle (EV) Policies Have Shaped the EV Market. Available online: https://www.sustainalytics.com/esg-blog/how-chinas-electric-vehicle-policies-have-shaped-the-ev-market/ (accessed on 26 November 2020).

26. Norwegian EV Policy. Available online: https://elbil.no/english/norwegian-ev-policy/ (accessed on 26 November 2020).

27. National Conference of State Legislatures. State Policies Promoting Hybrid and Electric Vehicles. Available online: https://www.ncsl.org/research/energy/state-electric-vehicle-incentives-state-chart.aspx (accessed on 26 November 2020).

28. The Netherlands Goes for EV Purchase Subsidies. Available online: https://www.electrive.com/2020/03/05/the-netherlands-goes-for-ev-purchase-subsidies/ (accessed on 26 November 2020).

29. Environmental and Energy Study Institute. Comparing U.S. and Chinese Electric Vehicle Policies. Available online: https://www.eesi.org/articles/view/comparing-u.s.-and-chinese-electric-vehicle-policies (accessed on 26 November 2020).

30. Steinbacher, K.; Goes, M.; Jörling, K. Incentives for Electric Vehicles in Norway. Available online: https://www.euki.de/wp-content/uploads/2018/09/fact-sheet-incentives-for-electric-vehicles-no.pdf (accessed on 8 April 2021).

31. Elektroautos und E-Mobilität–Förderungen und weiterführende Links. Available online: https://www.oesterreich.gv.at/themen/bauen_wohnen_und_umwelt/elektroautos_und_e_mobilitaet/Seite.4320020.html (accessed on 26 November 2020).

32. Richardson, J. The Incentives Stimulating Norway's Electric Vehicle Success. Available online: https://cleantechnica.com/2020/01/28/the-incentives-stimulating-norways-electric-vehicle-success/ (accessed on 26 November 2020).

33. Ajanovic, A.; Haas, R. Economic and environmental prospects for battery electric- and fuel cell vehicles: A review. *Fuel Cells* **2019**, *19*, 515–529. [CrossRef]

34. Ajanovic, A.; Haas, R. Driving with the sun: Why environmentally benign electric vehicles must plug in at renewables. *Sol. Energy* **2015**, *121*, 169–180. [CrossRef]

35. Ajanovic, A.; Haas, R. The impact of energy policies in scenarios on GHG emission reduction in passenger car mobility in the EU-15. *Renew. Sustain. Energy Rev.* **2017**, *68*, 1088–1096. [CrossRef]

36. International Energy Agency. *Global EV Outlook 2020*; IEA: Paris, France, 2020.

37. European Automobile Manufacturers' Association. Passenger Cars. Available online: https://www.eafo.eu/vehicles-and-fleet/m1 (accessed on 17 January 2021).

38. IEA. Global Electric Car Sales by Key Markets, 2010–2020. Available online: https://www.iea.org/data-and-statistics/charts/global-electric-car-sales-by-key-markets-2015--2020 (accessed on 20 January 2021).

39. Benzinpreise auf der Ganzen Welt, 11-Jan-2021. Available online: https://de.globalpetrolprices.com/gasoline_prices/ (accessed on 16 January 2021).

40. Eurostat. Electricity Prices for Household Consumers–Bi-Annual Data (from 2007 Onwards). Available online: https://ec.europa.eu/eurostat/databrowser/product/page/NRG_PC_204 (accessed on 20 January 2021).

41. European Automobile Manufacturers' Association. *ACEA Tax Guide 2020*; ACEA: Brussels, Belgium, 2020.

42. Mock, P.; Yang, Z. *Driving Electrification: A Global Comparison of Fiscal Incentive Policy for Electric Vehicles*; International Council on Clean Transportation: Washington, DC, USA, 2014.

43. Presse- und Informationsamt der Bundesregierung. So Funktioniert der Neue Umweltbonus. Available online: https://www.bundesregierung.de/breg-de/themen/klimaschutz/umweltbonus-1692646 (accessed on 7 January 2021).

44. Bardal, K.G.; Gjertsen, A.; Reinar, M.B. Sustainable mobility: Policy design and implementation in three Norwegian cities. *Transp. Res. Part D Transp. Environ.* **2020**, *82*, 102330. [CrossRef]

45. Team, T.W. Discover Norway's Unique EV and EV Chargers Perks. Available online: https://blog.wallbox.com/en/norway-ev-incentives/#index_1 (accessed on 8 January 2021).

46. Lambert, F. China Boosts Electric Car Sales by Removing License Plate Quotas. Available online: https://electrek.co/2019/06/06/china-boost-ev-sales-license-plate-quotas/ (accessed on 7 January 2021).

47. He, H.; Jin, L.; Cui, H.; Zhou, H. *Assessment of Electric Car Promotion Policies in Chinese Cities*; International Council on Clean Transportation: Washington, DC, USA, 2018.

48. Western Automation. Government Incentives Encourage Adoption of Electric Vehicles. Available online: https://www.westernautomation.com/technology-innovations/government-incentives-encourage-adoption-of-electric-vehicles/?doing_wp_cron=1610048496.4899899959564208984375 (accessed on 7 January 2021).
49. Navarro, M. A Complete Guide to EV & EV Charging Incentives in the UK. Available online: https://wallbox.com/en_us/ev-incentives-uk (accessed on 8 January 2021).
50. Foley & Lardner LLP. Japan Continues to Offer Electric Vehicle Incentives. Available online: https://www.foley.com/en/insights/publications/2013/09/japan-continues-to-offer-electric-vehicle-incentiv (accessed on 6 January 2021).
51. Berger, R. *E-Mobility Index 2019*; Roland Berger: Munich, Germany, 2019.
52. Statista Mobility Market Outlook. *InDepth: eMobility 2020*; Statista: Hamburg, Germany, 2020.
53. Petrauskiene, K.; Dvarioniene, J.; Kaveckis, G.; Kliaugaite, D.; Chenadec, J.; Hehn, L.; Pérez, B.; Bordi, C.; Scavino, G.; Vignoli, A.; et al. Situation analysis of policies for electric mobility development: Experience from five European regions. *Sustainability* **2020**, *12*, 2935. [CrossRef]

Article

Electric Mobility in Cities: The Case of Vienna

Amela Ajanovic *, Marina Siebenhofer and Reinhard Haas

Energy Economics Group, Vienna University of Technology, Gußhausstraße 25-29, 1040 Vienna, Austria; siebenhofer@eeg.tuwien.ac.at (M.S.); haas@eeg.tuwien.ac.at (R.H.)
* Correspondence: ajanovic@eeg.tuwien.ac.at; Tel.: +43-(0)1-58801-370-364

Abstract: Environmental problems such as air pollution and greenhouse gas emissions are especially challenging in urban areas. Electric mobility in different forms may be a solution. While in recent years a major focus was put on private electric vehicles, e-mobility in public transport is already a very well-established and mature technology with a long history. The core objective of this paper is to analyze the economics of e-mobility in the Austrian capital of Vienna and the corresponding impact on the environment. In this paper, the historical developments, policy framework and scenarios for the future development of mobility in Vienna up to 2030 are presented. A major result shows that in an ambitious scenario for the deployment of battery electric vehicles, the total energy demand in road transport can be reduced by about 60% in 2030 compared to 2018. The major conclusion is that the policies, especially subsidies and emission-free zones will have the largest impact on the future development of private and public e-mobility in Vienna. Regarding the environmental performance, the most important is to ensure that a very high share of electricity used for electric mobility is generated from renewable energy sources.

Keywords: battery electric vehicles; public transport; emissions; policies

check for **updates**

Citation: Ajanovic, A.; Siebenhofer, M.; Haas, R. Electric Mobility in Cities: The Case of Vienna. *Energies* **2021**, *14*, 217. https://doi.org/10.3390/en14010217

Received: 8 December 2020
Accepted: 28 December 2020
Published: 4 January 2021

Publisher's Note: MDPI stays neutral with regard to jurisdictional claims in published maps and institutional affiliations.

1. Introduction

Globally, cities generate about 80% of the GDP, consume about 75% of global primary energy, and cause more than 70% of energy-related carbon dioxide emissions [1]. The number of the urban population is rapidly increasing worldwide, from 751 million in 1950 to 4.2 billion in 2018. It is likely to reach about 5.1 billion by 2030 [2]. In Europe, the majority of citizens live in urban areas, and for their daily life, they require some kind of private or public mobility. Urban mobility is an important facilitator for economic growth, employment and development. In the EU, urban mobility already accounts for 40% of all CO_2 emissions of road transport and up to 70% of other pollutants from transport [3]. With increasing urbanization, problems such as congestion and pollution are becoming more and more evident, especially in larger cities. Hence, one of the key challenges in the transition towards more sustainable development of the energy system is the transport sector. Moreover, there are two trends, urbanization and electrification, which are deeply transforming energy systems globally [4]. However, most of the cities have car-centric infrastructures built in the 20th century, which occupy large territories and cause high rates of urban air pollution and congestion. Especially in urban areas, electric mobility is seen as an important means to cope with local environmental problems as well as to combat global warming.

The shift from private cars to public transport and the electrification of mobility are often seen as the right ways towards sustainable development of urban areas. However, it is important to notice that electric mobility in both private and public transport has a long history.

In particular, the capital of Austria has a long history of electric mobility. In Vienna, trams powered by electricity, at 600 V DC, have been used since 1897. Due to their lower noise and smell compared to horse-drawn and steam trams, they very quickly became the favorable option. Horse-driven trams were removed from use in 1903 and steam trams in

1922. In the first half of the 20th century, since private cars were too expensive for most of the population, trams were a major transport mode in Vienna. Still today, electric trams are an essential part of the public transport system in Vienna in spite of the broader portfolio of other mobility options. Over time, some parts of the tramlines were replaced with buses and metros, but the Viennese trams network comprises around 220 km, which makes it the sixth largest in the world [5].

Today, the metro is the backbone of the public transport system in Vienna. After the first test operations in 1976, the modern metro opened in 1978. Over time, the number of lines increased, as did the total length of the network. The current underground network extends to 83 km.

Besides the tram and metro network, there is also a bus line network in Vienna of about 850 km. Although most of the buses are still powered by fossil fuels, there are different initiatives for purchasing electric buses. Since 2013, fully electric eight-meter minibuses have been used in the city center. Currently, 12-m battery-electric buses are in the test stage and should be included in passenger service from 2023. Moreover, the first hydrogen bus is in test operation this year, 2020 [6–8].

The electrification and historical development of the public transport system in Vienna is depicted in Figure 1.

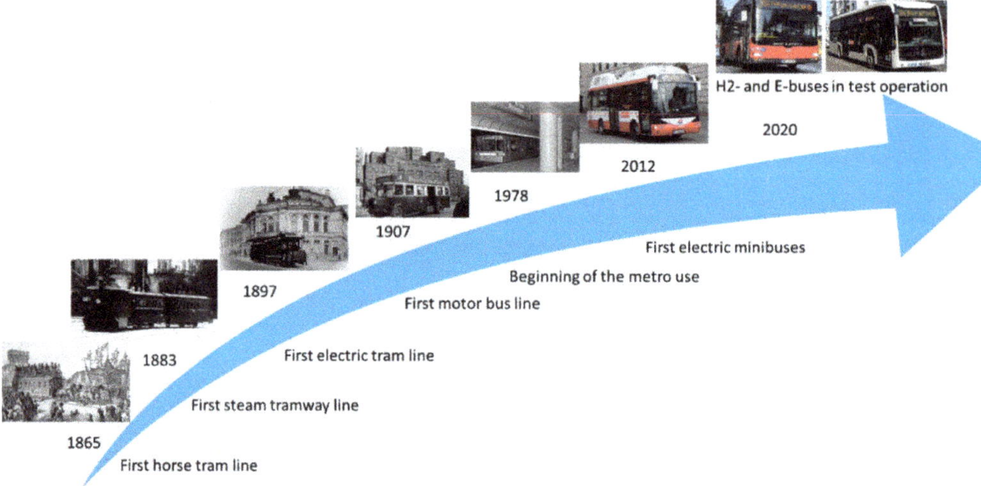

Figure 1. Development of electrification of the public transport system in Vienna.

Reorganizing urban mobility towards more sustainability is a common challenge to all major cities. Efficient and effective urban transport is crucial for the achievement of sustainability goals. However, the development of mobility is very dependent on policies implemented and policy objectives set for the future. Already in 2011, the Transport White Paper [9] sets the goal to reduce the use of conventional fossil-fueled cars in urban transport by 50% by 2030 and to phase them out completely by 2050. As a follow-up to this document, the European Commission came up in 2013 with an Urban Mobility Package with a special focus on urban road charging and access restriction schemes, monitoring and management of urban freight flows and financial support mechanisms for the preparation of Urban Mobility Plans.

However, the success of the European policies and goals, which are agreed upon at the EU level, depends also on actions taken and policies implemented by national, regional and local authorities. This is the reason that the development of urban mobility is quite different from city to city. Although the developments are different, the increasing use of alternative fuels and alternative automotive technologies can be noticed in most cities. Over the last decade, special interest in electrification of mobility has been rapidly increasing.

In the literature, different aspects of electrification of mobility have been discussed in recent years. Held and Geritts [10] conducted a qualitative comparative analysis of urban e-mobility policies in 15 European cities. Liberto et al. [11] analyzed the impact of electric mobility in Rome. Ajanovic and Haas [12] examined the major impact factors on the broader dissemination of electric vehicles in urban areas. Assessment of the electrification of urban mobility with the special focus on buses was conducted by Scarinci [13]. Since public transport is one of the backbones of sustainable transport strategy in the EU, further electrification of public transport is essential for the reduction in emissions in this sector. In the scope of the ELIPTIC project, 20 showcases with variations of electrified public transport under different operational, geographical and climate conditions are demonstrated and analyzed [14]. The development of e-mobility in urban areas is very different from city to city, and it is important to exchange lessons learned. However, in the literature, only a few contributions exist dealing with e-mobility in all its facets. Currently, there are many papers focusing on electric cars despite their minor relevance compared to public transport, e.g., underground and trams.

The major contribution of this paper is the comprehensive analysis of electric mobility in the city of Vienna from an economic and environmental point of view considering the individual as well as public mobility powered by electricity. An additional aim is to assess CO_2 emissions taking into account various electricity generation portfolios as well as the hidden and embedded GHG emissions from production, assembling and scrappage of the electric vehicles. The paper documents the major historical developments, shows the current situation and analyzes possible future scenarios for the electrification of mobility considering existing targets and policy framework, as well as the effects of stronger use of renewable energy in electricity generation.

This paper starts with a brief history of electric mobility in Vienna. In the next section, the recent developments and current situation of mobility in Vienna are described. In Section 3 policy framework is documented. Section 4 describes the methods used for economic and environmental assessments. In the next sections, the results of our economic and environmental analysis are provided. The scenarios for future development are presented in Section 7, and major conclusions are derived at the end.

2. Background: Recent Developments and the Current Situation of Mobility in Vienna

Vienna is a city with a continuously growing population and consequently an increasing demand for energy services, including mobility. The development of overall energy consumption in transport by fuel in Vienna is depicted in Figure 2. The increase in energy consumption between 1990 and 2018 was about 63%. The largest amount of this energy consumption was covered by fossil fuels. For example, in 2018, the total final energy consumption was about 14 TWh, with only a small amount, 644.3 GWh, covered by electrical energy. In total, the highest energy consumption was reached in 2005 with about 15 TWh.

Over time, different fuel mixes have been used in the transport sector. The fuel used most in 1990 was petrol and in 2018 diesel. For example, in 1990, the share of diesel fuel in the energy mix was 35%, but this had risen to 70% by 2018; see Figure 3. It is interesting to note that in spite of all supporting measures implemented with the goal of accelerating electrification of mobility, the share of electricity in the energy mix in the transport sector in 2018 was almost the same as in 2000 at about 5%.

Over the years, interest in different transport modes has also changed. Between the 1970s and the early 1990s, the use of public transport steadily decreased while the use of private passenger cars increased. However, due to the more evident environmental problems related to the transport sector, as well as reduced service prices and improved quality in services, the use of public transport in Vienna has increased remarkably in recent years; see Figure 4. Moreover, like in other cities, cycling is becoming a more and more popular mobility option. Electric mobility increased in total from about 17% in 1970 to about 23% in 2019 with a historical low in 1993.

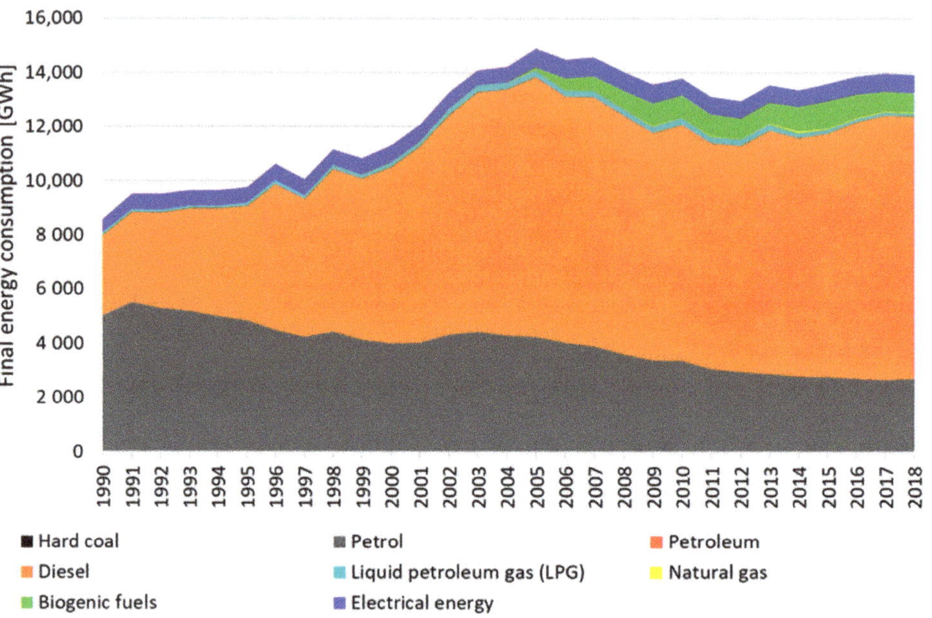

Figure 2. Development of overall energy consumption in transport by fuel (data source [15]).

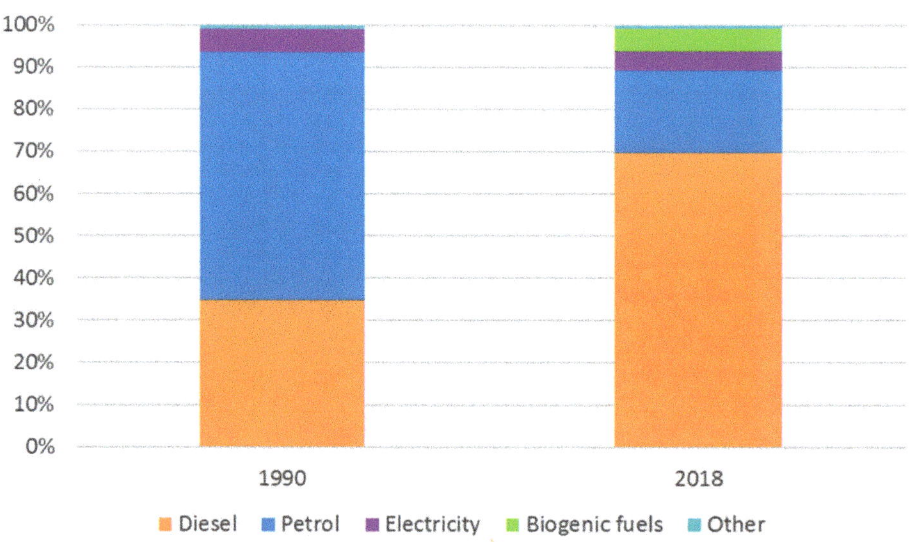

Figure 3. Share of fuels used for mobility in Vienna, 1990 vs. 2018 (data source [15]).

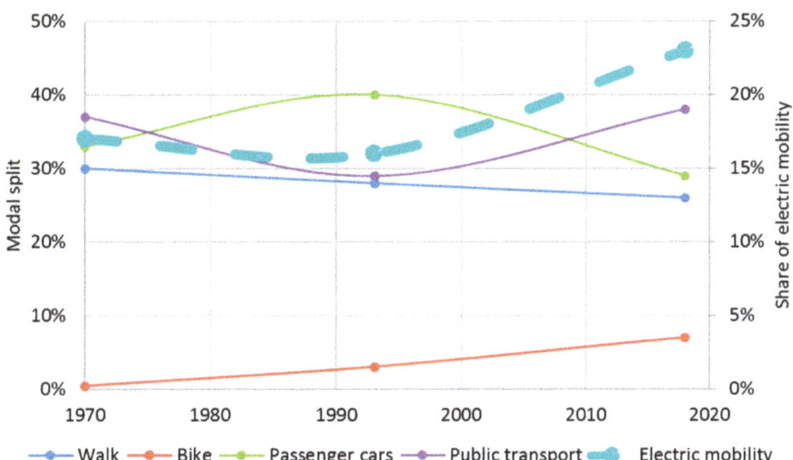

Figure 4. Development of modal split in Vienna, 1970–2018 (data source [15,16]).

Currently, the most kilometers driven in Vienna are covered by public transport. In 2018 in total, 153 million vehicle kilometers (vkm) were traveled by public transport. The underground had the biggest share, with 70.7 million vkm; see Figure 5. In Vienna, most of the buses still operate on fossil fuels, but there is a significant effort to increase the use of electric buses, with electric minibuses used currently in the city center.

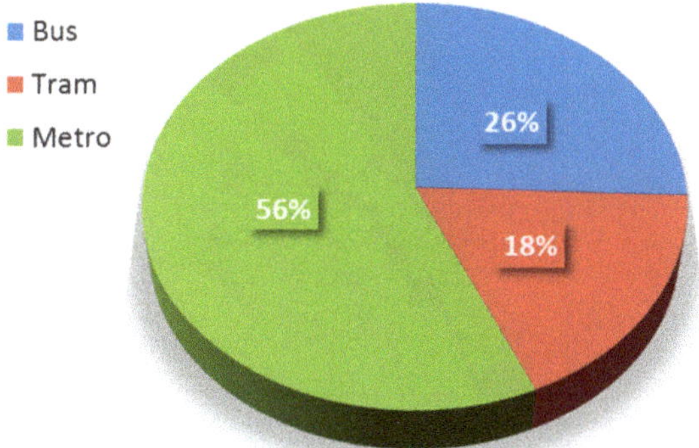

Figure 5. Share of vehicle kilometers of public transport, 2018 (data source [17]).

The development of electricity consumption in transport is depicted by mode in Figure 6. The results combine a top-down and a bottom-up approach. The electrical energy consumption of railway and underground is available in statistical data [15]. The electricity demand of city buses, trams and passenger cars is calculated using data for the vehicle's fuel intensity (FI) and vehicle kilometers driven (d_{vkm}):

$$E_i = FI_i \cdot d_{vkm_i} \tag{1}$$

with $i \in \{\text{city bus}; \text{tram}; \text{passenger cars}\}$.

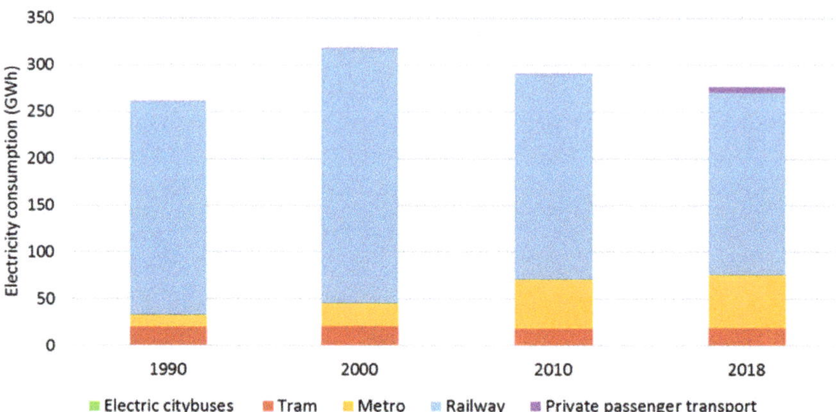

Figure 6. Electricity consumption in the transport sector by transport mode in the period of 1990–2018 (data source [17]).

Electricity consumption in tram transport was stable in the period of 1990–2018, whereas the electricity consumption of underground transport more than quadrupled in the same time.

The use of electricity for private passenger mobility is still very low, although a broad portfolio of policy measures is provided with the goal of accelerating the use of electric vehicles. Figure 7 shows the development of the passenger car stock from 1990 to 2019. Private passenger mobility is dominated by conventional internal combustion engine vehicles, powered by petrol and diesel.

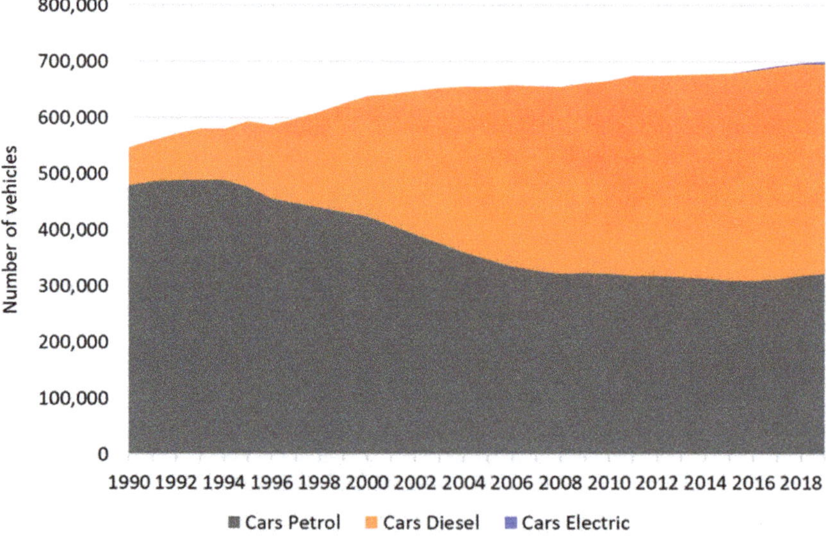

Figure 7. Stock of passenger cars from 1990–2019 (data source [18]).

However, the total number of electric vehicles used for private and public mobility is rapidly increasing over time; see Figure 8. Since 2010, electric cars, motorcycles and trucks have shown significant growth in their stock numbers.

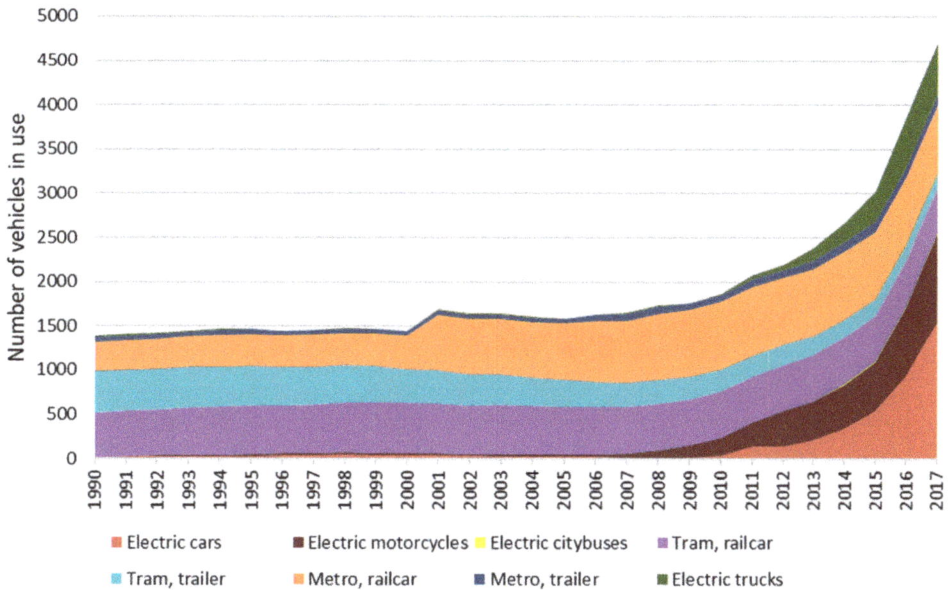

Figure 8. Development of the stock of private and public electric vehicles by mode from 1990–2017 (data source [17,18]).

In 2001, Wiener Linien, Vienna's public transport operator, fundamentally changed its counting method of underground trains. This explains the immediate growth of metro railcar stock between 2000 and 2001.

The number of underground railcars is significantly higher than the number of underground trailers. The standard U-type underground train has six railcar wagons (three twin railcars). The type T has mostly four railcars. Type V consists of six wagons, with the end-wagons as trailers without a propulsion system, meaning a type V metro train regularly consists of two trailers and four railcars.

3. Policy Framework

With the oil crisis in the 1970s, it became apparent that traffic could not continue to grow only based on fossil fuels. At first, measures such as restrictions in car use, e.g., leaving the car at home one day a week were implemented. In addition, in the 1980s, the extension of local public transport was accelerated. Due to environmental problems, freight transport was transferred from road to rail. The Transport Master Plan [19] was published in 1991 with the goal to improve the attractiveness of public transport, to reduce road-traffic-related emissions (noise and air pollution) and to introduce the concept of true costs in the transport sector.

The Traffic Concept for Vienna was first published in 1969. Since then, new transport and mobility concepts have been adopted every ten years. Currently, the policy framework in Vienna is based on policies and targets set on the EU and Austrian national level. The most relevant documents on the EU level are the White Paper on Transport [9], and the guidelines for Sustainable Urban Mobility Plans [20]. On the national level, the most important documents are the Energy Strategy Austria from 2010 and the Transport Master Plan from 2013. With the Smart City Vienna Framework Strategy (2014), Vienna has committed to the European energy and climate targets [21,22]. the documents such as the Urban Development Plan (STEP 2025) from 2014, Urban Mobility Plan Vienna (2015) and E-Mobility Strategy (2016), the goals for the future development of mobility in Vienna are also clearly set.

The essential strategies relevant for the implementation of e-Mobility are [22]:

- Development of a master plan for electric mobility;
- Development of innovative business models, creating market rules for the provision and use of electricity and charging infrastructure;
- Creation of tax incentives to promote the procurement and distribution of electric vehicles;
- Information, motivation, training and awareness-raising.

There are different tax benefits and incentives provided for e-mobility. They can be divided into three categories: federal subsidies for private individuals, subsidies for companies and tax advantages.

Currently, individuals who purchase electric cars, e-mopeds, e-motorcycles and e-transport bikes receive subsidies. Prerequisites are the use of electricity from renewable energies with a range of at least 50 km and a gross list price of a maximum of €50,000. Electric and fuel cell vehicles are subsidized with €1500, plug-in hybrid and range extender with €750, e-motorcycles with €500, e-mopeds with €350 and e-transport bikes with €200. The purchase of e-charging stations is also subsidized. In addition, there are federal-state subsidies (e.g., €1000 per e-car in Lower Austria) [23]. The same subsidies are also available for companies. However, the gross list price may not exceed €60,000.

Moreover, e-vehicles are exempt from the standard consumption tax (NoVA), the motor vehicle tax above 3.5 tons, and the motor-related insurance tax up to 3.5 tons. In addition, they are exempt from mineral oil tax (MÖSt), and only minor energy taxes (electricity) are payable [24]. If an e-vehicle (max. 80,000 purchase price) is used for business purposes, there is the possibility of an input tax deduction (purchase costs, leasing expenses, operating costs) [23].

Besides different tax benefits and incentives provided for e-mobility, there are also subsidiary programs provided with the goal to support implementation of various pilot and demonstrations projects, e.g., purchase of different types of electric vehicles (electric scooters, mopeds, motorcycles, electric bicycles), development of infrastructure, e-mobility management, e-fleets and e-logistics.

The City of Vienna supports expansion measures and enacts laws relevant to e-mobility. The Vienna Parking Garage Law stipulates the provisioning of empty cable ducts for future charging stations in new parking garages. Furthermore, it is recommended to provide charging infrastructure in semi-public areas such as parking slots and petrol stations. Strategic locations such as multimodal hubs are particularly suitable for the installation of charging infrastructure [22].

E-vehicles are increasingly used in the logistics sector. This reduces emissions as well as noise and increases the performance coefficient. With the framework strategy "Smart City Vienna", the City of Vienna has committed itself to cooperate with the logistics sector [25].

Energy suppliers should follow existing systems like the Vienna Energy "tank" system. They should offer tailor-made business models and integrated solutions for private and business customers from a single source that are economically viable, suitable for everyday use and user-friendly (regarding the existing electrical power supply infrastructure, the installation of charging stations and energy billing) [22].

One of the most important research projects in the field of e-Mobility is the E-Delivery on Demand' (Tranform+) project, a mobility lab that has been running since 2015. At the Liesing Industrial Park, it tests how delivery services, car rentals and carpools can be organized more efficiently. The aim is to generate a pooled, demand-oriented and cost-efficient model with the use of e-vehicles [22]. Another mobility lab is the aspern.mobil LAB, which deals with active mobility, mobility as a service as well as first- and last-mile logistics. The mobility lab thinkport VIENNA deals with freight logistics solely. The research program Mobility of the Future should also be mentioned here: it was started in 2012 and focuses on the sustainable and environmentally friendly development of alternative and innovative drive technologies and mobility solutions.

Furthermore, there is a big focus on raising the awareness of the general public regarding e-Mobility via campaigns. Advancements in e-mobility bring new jobs and thus a need for training and further education.

4. Method of Approach

In this section, the method of approach for the economic and environmental analyses conducted in this paper is documented. This approach is based on the Total Cost of Ownership calculation method of the total mobility costs [26,27], well-to-wheel methodology [28,29], as well as on the assessment framework developed in our previous works [30–32].

For the economic assessment investment costs, energy costs and other operating and maintenance costs, as well as relevant taxes and incentives, are considered. The total costs per km driven C_{km} are calculated as:

$$C_{km} = \frac{IC \cdot \alpha}{skm} + P_f \cdot FI + \frac{C_{O\&M}}{skm} \; [\text{€}/100 \text{ km driven}] \tag{2}$$

where:
IC: investment costs [€/car]
α: A capital recovery factor
skm: specific km driven per car per year [km/(car.yr)]
P_f: energy price incl. taxes [€/kWh]
$C_{O\&M}$: operating and maintenance costs
FI: energy intensity [kWh/100 km].

The environmental assessment is divided into assessment of well-to-tank (WTT) and tank-to-wheel (TTW) emissions. In addition, emissions embedded in the car (TTW$_{car}$) are included in the analysis. WTT emissions in the fuel cycle are calculated using the following equation:

$$WTT_{fuel} = f_{PrCO_2} \cdot E_{pr} \tag{3}$$

where:
f_{prCO2}: CO_2 emission factor of primary energy
E_{pr}: primary energy input.

TTW emissions in the fuel cycle are calculated using data for the energy intensity of vehicles and the specific CO_2 emission factors of energy used, as well as depending on the specific number of kilometers driven per year:

$$TTW_{fuel} = FI \cdot skm \cdot f_{CO_2} \tag{4}$$

where:
f_{CO2}: CO_2 emission factor of fuels
FI: fuel intensity
skm: specific km driven per year and car.

The embedded emissions of the car are calculated per year and kilometers driven as:

$$TTW_{car} = \frac{CO_{2_{car}}}{skm \cdot LT} \tag{5}$$

where:
CO_{2car}: CO_2 emissions of car manufacturing including materials
LT: lifetime of car.

Finally, the total CO_2-emissions are calculated as the sum of all emissions related to energy use including the embedded emissions of the car:

$$CO_2 = WTT_{fuel} + TTW_{fuel} + TTW_{car} \tag{6}$$

5. Costs of Private and Public Mobility

One of the major reasons for the low number of battery electric vehicles (BEV) is related to their higher mobility costs in comparison to conventional vehicles.

Figure 9 depicts the structure of total mobility costs per 100 km driven for conventional petrol and diesel cars in comparison to battery electric vehicles, assuming 16,000 km driven per year for all car types and 70% of the total travel activity in Vienna. It can be seen that the advantage of lower energy costs in the case of electric vehicles is currently more than compensated by higher capital costs; see Figure 9. They are currently not economically competitive with conventional cars. Moreover, considering the same travel activity, it is obvious that public transport is by far the cheapest mobility option in Vienna.

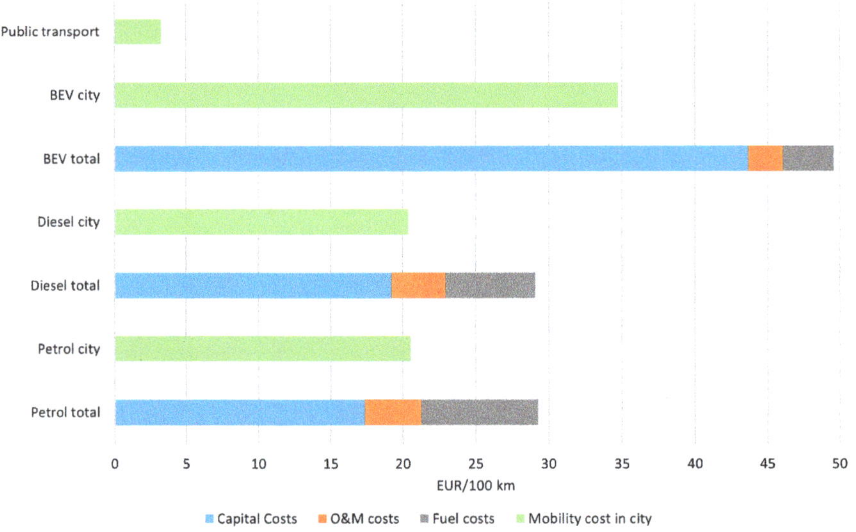

Figure 9. Total mobility costs of passenger cars and public transport per 100 km driven and year in 2018 (average car size: 80 kW).

The major assumptions for the calculation of the total mobility costs of passenger cars per 100 km driven in 2018 are given in Table 1. In the case of private cars, there is a significant price difference in mobility depending on the vehicle type and energy used. However, in the case of public transport, the situation is completely different. In Vienna, all public transport means have the same price. Nevertheless, the most expensive is to purchase single tickets, and cheapest to have an annual ticket for public transport. The annual ticket is valid on all means of public transport within Vienna for exactly 365 days. It costs on average 1 EUR per day. This is the reason that the number of such tickets is rapidly increasing. In 2018, 822,000 annual tickets were sold.

Table 1. Assumptions for calculating the total mobility costs of passenger cars and public transport in Vienna per 100 km driven and year in 2018.

	Investment Costs (Incl. VAT and Registration Taxes)	Fuel Price	O&M Costs	Fuel Intensity	Ticket	Total Distance Driven
	EUR/Car	EUR/Unit	EUR/Car/Year	kWh/100 km	EUR/Year	km/Year
Passenger cars						
Petrol	14,000	1.3	624	53.8		16,000
Diesel	15,500	1.2	600	50.8		16,000
BEV	37,000	0.18	382	19.7		16,000
Public transport					365	

It is of course clear that the number of km driven is not the same for all transport technologies and modes. Yet for the sake of comparison, the same number of km driven is assumed for all vehicles, and 70% of the total travel activity is in the city.

6. Environmental Assessment

In the following, at first, the current CO_2 emissions of battery-electric and conventional vehicles are compared, considering different primary energy sources; see Figure 10. Note that in overall CO_2 emissions, the embedded CO_2-emissions of car and battery production as well as of the materials used (TTW_{car}) are included. For BEV, two different cases are analyzed, one with electricity produced from renewable energy sources (RES), and one with the electricity generation mix in Vienna in 2018, which is largely dominated by natural gas-fired power plants. A major perception of this figure is that despite the fact that BEVs do not emit CO_2 at the point of use (TTW_{fuel}), their full environmental benefits are possible if the electricity used is generated from renewable energy sources and not in fossil power plants.

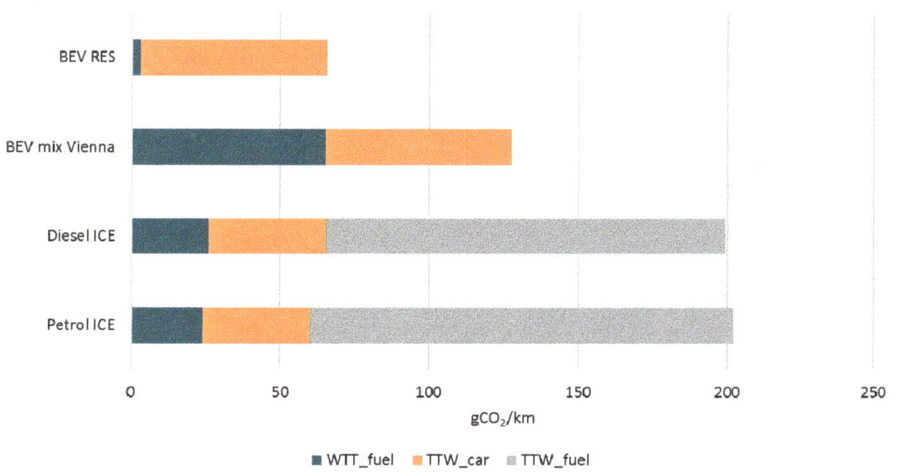

Figure 10. Overall CO_2 emissions of conventional and BEVs with various energy sources, (Car size: 80 kW). Abbreviations: ICE, internal combustion engine; BEV, battery electric vehicle.

WTT emissions of BEVs are very dependent on the primary energy sources used for electricity generation. The historical development of the CO_2-emission factor of electricity used in Vienna is shown in Figure 11. For electricity, the emission factor of primary energy f_{prCO2} is calculated for every year based on the stock of Viennese power plants and the electricity imports. It can be seen in Figure 11 that over time, a continuous decrease took place. However, in Vienna, this factor is still very high compared to the Austrian average (it was 264 kg CO_2/MWh in 2018). The possible future developments of this factor up to 2030 are sketched at the end of Section 7.

Based on this emission factor, it is possible to compare the emissions of public transport, Austrian railways and passenger cars. Figure 12 shows a comparison of CO_2 emissions of various transport modes in Vienna in 2018. For BEV, CO_2 emission factor in Vienna is used, according to Figure 11. The figure of Austrian railways is from their own electricity generation portfolio, which almost solely consists of hydropower plants. The number for BEV Austria builds on 72% RES in the Austrian electricity mix. Public transport in Vienna builds on the same electricity generation mix as BEV in Vienna, yet with a higher occupation.

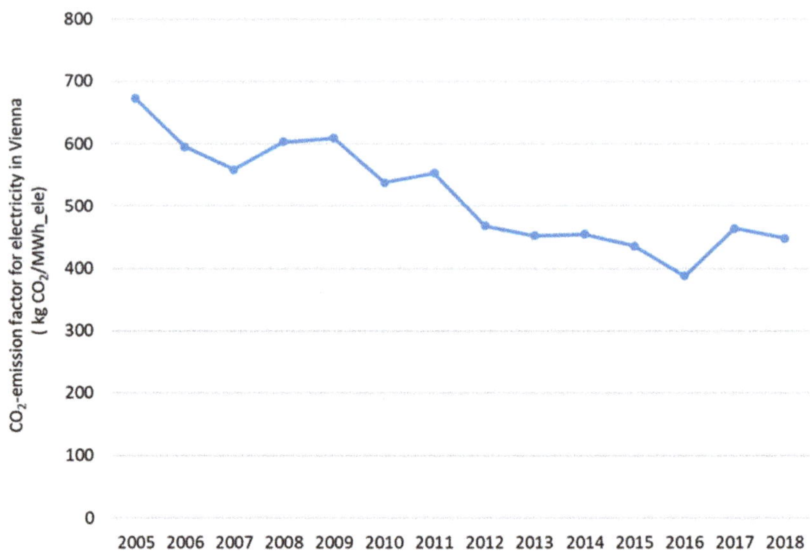

Figure 11. Development of CO_2 emission factor of primary energy of electricity in Vienna, 2005–2018.

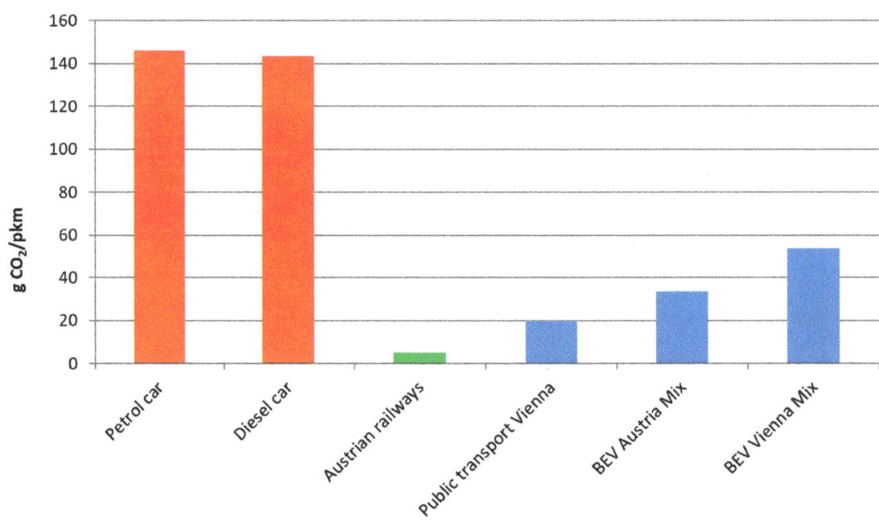

Figure 12. Comparison of CO_2 emissions of various transport modes in Vienna in 2018 (own calculation based on statistical data [15]).

7. Scenarios for the Future of Electric Mobility in Vienna

Since the development of public transport in Vienna is already planned up to 2030, there is limited freedom in the development of scenarios. The only segment where significant changes are possible is passenger car transport, where the shift to BEV could be intensified. In the following, two possible scenarios for the deployment of BEVs in passenger car transport are presented.

The first scenario is a business-as-usual (BAU) scenario created based on recent trends and developments. In the BAU scenario, it is assumed that the growing rate of BEVs, which was on average 63% over the last five years, will remain up to 2025 and then will be reduced to 40% by 2030.

The second scenario is an ambitious (AMB) scenario created assuming stronger promotion policies such as higher subsidies and other benefits for electric vehicle drivers. Here, the growing rate of BEVs is reduced from 63% in 2018 to only 50% in 2030. Of course, the decrease in conventional vehicles will be steeper in the AMB scenario including the target of an overall reduction in the total number of cars (-10%) by 2030.

The vehicle stock (VST) is modeled as:

$$VST_{jt} = VST_{jt-1} \cdot f_{g\_jt} \tag{7}$$

where:

VST_{jt}: stock of vehicle type j in year t

f_{g_jt}: growth factor of car type j in year t.

The magnitude of f_{g_jt} is 10% in the BAU scenario and 15% in the AMB scenario.

In addition, the total vehicle stock in 2030 must be lower than the maximum (VST_{Max_2030}) allowed by current policy restrictions:

$$VST_{Tot\_2030} \leq VST_{Max\_2030} \tag{8}$$

VST_{Tot_2030}: total car stock in the year 2030.

Figure 13 depicts the development of the car stock in the BAU and AMB scenario. In both scenarios, an increase of BEVs and a decrease in the number of internal combustion engine (ICE) vehicles, as well as in total vehicle stock, can be seen. The curves on the bottom represent the scenarios for BEVs. In the AMB scenario, the number of BEVs could be about 110,000 in 2030. Due to policies and measures already implemented, the number of overall car is decreases even in the BAU scenario. The curve on the top shows the development of overall car stock in the BAU scenario.

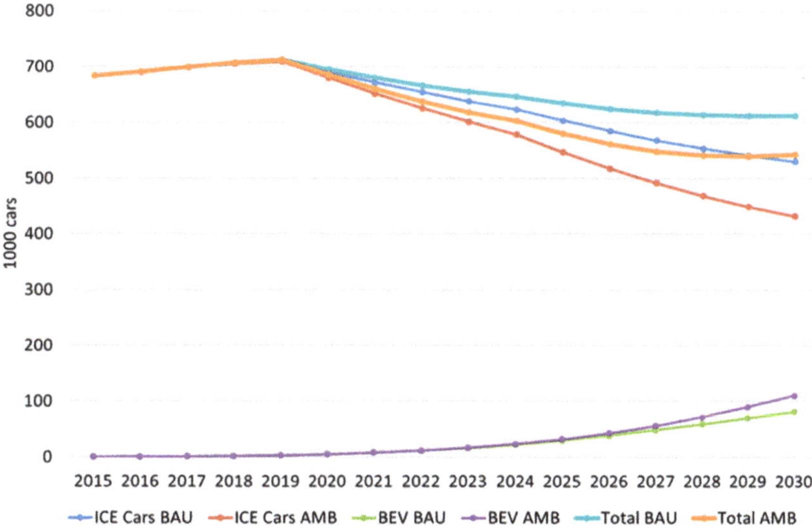

Figure 13. Development of number of conventional and electric cars in the business-as-usual (BAU) and the ambitious (AMB) scenarios.

The major results of the scenarios with respect to energy demand are shown in Figures 14 and 15. Most important is that in the AMB scenario, through the introduction of BEV, the total energy demand in road transport can be reduced by about 60% in 2030 compared to 2018. In the BAU scenario, energy demand also decreases but just by 35%, mainly because of the mandatory reduction of the overall vehicle stock. As Figure 14 shows, already in the BAU scenario, overall energy demand in passenger car transport can be reduced by 50% given an overall reduction of the car stock.

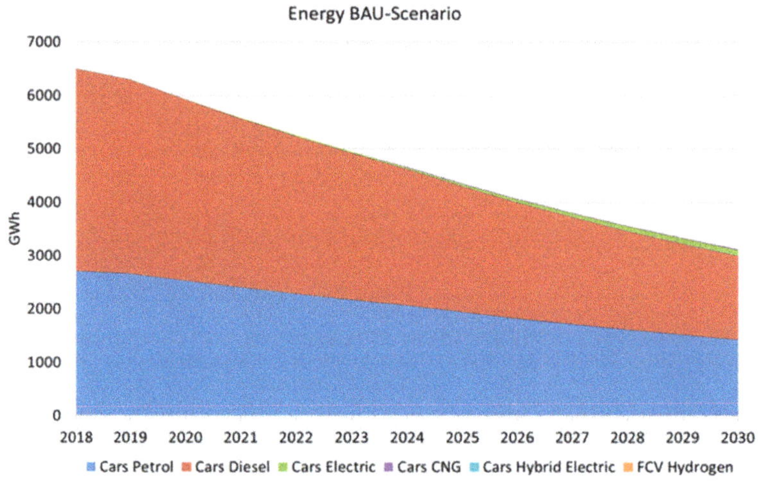

Figure 14. Development of overall energy demand in passenger car transport in the BAU scenario.

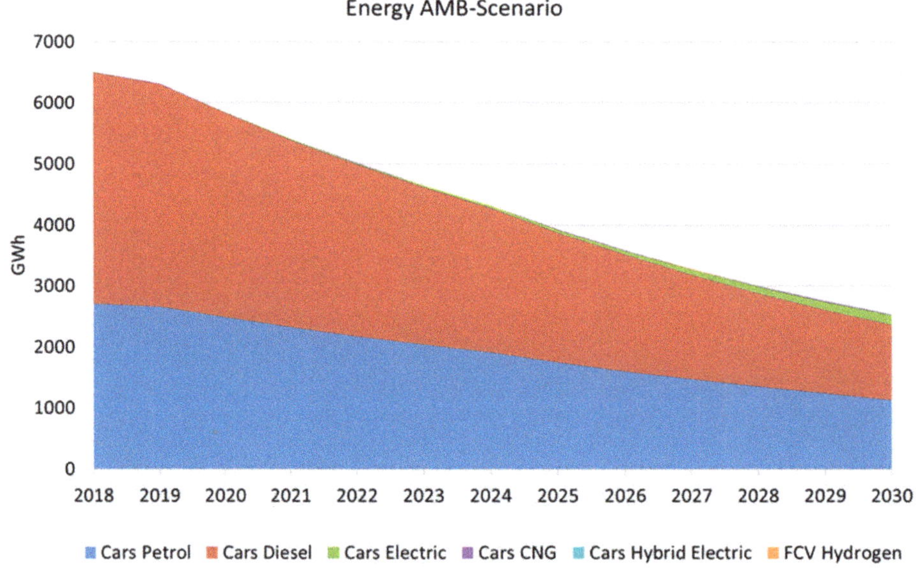

Figure 15. Development of overall energy demand in passenger car transport in the AMB scenario.

Although the number of BEVs in the AMB-scenario is higher than in the BAU scenario, the corresponding increase in energy demand is small, since BEVs have a high efficiency and hence can be seen as an energy conservation technology.

The major finding is that until 2030, the total energy demand was reduced from 3.1 TWh in the BAU-scenario to 2.5 TWh in the AMB-scenario.

The development of specific CO_2-emission factors of the electricity mix is depicted in Figure 16. The figure shows two possible scenarios: a BAU electricity mix scenario and a progressive RES scenario. The mix scenario considers that the district heating share requires more combined heat and power production, which is in principle favorable to the current mix but worse than electricity generation from RES. The RES scenario assumes that 100% RES electricity generation is reached in Austria by 2030.

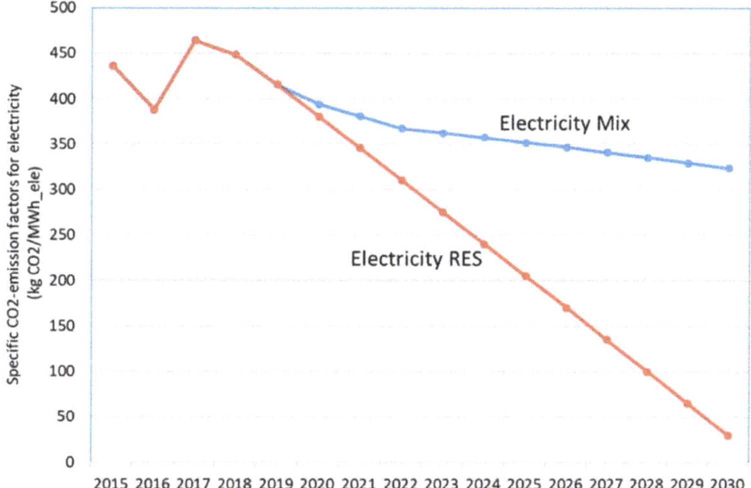

Figure 16. Development of specific CO_2-emission factors of electricity in a BAU electricity mix scenario and in a progressive electricity RES scenario.

The total CO_2-emissions (CO_{2Tot}) in these scenarios are calculated as:

$$CO_{2Tot_t} = \sum_{j=1}^{n} f_{CO_{2j_t}} \cdot E_{j_t} \ (\text{Mill tons } CO_2) \tag{9}$$

with

E_{jt}: Energy consumption of energy carrier j in year t (MWh)

$f_{CO2\_j\_t}$: Overall CO_2 emission factor of fuel j.

Figure 17 shows the development of overall CO_2-emissions in passenger car transport in the BAU and the AMB scenario for BEV development and for electricity in an electricity mix and a progressive electricity RES scenario. Four possible paths are considered. It can be noticed that the electricity mix used in BEVs has a considerable impact on total emissions and that use of electricity from RES in the AMB scenario can significantly reduce emissions from passenger car transport.

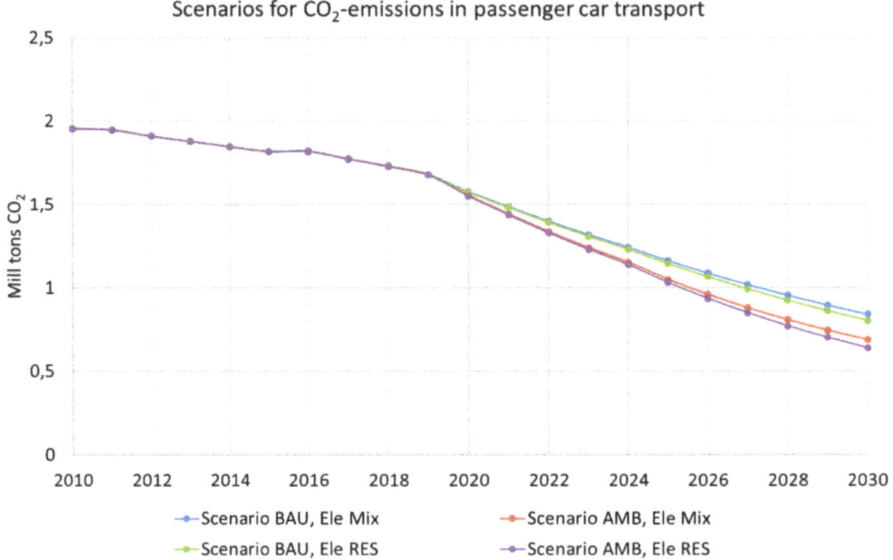

Figure 17. Development of overall CO_2-emissions in passenger car transport in the BAU and the AMB scenarios for BEV development and for electricity in an electricity mix and a progressive electricity RES scenario.

8. Conclusions

The major conclusion is that the policy of the city will have the largest impact on the future development of e-mobility in Vienna. This concerns public transport as well as private BEVs. With respect to the promotion of BEVs, two major options are subsidies and emission-free zones. In addition, announced bans of internal combustion vehicles, especially diesel vehicles, could also become a strong incentive to purchase BEVs.

However, the following important questions with respect to the development of electricity generation and electricity demand remain open. First, regarding the environmental performance, by far the most important is how the electricity generation mix for the electricity used for e-mobility will develop. Second, specifically for the city of Vienna, it is important how the demand for district heating will develop. A higher district heating share requires more combined heat and power production, which is in principle favorable to the current electricity mix but worse than electricity generation from RES. Finally, it is also important how the overall electricity demand will develop. The lower it is in general, the easier it will be to have a larger share of renewable electricity.

Another important issue is the development of the corresponding infrastructure. The deployment of the proper infrastructure such as overhead lines or other grids for electricity, installation of rapid and regular charging stations is of utmost importance. The development of infrastructure is in principle always dependent on regulation, which means it highly depends on the corresponding policies.

Furthermore, the national policies of Austria will play an important role. A current official overall target of Austrian energy policy is to reach 100% RES electricity generation by 2030 in a balanced system. If this could be achieved, of course, it would be a huge jump forwards towards a sustainable energy system. Consequently, electricity used in transport would become much more environmentally benign, and the corresponding CO_2-emissions could be reduced significantly.

However, there are also other options for emission-free energy carriers in transport such as hydrogen or green gases. Today, for short distances and smaller cars, hydrogen is not an economically viable solution. However, for larger buses, it could at least be a possible option.

An outlook for the future emphasizes the core role of generating electricity from RES. This is the core crucial issue for environmentally benign electric mobility regardless of which mode is chosen. In addition, there is the question of whether the city will introduce a more ambitious strategy with respect to the reduction of private cars in general and ambitious towards more environmentally benign mobility modes. In any case, from the individual and societal points of view, economics will play a crucial and predominant role.

In summary, concrete steps recommended for future policies are to (i) introduce and extend emission-free zones in urban areas; (ii) force electricity generation from RES; (iii) ensure that mobility costs of all transport modes reflect full life cycle emissions are as correct as possible. Finally, it can be stated that these findings are generalizable virtually worldwide.

Author Contributions: Conceptualization, A.A. and R.H.; methodology, A.A. and R.H.; formal analysis, A.A.; investigation, A.A. and M.S.; resources, A.A. and M.S.; data curation, A.A. and M.S.; writing—original draft preparation, A.A.; writing—review and editing, A.A., R.H. and M.S.; visualization, A.A.; supervision, A.A.; project administration, A.A.; funding acquisition, A.A. All authors have read and agreed to the published version of the manuscript.

Funding: The present work was funded by the Vienna Science and Technology Fund (WWTF) through the TransLoC project ESR17-067.

Institutional Review Board Statement: Not applicable.

Informed Consent Statement: Not applicable.

Data Availability Statement: MDPI Research Data Policies.

Acknowledgments: The authors are grateful for the data collected and inputs provided by A. Glatt.

Conflicts of Interest: The authors declare no conflict of interest.

References

1. The Sustainable Urban Systems Subcommittee. Sustainable Urban Systems Report. NSF; 2018. Available online: https://www.nsf.gov/ere/ereweb/ac-ere/sustainable-urban-systems.pdf (accessed on 1 October 2020).
2. Davis, S.J.; Lewis, N.S.; Shaner, M.R.; Aggarwal, S.; Arent, D.; Azevedo, I.L.; Benson, S.M.; Bradley, T.H.; Brouwer, J.; Chiang, Y.-M.; et al. Net-zero emissions energy systems. *Science* **2018**, *360*, eaas9793. [CrossRef] [PubMed]
3. European Commission, Urban Mobility. Available online: https://ec.europa.eu/transport/themes/urban/urban_mobility_en (accessed on 1 June 2020).
4. Romero-Lankao, P.; Wilson, A.; Sperling, J.; Miller, C.; Zimny-Schmitt, D.; Bettencourt, L.; Wood, E.; Young, S.; Muratori, M.; Arent, D.; et al. *Urban Electrification: Knowledge Pathway toward an Integrated Research and Development Agenda*; Whitte Paper. NREL's Transportation and Hydrogen Systems Center and the Mansueto Institute for Urban Innovation of the University of Chicago, 2019. Available online: https://papers.ssrn.com/sol3/papers.cfm?abstract_id=3440283 (accessed on 5 June 2020).
5. Wiener Linien. About Wiener Linien. Available online: https://www.wienerlinien.at/eportal3/ep/channelView.do/pageTypeId/66533/channelId/-2000622 (accessed on 1 June 2020).
6. VIENNA.AT. Neue Elektrobusse der Wiener Linien Wurden Präsentiert. Available online: https://www.vienna.at/neue-elektrobusse-der-wiener-linien-wurden-praesentiert/3354064 (accessed on 1 June 2020).
7. Inno4sd.net. Innovative Electric Buses in Vienna. Available online: https://www.inno4sd.net/innovative-electric-buses-in-vienna-548 (accessed on 1 June 2020).
8. Eltis. A Cleaner City: Electric Buses in Vienna (Austria). Available online: https://www.eltis.org/discover/case-studies/cleaner-city-electric-buses-vienna-austria (accessed on 1 June 2020).
9. European Commission, European Strategies—White Paper 2011. Available online: https://ec.europa.eu/transport/themes/strategies/2011_white_paper_en (accessed on 1 June 2020).
10. Held, T.; Gerrits, L. On the road to electrification—A qualitative comparative analysis of urban e-mobility policies in 15 European cities. *Transp. Policy* **2019**, *81*, 12–23. [CrossRef]
11. Liberto, C.; Valenti, G.; Orchi, S.; Lelli, M.; Nigro, M.; Ferrara, M. The Impact of Electric Mobility Scenarios in Large Urban Areas: The Rome Case Study. *IEEE Trans. Intell. Transp. Syst.* **2018**, *19*, 3540–3549. [CrossRef]
12. Ajanovic, A.; Haas, R. Dissemination of electric vehicles in urban areas: Major factors for success. *Energy* **2016**, *115*, 1451–1458. [CrossRef]
13. Scarinci, R.; Zanarini, A.; Bierlaire, M. Electrification of urban mobility: The case of catenary-free buses. *Transp. Policy* **2019**, *80*, 39–48. [CrossRef]

14. Glotz-Richter, M.; Koch, H. Electrification of Public Transport in Cities (Horizon 2020 ELIPTIC Project). *Transp. Res. Procedia* **2016**, *14*, 2614–2619. [CrossRef]
15. Statistik Austria. Energiebilanzen Wien 1988 Bis 2018. Available online: https://www.statistik.at/web_de/statistiken/energie_umwelt_innovation_mobilitaet/energie_und_umwelt/energie/energiebilanzen/index.html (accessed on 10 October 2020).
16. Wiener Linien. Zahlen Daten Fakten. Available online: https://www.wienerlinien.at/media/files/2018/facts_and_figures_2017_243486.pdf (accessed on 1 October 2020).
17. Wiener Linien: Facts and Figures. Available online: https://www.wienerlinien.at/media/files/2019/betriebsangaben_2018_engl_310520.pdf (accessed on 1 October 2020).
18. Statistik Austria (1991–2020): KFZ-Bestand nach Kraftstoffarten bzw. Energiequelle und Bundesländern. Available online: https://www.statistik.at/web_de/statistiken/energie_umwelt_innovation_mobilitaet/verkehr/strasse/kraftfahrzeuge_-_bestand/index.html (accessed on 10 October 2020).
19. *Bundesministerium für Öffentliche Wirtschaft und Verkehr, Mensch, Umwelt, Verkehr: Das Österreichische Gesamtverkehrskonzept 1991 (GVK-Ö 1991)*; Bundesministerium für Öffentliche Wirtschaft und Verkehr, 1991; pp. 1–308. Available online: http://worldcat.org/identities/lccn-n85160158/ (accessed on 1 October 2020).
20. European Commission. Sustainable Urban Mobility Plans (SUMPs). Available online: https://ec.europa.eu/transport/themes/urban/urban-mobility/urban-mobility-actions/sustainable-urban_en (accessed on 1 October 2020).
21. Vienna City Administration, MA 18—Urban Development and Planning. STEP 2025—URBAN Mobility Plan. Vienna, Austria, 2014. Available online: https://www.wien.gv.at/stadtentwicklung/studien/pdf/b008379b.pdf (accessed on 10 October 2020).
22. Vienna City Administration, MA 18—Urban Development and Planning. STEP 2025—E-Mobility Strategy. Vienna, Austria, 2016. Available online: https://www.wien.gv.at/stadtentwicklung/studien/pdf/b008465.pdf (accessed on 10 October 2020).
23. BDO Austria Holding Wirtschaftsprüfung GmbH. Newsletter: E-Mobilität—Förderungen und steuerliche Anreize. 2019. Available online: https://www.bdo.at/de-at/publikationen/tax-news/tax-news-3-2019/e-mobilitat-%E2%80%93-forderungen-und-steuerliche-anreize (accessed on 1 October 2020).
24. Austria Tech. Anreize für Effiziente und Umweltfreundliche Fahrzeuge. Vienna, Austria, 2014. Available online: https://www.austriatech.at/assets/Uploads/Publikationen/PDF-Dateien/1670cc5a35/ATE-PolicyBrief-012014_Anreizsysteme-fuer-E-Mobilitaet.pdf (accessed on 15 October 2020).
25. Vienna City Administration, MA18—Urban Development and Planning. Smart City Wien—Framework Strategy. Vienna, Austria, 2014. Available online: https://smartcity.wien.gv.at/site/files/2019/07/Smart-City-Wien-Framework-Strategy_2014-resolution.pdf (accessed on 15 October 2020).
26. Wu, G.; Inderbitzin, A.; Bening, C. Total cost of ownership of electric vehicles compared to conventional vehicles: A probabilistic analysis and projection across market segments. *Energy Policy* **2015**, *80*, 196–214. [CrossRef]
27. Hagman, J.; Ritzén, S.; Stier, J.J.; Susilo, Y. Total cost of ownership and its potential implications for battery electric vehicle diffusion. *Res. Transp. Bus. Manag.* **2016**, *18*, 11–17. [CrossRef]
28. Moro, A.; Helmers, E. A new hybrid method for reducing the gap between WTW and LCA in the carbon footprint assessment of electric vehicles. *Int. J. Life Cycle Assess.* **2017**, *22*, 4–14. [CrossRef]
29. Qiao, Q.; Zhao, F.; Liu, Z.; He, X.; Hao, H. Life cycle greenhouse gas emissions of Electric Vehicles in China: Combining the vehicle cycle and fuel cycle. *Energy* **2019**, *177*, 222–233. [CrossRef]
30. Ajanovic, A.; Haas, R. Economic and Environmental Prospects for Battery Electric- and Fuel Cell Vehicles: A Review. *Fuel Cells* **2019**, *19*, 515–529. [CrossRef]
31. Ajanovic, A.; Haas, R. On the economics and the future prospects of battery electric vehicles. *Greenh. Gas Sci Technol.* **2020**, *10*, 1151–1164. [CrossRef]
32. Ajanovic, A.; Haas, R. On the Environmental Benignity of Electric Vehicles. *J. Sustain. Dev. Energy Water Environ. Syst.* **2019**, *7*, 416–431. [CrossRef]

Article

Quantification of the Flexibility Potential through Smart Charging of Battery Electric Vehicles and the Effects on the Future Electricity Supply System in Germany

Felix Guthoff *, Nikolai Klempp and Kai Hufendiek

Institute of Energy Economics and Rational Energy Use (IER), University of Stuttgart,
DE-70565 Stuttgart, Germany; nikolai.klempp@ier.uni-stuttgart.de (N.K.);
kai.hufendiek@ier.uni-stuttgart.de (K.H.)
* Correspondence: felix.guthoff@ier.uni-stuttgart.de

Abstract: Electrification offers an opportunity to decarbonize the transport sector, but it might also increase the need for flexibility options in the energy system, as the uncoordinated charging process of battery electric vehicles (BEV) can lead to a demand with high simultaneity. However, coordinating BEV charging by means of smart charging control can also offer substantial flexibility potential. This potential is limited by restrictions resulting from individual mobility behavior and preferences. It cannot be assumed that storage capacity will be available at times when the impact of additional flexibility potential is highest from a systemic point of view. Hence, it is important to determine the flexibility available per vehicle in high temporal (and spatial) resolution. Therefore, in this paper a Markov-Chain Monte Carlo simulation is carried out based on a vast empirical data set to quantify mobility profiles as accurately as possible and to subsequently derive charging load profiles. An hourly flexibility potential is derived and integrated as load shift potential into a linear optimization model for the simultaneous cost-optimal calculation of the dispatch of technology options and long-term capacity planning to meet a given electricity demand. It is shown that the costs induced by BEV charging are largely determined by the profile costs from the combination of the profiles of charging load and renewable generation, and not only by the additional energy and capacity demand. If the charging process can be flexibly controlled, the storage requirement can be reduced and generation from renewable energies can be better integrated.

Keywords: electricity sector; flexibility; electric mobility; demand side integration; total system costs; decarbonization

Citation: Guthoff, F.; Klempp, N.; Hufendiek, K. Quantification of the Flexibility Potential through Smart Charging of Battery Electric Vehicles and the Effects on the Future Electricity Supply System in Germany. *Energies* **2021**, *14*, 2383. https://doi.org/10.3390/en14092383

Academic Editor: Amela Ajanovic

Received: 30 November 2020
Accepted: 15 April 2021
Published: 22 April 2021

Publisher's Note: MDPI stays neutral with regard to jurisdictional claims in published maps and institutional affiliations.

1. Introduction

In order to achieve climate targets specified in international agreements, countries set emission reduction targets and action plans in all sectors [1]. The strong expansion of renewable energies plays an important role for the target achievement, but also the rise of the electrification of the heat supply and the mobility sector [2] which, however, leads to a higher electricity demand. Therefore, a more fluctuating feed-in has to meet a demand which is varying in time and space, which increases the need for flexibility [3]. As a result, new problems arise for the dimensioning of the necessary grid infrastructure or generation and storage capacities.

The integration of technologies for sector integration, such as heat pumps or battery electric vehicles (BEV), and the use of direct load control, e.g., for charging vehicles, are options to provide flexibility and therefore should be part of the solution. Their actual flexibility potential, however, is subordinated to given boundary conditions from reality. For BEV in particular, there are restrictions resulting from individual mobility behavior. The batteries of electric vehicles cannot be assumed as a large storage capacity available at times when impact of additional flexibility potential is highest from a systemic point of view. An efficient use of flexibility options can reduce the need for additional technology investments and thereby the costs for the electricity supply.

The question arises: how can the demand and the flexibility potential of *BEV* be determined adequately based on empirical mobility data? Subsequently, the impact of the use of an intelligent charging control on total system costs as well as on necessary generation capacities has to be analyzed in order to quantify the effects within a systemic evaluation. Empirical household mobility data, which, amongst others, include duration, start and ending location as well as standing times, are available for Germany provided from the Federal Ministry of Transport and Digital Infrastructure (BMVI) [4]. The key findings of the study, in which the data collection was conducted, are presented in [5], while a more detailed overview on the data can be found in [6]. Taking only the averages of all rides in the data leads to a mobility pattern which does not reflect the variety of individual mobility needs and therefore additional assumptions to subsequently derive realistic charging profiles are necessary. In [7], for example, a clustering is used to construct representative driving patterns based on historical data. Data-samples on real charging behavior are published in several studies and research projects and often used to derive only generic load profiles, as is done in [8].

The flexibility potential based on charging curves can be determined with several methods and can be described differently. The timeframe of the use of the flexibility of BEV can be considered as short-term when it is compared with the definition in [9], because the mobility is assumed to have a daily pattern with its resulting restrictions. Approaches for modeling differ fundamentally in whether Vehicle-to-Grid (*V2G*) is enabled or only the pure shifting of the loading process is considered. In addition, further boundary conditions can be assumed, such as blocking times during which the charging process cannot be postponed or a minimum charge level of the vehicle. In the literature, the willingness of *BEV* owners to allow external access at all as well as possible financial incentives, which would be necessary, is discussed [10].

Above all, in the optimization of the renewable self-consumption or the net purchase, intelligent load strategies play a large role in the future. Partly they are already implemented in real demonstrators (for example in [11]) and are part of research (for example in [12]) as well as being implemented in reality. In many cases, a fixed electricity price curve is specified, the own consumption is maximized or the peak of the power supply is minimized. The use of flexibility to stabilize the power grid is also investigated and can be useful if sufficient knowledge of the locally available potential is available. For example in [13], optimal operation points of an existing network are identified using a Monte Carlo simulation under consideration of a given load (including *BEV*) and renewable energy production. These local analyses often neglect interaction with the rest of the energy system and therefore don't have a focus on future effects on the energy system regarding capacity expansion.

In analyses for the long-term planning of the energy system, the effects of increasing electrification of the transport sector on the energy system have already been analyzed, e.g., in [14]. The amount of energy required for motorized private transport and freight transport is determined and integrated into the respective model by means of a load curve. The studies differ in the spatial resolution or other aspects as well as the system boundary and the level of detail in the transport sector. Research studies focus, amongst other factors, on the necessary grid extension [15] or impacts on household prices [16] on an actor's perspective by a rising penetration of *BEV*. Effects on a systemic level concerning the flexibility use in high temporal resolution, the necessary capacity investments and the resulting system cost components are not analyzed extensively with a focus on *BEV*. The difficulty in all these investigations remains, however, to determine the individual mobility behavior in general in order to determine a realistic availability of flexibility and to quantify the benefit in interactions with the dynamic energy system.

Therefore, this paper presents an approach to adequately determine the load profiles and the flexibility potential of *BEV* and their integration, as well as its use, within an electricity market model. The question of the optimal technology mix for cross-sectoral decarbonization of the energy system is not the focus of attention.

2. Materials and Methods

In the given paper, a methodology is presented to adequately determine the flexibility potential of *BEV* and to calculate its use within an electricity market model considering feedback effects. *BEV*-specific input data are determined based on empirical mobility data for Germany within a mobility tool. In this tool, mobility profiles per car are determined using a Markov-Chain Monte Carlo simulation. Based on these profiles, the resulting load curve of *BEV*'s, and subsequently the flexibility potential, is determined as input for the electricity market model. In order to evaluate the benefits of a flexibilization of the *BEV* charging process by demand side integration (*DSI*), we determine the flexibility requirements and calculate total system costs in a scenario with ambitious climate targets with the electricity market model in which different technologies and flexibility options are implemented. An overview of the methodology is shown in Figure 1. The mentioned boxes and its key points, listed under the headlines "(*EV*-)Mobility Simulation Tool" (Empirical Mobility Data, Mobility Profiles, *BEV* Charging Profiles, Flexibility Potential of intelligent *BEV*-charging) and "Electricity Market Model E2M2", are addressed in more detail in the following methodology presented in this section.

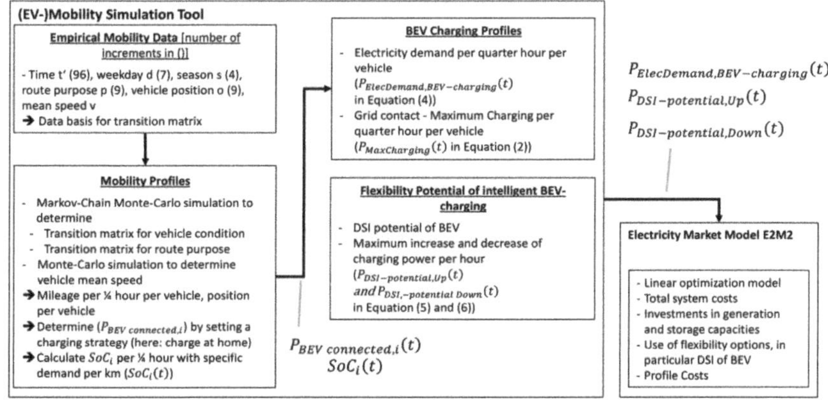

Figure 1. Overview of the approach presented in this paper.

2.1. Determination of Mobility and Load Curves

An empirical data set on the mobility behavior of private individuals in Germany with more than 316,000 respondents is available as the basis for determining mobility profiles [4]. It contains, amongst others, information on the time of departure, the purpose of the journey, the duration of the journey and the standing times in a quarter-hourly resolution. Based on these data, transition probabilities are determined in order to use them in a Markov-Chain Monte Carlo simulation and thereon develop mobility curves.

A probability for the vehicle status results from the sum of the trips at the respective point in time and the following driving status in the next time segment. The stochastic process of the time a vehicle starts to arrival at its destination can be reduced to the states "en route" or "standstill" and described as solely dependent on the initial state for every time step. Following [17], it can subsequently be modelled as a discrete-time stochastic process $(X_n)_{n \in N0}$ with finite state space Z and is called a Markov Chain if it satisfies the Markov property, i.e., if the following holds:

$$P(\{X_{n+1} = Z_{n+1}\}|\{X_0, \ldots, X_n) = (z_0, \ldots, z_n)\}) = P(\{X_{n+1} = Z_{n+1}\}|\{X_n = z_n\}) \quad (1)$$

For all $n \in N$ and $z_0, \ldots, z_{n+1} \in Z$ with $P(\{(X_0, \ldots, X_n) = (z_0, \ldots, z_n)\}) > 0$ [18].

The conditional probability $P(\{X_{n+1} = Z_{n+1}\} | \{X_n = z_n\})$ is called the transition probability from the state z at time n to the state z at time $n + 1$. Each theoretically possible state at time $n + 1$ is assigned a transition probability depending on all possible initial states at time n. These transition probabilities can be summarized in a quadratic matrix, which is called a transition matrix [18].

Such transition matrixes are generated for every time step and state using the vast data set of MiD 2017, which comprises empiric mobility profiles of 25,922 households with, in total, 60,713 persons and 193,290 routes. The Markov-Chain Monte Carlos simulation is used based on the transition matrix for the vehicle conditions depending on the time-step (96 quarter hours per day), the weekday (7 days), the season and the last route's purpose (9 different purposes). An example of such a transition matrix for the cars condition is given in Table 1.

Table 1. Structure of the transition matrix for the condition of the car with exemplary entries.

Season	Weekday	Time Step	Last Routes Purpose	Initial State	Final State		Σ
					En Route	Standstill	
...	...	...	...	en route		...	1.00
				standstill			1.00
Autumn	Monday	7:00 to 7:15	...	en route	0.93	0.07	1.00
				standstill	0.00	1.00	1.00
...	...	...	...	en route		...	1.00
				standstill			1.00

The implementation of the simulation is described in detail by [17]. The advantage of this procedure is that each of the single runs recorded in the data set are taken into account to determine the transition matrix and therefore the synthetic motion profiles. For every car, the current status and the initial state of the car leads to the probability of the transition matrix for the next time step, the final state of each simulation.

Following this concept, another transition matrix is created based on the empirical data to aquire the probabilities for the purpose of the route when the car's status is changing from "standstill" to "en route". The purposes of the route include, for example, "way to work", "leisure" or "drive home". The data set includes a total of 9 purposes, which are dependent on the season and the weekday. An example for the transition matrix for the route purposes is given in Table 2.

Table 2. Structure of the transition matrix for the route purposes of the car with exemplary entries.

Season	Weekday	Final State (9 Route Purposes)				Σ
		Way to Work	Leisure	...	Drive Home	
...	...			...		1.00
Autumn	Monday	0.23	0.17	...	0.12	1.00
...	...			...		1.00

The Markov-Chain is processed for each of the assumed cars within the Monte Carlo simulation until the result converges again to the expected value.

Furthermore, the vehicle mean speed $v(t)$ is determined by means of a conditional probability distribution of vehicle mean speed depending on the time, weekday and season in the same way using a Monte Carlo simulation. As a result of the simulation, we get the information about the mileage per quarter hour and the position per hour. Assuming a specific demand per km, we get the energy consumption and so the state of charge (*SoC*) per car and quarter hour. Consumption only occurs if the vehicle condition is declared as "en route".

If the state of the car changes to "standstill" and "at home" in the next time step, the car is stated to be connected to the grid as a specific user behavior, and immediate uncontrolled charging at home after each journey is assumed. The sum of all cars' simulation results leads to the mobility profile and the charging profile from *BEV*. In the following equations to describe the interface between the mobility simulation and the optimization model, parameters are annotated with "*P*", and variables with "*V*". By aggregating the quarter-hourly simulation results for each hour t, the maximum charging power of all connected cars, which is further referred to as the "grid contact curve" can be described by

$$P_{MaxCharging}(t) = P_{BEV\ connected}(t) * P_{charging\ capacity} \tag{2}$$

A charging infrastructure with a charging capacity of 3.7 kW, as installed in most households, is assumed [19]. The charging process in the simulation tool starts with maximum power until the vehicle is fully charged. The number of charging car at each time step can be described by

$$P_{BEV\text{-}charging}(t) = \sum_i P_{BEV\ connected,i}(t) \mid i \text{ with } SoC_i < SoC_{max} \tag{3}$$

An initial state of charge of the batteries based on the final state of the previous day is also taken into account. Within the tool other options, e.g., charging at work, are possible but not part of this study.

The consumption of the cars, the connection to the grid and the charging capacity results in the user controlled charging-curve of all *BEV* for each time step and is stated by

$$P_{ElecDemand,BEV\text{-}charging}(t) = P_{BEV\text{-}charging}(t) * P_{charging\ capacity} \tag{4}$$

In the work presented here, the mobility data are summarized for the core week from Monday to Thursday and for each season to reduce complexity. Further on, the Monte Carlo simulation is processed 750 times (which corresponds to 750 vehicles) to reach a high statistical accuracy of results while having a tolerable calculation time. Within a season, the mobility behavior for a Friday, for example, is always assumed the same. This results in a typical week for each season, which describes all weeks in this season. By aggregating all seasons, the course for a whole year is then obtained.

At the moment, Germany mainly has vehicles with combustion engines and thus these also form the basis for the majority of data on mobility behavior. For the derivation of the charging curve of *BEV*, it is assumed that the mobility requirements and user behavior will remain the same in the event of a greater penetration of electric vehicles in the future. Therefore, we use the current mobility behavior data for combustion engines for our mobility demand calculation without considering different user types.

A representative *BEV* is assumed for the calculation of the load curve. The specific consumption and battery capacity is calculated by quantity weighting for each vehicle segment and the electric vehicles newly registered in Germany. The data and the derivation of a typical *BEV* are given in the Appendix A.

2.2. Determination of the Flexibility Potential of Intelligent Charging Control

To determine the flexibility potential of an intelligent charge control, further assumptions are made. The use of *V2G* is excluded, since the technology is currently still in the experimental stage and no solutions have yet been found regarding the compensation payment for the expected reduction in service life due to aging of the batteries as a result of the additional number of cycles. Furthermore, it is assumed that the charging capacity can only be postponed to a later point in time, since every vehicle is charged in uncontrolled mode immediately after plugging in. A shifting forward of the load would be possible in regard to the aggregated charging curves, but does not reflect the results of the modeling and the spatial distribution of the cars in reality.

The goal is to describe and quantify the load shift potential based on the determined load curves. For this purpose, a range of the maximum possible load is defined, which corresponds to the power of all vehicles with full power supply connected at that time: the grid contact curve. It includes vehicles that are already fully loaded and currently not receiving any power. The difference between the maximum possible charging power $P_{MaxCharging}(t)$ (the grid contact curve) and the actual charging power at a given point in time $P_{ElecDemand,BEV\text{-}charging}(t)$ corresponds to the positive flexibility potential through power increase $(P_{DSI\text{-}potential,Up}(t))$, as shown in Equation (5).

$$P_{DSI\text{-}potential,Up}(t) = P_{MaxCharging}(t) - P_{ElecDemand,BEV\text{-}charging}(t) \tag{5}$$

This is also referred to as "*PowerUp*" or "*PowerUp-Potential*" in the following. The negative flexibility potential $P_{DSI\text{-}potential,Down}(t)$ results from the difference between the actual charging power and the zero line, no charging process, therefore called "*PowerDown*" or "*PowerDown-Potential*".

$$P_{DSI\text{-}potential,Down}(t) = P_{ElecDemand,BEV\text{-}charging}(t) \tag{6}$$

The relation is also depicted in Figure 2. It shows the maximum possible charging and the realized charging as well as the possible increase or decrease of the process by specifying the PowerUp or PowerDown for an exemplary day, each normalized to the grid contact of all vehicles.

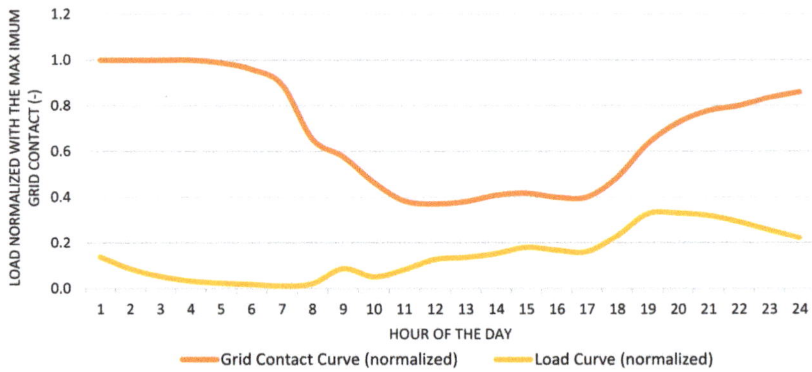

Figure 2. Depiction of the simulated realized charging profile for an exemplary day (yellow line), the maximum possible charging of all cars connected to the station (orange line), the maximum positive flexibility (blue arrow) and maximum negative flexibility (black arrow), normalized values.

As a further assumption to limit the flexibility potential, a maximum duration of the shift is given. We assume that owners of electric cars are willing to allow external access to control their vehicles if they are not restricted in their everyday life. An important point here is the specification of a time at which the vehicle is fully or partially charged. Therefore, the present model of the flexibility potential also specifies a point in time when all vehicles must be fully loaded.

Since the mobility curves are determined based on historical data from moving vehicles and the part of the vehicle fleet that is not moved (on average 41.2%) on a day is not included, the actual demand is overestimated (of about 72.3%) in the modeling when the number of simulated vehicles is scaled to the existing fleet. The average mileage of the vehicles in the given data [4] is 24,054 km and is thus significantly higher in comparison to literature data (13,931 km in 2017 [20]). Therefore, the average mileage of the empirical data is scaled to the average mileage of a passenger car and thus the load demand is reduced accordingly.

2.3. Implementation in the Linear Optimization Model

To work out the effects of an intelligent charge control of *BEV* in a nearly decarbonized energy system with a high share of renewable energies and to analyze the interaction and feedback with other parts of the energy system, the hourly charge performance curves and the flexibility potential are integrated in the electricity market model E2M2 (European Electricity Market Model). It is used to quantify the system effects of the additional flexibility potential created by the intelligent control of the *BEV* charging process.

The model, based on [21], simultaneously optimizes the necessary investments in power plant capacity and flexibility options and their use to meet a given electricity demand in hourly resolution while minimizing the total system costs C_{total}. The objective function includes fixed costs, such as annualized investment costs (C_{Invest}) or annual operating costs (C_{fixOaM}), as well as variable costs (Operation and Maintenance (C_{varOaM}), Fuel (C_{Fuel}), StartUp ($C_{StartUp}$)) for power plant deployment, as shown in Equation (7).

$$C_{total} = C_{Invest} + C_{fixOaM} + C_{varOaM} + C_{StartUp} + C_{Fuel} \tag{7}$$

The electricity demand specified in the model must be met at every hour (see Equation (8)). Various technology options, which are subject to specific techno-economic restrictions, are available for this purpose. In addition to feed-in from renewable energies and conventional power plants, the balance can also be covered by imports or excess electricity can be exported. The production of renewable energy is, different than described in [21], based on a discrete synthetic profile, as it is described in [22]. Furthermore, a temporal offset of the energy demand can be achieved by storages or load management ($V_{DSI,up}$, $V_{DSI,down}$), which are subject to restrictions across time steps. The entire demand equation for each time step then results in:

$$P_{ElecDemand} = V_{ElecProduction,conventional} + V_{ElecProduction,renewable} - C_{curtailment,renewables}$$
$$+ V_{ElecProduction,storages} - V_{Pumping,storages} + V_{DSI,down} - V_{DSI,up} + V_{Import} - V_{Export} \tag{8}$$

In addition, framework conditions, such as minimum shares of renewable energies or a CO_2-limit, can be taken into account and, depending on the issue at hand, various flexibility options or sector-integrating technologies, e.g., for heat supply, can be considered. The possible flexibility options are described in [23].

Electric mobility is incorporated into the model on the one hand via the load profile determined in the model and on the other hand via the resulting demand-side flexibility. Therefore, the implementation of load management according to [24] is used. The possibility of using load management results from the short-term reduction of the load and thus reduction of demand. The reduced demand has to be covered after a period of time or up to a point in time, depending on the specification. In any case, it must be balanced energetically, which may only be done after the load shift. In the case of electric mobility, in order not to cause a loss of comfort for the *BEV* drivers, it is assumed that the vehicles must be charged at a uniform time. This time is determined after the analysis of the mobility data.

The maximum positive or negative available shifting power is limited by a given hourly potential $P_{DSI\text{-}potential,Down}(t)$ and $P_{DSI\text{-}potential,Up}(t)$ (Equations (9) and (10)). The potential limitation is derived by the procedure described above (see Equations (5) and (6)).

$$V_{DSI,down}(t) \leq P_{DSI\text{-}potential,Down}(t) \text{ and} \tag{9}$$

$$V_{DSI,up}(t) \leq P_{DSI\text{-}potential,Up}(t) \text{ with } V_{DSI,j}(t) \geq 0 \mid j \in down, up \tag{10}$$

The amount of energy shifted by the realized load reduction ($V_{DSI,down}$) or load increase ($V_{DSI,up}$) is balanced in a negative, loss-free storage (Equation (11)). For this, the specification applies that after a certain period or at a certain point in time, the storage level ($V_{FillLevelDSI}$) must be at zero.

$$V_{FillLevelDSI}(t+1) = V_{FillLevelDSI}(t) + V_{DSI,down}(t) - V_{DSI,up}(t) \tag{11}$$

The technical restrictions and the resulting shift potential are implemented per technology. A detailed description, as well as the basic principles of the flexibility option, can be found in [24]. In the present study, a version is used that specifies a point in time when the energy demand must have been finally covered in order to adequately capture the shift potential of intelligent charge controls.

The chosen representation makes it possible to realistically consider the flexibility from electric mobility in the electricity market model. The implemented restrictions are the potential that varies over time, which was determined from empirical data, and the specification of a vehicle that is charged regularly. The electricity market model continues to assume no costs or losses for the use of the flexibility from *BEV*, as it is assumed that no additional costs are incurred by the mere shifting. The use of other *DSI* options, e.g., shiftable loads in industry, which are associated with costs, are not considered in this study.

2.4. Scope

The newly developed approach is demonstrated in a scenario for Germany with ambitious CO_2-reduction targets for the energy sector. The chosen framework is the scenario "KS95", which was developed in the study "Klimaschutzszenario 2050" [25]. It assumes a CO_2-reduction of 95% for the year 2050 compared to 1990. This means that in 2050 only 18.6 mt of CO_2 may be emitted to cover the demand for electricity. Furthermore, a path for the penetration of electric vehicles is assumed. For 2050 in total, 21.6 million *BEV* are assumed. The hourly resolved charge power demand from the previous simulation is scaled to the number of cars, which results in a demand of 61 TWh. Together with a time series on conventional demand of 375 TWh, this leads to a total electricity demand of 436 TWh.

In the present study, an investment calculation in power plants and flexibility options, without existing power plants, is carried out to cover the hourly demand. This so-called greenfield calculation serves to analyze effects detached from an existing power plant park.

Possible technology options are gas combined cycle plants (*GasCC*), Gas turbines (*GT*) and renewables (*PV*, wind onshore, wind offshore), which can be curtailed. Furthermore, an investment in storage with different energy-to-power (*E2P*) ratios is possible (short-term storages with an *E2P* equal to 2 and mid-term storages with an *E2P* equal to 7, such as battery storage systems in different setups, and long-term storage with an *E2P* equal to 500, such as hydro pump storage). Nuclear power plants as well as hard coal and lignite power plants are not included due to political decisions of a phase out [26]. The data for the investment options and further model data are given in the Appendix B.

For the year 2050, a certain degree of expansion of the renewable technologies *PV* and wind on- and offshore is stipulated in the given scenario, but is calculated endogenously as part of the optimization problem. Since the profiles of these technologies affect the model results and thus the investment decision in different assumptions in sensitivity calculations, the following calculations assume a fixed investment ratio of *PV*, wind offshore and wind onshore to receive robust results, as also done in [22]. The scenario specifies that the ratio should be 3.3 times the invested capacity of wind onshore and 2.9 times the invested capacity of *PV* compared to wind offshore.

Potential restrictions for renewable energies or storage technologies are not specified. The scope for the study is limited to Germany; imports and exports from neighboring countries are not considered.

2.5. Determination of Profile Cost

In order to quantify the costs of integrating electric mobility into the electricity supply task and to show the possibilities offered by the flexibility potential, this paper also focuses on profile costs as part of the system costs.

Profile costs of renewable energies arise from a profile that does not match the demand to be covered. For example, a photovoltaic system can provide a large amount of energy, but due to the inappropriate generation profile dependent on the solar radiation, it may not meet the demand at a given time. There is a time lag between supply and demand. In the research field of energy system analysis and electricity market modeling, profile costs have been part of various studies for several years. Due to the increasing share of renewable energies in power generation, this part is becoming more relevant. The authors of [27] characterize the profile costs as part of integration costs by analyzing different cost components of renewable energies. They split profile costs in backup costs, full-load hour reduction and overproduction costs and propose an approach to quantify them as well as a graphical interpretation via a residual load duration curve (*RLDC*). Furthermore, profile costs are described as "the marginal costs of the temporal variability of variable renewable energy output" [28].

Furthermore, the work presented in [22] analyzes profile costs according to the procedure of the classical system analysis, with and without the respective technology under investigation. It examines the total system costs with all profiles and without the residual profile and thereby determines the additional costs of the residual profile. It further states that it is only possible to analyze all profiles together in a residual profile [22].

Due to the demand for electric mobility, a further profile becomes relevant for the supply task. In order to quantify the profile effect and the additional costs caused by this profile analogous to [22], calculations are performed including and excluding the demand from electric mobility, each with the given and with flat profiles for the demand and the renewables feed-in. The four results are then compared to evaluate the profiles.

Since there is no effect of *BEV-DSI* without existing demand or generation profiles, the expected reduction of total system costs with profiles also defines the difference in profile costs with and without intelligent charging. An evaluation of the flexibility is therefore not possible. Hence, the consideration of the *RLDC* is supportively used to evaluate the effect of *DSI* on profile cost qualitatively.

3. Results

In this chapter, the results of the mobility tool are presented at the beginning. The charging behavior and flexibility potential derived from the mobility curves are presented exemplarily. These serve as input for the calculations in the electricity market model. The results for the total system costs, the installed capacity and the flexibility input are determined and compared in various sensitivity calculations.

3.1. Results of the Simulation of Mobility and BEV Load Curves

Figure 3 shows the average mobility curve for a Monday determined for the period in spring. Since the mobility tool is simulating the mileage per quarter hour, the values are summed up to hourly values. The mileage is furthermore normalized by all *BEVs*, resulting in values between 0 and 6 km. It is shown that the first journeys during the course of the day take place after 4 a.m. The mobility need on the Monday is highest in the morning and evening, while on weekends, the greatest demand usually is at midday.

If the charging behavior is derived from the determined mobility behavior, the charging curve for the assumed 21.619 million *BEVs* shown in Figure 4 is obtained exemplarily for Monday and Saturday in winter.

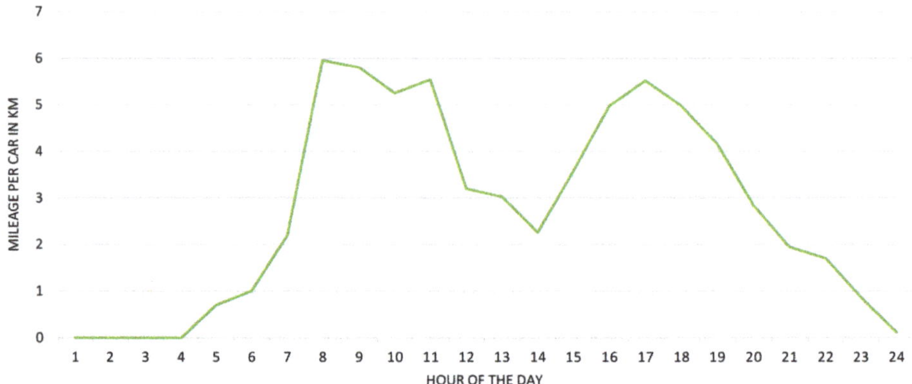

Figure 3. Mobility profile for a Monday in spring, normalized by all *BEVs*.

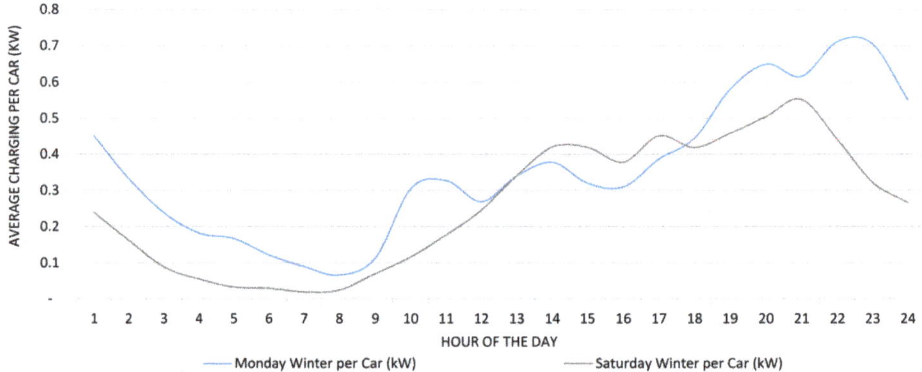

Figure 4. Load per *BEV* for a Monday (blue line) and a Saturday (grey line) in winter season.

The illustration shows that the maximum charging power occurs in the evening hours and flattens out with time until early in the morning. In the morning hours there is a low charging power demand. This can also be seen on Saturday. The charging power has a lower peak there, but is already at a relatively high level from midday until the evening. To validate the results, a qualitative and quantitative comparison with other studies, for example ([29], pp. 16–17), shows that the simulated hourly electricity demand profile from uncontrolled charging is realistic.

All vehicles are connected to the network until the beginning of their first journeys. After their return, they are reconnected for charging. The grid contact curve is derived from the driving profiles and the charging power requirements. As described in the methodology section, the load increase potential is determined from the difference between the grid contact curve and the user-controlled charging curve. The result is shown in Figure 5 as an example for a Monday in spring and in winter. The figure shows that there is a great potential for increasing load, especially in the morning hours until about 7:00 a.m. In the midday hours, the low maximum charging power and the grid contact curve ensure a low *PowerUp potential*. In the course of the day, the load reduction (or *PowerDown*) potential increases. User-controlled charging in the evening hours reduces the positive load shift potential there, whereas the load reduction potential is more available.

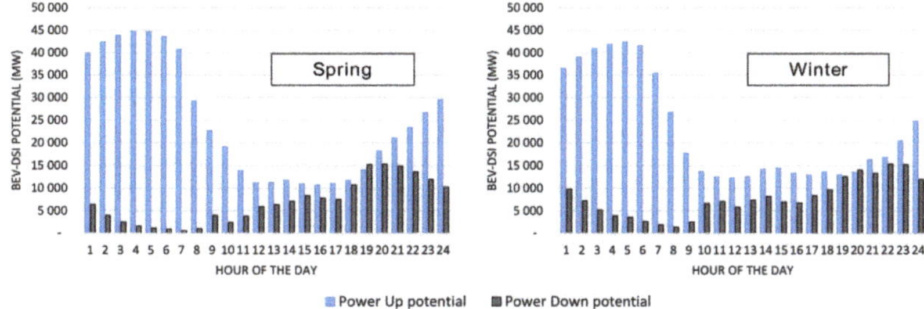

Figure 5. PowerUp (blue) and PowerDown (dark grey) flexibility potential of all *BEVs* for a typical Monday in spring and winter terms.

In the electricity market model as described in Section 2, the flexibility potential presented is subject to the additional restriction that the energy of the postponed charging process must be compensated at a fixed time every day. After analysis of the result of the mobility behavior, this point in time is set to 6 a.m., since most of the distances covered during the day are not covered until this hour (see also Figure 3).

3.2. Results of the Energy System Modelling

In this section, the results from the electricity market model are presented. First, the use of the newly implemented flexibility option, the shift of the charging load, is qualified. As a basis for the evaluation of the system effects, a reference result without flexibility of *BEV* is shown. In comparison, the scenario with additional flexibility from electric mobility is evaluated. Further sensitivity analyses are carried out. The effects of a stronger penetration and readiness for intelligent control of the charging process are analyzed. In addition, the changed use of flexibility and the possible benefit from a different compensation time point (*CTP*), the moment when the charging process must be completed, are determined.

3.2.1. Use of Flexibility in the Market Model

Figure 6 shows, for an exemplary period of 36 h, starting at midnight, the residual load (*ResLoad*, green area) and the resulting use of the flexibility of load shifting from *BEV* loads (*BEV PowerDown/Up*, in blue or light blue). Additionally, the corresponding fill level of the *DSI*-storage is shown as black dashed line on the secondary axis.

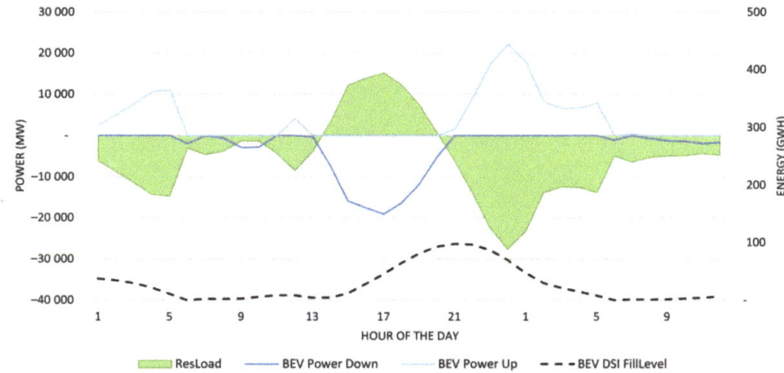

Figure 6. Use of *BEV-DSI* potential (PowerUp in light blue, PowerDown in blue) for 36 h to meet the residual load demands (ResLoad, green area). The corresponding fill level of the *DSI* storage is depicted in black dashes on the secondary axis.

On a day after 6 a.m., the load is first reduced, i.e., the flexibility storage (black dashed line, secondary axis) must be charged before the demand can be met and the charging capacity can be increased. It is shown how at times of positive residual load the demand is reduced by *BEV*-loading and at times of negative residual load the demand is increased. It can also be seen that the flexibility is also used to reduce demand during times of negative residual load in order to be able to reduce demand later with higher negative residual load. The shift is also reflected in the resulting demand after loading the *BEV* (see Figure 7).

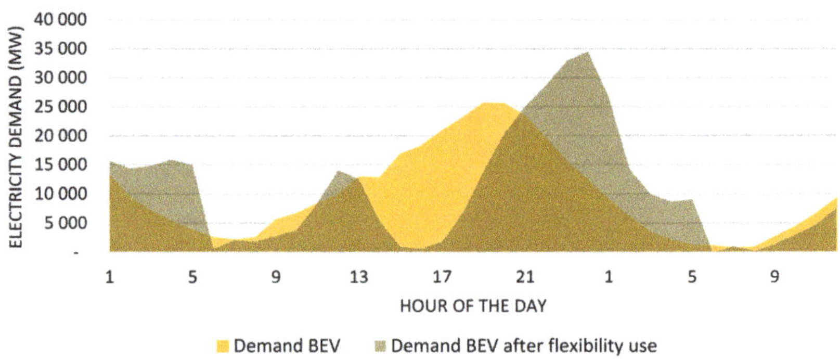

Figure 7. Electricity demand due to the *BEV*-charging process before (yellow) and after the flexibility use (brown, transparent) for an exemplary day.

The peak of the original demand by *BEV* charging (in yellow) changed from the evening hours into the night. Due to the limitation of being fully charged at 6 a.m. there is no potential to increase the load at this time, only to reduce it. Therefore, the resulting charge power (in brown, transparent) is always smaller or equal to the original charge power, so the shift is limited here and the resulting curve drops sharply.

Figure 8 shows how the load shift interacts with other flexibility options in the model. In this case we show the interaction storages with different *E2P* ratios to address short-, medium- and long term flexibility behavior. The use of load management is loss-free and without costs, which is why it is preferred to the use of storage. However, the figure clearly shows how the medium-term category 2 storage (Storage Cat 2) is used in a complementary way. The negative residual load and the resulting excess energy is stored by the storage system (Storage Cat 2 Pump, in yellow) and made available again at a time of positive residual load (Storage Cat 2 Prod, light yellow), in which, together with the reduction of *BEV* demand (*BEV PowerDown*, blue), the remaining residual load is covered. The long-term category 3 storage (Storage Cat 3 charging in red, discharging in light red), on the other hand, is mainly used for balancing over longer periods and is not used here for short-term operations.

The use of load management in comparison with the use of storage and the residual load curve shows the assumed behavior that demand is shifted from the evening hours to the night within the scope of the flexibility potential. A cost-efficient use of load shifting is possible, which is why in the following the effects on the system are worked out on the basis of the effects in the selected scenario by comparing the selected scenario with and without the possibility of using *DSI*.

3.2.2. Results of the Optimization Model on the Effects of DSI on Total System Costs and Capacity Investments

The reference calculation without the possibility of intelligent charging results in system costs of 59.1 billion €. In terms of power plant investments, this results in an installed renewable energy capacity of 220 GW and a conventional capacity (*GT* and *GasCC*) of 36 GW. Storage capacities are necessary to integrate renewable energies. These include 5 GW of seasonal storage (category 3) and 31 GW of medium-term storage (category 2), which contribute 40 TWh to load coverage.

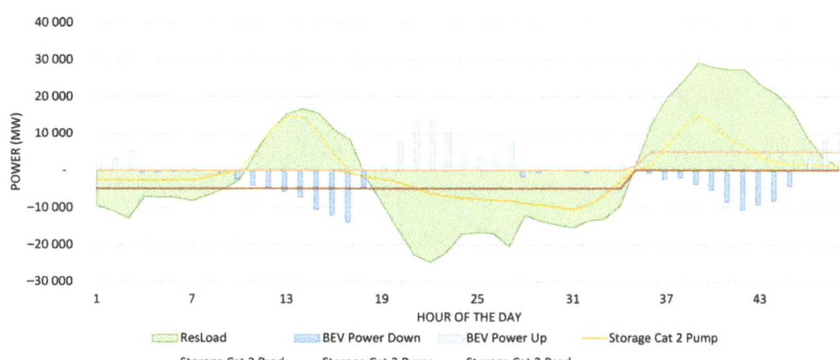

Figure 8. Interaction between the *DSI*-flexibility (blue/light blue) and the storages (Cat 2 in yellow/light yellow, Cat 3 in red/light red) to meet the given residual load (green area) for two exemplary days.

By making the charging process of all *BEVs* flexible, an additional flexibility option is available to the model, which can be used for cost-optimal demand coverage. The total system costs are thus reduced by 623 million € or 1.05% to 58.5 billion €, which is mainly due to the reduced fixed costs, as can be seen from Table 3, which compares the runs with and without the possibility for *BEV-DSI* usage.

Table 3. Comparison of the resulting costs for the scenario with and without the possibility for *BEV-DSI*-usage.

Costs and Cost Reduction	No BEV-DSI	With BEV-DSI
Total System Costs (bil. €)	59.132	58.508
Fix Cost (bil. €)	54.320	53.711
Variable Cost (bil. €)	4.511	4.797
Reduction of System Costs (bil. €)		0.623
Reduction of System Costs (%)		1.05

In terms of power plant investments (see Table 4), it is apparent that significantly less storage of category 2 (in total −3.2 GW) is being build. Slightly less renewable energies and combined-cycle power plants are built while slightly more gas turbines, which have comparatively low investment costs, are required.

Table 4. Comparison of the resulting power plant investments for the scenario with and without the possibility for *BEV-DSI*-usage.

Power Plant Invest (GW)	No *BEV-DSI*	With *BEV-DSI*	Difference
GasCC	28.0	28.0	+0.0
GasGT	8.3	8.4	+0.1
Storage Cat 2	30.8	27.6	−3.2
Storage Cat 3	5.0	5.0	−0.0
PV	88.8	88.3	−0.5
Wind Offshore	30.6	30.4	−0.2
Wind Onshore	101.0	100.5	−0.5

Table 5 shows the amount of generation from the invested capacities as well as the dispatch of flexibility options (storages and *DSI*). With *BEV-DSI*, less energy (7 TWh) is temporarily stored in the medium-term storage facilities, which leads to lower losses. In total, 13.7 TWh of the total power demand of electric vehicles, amounting to 61 TWh, are shifted. Calculating the system cost savings per shifted MWh results in 46 €/MWh. Converted to the number of *BEVs* involved, this results in a saving of 29 € per vehicle.

Table 5. Comparison of the resulting energy production and the use of flexibility options for the scenario with and without the possibility for *BEV-DSI*-usage.

Energy Production and Flexibility	No *BEV-DSI*	With *BEV-DSI*	Difference
Energy Production (TWh)			
GasCC	53.8	53.9	+0.1
GasGT	1.1	1.1	−0.0
PV	76.8	76.0	−0.8
Wind Offshore	101.6	102.4	+0.8
Wind Onshore	212.1	210.7	−1.4
Stored Energy (TWh)			
Storage Cat 2	30.5	23.5	−7.0
Storage Cat 3	20.0	19.7	−0.3
Flexibility Use (TWh)			
DSI Use	-	13.7	+13.7

3.2.3. Analysis of the Influence of a Varying Share of Participating BEV and a Different Time of Compensation at the Results of the Market Model

Since it cannot necessarily be assumed that all vehicles will participate in the *DSI* measures and this may not be necessary, this section will examine the issue in this context. In Figure 9, the course represents the system cost reduction over the share of *BEVs* participating in the flexibilization measures. It is shown that a large part (about 72%) of the achievable system cost reduction by smart charging is already achieved with a participation of 50% of the *BEVs*. At this share, this results in savings of 42 € per car. With a higher share of vehicles participating, the relative cost reduction is reduced.

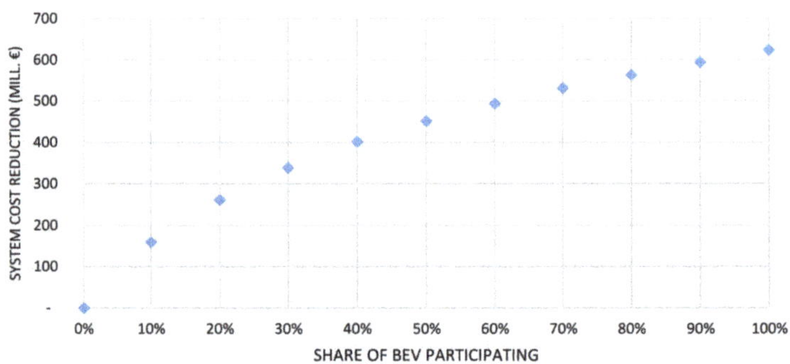

Figure 9. System cost reduction due to *BEV-DSI* in relation to the share of *BEV* participating in the measures.

The specification of a point in time when the vehicle must be fully loaded limits the flexibility of the application. As shown above, *PV* power generation at midday hours can hardly be integrated by an intelligent charge control system from *BEV*, as there is only a small total charge power requirement at this time (see Figure 5). Therefore, it is investigated what a shift of the compensation time point to other times can cause under the same assumptions. The varying effect of *BEV-DSI* by comparing the resulting total system costs with different *CTP* is shown in Figure 10. It shows, on the left side, the total system costs for scenarios without *BEV-DSI* and with *BEV-DSI*, but with changing *CTP* for four different moments during a day. On the right side, the relative cost reduction is depicted.

The figure shows that a *CTP* at 6 p.m. has the highest reduction in terms of system cost reduction, up to nearly 3 percent, while a shift of the *CTP* from midnight to 6 a.m. does not change a lot.

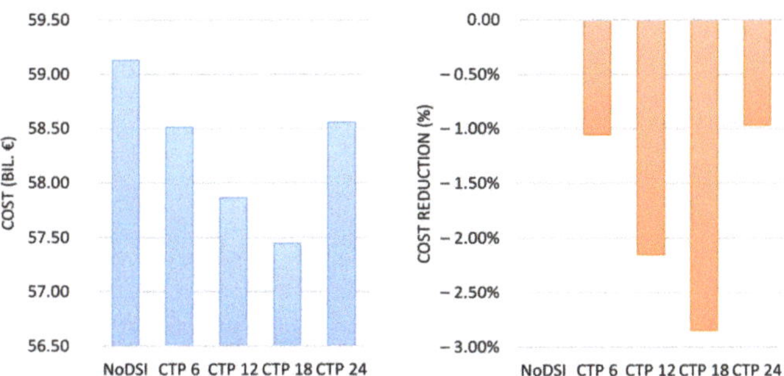

Figure 10. Total system costs (**left graph**) and relative cost reduction (**right graph**) by changing the preset compensation compared to the reference scenario without *BEV-DSI*.

3.2.4. Analysis of Profile Costs

For the analysis of profile costs, five runs are compared, with and without profiles, with and without *BEV* and one with *DSI*. Table 6 gives an overview of them.

Table 6. Overview of the model runs which are used to analyze profile costs.

Name of Model Run	BEV Demand	Profiles	BEV DSI
Flat	No	No	No
Profile	No	Yes	No
BEV-Flat	Yes	No	No
BEV-Profile	Yes	Yes	No
BEV-DSI	Yes	Yes	Yes

Figure 11 shows the difference in the demand curve for one day between the setups without profiles, with (right graph) and without *BEV* (left graph) demand. It is obvious that the profile with *BEV* increases the relative demand peak (scaled to the demand peak of the day, which is 40.6 GW in the no*BEV* and 65.0 GW in the *BEV* scenario).

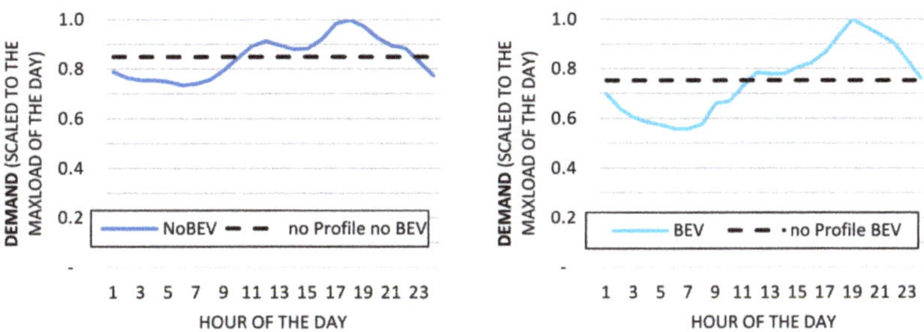

Figure 11. Exemplary depiction of the demand curves for the model runs without *BEV* (**left graph**) and with *BEV* demand (**right graph**) one time without profiles (black dashed lines) and with profiles (blue lines) for an exemplary day.

The total system costs of the model runs without profiles are significantly lower than those with profiles (see Table 7). The difference is 14.5 million € for the scenario without *BEV* and 18.7 million € for the scenario with *BEV* respectively. The share of the hereby calculated profile costs on total system costs is higher in the *BEV*-scenario higher than in the scenario without *BEV* demand (31.9% vs. 29.6%).

Table 7. Total system costs and determined profile costs for the scenarios with and without *BEV*.

Cost Component	Flat	Profile	*BEV*-Flat	*BEV*-Profile	*BEV-DSI*
Total System Costs (mill. €)	34.637	49.172	40.265	59.132	58.508
Profile Costs (mill. €)		14.535		18.867	18.243
Share of Total Costs		29.6%		31.9%	31.1%

As shown in Section 3.2.2, *DSI* of *BEV* can reduce the total system costs, needs nearly the same amount of renewable capacities but reduces medium-term storage needs. The impact on the *RLDC* is rather low because of the equal renewable production (before curtailment), which can be seen in Figure 12. On the left graph, the figure shows on the left graph the *RLDC* for only the *BEV*-Profile Scenario, because it does not differ graphically to the results of the *BEV-DSI* model run. The important parts of the graphic are marked, oriented on the representation of [27], Here "A" marks the back-up demand, defined by the maximum residual load, "B" addresses the full load hours of conventional generation and storage capacities while "C" represents overproduction. The back-up need for the run with *DSI* is slightly higher (17 MW). For the further evaluation of the results, on the right side of Figure 12, the difference of the absolute *RLDC* values between the runs without and with *BEV-DSI* are shown. The figure shows that the full load hours (area B) are higher in the run with the *DSI*-option, and therefore the specific costs of the energy production are reduced. Overproduction is higher without *DSI* (area C), which accounts for 1.75 TWh in total. The sum of these effects leads to higher total system costs and therefore to the higher profile costs in the scenario without *BEV-DSI*.

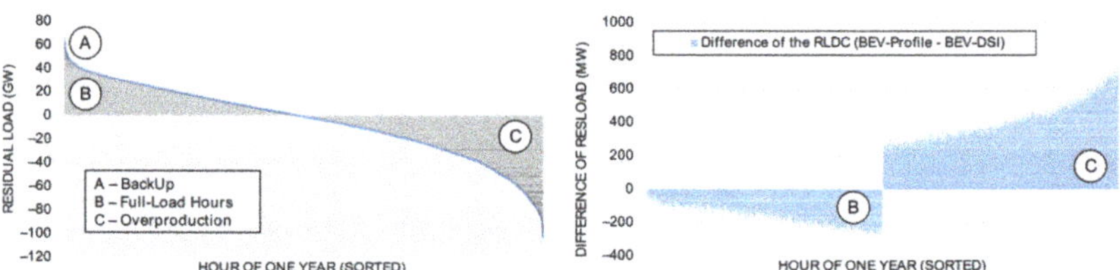

Figure 12. Residual load duration curve, congruent for the *BEV* scenarios (left graph), with the areas for the back-up need (**A**); full-load-hours (**B**) and overproduction (**C**). The right graph shows the resulting differences between the scenario without and with *BEV-DSI* concerning the FLH-reduction (**B**) and the overproduction (**C**). Concept of the figure based on [27].

4. Discussion

The approach displayed in this paper shows a methodology to determine the flexibility potential by load shifting during the charging process of *BEV* and to consider it in an electricity market model. Due to the chosen method of the Markov-Chain Monte Carlo simulation for the determination of the mobility and charging power demand, as well as the chosen definition of flexibility, it is possible to consider the restrictions of the driving behavior of individuals and at the same time to determine the system-optimal use of the flexibility potential.

The derived mobility curves lead to an hourly demand for electricity through the charging process of the electric vehicles. Here, high simultaneities are evident, resulting in high load peaks in the late evening and thus, with a high penetration of electric vehicles, significantly changing the profile of the electricity generation required to meet demand. This also leads to an increased demand for flexibility options.

The batteries available in the electric vehicles can, if used correctly, be systemically useful and represent a possible flexibility. As shown in this paper, this flexibility is considerably confined by restrictions from user behavior and availability is highly time-dependent.

In order to determine the benefit of the flexibility potential integrated with the time-varying generation from renewable energies and the conventional electricity demand, this paper implements flexibility as a *DSI* option in an electricity market model. In the analyses shown it becomes clear that the implementation of an intelligent charge control has a cost-reducing effect on the design of a decarbonized energy system with a high share of renewable energies.

The majority of the energy supply in the scenarios is provided by renewable energies, which, however, require a high degree of flexibility in order to compensate for supply-dependent fluctuations on the one hand and to enable cost-efficient integration on the other. By implementing *BEV* flexibility storage, requirements are reduced while the investments in renewables are not reduced. Since renewables are often questioned due to the high capacities and the associated space requirements, intelligent charge controls cannot make a major contribution in this area, but *BEV* can nevertheless reduce the (partly self-induced) system costs.

The analyses also show that it is not necessary to upgrade all vehicles or charging stations for flexibility provision. With the participation of 50% of *BEVs*, enormous positive effects are achieved, which increase at a slower rate as the share of participating *BEVs* increases. Since no costs (neither for development nor for deployment) were assumed in the analysis carried out, an optimal ratio must be evaluated in comparison with the additional expenditures in further investigations.

Quantifying the cost reductions per participating vehicle or energy used shows what possible financial incentives could be created. A player in the electricity market could use the flexibility to make profits in the electricity market. For this, the vehicle owner must be offered a financial (or even qualitative) incentive and the costs for the (additional) intelligent infrastructure must also be offset.

Considering the change in the specified loading time and the additional cost reductions that can be achieved as a result, it becomes clear that a point of completed charging in the evening at 6 p.m. is particularly advantageous. This suggests that (intelligent) charging stations should be set up at the workplace to integrate these vehicles. The implementation could either be carried out by a service provider or the company itself. The actual proportion of vehicles that can be used for this purpose is not explicitly stated here and determines the actual flexibility available.

From a systemic point of view, the advantage of a later point in time can be explained by the integration of *PV* power. The charging power demand from the morning hours can be shifted to the afternoon and thus use the peak generation from the *PV* systems. In addition, the load peaks that occur in the evening can be flattened out by classic electricity demand and electric mobility.

Profiles of electric mobility charging contribute significantly to the total system costs due to the necessary flexibility options and the different share of renewable power generation that can be integrated. They should therefore be explicitly identified when quantifying the costs. A possible method offers the evaluation of the *RLDC*, which gives insights on the temporal matching of electricity demand and renewable generation (and overproduction). Profile costs do not only arise from the expansion of renewable energies, but also from new consumers such as *BEVs*. The decisive factor is the combination of profiles and the availability of flexibility options. At this point, the analysis of the electricity market may neglect the grid costs that arise from regional differences in the profiles.

5. Conclusions

The study shows a new approach to derive the flexibility potential of electric vehicles available for the electricity market on the basis of simulated charging curves taking into account the real driving behavior of individual vehicles. The charging demand and the identified flexibility potential are integrated into an electricity market model to quantify the effects on total system costs as well as the necessary flexibility demand in a scenario with ambitious climate targets. It is shown that the high demand for electricity from electric vehicles leads to a considerable increase in the need for renewable energy capacities, which in turn increases system costs. The flexibility of intelligent charging stations can support the integration of renewable power generation and reduce the necessary storage requirements. The assumption for the time when the *BEV* has to be completely charged influences the cost reduction potential, with the best use of flexible charging during the afternoon. Furthermore, it is shown that the profile of the electricity demand in combination with the generation profile of the weather-dependent renewables feed-in, and additionally the availability of flexibility options, has a great influence on the design of the energy supply and thus on the total system costs.

In the modelling process some limitations have been identified, and a higher level of detail in these areas may provide further insights. Since the actual charging behavior cannot yet be foreseen in the future, further charging patterns, such as charging only at the weekend or only when the battery level is below 50%, could be investigated here. In addition, the assumptions that all users have the same behavior and that the charging power is maximum 3.7 kW could be revisited. Some vehicles are going to be charged at work and, as there is a strong dependency on the charging time, it might therefore be useful to add more than one user group to increase the level of detail. In the coming years, further experience and data from users of electric vehicles will provide more insights.

In this study, we do not analyze effects on the electricity grid, even though there will be issues concerning peak loads (especially in the distribution network) induced by technologies with significant profiles such as *BEV* or *PV* and this will therefore will make flexibility on a regional level necessary. Aggregators can combine the flexibility of intelligent charging of *BEVs* with other options with complementary limitations such as heat pumps or other technologies to offer it on relevant markets.

To what extent users are willing to leave the control of their charging process to third parties is also a question of acceptance, which, like legal issues, still needs to be clarified. The financial incentive for private households (or aggregators) to sell the charging flexibility on electricity markets would therefore have to be correspondingly high.

The modelling of mobility behavior, *BEV*-charging curves and flexibility potential shown in this paper provides a basis for discussion to evaluate the impact of a rising penetration of electric vehicles with an additional energy demand and changes in the electricity system and its costs.

Author Contributions: Conceptualization, F.G., N.K. and K.H.; methodology, F.G. and N.K.; software, F.G. and N.K.; validation, F.G., N.K. and K.H.; formal analysis, F.G., N.K. and K.H.; investigation, F.G. and N.K.; resources, F.G. and N.K.; data curation, F.G. and N.K.; writing—original draft preparation, F.G.; writing—review and editing, N.K. and K.H.; visualization, F.G.; supervision, N.K. and K.H.; project administration, F.G. and N.K.; funding acquisition, F.G., N.K. and K.H. All authors have read and agreed to the published version of the manuscript.

Funding: This research was funded by the Open Access Publication Fund of the University of Stuttgart.

Data Availability Statement: Not applicable.

Acknowledgments: The authors gratefully acknowledge the support of the German Federal Ministry for Economic Affairs and Energy, BMWi, the Federal Ministry of Transport and Digital Infrastructure (BMVI) as well as the support of the Open Access Publication Fund of the University of Stuttgart.

Conflicts of Interest: The authors declare no conflict of interest. The funders had no role in the design of the study; in the collection, analyses, or interpretation of data; in the writing of the manuscript, or in the decision to publish the results.

Appendix A. Data for the Mobility Tool

Table A1. Data for the simulation of the mobility and charging load, data source: [30–32].

Parameter	Parameter Value	Comment
Vehicle-to-Wheel-consumption (kWh/100 km)	18.0	Quantity weighted consumption
Charging Power (kW)	3.7	
Charging Losses (%)	13.4	
Grid-to-Wheel consumption (kWh/100 km)	20.8	
Range (km)	345	Prognosis for 2020, assumption, that 20% less range than with NEDC
Battery Capacity (kWh)	71.8	

Appendix B. Data for the Electricity Market Model

Table A2. Data for power plant investment options, data source [22,25].

Power Plant	Efficiency (%)	Lifetime	Invest Cost (€/kW)	Annual OaM Costs (€/kW)	VarOaM Cost (€/MWh)	Full Load Hours
GasCC	61	30	780	22	1.5	
GasGT	41	30	400	15	1.5	
Storage Cat1 E2P 2	81	20	714	10	2.5	
Storage Cat2 E2P 7	81	20	1059	10	2.5	
Storage Cat3 E2P 500	81	20	9219	10	2.5	
PV		25	1150	34	0	949
Wind Offshore		20	2850	120	0	4000
Wind Onshore		20	1400	50		2598

Assumed Interest Rate: 7.5%, Primary energy prices are based on the KS 95 [25]. Synthetic profiles for renewables and demand are taken from [33], based on the year 2018.

References

1. Federal Ministry for the Environment, Nature Conservation, Building and Nuclear Safety. *Climate Action Plan 2050—Principles and Goals of the German Government's Climate Policy*; Federal Ministry for the Environment, Nature Conservation, Building and Nuclear Safety: Berlin, Germany, 2016.
2. Maximilian Schulz, M.S.; Kai Hufendiek, K.H. Discussing the Actual Impact of Optimizing Cost and GHG Emission Minimal Charging of Electric Vehicles in Distributed Energy Systems. *Energies* **2021**, *14*, 786. [CrossRef]
3. Van Nuffel, L.; Dedecc, J.G.; Smit, T.; Rademaekers, K. *Sector Coupling: How Can It Be Enhanced in the EU to Foster Grid Stability and Decarbonise?* European Parliament, Policy Department for Economic, Scientific and Quality of Life Policies: Brussels, Belgium, 2018. [CrossRef]
4. Clearingstelle Verkehr, Deutsches-Zentrum für Luft- und Raumfahrt (DLR). *Mobilität in Deutschland 2017*; Zeitreihendatensatz: Berlin, Germany, 2017; Available online: https://daten.clearingstelle-verkehr.de/279/ (accessed on 22 March 2021).
5. Follmer, R.; Gruschwitz, D. *Mobility in Germany, Short Report*; Edition 4.0 of the study by Infas, DLR, IVT and Infas 360 on behalf the Federal Ministry of Transport and Digital Infrastructure (BMVI) (FE no. 70.904/15); Federal Ministry of Transport and Digital Infrastructure (BMVI): Berlin, Germany, 2019.
6. Follmer, R. *Mobilität in Deutschland—Tabellarische Grundauswertung*; Infas, DLR, IVT and Infas 360 on behalf the Federal Ministry of Transport and Digital Infrastructure (BMVI); Federal Ministry of Transport and Digital Infrastructure (BMVI): Berlin, Germany, 2018.
7. Kristoffersen, T.K.; Capion, K.; Meibom, P. Optimal charging of electric drive vehicles in a market environment. *Appl. Energy* **2011**, *88*, 1940–1948. [CrossRef]
8. Schäuble, J.; Kaschub, T.; Ensslen, A.; Jochem, P.; Fichtner, W. Generating electric vehicle load profiles from empirical data of three EV fleets in Southwest Germany. *J. Clean. Prod.* **2017**, *150*, 253–266. [CrossRef]
9. Mladenov, V.; Chobanov, V.; Zafeiropoulos, E.; Vita, V. Characterisation and evaluation of flexibility of electrical power system. In Proceedings of the 10th Electrical Engineering Faculty Conference (BulEF), Sozopol, Bulgaria, 11–14 September 2018. [CrossRef]
10. Dütschke, E.; Paetz, A.G.; Wesche, J. Integration Erneuerbarer Energien durch Elektromobilität—Inwieweit sind Konsumenten bereit, einen Beitrag zu leisten? *UWF UmweltWirtschaftsForum* **2013**, *21*, 233–242. [CrossRef]

11. Project Fellbach ZEROplus—Intelligent Usage of Electric Vehicles in the Overall Energy Concept of an Energy Plus House Group in Fellbach. Available online: https://www.ise.fraunhofer.de/en/research-projects/fellbach-zeroplus.html (accessed on 17 November 2020).

12. Blasius, E. Ein Beitrag zur Netzintegration von Elektrofahrzeugen als Steuerbare Lasten und Mobile Speicher Durch Einen Aggregator. Ph.D. Thesis, Brandenburgischen Technischen Universität Cottbus-Senftenberg, Cottbus, Germany, 2016.

13. Lazarou, S.; Vita, V.; Christodoulou, C.; Ekonomou, L. Calculating operational patterns for electric vehicle charging on a real distribution network based on renewables' production. *Energies* **2018**, *11*, 2400. [CrossRef]

14. Heinrichs, H.U.; Jochem, P. Long-term impacts of battery electric vehicles on the German electricity system. *Eur. Phys. J. Spec. Top.* **2016**, *225*, 583–593. [CrossRef]

15. Robinius, M.; Linssen, J.; Grube, T.; Reuß, M.; Stenzel, P.; Syranidis, K.; Kuckertz, P.; Stolten, D. *Comparative Analysis of Infrastructures: Hydrogen Fueling and Electric Charging of Vehicles*; Forschungszentrum Jülich GmbH, Zentralbibliothek: Jülich, Germany, 2018.

16. Kühnbach, M.; Stute, J.; Gnann, T.; Wietschel, M.; Marwitz, S.; Klobasa, M. Impact of electric vehicles: Will German households pay less for electricity? *Energy Strategy Rev.* **2020**, *32*, 100568. [CrossRef]

17. Richardson, I.; Thomson, M.; Infield, D. A high-resolution domestic building occupancy model for energy demand simulations. *Energy Build.* **2008**, *40*, 1560–1566. [CrossRef]

18. Müller-Gronbach, T.; Novak, E.; Ritter, K. *Monte Carlo-Algorithmen*; Springer: Berlin/Heidelberg, Germany, 2012.

19. Propfe, B. Marktpotenziale Elektrifizierter Fahrzeugkonzepte unter Berücksichtigung von Technischen, Politischen und Ökonomischen Randbedingungen. Ph.D. Thesis, University of Stuttgart, Stuttgart, Germany, 2016.

20. Kraftfahrt-Bundesamt. *Kurzbericht Verkehr in Kilometern (VK)—Inländerfahrleistung*; Kraftfahrt-Bundesamt: Flensburg, Germany, 2019.

21. Sun, N. Modellgestützte Untersuchung des Elektrizitätsmarktes. Kraftwerkseinsatzplanung und -Investitionen. Ph.D. Thesis, University of Stuttgart, Stuttgart, Germany, 2013.

22. Fleischer, B. Systemeffekte von Bioenergie in der Elektrizitäts- und Fernwärmewirtschaft—Eine Modellgestützte Analyse Langfristiger Energiewendeszenarien in Deutschland. Ph.D. Thesis, University of Stuttgart, Stuttgart, Germany, 2019.

23. Bothor, S.; Steurer, M.; Eberl, T.; Brand, H.; Voß, A. *Bedarf und Bedeutung von Integrations- und Flexibilisierungsoptionen in Elektrizitätssystemen mit Steigendem Anteil Erneuerbarer Energien*; 9 Internationale Energiewirtschaftstagung Wien; University of Vienna: Vienna, Austria, 2015.

24. Steurer, M. Analyse von Demand Side Integration im Hinblick auf Eine Effiziente und Umweltfreundliche Energieversorgung. Ph.D. Thesis, University of Stuttgart, Stuttgart, Germany, 2017.

25. Öko-Institut e.V. and Fraunhofer Institute for Systems and Innovation Research ISI. *Klimaschutzszenario 2050 2. Endbericht*; German Federal Ministry of the Environment, Nature Conservation and Nuclear Safety, BMU: Berlin, Germany, 2015.

26. BMWi. Press Release: Final Decision to Launch the Coal-Phase Out—A project for a Generation. In *German Federal Ministry for Economic Affairs and Energy*; BMWi: Berlin, Germany, 2020; Available online: https://www.bmwi.de/Redaktion/EN/Pressemitteilungen/2020/20200703-final-decision-to-launch-the-coal-phase-out.html (accessed on 17 November 2020).

27. Ueckerdt, F.; Hirth, L.; Luderer, G.; Edenhofer, O. System LCOE: What are the costs of variable renewables? *Energy* **2013**, *63*, 61–75. [CrossRef]

28. Hirth, L.; Ueckerdt, F.; Edenhofer, O. Integration costs revisited, An economic framework for wind and solar variability. *Renew. Energy* **2015**, *74*, 925–939. [CrossRef]

29. Gnann, T.; Plötz, P.; Globisch, J.; Schneider, U.; Dütschke, E.; Funke, S.; Wietschel, M.; Jochem, P.; Heilig, M.; Kagerbauer, M.; et al. *Öffentliche Ladeinfrastruktur für Elektrofahrzeuge: Ergebnisse der Profilregion Mobilitätssysteme Karlsruhe*; KITopen-ID: 1000093821; Fraunhofer-Institut für System-und Innovationsforschung ISI: Karlsruhe, Germany, 2017.

30. Held, M.; Graf, R.; Wehner, D.; Eckert, S.; Faltenbacher, M.; Weidner, S.; Braune, O. *Abschlussbericht: Bewertung der Praxistauglichkeit und Umweltwirkungen von Elektrofahrzeugen*; German Federal Ministry of Transport and Digital Infrastructure BMVI: Berlin, Germany, 2016.

31. Kraftfahrt-Bundesamt, KBA. Statistics: Monthly New Registrations—July 2018. Available online: https://www.kba.de/DE/Statistik/Fahrzeuge/Neuzulassungen/MonatlicheNeuzulassungen/fz_n_MonatlicheNeuzulassungen_archiv/2018/201807_GV1monatlich/201807_uebersicht.html?nn=2601598t (accessed on 17 November 2020).

32. Greiner, O. *Faktencheck Mobilität 3.0*; Horvath & Partners Management Consulting: Stuttgart, Germany, 2018.

33. Open Power System Data. Data Package Time Series. Version 2020-10-06. (Primary Data from Various Sources, for a Complete List See URL). 2020. Available online: https://data.open-power-system-data.org/time_series/2020-10-06 (accessed on 27 April 2020).

Article

Discussing the Actual Impact of Optimizing Cost and GHG Emission Minimal Charging of Electric Vehicles in Distributed Energy Systems

Maximilian Schulz * and Kai Hufendiek 

Institute of Energy Economics and Rational Energy Use (IER), University Stuttgart, 70565 Stuttgart, Germany; kai.hufendiek@ier.uni-stuttgart.de
* Correspondence: maximilian.schulz@ier.uni-stuttgart.de

Abstract: Electric vehicles represent a promising opportunity to achieve greenhouse gas (GHG) reduction targets in the transport sector. Integrating them comprehensively into the energy system requires smart control strategies for the charging processes. In this paper we concentrate on charging processes at the end users home. From the perspective of an end user, optimizing of charging electric vehicles might strive for different targets: cost minimization of power purchase for the individual household or—as proposed more often recently—minimization of GHG emissions. These targets are sometimes competing and cannot generally be achieved at the same time as the results show. In this paper, we present approaches of considering these targets by controlling charging processes at the end users home. We investigate the influence of differently designed optimizing charging strategies for this purpose, considering the electrical purchase cost as well as the GHG emissions and compare them with the conventional uncontrolled charging strategy using the example of a representative household of a single family. Therefore, we assumed a detailed trip profile of such a household equipped with a local generation and storage system at the same time. We implemented the mentioned strategies and compare the results concerning effects on annual GHG emissions and annual energy purchase costs of the household. Regarding GHG emissions we apply a recently proposed approach by other authors based on hourly emission factors. We discuss the effectivity of this approach and derive, that there is hardly no real impact on actual GHG emissions in the overall system. As incorporating this GHG target into the objective function increases cost, we appraise such theoretical GHG target therefore counterproductive. In conclusion, we would thus like to appeal for dynamic electricity prices for decentralised energy systems, leading at the same time to cost efficient charging of electric vehicles unfolding clear incentives for end users, which is GHG friendly at the end.

Keywords: smart charging; electric mobility; distributed energy systems; cost- and GHG emission minimization

Citation: Schulz, M.; Hufendiek, K. Discussing the Actual Impact of Optimizing Cost and GHG Emission Minimal Charging of Electric Vehicles in Distributed Energy Systems. *Energies* **2021**, *14*, 786. https://doi.org/10.3390/en14030786

Academic Editor: Amela Ajanovic
Received: 30 November 2020
Accepted: 29 January 2021
Published: 2 February 2021

1. Introduction

Germany is aiming for ambitious greenhouse gas (GHG) reduction goals to contribute adequately to the targets resulting from the Paris Climate Agreement. Germany committed itself to reduce GHG emissions by 80% compared to 1990 and increase the share of renewable energies (RE) to 60% of energy consumption by 2050 [1]. These targets seem to be increased further by the recently proposed "green deal" of the EU.

Due to the low energy density of RE as solar or wind power, the generation needs to spread into the plain calling for small distributed generation especially by photovoltaics (PV). Therefore it is expected, that the number of distributed installed generation and storage technologies, especially PV systems (PV) and battery storage (BS) will increase strongly. This expectation is supported by recent figures showing increased use of all this technologies already (Figure 1). This means, that individual households (HH), living in

113

single or multifamily houses, can integrate their heat and mobility demand resulting from electric heat pumps and battery electric vehicles (EV) with their local RE production by smart distributed energy systems.

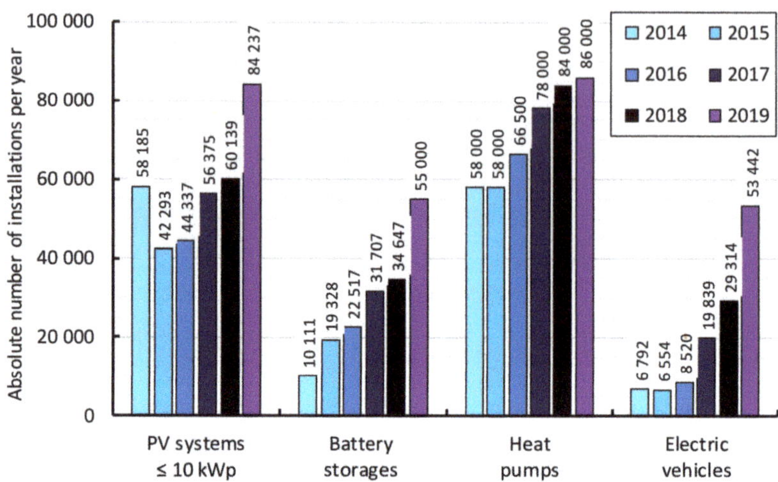

Figure 1. Overview of the annually installed number of PV systems [2], battery storages [3] (due to missing values for battery storage for 2019 figures are assumed by the previous year's ratio of 0.5088 battery storage per installed PV system) [4], heat pumps [5] and electric vehicles [6] from 2014 to 2019 in Germany.

First, the increasing number of distributed technologies as well as increased demand by electric heating and charging loads of EV results in a challenge for the distribution grid, e.g., inverted load flow and peaking loads. In contrast, the high degree of flexibility within the operation of these technologies allows for different operation strategies offering flexibility for the overall system or optimising different objectives for the individual household. Based on the number of EVs in Figure 1 from 2014 to 2019, a flexibility potential of 0.46 GW and 4.43 GWh results for Germany, in case Volkswagen eGolf parameters are assumed as presented later. This potential is expected to grow in line with increasing EV fleet dramatically in the future. In Germany vehicles remain parked on average 16 h per day without a preceding or following trip [7]. To exploit this energetic and temporal flexibility and enable optimum system integration, a smart charging control is required.

This smart control system simultaneously enables the optimal utilization of the exploited flexibility. The criterion for optimization depends on the target to be achieved. We analyse in general two objectives, which for us appear to be the most relevant for the perspective of an individual HH:

- Minimisation of power cost of individual HH:

The target is to achieve a power supply of the EV and the entire local energy system of the HH at minimum cost. The target function contains the total operation and purchase cost of the individual HH, including the cost of local power generation or purchase as well as detailed cost components like the cost of degradation of the different system components, as well as the refund for power fed into the grid.

- Minimisation of caused GHG emissions of the entire power system:

This approach is proposed by several researchers discussed in Section 2 recently, aiming for improving climate protection. The target value is derived from specific GHG factors of each hour of the entire power system, shifting demand to times with high RE production and therefore lower GHG emissions.

At this point, it could also be argued that ensuring the comfort of the local HH should be considered as an objective as well. However, as we assume that comfort of the HH is rather a basic requirement, it has to be met in any case. Therefore, it is integrated as a restriction being observed at any time. Focusing the charging processes of EV, it must be guaranteed that the charging level of the EV is always sufficient to meet the upcoming mobility demand of the HH.

Although it is an important issue for reaching overall targets of integrating high fractions of RE in the entire system and reducing GHG emissions, we have not analysed the optimal application of local flexibility for the entire system as a further control objective. However, the effects will be discussed later and an analysis can be provided by the same approaches shown in this paper.

Most of the work in literature on integration of EV charging covers two main topics: The first topic targets the effect of applying the flexibility of EV charging, most often of a fleet of EVs, on the optimal system integration in the entire energy system. As a second topic, the effects of controlled vs. uncontrolled EV charging processes are compared, again from the perspective of the overall energy system integration. To our knowledge, there is just little publication focusing on the minimum charging cost of electric vehicles in a distributed energy system in combination with the minimisation of the locally assigned GHG emissions. In our opinion, these two objectives represent the owners basic motivation for the purchase and usage of EVs. Therefore, our approach aims at the optimising of the corresponding charging processes regarding these two objectives. However, the assumed cost components can certainly not yet be realised at all at the moment, but it is expected, that with increasing awareness of distributed energy systems and their flexibility, the regulatory framework will change in a likewise way. There is evidence already that individual households are aware of GHG effects and show a willingness to pay for GHG reduction as they contract green electricity tariffs. Therefore, we assume that the issue analysed is relevant for implementation in the real world.

The paper is organised as follows: We first provide an overview of related publications with a special focus on EV charging strategies for distributed energy systems in Section 2. Following, we present our approach to optimise these charging processes in a cost- and simultaneously GHG emission minimum way in Section 3. Then we developed three different smart charging strategies and, as a benchmark, a heuristic direct charging approach, which we apply to a representative individual HH with a distributed energy system. We describe this energy system as well as its interface to the entire system in more detail in Section 4. In Section 5, we present the results of the comparison of the three charging strategies with the heuristic direct charging approach. Hereby, we provide an overview of the resulting yearly cost savings and calculated emissions reductions of each charging strategy and discuss whether this achieves an effective reduction of the GHG emissions in the overall system. Finally, we summarize our key findings in Section 6 and provide an outlook on further research issues.

2. Related Work on Smart Charging of Electric Vehicles

In the literature there is a huge number of publications focusing on smart charging of EVs: A detailed and comprehensive introduction to the topic of smart charging of EVs is given in [8]. The authors show a large number of relevant publications and cluster them into different subcategories, but do not subdivide them according to the design of the charging process (uni- vs. bidirectional, static vs. mobility-aware, centralised vs. distributed) or the specific objective of the optimisation of the charging process (integration of REs, providing ancillary services). A deeper analysis of central charging control by an aggregator and distributed smart charging control, e.g., by the owner, is offered in [9]. However, the comparison of the control approaches focuses exclusively on the charging of EVs: Other components involved in the distributed energy system of a household, such as battery storage or heat pumps are not considered, only a small outlook on the operation of EVs in virtual power plants or microgrids is given. The (information) technical integration of EVs is dealt with intensively in [10]. In this context, the potential offered by smart charging processes is highlighted in particular, but no corresponding charging approaches are listed. Exemplary approaches to control the corresponding multitude of public EV charging processes while ensuring the distribution grid-side requirements are described in a highly detailed and comprehensive way in [11,12]. Reference [13] classifies the benefits of system integrated EV charging processes into three categories: Improved technical grid efficiency, financial, for example for different stakeholders like the owner or the supplying utility, and socio-environmental benefits like the reduction of GHG emissions or an improved integration of renewable energies. These identified advantages show a strong alignment with our objectives, but are not applied as objectives for charging control. In conclusion, we can classify our smart charging approach according to this categorisation as a distributed, unidirectional and mobility-oriented charging strategy.

Following this overview of different publications on the classification of the smart integration of electric vehicles, we want to take a closer look at the charging strategies reported in literature aiming in particular at minimising costs and GHG emissions. The approach in [14], comparable to our approach, intends to minimise costs as well as the GHG emissions of the charging processes, although these are centrally managed by the distribution system operator (DSO). The advantage is the immediate knowledge of the DSO about the feed-in of volatile renewable energies, in this case the local wind power production, which leads to it being maximally expoited for the charging of the EVs. However, with regard to the regulatory requirements of unbundling, such a scenario is difficult to realise under market conditions. An extensive comparison between conventional and electric vehicles is provided by [15], but from the perspective of the overall energy system. With regard to EV charging, however, two simple charging strategies, uncoordinated and off-peak charging, are considered only. A similar approach is pursued by the main focus on smoothing the peaks of the load of a multitude of EVs [16,17]. An interesting approach to estimate the reduction of emissions by smart charging of EVs is reported in [18], in which a centrally managed charging strategy is compared to a distributed charging strategy with a focus on the California electric grid. In this case, first the residual load, i.e., the absolute load minus the generation based on RE, is considered. The charging of the EVs is then determined and, in conjunction with the residual load, the conventional generation is scheduled to compare the benefits by means of the investigated emissions of CO_2 and NO_x, power plant requirements and levelized cost of electricity. At this point we can state that a large part of the relevant publications evaluates and optimises the charging processes from the overall energy perspective. However, since we focus on the optimization of distributed charging processes at the household level, we will discuss recent charging approaches from the perspective of an individual household.

This focus is rather described in other papers: A detailed analysis of the effects of three charging strategies (direct charging, uni- and bidirectional charging) on different user groups with varying driving profiles is carried out in [19]. This analysis though, focuses on the costs of the loading processes only and does not consider any further objectives. In contrast, the GHG emissions of the charging process are minimized in [20] applying a regional specific marginal emission factor and in [21] considering probabilistic input data. Another possibility for GHG emission-minimised charging of EVs is given in [22]. Here, emission factors are predicted for the potential charging time of the EV and the points of time with the lowest emission factors are chosen for the charging.

In [23] two different charging strategies are investigated to achieve a maximum consumption of locally produced PV energy: A heuristic rule-based strategy examines the charging process, whether there is a surplus of PV energy or whether the state of charge falls below a required amount. The influence of four different emission factors as input variables for GHG emission minimization using the example of battery storage systems in industrial plants is intensively analysed in [24]. The optimisation of the battery storage dispatch is executed in two steps: First, the cost-optimum capacity and dispatch of the battery storage is identified. Then, this dispatch is redefined to the most climate friendly way.

3. Methodology

To optimize the EV charging process, we apply the optimisation model E2M2_DES. Here, we present the general structure of the model with a special focus on the modelling of electric mobility. We then continue to explain the different charging strategies in order to quantify the potential of EV charging processes in more detail and allow for the comparison of the different strategies. We hereby intentionally refer to the term charging strategy instead of charging control because we only investigate a theoretical charging process. Focusing a real charging process, deviations of our calculated charging schedule occur: First, they result from the simplification of our modelling approach. Second, we assume a perfect foresight over the investigated energy system which in reality will never be attained.

3.1. The General Optimisation Model E2M2_DES

The optimization model E2M2_DES represents a mixed-integer linear programming (MILP) optimisation model formulated by GAMS minimising energy cost for a defined energy system by computing optimum schedules for its technologies. We have already introduced the model in [25] for application to the heat supply of a single family household by a heat pump system. For the following analysis, we enhanced its functionalities, especially regarding its applicability to optimize EV charging processes and to consider different price incentives and GHG emission targets, which we will present in more detail below.

3.1.1. Objective Function

Applying the model for the optimum operation of a distributed energy system, we introduced the following objective function to minimise the total cost $V_{CostTotal}$ of all technologies TEC of the energy system over all time steps $t \in T$ as:

$$\text{Min } V_{CostTotal} = \sum_{t \in T} \sum_{tec \in TEC} \begin{bmatrix} V_{C,StartUp_{t,tec}} \\ V_{C,FeedIn_{t,tec}} \\ V_{C,ElecPurch_{t,tec}} \\ V_{C,CO2Em_{t,tec}} \\ V_{C,Charging_{t,tec}} \\ V_{C,VarOaM_{t,tec}} \\ V_{C,FO_{t,tec}} \end{bmatrix} \tag{1}$$

with:

$$V_{C,StartUp,t,tec} \text{ start-up costs of a technology } tec,$$
$$V_{C,FeedIn,t,tec} \text{ eed-in remuneration for power fed into the grid,}$$
$$V_{C,ElecPurch,t,tec} \text{ costs of the power withdrawn from the grid (power purchase),}$$
$$V_{C,CO2Em,t,tec} \text{ costs of GHG emissions of this purchased power,}$$
$$V_{C,Charging,t,tec} \text{ costs of the EV charging,}$$
$$V_{C,FOaM,t,tec} \text{ fixed O\&M costs and}$$
$$V_{C,FO,t,tec} \text{ variable operation costs.}$$

Since the local energy system of the HH operates a PV generation which has not got any variable cost nor GHG emissions, we will focus on the cost of the power purchase and its GHG emissions in particular. We describe their evaluation in more detail in the following.

To cope with such different tariff structures in one modelling approach we apply time-specific energy purchase cost $COST_{ElecPurch,t}$. Therefore $V_{C,ElecPurch,t}$ results per time step t as:

$$V_{C,ElecPurch,t} = V_{el;Purch,t} \bullet COST_{ElecPurch,t} \tag{2}$$

with:

$$V_{el;Purch,t} \quad \text{purchased power and}$$
$$COST_{ElecPurch,t} \quad \text{power purchase price at HH level at timestep } t.$$

This parameter $COST_{ElecPurch,t}$ enables us to depict a constant or a time-dynamic electricity tariff. To minimise the GHG emissions of the energy system as well, we assign cost to them by:

$$V_{C,CO2Em,t} = V_{el;Purch,t} \bullet EMIS_{ElecPurch,t} \bullet COST_{CO2Em,t} \qquad (3)$$

with:

$$V_{el,Purch,t} \quad \text{power purchased and withdrawn from grid,}$$
$$EMIS_{ElecPurch,t} \quad \text{time-specific emissions factor of the grid purchase and}$$
$$COST_{CO2Em,t} \quad \text{emission price of time step } t.$$

This term can be understood as a penalty term and increases influence especially in case of a high choice of the CO_2-price, as CO_2-costs are generally minimised.

3.1.2. Restrictions

To meet the electrical demand of the HH and to charge the EV, we define appropriate constraints to ensure the basic physical properties of the technical components. The electrical balance equation is therefore given for each time step t as

$$LOAD_{el,Sys,t} + \Sigma_{tec \in TEC}\, V_{el,Cons_{tec},t} + V_{el,FeedIn,t}$$
$$\leq \Sigma_{tec \in TEC}\, Gen_{el,tec,t} + \Sigma_{tec \in TEC}\, V_{el,Gen_{tec},t} + V_{el;Purch,t}, \qquad (4)$$

with:

$$LOAD_{el,Sys,t} \quad \text{electrical load of the HH,}$$
$$V_{elCons,t,tec} \quad \text{power consumption of component } tec \text{ of the local energy system of the HH,}$$
$$V_{el,FeedIn,t}. \quad \text{power fed into the grid,}$$
$$Gen_{El,tec,t} \quad \text{fluctuation power generation of component } tec \text{ of the local energy system,}$$
$$V_{el,Gen,t} \quad \text{controllable power generation of component } tec \text{ of local energy system and}$$
$$V_{el,Purch,t} \quad \text{power purchased and withdrawn from grid at time step } t.$$

An electrical energy storage unit is included in this general formulation as it can either be a consumer of power (charging of the storage unit) or a generator of power (discharging of the storage unit) per time step t.

The charging of the EV is also modelled as an electrical storage unit. However, as we focus on unilateral charging only, it cannot be discharged. For accounting the availability of the EV at the charging point, a binary time series $AVAIL_{EV,t}$ is introduced. The availability of the EV is derived from the individual trips per day presented in the following section. If the EV undergoes such a trip, the value 0 is assumed for $AVAIL_{EV,t}$, otherwise 1 for the respective time step t.

The balance equation for the EV for each time step t result as

$$V_{EV,FL,t} \leq V_{EV,FL,\,t-1} + V_{EV,Load,t} - LOAD_{EV,cons,t} - V_{EV,StoLosses,t} \qquad (5)$$

With:

$$V_{EV,FL,t} \quad \text{charging condition of EV,}$$
$$V_{EV,Load,t} \quad \text{charging power during charging of EV,}$$
$$LOAD_{EV,cons,t} \quad \text{power consumption of EV resulting from trip starting in time step } t \text{ and}$$
$$V_{EV,StoLosses,t} \quad \text{storage losses of EV at time step } t.$$

3.1.3. Application of the Optimisation Model

Aiming for the optimum schedule of the different components of the energy system, the schedule for the charging of the EV as well as the schedule for the charging and discharging of the local battery storage are determined. As input for these schedules, we assume perfect information about the required parameter time series we mentioned above. Applying this information, we calculate the optimum schedule for the defined control period by the Cplex optimizer.

In the following, we analyse the charging of the EV over the investigation period of one year. With regard to the computing time of the optimisation problem, we execute a rolling time horizon approach with a time period of one week as optimisation period. This seems adequate, as the utilisation of the EV mainly exhibits daily and weekly patterns and the capacity of the local battery storage is as small, that it cycles in this period already different times. The optimisation model repeatedly calculates the optimum schedule for each weekly period until the end of the investigation period is reached documenting each schedule (among other data). Since the optimisation model targets the minimum of the storage levels at the end of each optimisation period due to cost optimality, we apply an overlap of one day, i.e., we extend the optimisation period by one day which is cut off later in the documentation afterwards. After each model run for a certain week, the fill level of each storage is transferred as the new starting value for the next week. This process is described in more detail in Figure 2.

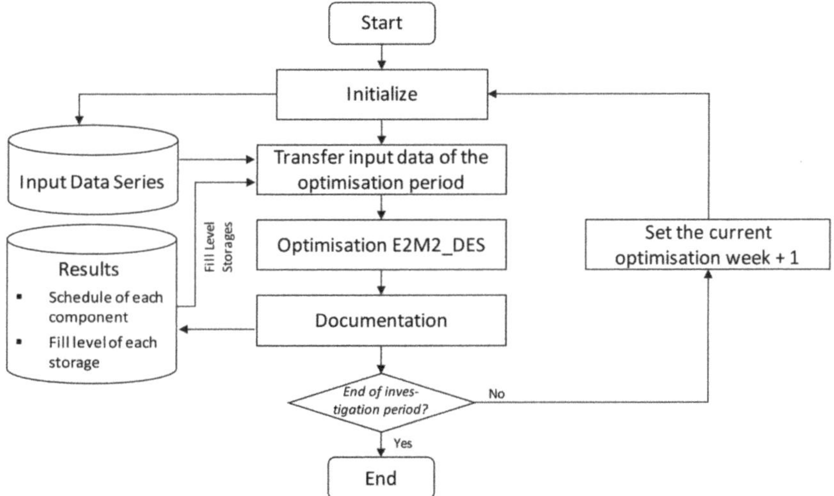

Figure 2. Overview of the optimisation model application per flowchart diagram.

The characteristics of the input variables vary depending on the charging strategy, e.g., with regard to the representation of a constant or dynamic electricity tariff. We will now present these different charging strategies in more detail in the following section.

3.2. Charging Strategies

To quantify the EV charging potential of a HH with local PV generation in more detail, we developed three different charging scenarios combined with appropriate charging strategies. All strategies are based on the previous presented optimisation model, but consider different targets and/or regulatory frameworks. These result are as follows:

(1)   Charging strategy 1 (CS 1) considers a constant power purchase price at HH level and no GHG emissions effects of the grid purchase are taken into account, representing the typical regulatory framework in Germany at present.

(2) Charging strategy 2 (CS 2) assumes a dynamic power purchase price at HH level similar to the price at the power whole sale market, while GHG emissions will still not have to be remunerated and are therefore not considered in the optimisation.

(3) Charging strategy 3 (CS 3) continues to apply a dynamic power price at HH level and furthermore GHG emissions are to be minimised. Therefore, the price for $COST_{CO2Em,t}$ is set at 1000 Euro/kWh constantly. This refers to the basic approach of multi-objective optimisation, where two objectives are combined in one objective function by applying weighting factors. The level of GHG price compared to power purchase prices corresponds to such a weighting factor and both objectives are optimised to a certain extent determined by this weighting.

As self-consumption of PV is not associated with variable cost and fed in remuneration is typically lower than power purchase prices at HH level, cost minimisation maximises self-consumption in each of the three scenarios. The scenario 3 was implemented to analyse the effect of including GHG reduction into the target function of the individual HH as this is proposed at present by some authors. However, there are two aspects to investigate: (1) we will analyse how much including such target leads to different charging schedules compared to scenario 1 and 2 and (2) whether it generates an effective GHG reduction effect in the overall system.

In order to get a better understanding of these three scenarios on the basis of an optimisation module, we have implemented as a 4th scenario an heuristic direct charging approach as a reference. This approach represents a simple rule-based charging strategy for each time step. Since each time step covers 15 min, we convert electrical energy with a factor of 60 min/15 min to power and vice versa. The rules for the heuristic approach are defined in the following. Since this algorithm only serves as a benchmark for the results of our optimum charging strategies, we have attached a detailed description in the form of a flowchart diagram in Appendix A.

(1) Charging restrictions: If the EV is connected to the charging station, charge with the maximum charging power. If the maximum battery level is exceeded, charge the battery only up to the maximum level.

(2) PV self-consumption: If there is a surplus of power generated by the PV system, supply it to the power demand of the HH as well as the charging demand as far as possible. If there is not enough energy, split it on the demand of the HH and the EV charging demand according to their respective share.

(3) Self-consumption discharging batterie storage: If there is no local power production, discharge the power required from the battery storage as long as the fill level of the battery is sufficient.

(4) Grid purchase: If the available power is not adequate, purchase the remaining power from the grid.

(5) Charging batterie storage with PV surplus: If there is still a surplus of power generated by PV, while load of HH and EV charging is supplied, charge the batterie storage.

(6) Grid Feed-In: If this power exceeds the maximum charging power or the maximum capacity of the battery system, feed the surplus into the grid.

4. Investigated Local Energy System

Key for charging strategies is the energy system in which the charging process will be integrated. The presented charging strategies are evaluated by their application to a typical grid-connected single-family household with an electrical demand and a demand for mobility, equipped with a distributed generation system based on PV and a stationary battery storage. First, we present this energy system by its technical and economical characteristics in more detail. Afterwards, we take a closer look at the representative HH mobility profile. Finally, we describe the interface of the distributed energy system with the overall system, the purchase from and feed-in of power via the public grid.

4.1. Structure of the Local Energy System of the Indivdual Household

The behaviour of the HH characterises the profile of the power as well as the mobility demand. The latter is described by various EV trips while the power demand without EV charging is described by a typical HH profile. For supplying the required power demand, the local energy system consists of a PV plant, an electrical battery storage unit and a connection to the power grid enabling power purchase and feed-in. The specific values of the purchase costs and feed-in revenues are predefined for each scenario to enable the analysis of the effects of the pricing structure. Focusing on EV charging processes, we don't consider a thermal demand in our investigation, leaving scope for future work.

The electrical demand profile of the HH and the PV generation profile stems from real measured data, corresponding to [25], presented in more detail in Figure 3. Both profiles show a time resolution of 15 min, referring to relevant time intervals for billing and balancing in the German system.

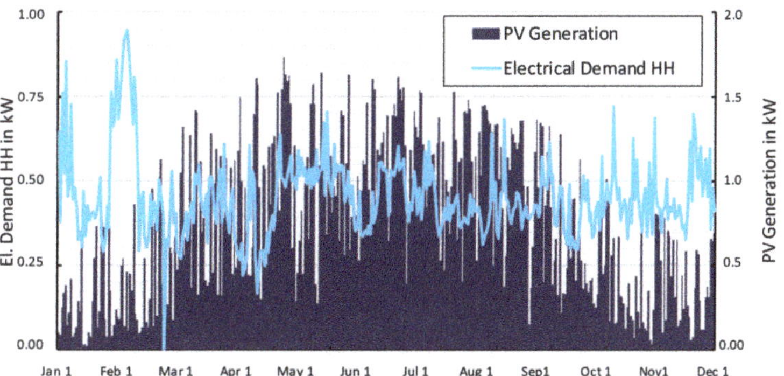

Figure 3. Main characteristics of the investigated energy systems as mean daily values: electric demand of the household (light blue) and PV generation (dark blue).

The dimensioning of the stationary battery storage corresponds to the average installation of a solar power battery storage packs in Germany in 2018 [2]. We have selected the e-Golf of Volkswagen as representative for the characteristics of the electric vehicle with a battery capacity of 35.8 kWh and an average EV power consumption of 15.8 kWh per 100 km [26]. The choice of this battery size may at first seem rather small as more up to date models already offer higher battery capacities. However, regarding the subsequent optimisation for unidirectional charging, the optimisation model will charge at maximum the energy, required for the next journey only. With our focus on a representative household profile, which predominantly consists of short journeys in the local area, as opposed to, e.g., longer journeys taken on holiday, this battery size with a maximum range of around 237 km already proves to be sufficient, though higher capacity will offer increased flexibility to the user of the EV rather than different charging patterns. The local charging facility is a wallbox with 3.7 kW maximum power, analogous to [27]. This is the minimum charging power you will find for sensible operation. However, as there is plenty of immobilisation time at home for the EV, even this power offers some flexibility already.

If higher charging power, like 11 kW, which seems to get the standard for charging at home, will be considered, the flexibility and therefore the advantage of an optimisation approach increases. In Table 1, we sum up the specific technical and economical properties of the investigated local energy system.

Table 1. Technical and economical properties of the energy system.

Component	Property	Value	Unit
Household Load	Electrical Demand	3995	kWh_{el}/year
Photovoltaic System	Electrical Power	6.8	kW_p
	Electrical Energy	6800	kWh_{el}/year
Batterie Storage Unit	Electrical Power	4.6	kW_{el}
	Electrical Capacity	7.2	kWh_{el}
Electrical Vehicle	Charging Power	3.7	kW_{el}
	Electrical Capacity	35.8	kWh_{el}
Supply/Feed-In via Power Grid	Purchase Price (average)	30.85	€ct/kWh_{el}
	Feed-In Tariff	10.00	€ct/kWh_{el}

4.2. Mobility Profile of the Household

The mobility demand of the household results from the trips of the household on each day. We characterise those trips by two factors: First the time slot the EV is on the trip, it is not available for charging at home ($V_{EV,Avail} = 0$) and second the energy consumed by the trip in kWh. Figure 4 displays the resulting mobility profile for a typical week.

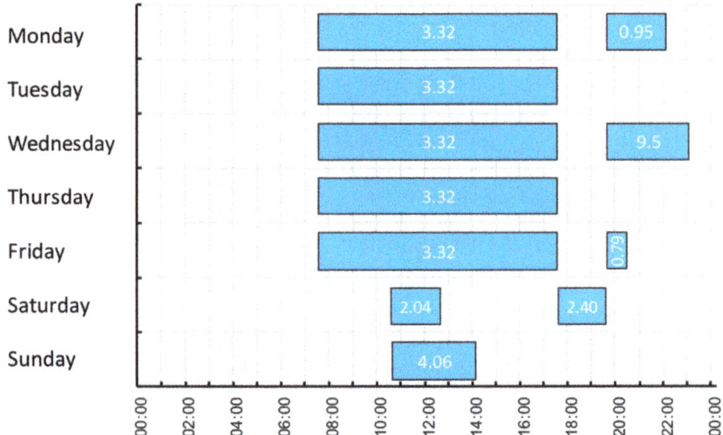

Figure 4. Trips over a week and their absolute electrical consumption of the trip in kWh.

Since empirical data were not available for the mobility demand, we synthesied this mobility demand considering individual trips. For this, we assume, that the EV user commutes to work every weekday like 77% of the working population in Germany. For most of the individuals surveyed, this commuting to work takes 30 min [28].

A full-time employee stays an average of 41 h per week, i.e., 8.2 h per day at his place of work [29]. The average distance of a trip to the workplace is 10.5 km per trip [30].Beyond this information regarding the journeys to work, there is no further information available on other trips by individual verhicles for other purposes. Therefore, we assumed that the EV is used on Mondays, Wednesdays and Fridays in the evening in addition. On Mondays a short trip is assumed (6 km), on Wednesdays a longer trip (60 km) and on Fridays a short trip (5 km). These short trips represent e.g., smaller purchases on site, like a trip to a supermarket, the longer distance a trip to a neighbouring town. The EV usage during weekends is described in detail in [31] for the average behaviour in Germany, from where we have deduced therefore two medium trips on Saturday (each 13 km) and one on Sunday (29 km).

4.3. Tariffs and Average GHG Emissions of Purchased Power

The interface to the overall energy system of our local energy system represents the connection to the power grid, where power can be purchased and fed-in due to defined tariff structures designed different for the different scenarios. This exchange of power via the public grid is connected on the one hand with costs or remunerations for the household and on the other hand with an impact of GHG emissions caused by change in generation in the overall system, which we allocate to the household by dynamic GHG emission factors of the power mix in a hourly time resolution.

Both parameters are derived by investigations of the project ENavi (part of the Kopernikus-projects funded by German Ministry for Education and Research (BMBF)) [32]. In this project, the electricity market model E2M2 has been applied and validated to model historic generation of the German power market. The model results show the hourly merit order of the unit commitment in Germany. The corresponding results have been taken up in [33], where the hourly GHG emissions have been investigated more closely. The model provides an annual hourly price for the power at the whole sale market and at the same time their average GHG emissions for each hour. Because [33] provides both time series, we do not provide a more detailed description at this point. Instead, we now focus on how we have further utilized both parameters in the present study.

Focusing the purchase cost, we converted the hourly generation prices to a representative household's power tariff. The German customer power price deviates significantly from the above mentioned whole sale price due to a large fraction of allocated taxes, duties and fees. The average power purchase price for a household in July 2020 amounts to 0.3171 Euro per kWh, whereby 52% is for taxes, duties and fees, 25% for grid fees (incl. measurement, metering point operation and billing) and 23% for procurement and sales [34].

As for our purpose comparing different charging strategies and pricing structures for the end consumer the exact price is not relevant, we assumed the purchase of our HH to be 10 times the average generation cost of 30.84 EUR/MWh, i.e., 0.3094 EUR/kWh which approximates the purchase price for households mentioned above. For one scenario we applied this price as constant price for power purchases of our HH. In another scenario we used the modelled wholesale power prices of E2M2 model in an hourly resolution and multiplied each by 10 for gaining a purchase price for HHs as well. Though this does not comply with present German regulation, such ideas where proposed by some researchers, e.g., in [19]. However, for the analysis reported in this work, this assumption is suitable, as we want to show the effect of dynamic pricing on charging strategies. We argue, that level of price differences will be relevant as well, but determining the minimum level of price fluctuation to gain impact is not in the scope of this work and left to future research.

With regard to the GHG emissions, the GHG emissions caused by our individual HH result from the households power purchase resulting from power generation in fossil plants. To keep analysis simple, we apply the pragmatic approach of considering a grid emission factor, e.g., analogous to [22,24]. However, we will discuss the validity and the impact of this approach later on in Section 5.3.

In order to guarantee comparability and reproducibility, we need to consider the GHG emissions related to the production, installation and removal of our local energy system and its components, which do not change with the charging strategy. We have therefore assumed specific emissions of 50 g CO_2/kWh for the electricity from the PV system [35].

In the literature, we could not find any specific emissions for the electricity from the battery storage, but only the absolut CO_2 emissions of 150–200 kg CO_2/kWh battery capacity, resulting from the production of the storage [36]. Here, we have chosen the more pessimistic value of 200 kg CO_2/kWh in order not to underestimate the CO_2 emissions, which we allocated over the first 10 years by means of a linear amortization with the annual amount of electrical energy to be discharged. This results in specific CO_2 emissions per discharged electrical energy of 80.03 g CO_2/kWh.

5. Comparison of Charging Strategies

In this section we present the results of the application of the different analysed charging strategies to the described energy system. Therefore, we first give a qualitative comparison between two exemplary schedules, the one of the direct charging and the other one of the optimised schedule considering the energy purchase costs as well as the GHG emissions. We then compare the different charging strategies quantitatively due to the absolute annual energy procurement costs and the estimated induced GHG emissions. Finally, we discuss the results with regard to GHG emission reduction effects and dispute, whether the GHG reductions will impact on the overall system.

5.1. Simulation of the Charging Strategies

We applied all 3 charging strategies based on the optimisation model as well as the heuristic reference charging strategy on the previously described energy system over the period of one year. Each simulation produces a charging schedule for the EV in this process. Figure 5 exemplarily shows the charging schedules of the heuristic charging scenario (HCS) as well as scenario 3 over the time of 3 days. For better analysis of the interdependencies in the local energy system, the feed-in or demand of the other components of the local energy system of the HH are shown as well.

In the Figure, it can be realised that there is no charging during the day except on the third day, though there is PV generation available as well as low purchase prices and low GHG emission factors. The reason for this is, that the EV is not connected to the charging station during daytime onthe other days. Comparing the light blue (CS 3) and the dark blue (HCS) line, it can be detected, that the charging in scenario 3 is moved to time slots offering low specific GHG emission factors. We see this for example in the morning of Feb 14th and in the afternoon of 16th February, where the optimisation of CS 3 selects times with lower GHG emissions and purchase prices for the charging process. Since GHG emissions are linked to very high prices of 1000 Euro/kg CO_2, the objective of GHG emissions is preferred. This can be realised by the shifting of the schedule into a higher-priced period of purchase-power price but lower GHG emissions in the morning of 15th February, in order to reduce GHG emissions.

Compared to HCS the charging strategy CS3 prefers the target value of reducing GHG emissions compared to minimisation of purchase cost due to the high prices for GHG emissions chosen by this scenario. The HSC, in contrast, is completely insensitive to the PV electricity generated on site, the purchase costs and the GHG emissions as they are not implemented in the HSC approach.

Following, we analyse the interaction with the energy system via the the exchange of electrical power, represented in the following Figure 6. Here, this exchange via the public grid is shown as a dotted line (CS 3 blue, HDC grey). Additionally, there are solid lines for charging (light blue/grey) and discharging (dark blue/grey) of the stationary battery unit of the local energy system.

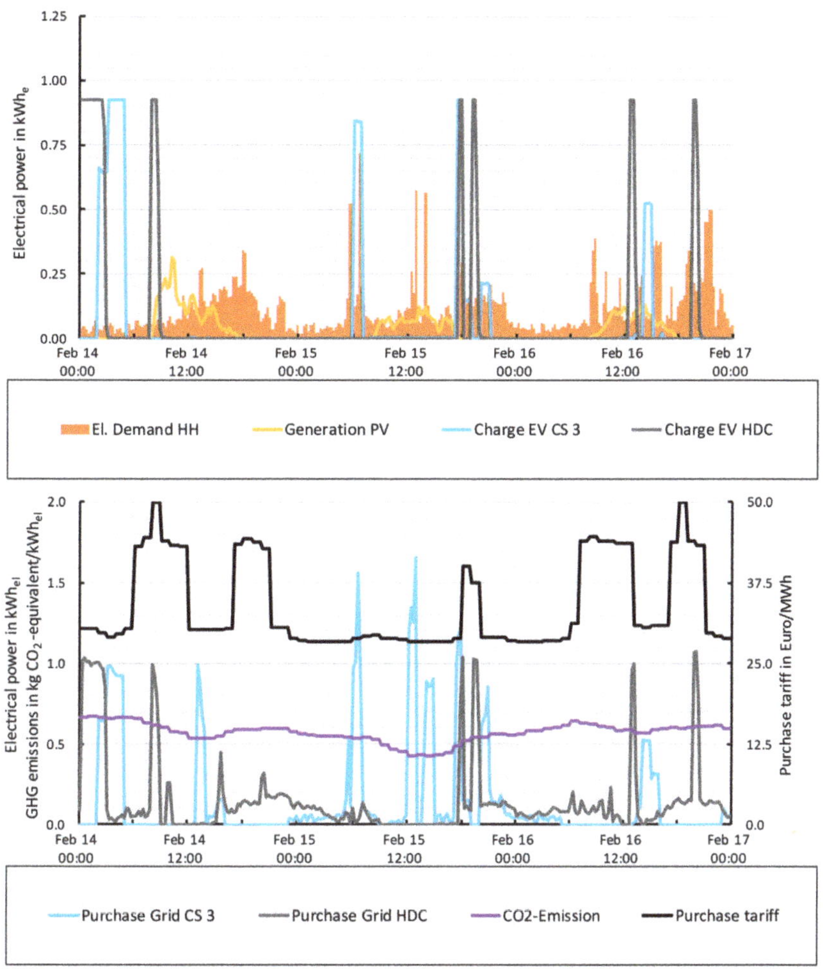

Figure 5. Resulting charging schedule of the charging strategy 3 (light blue) and the heuristic charging (HCS) strategy (light grey) at a dynamic electricity tariff (black lined) and dynamic CO_2 emissions (violet lined) from 14 to 17 February.

In the figure, it can be seen, that the smart charging strategy CS 3 covers almost the entire electrical load of the household by the stationary battery storage, although the PV system does not generate much power during this period. This demonstrates that the charging strategy CS 3 also optimises the flexibility of the stationary battery storage in times of low procurement costs and GHG emissions, e.g., on the 15th February. Simultaneously, we identify a potential disadvantage of the charging strategy CS 3: By optimising to the most beneficial time of day, the power supply via grid is concentrated on individual time steps, resulting in clearly visible power peaks. In order to prevent these peaks and to shape them more smoothly, we could consider them as a further objective in the optimisation. For this purpose, we could extend the objective function by costs for the maximum power purchase over the optimisation horizon referring to a peak power pricing and—in case it is appropriate—we could do the same for the feed-in. As typical for multi-objective optimisation, the weighting factor of this maximum power peak pricing refers to the level of peak power prices compared to the prices of the other objectives. Since our approach initially aims to minimize purchase costs and GHG emissions, we have not examined this approach in detail.

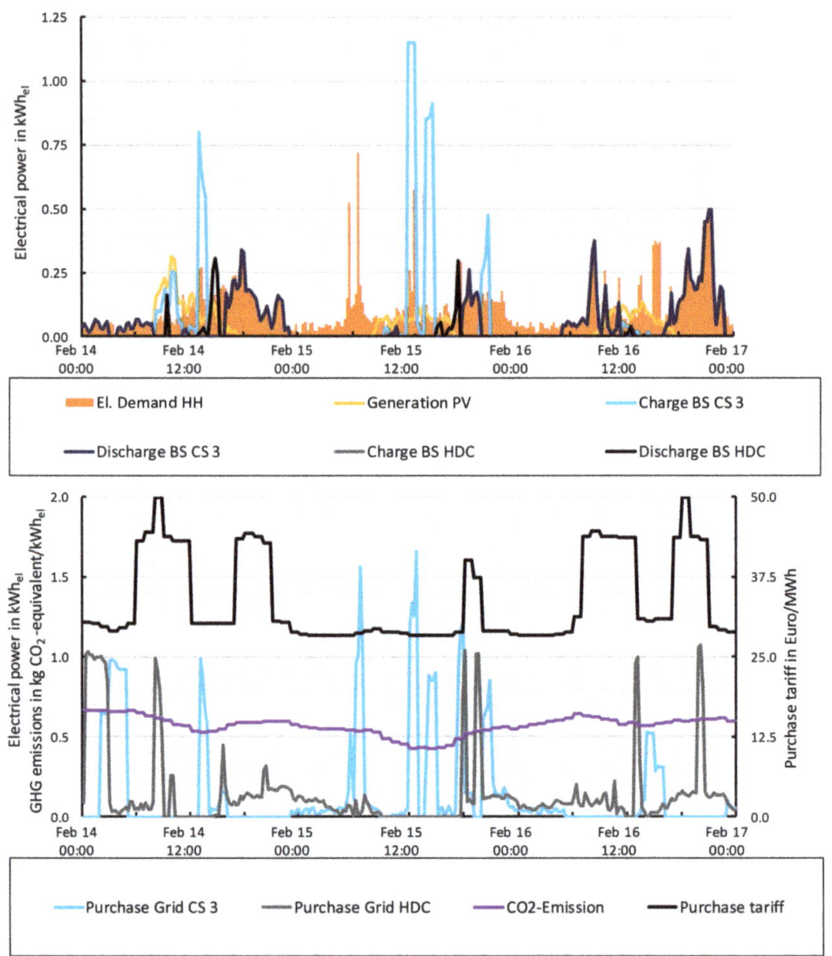

Figure 6. Resulting battery charging/discharging schedule of charging strategy 3 (light/dark blue) and heuristic charging (HCS) strategy (light/dark grey) at a dynamic electricity tariff (black lined) and dynamic CO_2 emissions (violet lined) from 14 to 17 February.

5.2. Comparison of the Charging Strategies

Figure 7 shows the annual results of the four charging strategies: Heuristic charging, optimised charging with minimised purchase cost and constant (CS 1) or dynamic purchase prices (CS 2) as well as conditions of CS 2 with adding the second objective of minimisation of assigned GHG emissions (CS 3).

It can be clearly seen, that applying a charging strategy based on an optimisation approach reduces yearly power cost significantly in the range from 10% to 20% depending on the pricing structure of power purchase pricing. In absolute terms we can show a cost saving of 45 EUR/a or 105 EUR/a respectively. In case of dynamic purchase prices, the higher cost saving can be achieved. Thereby, it is obvious, that introducing the second optimisation objective of minimising GHG emissions at the same time results in a cost increase, due to the reduced solution space concerning the cost minimisation in this case.

However, the increase in cost by applying CS 3 is rather small. This is not astonishing: the whole sale price itself contains GHG pricing induced by the EU ETS and the dynamic pricing structure relays on the whole sale prices, so that an additional (virtual) GHG pricing at the end customer level, as introduced for CS 3, does not change that much.

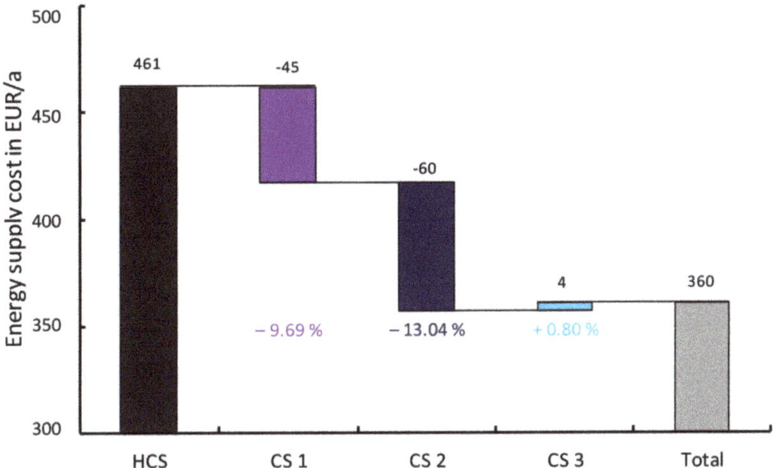

Figure 7. Resulting annual energy supply cost (Power purchase cost less feed-in remuneration) of each charging strategy.

In summary, we can conclude that the smart charging strategies can reduce the specific annual procurement costs for power by 22.73%, but the absolute savings seem to be rather challenging, as at present regulation there are just 45 EUR/a to be invested in such improved systems. And this is the breakeven case, i.e., if this sum is invested, there will be no incentive left for the HH. This figure increases to 105 EUR/a in CS 2, but the assumed dynamic of the purchase price structure is overly optimistic, as the high fluctuation of the assumed prices do not reflect causally determined cost structures. Within the power price for end customers, there are several significant inflexible components (e.g., refund of feed-in renumeration of renewable energy supporting scheme or grid fees), which do not reflect system cost appropriately, in case they are differentiated over time.

As the second objective analysed was GHG reduction, Figure 8 shows the assigned GHG emssions for each scenario resulting from estimated GHG emission changes in the overall system caused by power purchase or feed-in via the grid of our local energy system by the different scenarios.

It is not astonishing that the lowest assigned GHG emissions are related to CS 3, as this strategy is the only one including this target into the objective function. It achieves almost 9% less assigned GHG emissions compared to HCS. However, power cost minimisation of the HH leads to a reduction of assigned GHG emissions even in CS 1, due to the pricing structures in place in Germany. This assumed reduction of GHG emissions estimated by the chosen approach relies on the effect, that the objective function maximises local consumption of low GHG emission generation of the PV system, as feed-in remuneration of local produced PV power is less than the cost of power purchase at end customer level. CS 2 reduces the assigned GHG emissions by almost 7% (compared to HCS). This additional assigned GHG emissions are caused by the effect mentioned above, that the dynamic whole sale price includes a GHG pricing component already and drives the power purchase prices at end customer level in this scenario.

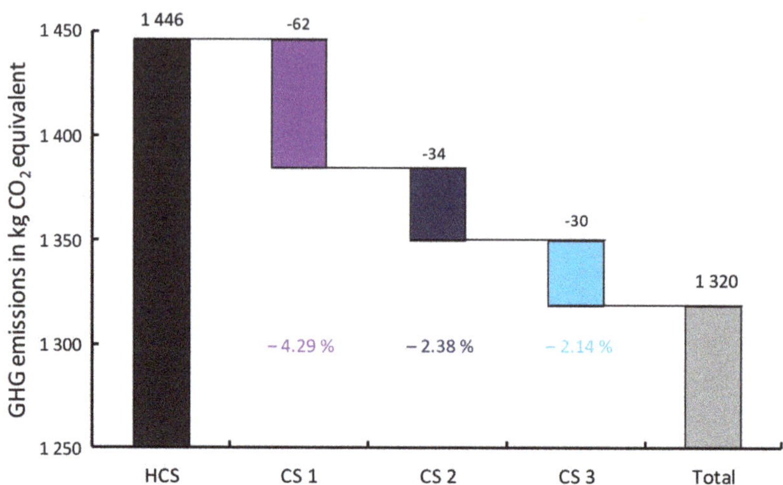

Figure 8. Resulting annual GHG emissions of each charging strategy.

5.3. Discussion of the Impact of the Charging Strategies

There are two aspects assumed to be different to up to date incentive structures in place in Germany at present. The present situation is reflected by CS 1 while CS 2 and CS 3 rely on assumed changes in pricing structures and an assignment of GHG emissions proposed in literature. However, it is important to investigate, whether it can be expected that those assumptions proposed in literature will reach the goals striving for. We will discuss this further in the following.

(1) Dynamic pricing at the end customer level: The assumed prices in CS 2 and CS 3 contain dynamic components for taxes, duties and fees as well as dynamic grid fees accounting for about 77% of the cost at end customer level and therefore the largest component while procurement cost relying on the dynamic prices of the whole sale market are included in the component accounting just for 23% of the end customer price. There is no good reason for assuming that the price dynamic of taxes (e.g., VAT), grid fees or duties (EEG remuneration) should exhibit the same price pattern than the whole sale market. From an economic point of view, prices should reflect cost structures and scarcities. The grid bottlenecks are not at all related to the bottlenecks in generation of the overall system as well as the generation of renewables. Low prices at whole sale markets relate especially to high renewable generation, which in consequence this leads to high supporting cash flows. Applying the assumed pricing at end customer level in CS 2 and CS 3, this would lead to especially low duty components for EEG remuneration at the same time, which is the polar opposite of supporting the situation at the same period. Therefore, in our judgement, the proposed structure for purchase prices at end customer level are contradictory to our main goal, designing a most cost-efficient overall system with less (or in the long run no) GHG emissions, as such correlations do not lead to the appropriate incentives governing the system in an economic efficient way. Restricting the price dynamic to the price component relying on the wholesale market, reduces the fluctuation in prices at end customer level and therefore reduces the savings shown in CS 2. However, there will be some additional saving never the less, but it is assumed to be quite small compared to CS 1.

(2) Assignment of GHG emissions at the end customer level by GHG emission factors: The assignment of GHG emissions at the end customer level by applying hourly GHG emission factors derived from a unit commitment model seems to be reasonable at first approximation. For increased simplicity, we applied factors representing the average

emission of the grid purchase and withdraw of electrical energy in each time step in this work. This does clearly not reflect the GHG reductions in case of any change of production linked to additional feed-in or power purchase by our local energy system. The more appropriate value would be the marginal GHG factor in each time step. This is somewhat more difficult to evaluate by a conventional unit commitment model as E2M2, but could be used in principle in the same way. Those figures were not available for this work, but could be generated for future work. Anyhow, it is assumed that those values will not differ significantly for a lot of time steps, as the overall system contains a lot of flexibility itself leading to the reduction of the impacts of shifting feed-in or demand along the time line in cost minimum production of fossil plants. In addition, in some more extreme situation (e.g., curtailment of renewable feed-in), there will definitely be some impact on marginal GHG emissions.

However, there is another issue we need to consider necessarily: Most of the GHG emissions originated from power production of the overall system are included in the EU emission trading system (EU ETS). That means, any reduced emission in any time step is linked to a reduced demand for an EU allowance (EUA) in this system in that time step. The saved up EUA is free to be used to cover additional GHG emissions in any other time step or sector (e.g., iron & steel). As the EU ETS is a cap and trade system, it ensures to achieve an absolute limit of GHG emissions during a defined period corresponding to the GHG emission targets set. As the cap is not affected by the behaviour of any player in the energy system, the estimated GHG reduction shown in this work, is just a figure depending on the assumptions, but does not reduce any GHG emission in fact as the cap in the EU ETS is not affected. Following our judgement, the approach of integrating the objective of GHG emission into any control strategy at local energy system level is therefore useless. Furthermore, we consider it as even counterproductive, as the HH is given the impression it could reduce GHG emissions further by another strategy, which does not reflect reality and then results in additional useless costs. The approach supports the motivation of climate-friendliness of the HH, though the intended target of the HH is not achieved in reality at all. For achieving an optimal GHG reduction overall, optimal system integration of the local energy system is crucial. This can be easily achieved by a cost minimisation of power cost of HH in case the appropriate incentives are provided in the pricing structures, which is not always achieved today [37]. However, this calls rather for adapting pricing structures and regulation as an appropriate way rather than changing objective functions for the operation of the local energy system.

6. Conclusions and Outlook

In the context of the German energy transformation, distributed energy systems are becoming increasingly important due to different sectors (power, heat, transport) being increasingly integrated on the end customer level. The high degree of flexibility in their operation results in an enormous flexibility potential for the overall energy system, which in view of the increase in electric vehicles has resulted in controllable flexibility of 0.46 GW and 4.43 GWh over the last six years. In this paper we have presented an approach to control the respective charging processes at minimum cost and we have tested an approach proposed by other authors to impact GHG emissions by adapted operation of such systems. This approach is based on an optimisation model, which can easily be extended by further objectives besides cost minimization and climate friendliness and is able to approach multi-objective targets as well (e.g., minimum cost and minimum GHG emissions).

For this purpose, we compared three different charging strategies, with minimisation of power cost for an exemplary household and different settings of structure of power purchase price at end customer level as well as explicit minimisation of GHG emissions as additional objective at HH level, with a simple heuristic charging approach. The results of our charging strategies undercut the costs of the heuristic charge control, as can be expected, byt 10% to 20%, though 10% should be achievable with the present regulatory framework in Germany already.

At the same time, we discussed the impact of such a charging strategy and the impact that can be reached in reality. This leads to the conclusion by deductive reasoning, that the 20% savings do not comply with cost savings in the entire system, as the applied dynamic pricing structure at end customer level does not reflect causality of cost and therefore leads to inappropriate incentives missing the most efficient system behaviour in the end. The estimated assigned GHG emission reduction on the other hand will not lead to any GHG reduction in reality, as the cap and trade system of EU ETS will keep emissions exactly in the cap level in any case. The introduction of the second objective into the optimisation of operation of the local energy system leads to increased system cost as well as increased cost for the HH while no GHG emission reduction is achieved in fact. This approach proposed in literature is, due to our judgement, counterproductive, as it increase the cost of energy transition while not achieving any significant GHG reduction.

However, we were able to show, that applying appropriate optimisation approaches to scheduling of local energy systems, including charging strategies of EVs, show clear potential to improve system operation from local as well as overall system perspective. Therefore research in this field should be intensified in several aspects. First, we recommend to improve data on characteristics of EV utilization and limitations of charging processes. For a better integration in real systems, the stochastic characteristic of the EV usage pattern has to be taken into account in addition. Second, additional components of local energy systems should be added like heat pumps and heat storage to consider the ongoing comprehensive electrification of distributed energy systems. Third, further research on the appropriate regulating framework should be intensified for generating appropriate incentives leading to an optimum for local HH as well as entire system.

However, we believe that this approach is one of the basic fundamentals for the comprehensive integration of distributed energy systems including EV charging into the overall energy system by means of smart control systems.

Author Contributions: Conceptualization, M.S. and K.H.; methodology, M.S. and K.H.; software, M.S.; validation, M.S. and K.H.; formal analysis, M.S. and K.H.; investigation, M.S.; resources, M.S. and K.H.; data curation, M.S.; writing, M.S. and K.H.; visualization, M.S.; supervision, K.H.; project administration, M.S.; funding acquisition, M.S. and K.H. All authors have read and agreed to the published version of the manuscript.

Funding: This research was funded by the Open Access Publication Fund of the University of Stuttgart.

Institutional Review Board Statement: Not applicable.

Informed Consent Statement: Not applicable.

Acknowledgments: The authors gratefully acknowledge the support of the Ministry of the Environment, Climate Protection and the Energy Sector Baden-Württemberg as well as the support of the Open Access Publication Fund of the University of Stuttgart.

Conflicts of Interest: The authors declare no conflict of interest.

Appendix A

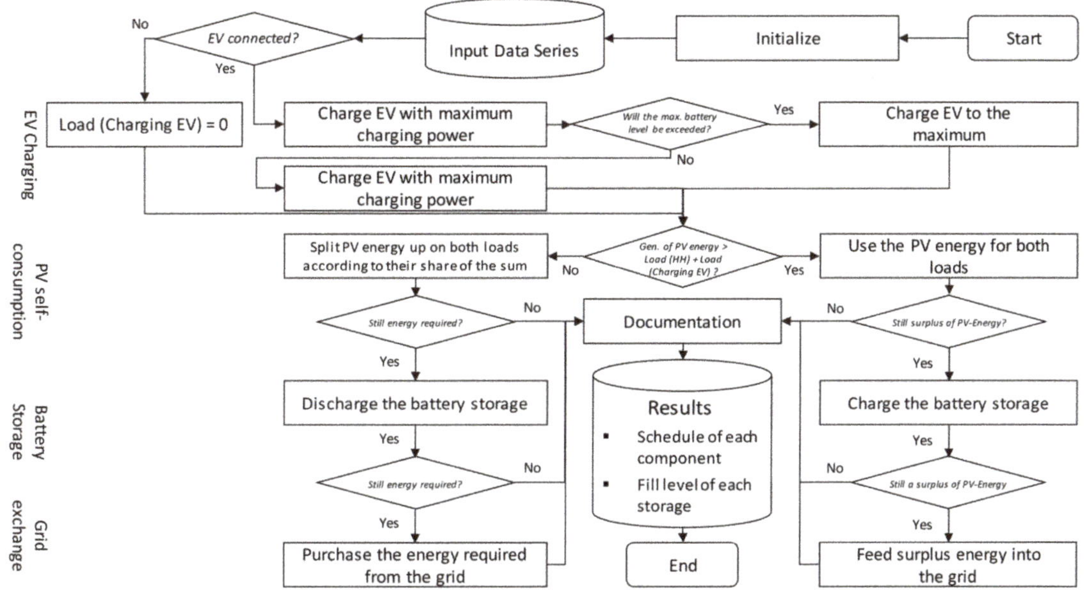

Figure A1. Overview of the heuristic direct charging approach per flowchart diagram.

References

1. Federal Ministry for the Environment, Nature Conservation, Building and Nuclear Safety. *Climate Action Plan 2050—Principles and Goals of the German Government's Climate Policy*; Federal Ministry for the Environment, Nature Conservation, Building and Nuclear Safety: Berlin, Germany, 2016.
2. Federal Network Agency for Electricity, Gas, Telecommunications, Posts and Railways. 2020. EEG Register Data and Reference Values for Payment. Berlin. Available online: www.bundesnetzagentur.de/EN/Areas/Energy/Companies/RenewableEnergy/Facts_Figures_EEG/Register_data_tariffs/EEG_registerdata_payments_node.html (accessed on 18 November 2020).
3. Figgener, J.; Haberschusz, D.; Kairies, K.P.; Wessels, O.; Tepe, B.; Sauer, D.U. *Wissenschaftliches Mess-und Evaluierungsprogramm Solarstromspeicher 2.0-Jahresbericht 2018*; ISEA Institut für Stromrichter-technik und Elektrische Antriebe RWTH Aachen: Aachen, Germany, 2018.
4. Deutsche Presse Agentur (dpa). 2020. Zahl Privater Solarspeicher Legt 2019 deutlich zu. Düsseldorf. Available online: https://www.handelsblatt.com/unternehmen/energie/energie-zahl-privater-solarspeicher-legt-2019-deutlich-zu/25718994.html?ticket=ST-16144660-PQ0Jz0DbRO2dE0H0Qxqp-ap5 (accessed on 18 November 2020).
5. Bundesverband Wärmepumpe e.V. 2020. Absatzzahlen. Berlin. Available online: www.waermepumpe.de/presse/zahlen-daten/absatzzahlen/ (accessed on 18 November 2020).
6. Kraftfahrt-Bundesamt. 2020. Bestand nach Umwelt-Merkmalen. Flensburg. Available online: https://www.kba.de/DE/Statistik/Produktkatalog/produkte/Fahrzeuge/fz13_b_uebersicht.html (accessed on 18 November 2020).
7. Pasaoglu, G.; Fiorello, D.; Martino, A.; Zani, L.; Zubaryeva, A.; Thiel, C. Travel patterns and the potential use of electric cars—Results from a direct survey in six European countries. *Technol. Forecast. Soc. Change* **2014**, *87*, 51–59. [CrossRef]
8. Mukherjee, J.C.; Gupta, A. A Review of Charge Scheduling of Electric Vehicles in Smart Grid. *IEEE Syst. J.* **2014**, *9*, 1541–1553. [CrossRef]
9. García-Villalobos, J.; Zamora, I.S.; Martín, J.I.; Asensio, F.J.; Aperribay, V. Plug-in electric vehicles in electric distribution networks: A review of smart charging approaches. *Renew. Sustain. Energy Rev.* **2014**, *38*, 717–731. [CrossRef]
10. Mwasilu, F.; Justo, J.J.; Kim, E.-K.; Do, T.D.; Jung, J.-W. Electric vehicles and smart grid interaction: A review on vehicle to grid and renewable energy sources integration. *Renew. Sustain. Energy Rev.* **2014**, *34*, 501–516. [CrossRef]
11. Hussain, S.; Ahmed, M.A.; Kim, Y.-C. Efficient Power Management Algorithm Based on Fuzzy Logic Inference for Electric Vehicles Parking Lot. *IEEE Access* **2019**, *7*, 65467–65485. [CrossRef]
12. Hussain, S.; Ahmed, M.A.; Lee, K.-B.; Kim, Y.-C. Fuzzy Logic Weight Based Charging Scheme for Optimal Distribution of Charging Power among Electric Vehicles in a Parking Lot. *Energies* **2020**, *13*, 3119. [CrossRef]

13. Sovacool, B.K.; Axsen, J.; Kempton, W. The Future Promise of Vehicle-to-Grid (V2G) Integration: A Sociotechnical Review and Research Agenda. *Annu. Rev. Environ. Resour.* **2017**, *42*, 377–406. [CrossRef]
14. Tashviri, M.H.; Ghaffarzadeh, N. Method for EV charging in stochastic smart microgrid operation with fuel cell and renewable energy source (RES) units. *IET Electr. Syst. Transp.* **2020**, *10*, 249–258. [CrossRef]
15. Van Vliet, O.; Brouwer, A.S.; Kuramochi, T.; Broek, M.V.D.; Faaij, A. Energy use, cost and CO_2 emissions of electric cars. *J. Power Sources* **2011**, *196*, 2298–2310. [CrossRef]
16. Yi, Z.; Scoffield, D.; Smart, J.; Meintz, A.; Jun, M.; Mohanpurkar, M.; Medam, A. A highly efficient control framework for centralized residential charging coordination of large electric vehicle populations. *Int. J. Electr. Power Energy Syst.* **2020**, *117*, 105661. [CrossRef]
17. Papadaskalopoulos, D.; Strbac, G.; Mancarella, P.; Aunedi, M.; Stanojevic, V. Decentralized Participation of Flexible Demand in Electricity Markets—Part II: Application with Electric Vehicles and Heat Pump Systems. *IEEE Trans. Power Syst.* **2013**, *28*, 3667–3674. [CrossRef]
18. Cheng, A.J.; Tarroja, B.; Shaffer, B.; Samuelsen, S. Comparing the emissions benefits of centralized vs. decentralized electric vehicle smart charging approaches: A case study of the year 2030 California electric grid. *J. Power Sources* **2018**, *401*, 175–185. [CrossRef]
19. Schuller, A.; Dietz, B.; Flath, C.M.; Weinhardt, C. Charging Strategies for Battery Electric Vehicles: Economic Benchmark and V2G Potential. *IEEE Trans. Power Syst.* **2014**, *29*, 2014–2022. [CrossRef]
20. Hoehne, C.G.; Chester, M.V. Optimizing plug-in electric vehicle and vehicle-to-grid charge scheduling to minimize carbon emissions. *Energy* **2016**, *115*, 646–657. [CrossRef]
21. Karan, E.; Asadi, S.; Ntaimo, L. A stochastic optimization approach to reduce greenhouse gas emissions from buildings and transportation. *Energy* **2016**, *106*, 367–377. [CrossRef]
22. Huber, J.; Lohmann, K.; Schmidt, M.; Weinhardt, C. Carbon efficient smart charging using forecasts of marginal emission factors. *J. Clean. Prod.* **2021**, *284*, 124766. [CrossRef]
23. Van der Kam, M.; van Sark, W. Smart charging of electric vehicles with photovoltaic power and vehicle-to-grid technology in a microgrid; a case study. *Appl. Energy* **2015**, *152*, 20–30. [CrossRef]
24. Braeuer, F.; Finck, R.; McKenna, R. Comparing empirical and model-based approaches for calculating dynamic grid emission factors: An application to CO2-minimizing storage dispatch in Germany. *J. Clean. Prod.* **2020**, *266*, 121588. [CrossRef]
25. Schulz, M.; Kemmler, T.; Kumm, J.; Hufendiek, K.; Thomas, B. A More Realistic Heat Pump Control Approach by Application of an Integrated Two-Part Control. *Energies* **2020**, *13*, 2752. [CrossRef]
26. Volkswagen. 2019. Der e-Golf:Technik und Preise—Gültig für das Modelljahr 2020. Wolfsburg. Available online: www.volkswagen.de/idhub/content/dam/onehub_pkw/importers/de/besitzer-und-nutzer/hilfe-und-dialogcenter/downloads/produktbroschueren/golf-(bq)/e-golf_preisliste.pdf (accessed on 18 November 2020).
27. Ebner, M.; Fattler, S.; Ganz, K. *Kurzstudie Elektromobilität. Modellierung für die Szenarienentwicklung des Netzentwicklungsplans*; Forschungsstelle für Energiewirtschaft e.V: München, Germany, 2019.
28. Stepstone GmbH. *Stepstone Mobilitätsreport 2018*; StepStone: Düsseldorf, Germany, 2019.
29. Statistisches Bundesamt (DESTATIS). *Qualität der Arbeit—Wöchentliche Arbeitszeit*; Statistisches Bundesamt: Wiesbaden, Germany, 2020.
30. Dauth, W.; Haller, P. *Berufliches Pendeln zwischen Wohn-und Arbeitsort: Klarer Trend zu Längeren Pendeldistanzen*; No. 10/2018; IAB-Kurzbericht: Nürnberg, Germany, 2018.
31. Bundesministerium für Verkehr und Digitale Infrastruktur (BMVI). *Mobilität in Tabellen*; BMVI: Köln, Germany, 2017.
32. Fahl, U.; Gaschnig, H.; Hofer, C.; Hufendiek, K.; Maier, B.; Pahle, M.; Pietzcker, R.; Quitzow, R.; Rauner, S.; Sehn, V.; et al. *Das Kopernikus-Projekt ENavi. Die Transformation des Strom-systems mit Fokus Kohleausstieg. Endbericht*; Institut für Energiewirtschaft und Rationelle Energieanwendung, Potsdam Institut für Klimafolgenforschung: Stuttgart/Potsdam, Germany, 2019. [CrossRef]
33. Seckinger, N.; Radgen, P. *Dynamic Prospective Average and Marginal GHG Emission Factors—Scenario Based Method for the German Power System Until 2050*; International Energy Agency: Stuttgart, Germany, 2020; in review.
34. Schwencke, T.; Bantle, C. *BDEW-Strompreisanalyse Juli 2020*; BDEW Bundesverband der Energie-und Wasserwirtschaft e.V: Berlin, Germany, 2020.
35. Eichmann, E. Klimabilanz Photovoltaik: Wie groß ist der CO2-Fußabdruck von Solarstrom. EnergieAgentur.NRW: Düsseldorf, Germany, 2017.
36. Emilsson, E.; Dahllöf, L. *Lithium-Ion Vehicle Battery Production: Status 2019 on Energy Use, CO2-Emissions, Use of Metals, Products Environmental Footprint, and Recycling*; IVL Swedish Environmental Research Institute: Stockholm, Sweden, 2019.
37. Schick, C.; Klempp, N.; Hufendiek, K. Role and impact of prosumers in a sector-integrated energy system with high renewable shares. *IEEE Trans. Power Syst.* **2020**, *1*. [CrossRef]

Article

Optimization of Electric Vehicle Charging Points Based on Efficient Use of Chargers and Providing Private Charging Spaces

Lukáš Dvořáček ***, Martin Horák** **, Michaela Valentová and Jaroslav Knápek**

Department of Economics, Management and Humanities, Faculty of Electrical Engineering, Czech Technical University in Prague, 16627 Prague, Czech Republic; horakm11@fel.cvut.cz (M.H.); valenmi7@fel.cvut.cz (M.V.); knapek@fel.cvut.cz (J.K.)
* Correspondence: dvoral14@fel.cvut.cz

Received: 23 November 2020; Accepted: 16 December 2020; Published: 21 December 2020

Abstract: Electric vehicles are a mobility innovation that can help significantly reduce greenhouse gas emissions and mitigate climate change. However, increasing numbers of electric vehicles require the construction of a dense charging infrastructure with a sufficient number of chargers. Based on the identified requirements for existing electric vehicle users and potential new customers, the paper proposes a charging point model for an urban area equipped with a local transformer station and a sufficient number of low-power chargers. In particular, the model focuses on efficient use of chargers throughout the day, considering private rental of chargers paid by residents in the evening. The model uses an optimization method that compares the non-covered fixed costs due to unsold electricity to nonresidents and the annualized costs of building an additional transformer. The proposed optimal charging point solution was tested in a case study using real data capturing users' habits and their arrivals in and departures from the car park. As our model results show, the great benefit of a park-and-ride car park equipped with chargers consists of a simple increase in car park efficiency, ensuring sufficient numbers of private charging lots, optimizing operating costs, and supporting the development of electromobility.

Keywords: shared parking; sharing economy; P + R car park; low-power chargers; electric vehicle

---

1. Introduction

Following the European Union (EU) 2030 targets [1], many subsidy programmes have been drawn up to support electric car sales and construction of charging stations [2]. Notably, support for sale of electric vehicles in many countries has caused rapidly growing numbers of electric vehicles (EV) in operation [3], which have swiftly overtaken the numbers of charging stations built [4]. To achieve a balanced ratio between electric vehicles in operation and charging stations, we need to step out of the vicious circle [5] where many investors wait with construction of charging stations because the number of EVs in operation has not achieved the required level. Conversely, many potential new EV owners wait for the numbers of charging stations to increase.

However, construction of public charging stations itself is not enough [1]. Due to the expected increase in the numbers of EVs, their owners will require similar parking and charging comfort that is typical of the owners' private garages or parking spaces [2]. Therefore, sharing them seems to be a smart and simple solution to ensure a sufficient number of parking and charging stations for electric cars.

Concerning the issue of sharing parking and charging places, some studies focus on peer-to-peer (P2P) sharing, cf. [2–4]. The disadvantage of the proposed P2P solutions can be seen mainly in the difficulty of ensuring the quality of the services provided. However, not all charging station owners are sufficiently conscientious and aim at ensuring high quality of the services. Usually, the charging station owner is responsible for their charging station failure, but not penalized in any way for it [4].

Another key issue associated with P2P sharing of charging stations is ensuring electricity grid stability. The electric grid can be in jeopardy mainly when the many shared charging stations in urban areas are fully occupied [5–7]. The methods used to smooth the peak load as a method of dividing the charging power between several charging stations cannot be used, because the charging station does not constitute one microgrid. Moreover, P2P charging station sharing does not provide a clear answer to the issue of procuring private charging points for EV users. Instead, it focuses on recharging EVs during the day. EV sharing assumes that the charging station will be used in the evening by the charging station owner who typically returns from work [6].

A good base for dealing with the issue of private charging stations is the idea of sharing parking places near commercial buildings [8,9]. Assuming that car parks with a high concentration of vehicles will be equipped with a sufficient number of charging stations, mainly due to the growing trend of electric cars, the possibility of sharing these places as private parking spaces for EV users offers itself directly. These places, equipped with charging stations, are mainly occupied during working hours. Beyond working hours, they are unused and thus offer the opportunity to create private charging points for EV owners.

Assuming these two conditions, i.e., charging stations connected to a microgrid and long EV parking times, the power division or charging shifting method [6] can be used to smooth the peaks in electricity consumption. In extreme cases, it is possible to supplement the microgrid with battery storage to further support the stability of the entire system and of the local electric grid.

The aim of the present paper is to offer a solution to one of the most common obstacles of potential EV users deciding whether to buy an electric car: the absence of a private charging space. Therefore, it is necessary to provide these users with similar charging comfort to what users with private home charging have, and to offer a price similar to that of home charging. We do so by defining the following research questions:

1. Can evening reservation of car parks equipped with low-power chargers in a P + R car park compete with home charging in terms of both costs and charging comfort?
2. Is it essential to equip P + R car parks with fast charging stations, due to the mode of using this type of car park and the time users park their EVs in this type of car park?
3. Is it possible to ensure that EV owners having no private garage and charger will have a fully charged EV battery before the next journey and the same comfort as users having a private garage and charger?
4. Is it possible to ensure for EV owners having no private garage and charger a price per kWh similar to users having a home charger?

The major contributions of this paper are as follows: (1) the problem of insufficient numbers of charging stations can be solved by effective use of existing car parks equipped with low-power chargers; (2) charging electric cars with low-power chargers contributes to efficient use of transformer station potential; (3) by reserving parking places in a car park during evening hours, it is possible to achieve higher efficiency of their use and simultaneously create charging conditions and a price per kWh comparable to home charging.

The rest of this paper is organized as follows. Section 3 analyses EV drivers' charging behaviour and proposes a model of sharing parking places in a P + R car park equipped with low-power chargers. Section 4 presents a case study based on both the real occupancy of a P + R car park and the real charging requirements of EV users, with the objective to optimize the installed power of transformer stations and the number of chargers used simultaneously from the technical and economic point of view. Section 5 presents the results of technical and economic optimization of the car park model, followed by discussion in Section 6 and conclusions with suggestions for future research in Section 7.

2. State of the Art

There are several aspects that potential customers consider when deciding whether to purchase an electric vehicle (EV). Firstly, the high investment costs of EVs are seen as the major barrier to EV scale-up [2,3,6,10]. Furthermore, the high initial investment in EVs currently sold correlates with the users' annual income. Xu et al. [6] stratified the percentage distribution of EVs registered in California into user groups with annual incomes above and below USD 150,000. The group of registered EVs in the group of users with an annual income exceeding USD 150,000 makes up 45%. However, the percentage of registered EVs in the larger group of users with an annual income lower than USD 150,000 is only 15%.

Individual countries try to compensate for this barrier using various financial instruments. The structure and scope of the financial instrument and its implementation varies from country to country. Additionally, not all countries provide subsidies for non-business vehicles. A questionnaire conducted in the Nordic region (Iceland, Sweden, Denmark, Finland, and Norway) shows that users would prefer a tax relief to a financial subsidy when purchasing EVs [3] with the exception of Sweden, where the government has taken several reform steps since 2006 to reduce the tax burden [11], which included the abolition of the most criticized taxes on real estate and property, also including cars. Conversely, Sweden has significantly increased its fuel tax since 2014 to support CO_2 reductions and also increased the number of EVs in operation [11]. Other financial benefits that users will receive from EV, cf. [2,3], are free parking in restricted areas, free use of motorways, and an exemption from road tax if required by a particular country.

The second important issue for potential EV users is a short range of the EVs due to the small capacity of the installed batteries. The study [4] states that the average daily distance travelled by a user living in Beijing is 50 km. Users living in California travelled less than 48 km by car during the day [6]. Respondents from the Nordic countries were also concerned about the short range of the EV, but finally, they stated that most of them were satisfied with an EV that would be able to achieve roughly 200 to 300 km per charge [3]. Only a minority of respondents would decide to purchase an EV only if the capacity of its battery allowed them to achieve a distance of at least 500 km [3].

Taking into account all currently available battery electric vehicle (EV) models, the average installed battery capacity is 61 kWh [12], and the average energy consumption is roughly 19 kWh/100 km [13]. The real range that EVs are able to achieve with one battery charge is approximately 314 km [14]. As can be presumed, the capacity of an installed electric vehicle battery strongly correlates with its purchase price [15]. However, EVs with a low purchase price will achieve 50 km per battery charge without difficulty. Even a two hundred-kilometre journey with a single charge is not unrealistic for them.

The third important issue for many respondents is charging electric cars and the question of a dense network of charging stations [2,3,6,10,16]. The questions in the questionnaire which focused on charging efficiency, charging station locations and network density are in line with current scientific issues addressed in the literature [1,4,8,9,17–23]. The issue of electric car charging can be principally divided into two main subcategories: home charging and public charging stations.

Home charging stations paved a new way for EV users, who are thus able to recharge their cars very comfortably. One of the most significant benefits of home charging of electric vehicles is a lower price per kWh compared to a business fast-charging station [24], together with a higher degree of comfort and less time spent charging on public charging sites, cf. [3,4,24,25]. On the other hand, the disadvantage of the home charging solution is the necessary upfront investment in the home charger set and, surely, also an investment in one's own parking place or garage, cf. [2,4,26].

However, not all users (especially in urban areas) have their own garage or parking place [4]. A dense network of charging stations is often a good bonus for EV users and a quick way to recharge their car during daily travel [3], but generally, EV users prefer home charging [3,6,27]. A private parking place or garage that provide EV users with the ability of charging an electric car was very important for potential new EV users, cf. [2–4]. For instance, Zhao et al. [4] describe the ratio between the number of registered EVs and the number of private charging stations in Beijing. In 2017, a total of 124,000 EVs were registered, and 67% had their private charging stations. The remaining 33% of EVs were dependent on recharging using public charging stations. Most of the population in cities live in flats without a private garage or a private parking place [26]. These larger groups of users leave their cars in public car parks near their apartments. They do not usually have the possibility of buying a home charging set and using it to charge EVs [4]. Furthermore, in these densely populated areas, a problem arises not only with a lack of private parking places and garages but also with a lack of free public car parks, cf. [1,2,4,8,9,23,28]. The sharp increase in the number of cars and the expected increase in the number of EVs [29] in future means that there will be a lack of new parking places [9]. Very often, it is not possible to build new car parks in the required location at all, or it is not economically viable [30].

From a global perspective, the lack of parking places is not caused by a physical shortage of car parks, but by inefficient use of existing ones. In particular, corporate car parks, parking spaces near commercial buildings, post offices, hospitals, schools, shopping centres and banks are busy mainly during working hours, cf. [8,9,23]. Cai et al. [9] estimate that the percentage utilization of these parking spaces in Beijing during the day is approximately 50%. At night, the utilization decreases further to 40%.

Increasing the efficiency of use of existing car parks is therefore crucial. Several optimization methods have been used so far. Perhaps the most straightforward and now well-established method for increasing the efficiency of car parks is to reserve a particular parking place for the required time interval [30]. Friesen and Mingardo [30] focus on increasing the efficiency of car parks using a method based on revenue management and dynamic parking. This method adjusts the price of parking to the car park location, the time of day when a customer parks their vehicle, and the day of the week. Friesen and Mingardo [30] and Lei et. al. [31] came to a very similar conclusion in their studies: the dependence of users' demand for parking places depends on changing prices of parking.

Cai et al. [9] present the optimum method of sharing the same area from the perspective of car parks and users at administrative buildings. To ensure the same occupancy of individual car parks, they recommend approximately equal distances of car parks from points of interest. Better management of the occupancy of individual car parks can also be achieved by changing the price of parking. A problem may occur in residential car parks of office buildings with highly sought-after locations, where, even after the increase in the parking fees, there are still problems with lack of free parking places [9]. These car parks cannot, however, be shared with nonresidents. The authors further state that a choice of car park is negatively influenced by the parking fee, travel time, risk of no parking space, and general obstacles to parking [9].

Similarly, Huang et al. [8] agree that an effective way to solve the problem of the lack of parking spaces is not their new construction, but more efficient use of existing ones. The main research objective was to determine the optimum number of shared parking places for the car park assumed. Assuming that the demand for parking spaces is constant and greater than the number of parking spaces provided, they conclude that the optimum number of reserved places is approximately equal to half the proportion of long-term parking users. However, it can be assumed that the number of parking places offered for sharing will depend mainly on the car park location related to the demand for the given location.

Similarly to the lack of car parks, the lack of charging stations for EVs is perceived as a significant obstacle, cf. [2,4,23,28].Sharing of existing charging stations could eliminate the vicious circle between public charging station infrastructure developers and potential EV owners [2]. From the economic point of view, the development of a dense network of public charging stations depends on increasing numbers of EVs in operation, and new EV users are often discouraged by small numbers of charging stations [2]. The incentive not only for the investor to build a dense network of charging stations is supported by the European Commission which, supports the Europe-wide electromobility initiative and Green eMotion projects with an approximate amount of EUR 41.8 million [32].

The use of existing private charging stations is inefficient, cf. [4]. EV owners usually use private charging stations in the evening and the station remains used inefficiently during the day [4]. Therefore, optimizing the use of charging stations by sharing them results in benefits not only for the charging station owners but also for other users of electric cars who could then use the dense network of charging stations [3,6,18,19].

Plenter et al. [2] dealt with the development of "CrowdStrom," representing a peer-to-peer (P2P) sharing and collaborative transportation consumption service for private charging infrastructure by using information technologies. These authors specify three main factors that an individual must ensure to provide his charging point for other EV users. The first important factor is the owner's willingness to share their charging station with other users. The second necessary factor is to ensure free access to the charging station (which tends to be an issue, especially in private underground garages of condominiums, where only renters have primary access). The third, final factor is internet connection, important for sharing information about the station occupancy. In addition to an IT solution, the authors also asked residents about their willingness to participate in a charging station sharing project. From a total sample of 327 respondents, 45% indicated above-average interest in becoming a provider regularly offering private charging stations to the public. However, a more detailed analysis of the questionnaire data showed that only 450 shared charging points would be created in a given locality with 300,000 inhabitants. The estimate was based on the answers of respondents who had the opportunity to build a parking place and add a charging station. Additionally, with above-average interest, they had to agree with purchasing an EV and providing the charging station for other EV users.

The results of the questionnaire [2] also show that people have a better chance of sharing private charging points in suburban areas because they are more likely to have their own parking place and to equip it with a charger. Zhao et al. [4] had a similar idea of sharing private charging stations with other users. The analysis found out that the utilization rate of private charging stations in Beijing was 4.5%. By offering these private charging stations to other users, their utilization rate has increased by 0.4%. The main limitation to the study by Zhao et al. is that the charging station owners provide their charger for other EV users during the daytime, but not in the evening, when the owner occupies the charging station. We take the opposite approach. Sharing of parking places in a P + R car park, which are unused at night-time, correlates with EV users' demand for charging in the evening, when they usually return home from work.

Xu et al. [23] approach sharing of parking places and charging stations from a different angle than the two abovementioned studies. They consider sharing parking places primarily in a car park in a commercial building. The pairing of parking place owners and customers then proceeds as follows. Each parking place owner sets a minimum price of renting in the time interval when they do not use it and each customer suggests the maximum price they are willing to pay for parking. An important factor of pairing is that the parking place owner is assigned a customer who offered the same or higher price of renting the parking place. Customers' fees consist of renting a parking space and the electricity consumed. However, the customers can get a financial benefit if they opt for demand-side management (DSM). The DSM method is used to optimize the parking costs by modifying EV charging/discharging behaviour. DSM aims at encouraging consumers to use less energy during peak hours, or to shift their energy use to off-peak times such as night-time and weekends [23].

Many studies [2–4] state that owning a private parking place is often an important condition for deciding to purchase an EV. In urban areas, where there is a limited number of public parking places, the possibility of sharing offers itself directly. Car parks near commercial buildings, post offices, hospitals, schools, and shopping malls are often the best possible solution for sharing, as the studies show. Especially beyond working hours, these car parks are used with low efficiency.

The research gap to which this article responds is effective use of parking spaces near commercial buildings, shopping malls, hospitals, etc. [2–4]. These car parks will be equipped with a sufficient number of chargers due to the expected increase in EVs, resulting in a high concentration of cars in the working hours when these institutions are most busy [2]. Furthermore, these car parks are very often located near settlements. Therefore, it offers itself directly to provide these car parks equipped with a charger for EV owners living close by. Rental of these charging points then offers private charging points for EV users without a garage or home charger.

Furthermore, our article solves the problem of the absence of a private parking place equipped with a charger, which often dissuades potential customers from buying an electric car [2]. This research presents a P + R car park model equipped with a sufficient number of chargers. It is assumed that P + R car parks are used mainly during the day and at least in the evening. Plenty of available charging points in the evening also correlates with EV users' demand for charging in the evening, when they usually return home from work. Thus, the proposed solution directly solves the problem of many potential EV users who do not have their own private charging points, a factor that often discourages them from buying an EV [2–4].

3. Model of EV Charging in Shared Parking Places

EV users come mainly from urban areas with a high proportion of apartment buildings, so they often face not only a lack of places for charging, but also places for private parking. Thus far, many charging models have focused on home and public charging, and similarly, there have already been many general ways of sharing parking places [2,4,8,23,28]. The proposed model combines the above approaches and models by building charging infrastructure in an existing car park and introduces the concept of sharing it. In order to increase the efficiency of use, the model contains parking places and charging infrastructure not only for residents living around but also for those randomly arriving in the daytime.

The model aims to design a technically and economically optimal solution for additional installation of 3.6 kW AC chargers and other necessary energy equipment in an existing car park. The model focuses on sharing parking places from the car park operator's perspective, trying to offer comfort and service for EV users comparable to home charging. Simultaneously, the model aims to guarantee low-power night-time charging for users' EVs, which will sufficiently compete with public fast-charging stations and achieve low-priced home charging.

3.1. User Description

A key issue in setting up a model of a car park equipped with chargers is to dimension and adapt its equipment to the requirements of individual groups of electric car users. Generally, car park users can be divided into two main groups. The first group consists of randomly arriving EV users who commute to the site for work or other purposes. The other group comprises users (residents) who live close to the car park and rent a parking place primarily for charging. Residents are offered a level of comfort similar to that of people living in family houses and having their own garage equipped with a charger.

Our model assumes that the first group of EV users (nonresidents) use the car park primarily in the daytime. The purpose of daytime parking might be the vicinity of a shopping centre or means of public transport. Therefore, nonresident users' priority is parking rather than charging. Although the price is close to home charging, charging is not mandatory for nonresidents. Before parking, each nonresident user chooses whether they want to charge the EV. After leaving the parking place, the user pays not only an appropriate hourly parking rate but also the energy bill. Since the car park is not primarily intended as a public charging site, nonresidents are not guaranteed recharging during their whole parking time. Parking only increases the energy level in the EV; the problem of full recharging is solved by home charging or other means.

For the second group (residents), the car park model offers the possibility of renting parking places with a charger. The parking places are reserved for the residents on weekdays during the night-time. At weekends, resident parking places are also reserved for the residents throughout the day. If the residents arrive no later than at the set evening time (T_l), they are guaranteed a sufficient number of charging hours correlated with recharging an appropriate amount of energy. However, the rules for the resident group require different settings. It is assumed that the residents want to rent a parking place mainly due to the absence of private parking and charging places.

Compared to nonresidents, residents are financially motivated to use car parks to charge their EVs. The principle of incentives we propose consists in the introduction of a minimum flat rate for residents including guaranteed charging hours, regardless of the actual use. Charging after the guaranteed hours is then calculated per kWh. The flat rate will ensure a reduction in the risk of non-covered fixed costs (NCFC) due to a blocking car with little need for charging. Another source of risk is mutual blocking of users; this is addressed in the following chapter.

3.2. Blocking Issues

To completely prevent charger blocking, additional installation of chargers in each parking place is assumed. The problem of sharing parking places between groups is addressed by resident membership booking for a specific day. Here, a resident does not have the option to rent a particular parking place but can book membership in a parking group. Since overlaps can occur in the case of a nonresident's or a resident's delayed departure, the calculation takes into account an overlap around the morning (X) and evening (Y) hours, when the overlaps are the most frequent. Above all, the model takes into account a capacity reserve (Q_r).

$$Q_r = Q_{pl} - Q \tag{1}$$

where Q_{pl} is the total number of parking places in a car park and Q is the number of parking places reserved for residents.

For a resident who decides not to leave, the resident daytime tariff is charged. In the case of a resident staying in the daytime, the minimum tariff rate ($R_{resident}$) is set as the product of the NCFC per unsold kWh ($L_{p\_kWh\_nonresident}$), the average occupancy of the charger (T_o), and the charger (P_{CH}).

$$R_{resident} = L_{p\_kWh\_nonresident} \cdot T_o \cdot P_{CH} \tag{2}$$

$$L_{p\_kWh\_nonresident} = LCOE_{nonresident} - N_{var} \tag{3}$$

where $LCOE_{nonresident}$ is the levelized costs of electricity for the nonresident group, and N_{var} is the marginal price per kWh.

For a nonresident who does not leave until the last daytime parking hour, the night-time nonresident tariff is charged. In the case of a nonresident staying overnight, the minimum tariff rate ($R_{resident}$) is set as the product of the NCFC per unsold kWh ($L_{p\_kWh\_resident}$), the guaranteed resident charging hours (T_g), and the power of one charger (P_{CH}).

$$R_{nonresident} = L_{p\_kWh\_resident} \cdot T_g \cdot P_{CH} \tag{4}$$

$$L_{p\_kWh\_resident} = LCOE_{resident} - N_{var} \tag{5}$$

where $LCOE_{resident}$ is the levelized costs of electricity for the resident group, and N_{var} is the marginal price per kWh.

3.3. Charging Scheme

Nowadays, building power infrastructure with a minimum charging output 50 kW DC is most common if short charging times are to be achieved. Furthermore, construction of charging infrastructure requires installation of either more local transformer stations or higher-power stations. It is also necessary to increase the capacity of the high voltage (HV) connection and increase the reserved power. The time of utilization of the installed capacity is short in this case. From the economic point of view, especially the costs associated with the increase in the reserved power from the HV power grid are not viable.

Contrarily, construction of charging infrastructure with low-power charging stations (usually up to 3.6 to 10 kW) avoids these issues. The time of utilization of the installed capacity will increase, and the smoothness of consumption ensures predictability for purchasing electricity from the trader. Installing a low-power charger in each parking place (Figure 1) then appears to be a more advantageous option for this model.

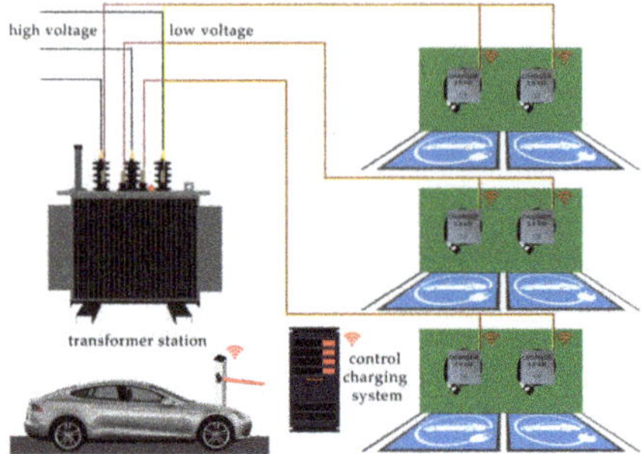

Figure 1. Car park model equipped with low-power chargers.

3.4. Model Input Parameters

The input parameters of the car park model equipped with low-power chargers are as follows:

- Maximum number of parking places in the car park assumed (Q_{pl});
- EV arrivals on weekdays with the highest demand for parking;
- Peak parking occupancy at the weekends (Q);
- Data describing the demand for electricity by arriving EV users;
- Maximum number of residential parking places (Q_r).

The model of charging infrastructure for an existing car park consists of a HV/LV transformer station, switch disconnectors and isolators, a circuit breaker, 16A fuses, cabling, a heat exchanger (cooler), a charging control system, and chargers.

3.5. Dimensioning the Transformer Station Power

The charging issue is always related to the maximum load of one or more locally installed transformers. The minimum power (P) of a transformer station is based on the total requirement for recharging the residents' EVs (E_r). Residents' recharging requirements are defined by the probability density function (PDF) of their daily mileages ($F(T_g)$). Based on the charger power of 3.6 kW AC and the given percentile of charging success (p), a minimum charging time (T_g) obtained should be guaranteed for residents. Simultaneously, information about the total energy needed for recharging (E_r) is obtained from the PDF of resident daily mileages.

To verify whether power of a group of installed transformers is sufficient, the model compares energy needed to charge the residents' EV (E_r) and the energy that the local transformers can provide in the time interval reserved for residents (E_{TR_YX}).

$$E_{TR\_YX} \geq E_r \tag{6}$$

$$E_r = \sum_{x=1}^{Q_r} E_r(x) \tag{7}$$

$$E_{TR\_YX} = P_{ch\_par} \cdot (Y - X)\,[\text{kWh}] \tag{8}$$

$$P_{ch\_par} = ch_{par} \cdot P_{CH}\,[\text{kW}] \tag{9}$$

$$ch_{par} \leq \frac{P}{P_{CH}} \tag{10}$$

where P_{ch_par} is the power consumption of the entire array of chargers used simultaneously ch_{par}. P_{CH} is the power of one charger.

The optimal number of local transformers should be the subject of economic analysis. The economic assessment also includes a comparison of NCFC due to electricity unsold to nonresidents, and annualized costs of building an additional transformer. If the NCFC exceed the annualized costs, it is advantageous to add another transformer to the transformer station.

3.6. Charging Power Sharing Method

An important issue in terms of charging is the distribution of the transformer station power (P) among the individual EVs. Compared to residents, nonresidents arrive in the car park randomly during the time span from (X) to (Y).

The occupancy curve strongly depends on the type of car park. Therefore, the estimate of the number of arriving nonresidents in a given period must result from long-term observation. The PDF of the length of stay (L) and the PDF of the quantity of arriving nonresidents in the given time intervals (A) are calculated from the observed data. As for the residents, we obtain information about the total energy needed to recharge their EVs (E_r) from the PDF of resident daily mileages.

The main difference in the division of the power of a local transformer station among the nonresident group is that nonresidents do not have a guaranteed energy supply (Figure 2). Therefore, the power of a local transformer station is divided among the EV being charged. The number of EVs being charged ($ch_{par}(t)$) and the number of new charging requests are monitored in the given time steps (t). The limiting factor for accepting a charging demand is the remaining unused power of the transformer stations ($P-P_{ch\_par}(t) > 0$). In a situation where there are several requests for charging ($N_w(t) > 0$) and recharging of all EVs is impossible due to transformer station overload ($P_{ch\_par}(t) \geq P$), the first-in, first-out (FIFO) basis is used. The charging of waiting EVs is postponed until they are guaranteed continuous charging. The EVs are then charged until either they have fully charged batteries, or their owners voluntarily interrupt charging and depart.

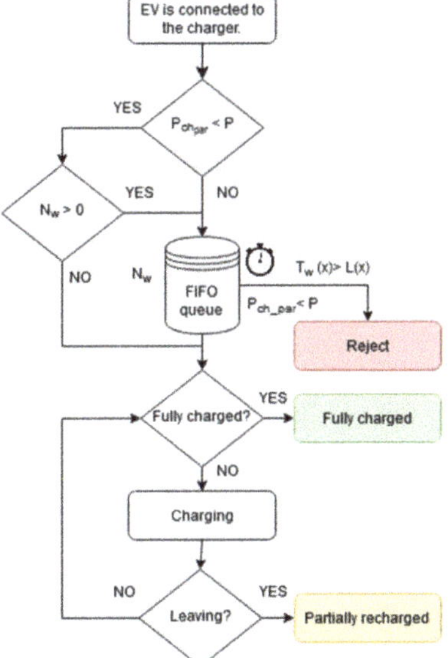

Figure 2. Nonresident charging diagram.

Contrarily, the residents have guaranteed (T_g) charging hours (Figure 3) as part of the parking rate, providing that they arrive between Y and the last morning hour (T_l). If the transformer station is not fully used and the battery of a resident's EV is not fully charged even after the guaranteed (T_g) hours, the battery is charged further. An interruption in the charging of this car will occur at the moment of the full battery capacity or on the arrival of another resident who has not exhausted the charging hours guaranteed and if the simultaneous charging of both EVs would exceed the maximum power of the transformer station.

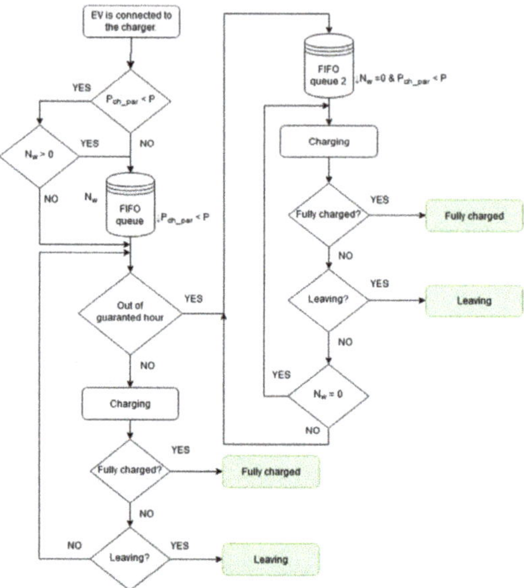

Figure 3. Resident charging diagram.

3.7. Economics of System Dimensioning

Apart from technical aspects, economic factors need to be taken into account when constructing a charging point. One of the key issues is the competitiveness of the project compared to other charging methods, such as using public fast-charging stations and home charging using a private charger. We assume that the main priority of the model is to create a sufficient number of shared parking places for residents and comfort and charging costs similar to home charging. Therefore, home charging is the main competitor for the car park with the charging infrastructure.

To assess the economic effectiveness, the model uses the levelized costs of electricity (LCOE) to recalculate individual capital expenditures (CAPEX) with different lifetimes and operating expenses (OPEX) converted to a unit price per kWh. CAPEX involve a HV/LV transformer station, switch disconnectors and isolators, circuit breaker, fuses, cabling, heat exchanger (cooler), installation material, chargers, connection fee, installation work, and charging control system. OPEX are generally divided into two items, the first one consisting of maintenance and the second one associated with electricity consumption.

Since the transformer station causes the highest losses, the empirical formula given by CSN 341610 standard is used for the estimation of annual losses in the transformer station.

$$T_L = \left[0.2 \frac{T_m}{T_0} + 0.8 \left(\frac{T_m}{T_0} \right)^2 \right] \cdot T_0 \tag{11}$$

where T_L is the annual time of full transformer losses, T_m is the utility factor, and T_0 is the annual operating time of the transformer station.

$$W_L = P_{nl} \, T_0 + P_{kn} \frac{S_m^2}{S_n^2} \, T_L \tag{12}$$

where W_L is the annual energy losses in the transformer, P_{nl} is the power transformer no-load loss, T_0 is the annual operating time of the transformer station, P_{kn} is the short-circuit transformer losses, S_m is the maximum apparent-power load of the transformer, S_n is the apparent power of the transformer station, and T_L is the time equivalent of losses at the maximum load of the transformer per year.

From a technical point of view, the power of one transformer installed in a transformer station may be sufficient for the full charging of residents' EVs, but the percentage of refused nonresidents and the associated unsold electricity can generate significant NCFC. A comparison of these NCFC with the annuity costs associated with an increase in the power of the local transformer station determines whether the increase is economically beneficial or not. The economics of system dimensioning aim at determining the levelized costs of electricity (LCOE) for both the nonresident and resident groups.

$$LCOE = \frac{sum\ of\ costs\ over\ lifetime}{sum\ of\ electrical\ energy\ sold} = \frac{N_i + \sum_{t=1}^{n}(M_t + N_{Et})(1+r)^{-t}}{\sum_{t=1}^{n} E_t(1+r)^{-t}} \tag{13}$$

where the variable N_i denotes the initial investment in the equipment; M is the annual maintenance expenditures, N_E is the electricity bill in the year t, r is the discount rate, and E denotes electricity sold in the year t [kWh].

The optimization method is based on economic cost evaluation. This method compares the NCFC due to electricity unsold to nonresidents and annualized costs of building an additional transformer. The optimization minimizes the number of refused nonresidents via installation of additional transformers. The algorithm applies a constraint that the overall costs of charging, including NCFC and annualized costs of building of another transformer, do not exceed 0.19 EUR/kWh (for both residents and nonresidents).

4. Case Study

The following chapter deals with an application of the model to real data. The chapter consists of four subchapters. The first and second present and analyse the real data entering the model. They are followed by technical evaluation of the case study data (third subchapter) and economic evaluation (fourth subchapter).

4.1. Parking Place Input Data Analysis

Data from a park and ride (P + R) car park with 650 parking places in Prague Letňany were selected for the survey [33]. The data capture the car park occupancy in five-minute intervals from the 1st quarter of 2013 to the 4th quarter of 2016. The car park occupancy data do not include data for holidays and public holidays, which would alter the occupancy information for the given days significantly. The parking data do not include any private residents.

For further processing, the car park occupancy information for individual five-minute intervals was broken down into weekdays (from Monday to Friday) and weekends (Saturday and Sunday). The average occupation level in the corresponding five-minute interval was calculated for each day based on this four-year study.

Subsequently, an extensive analysis of the data was performed in order to identify significant differences in the car park occupancy between individual weekdays (Monday to Friday), and weekdays and weekends. The analysis of weekdays shows that the user behaviour is almost identical for Monday, Tuesday, Wednesday and Thursday. On Friday, the parking places were occupied significantly less. Considering that the occupancy in the individual five-minute intervals has a normal distribution with a 64% probability, the difference in the occupancy in the time interval from 7 am to 7 pm on Monday, Tuesday, Wednesday, and Thursday is less than 3% (Figure 4).

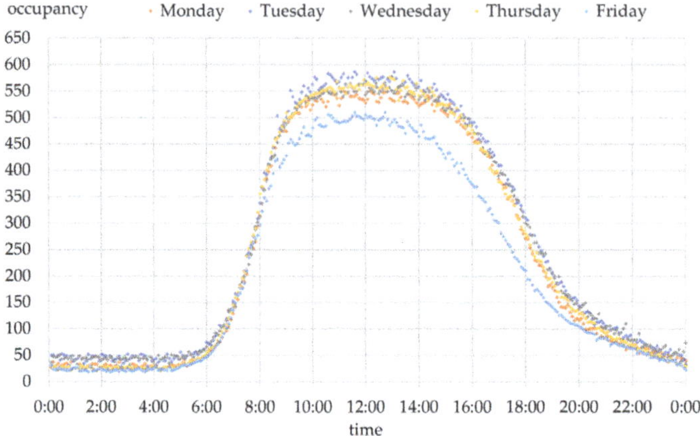

Figure 4. Occupancy on weekdays ([33], calculated by authors).

Moreover, an analysis was performed to identify changes in user behaviour depending on the quarter of the year. The analysis shows that the user behaviour remains almost unchanged throughout the year. For example, for Thursday, considering that the occupancy values in each five-minute interval have a normal distribution with a probability equal to 64%, the difference in occupancy from 7 am to 7 pm in the first, second, third and fourth quarters is less than 6% (Figure 5).

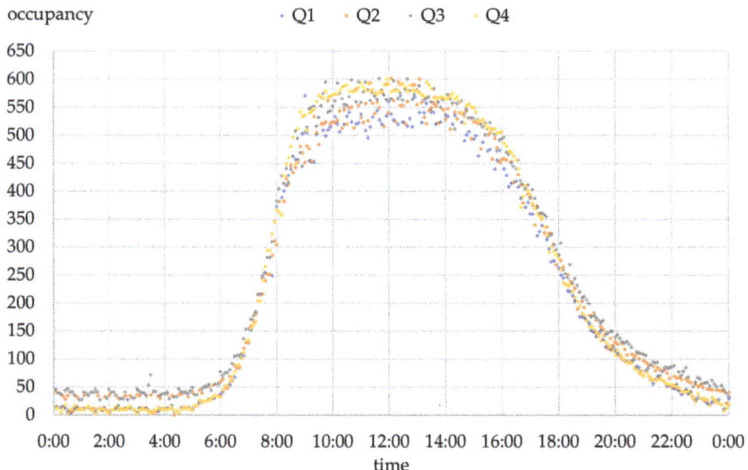

Figure 5. Thursday arrivals/departures ([33], calculated by authors).

Since the user behaviour in individual quarters and on weekdays, except Friday, was nearly the same, we selected the Thursday occupancy for further modelling. In terms of parking place occupancy, Thursday parking place occupancy has the character of the busiest weekday. Furthermore, the weekend car park occupancy analysis shows that the user behaviour on both Saturday and Sunday is the same in all the quarters. The average occupancy on these weekend days is constant, equal to 100 parking places.

The disadvantage of the provided data describing only the car park occupancy is the loss of information on arrivals and departures within the measured five-minute interval; for example, arrivals and departures of the same number of cars in the same five-minute interval show a zero change in occupancy. Therefore, the Poisson distribution using random arrivals of individual cars was used to approximate the real car park occupancy. The departure time, which correlates with the parking time for a car in a parking place, was modelled using a Gaussian distribution with a 4 h mean and a standard deviation of 1 h.

Data obtained from charging stations operated by ŠKO-ENERGO (Mladá Boleslav, Czech Republic) were used to simulate demand for recharging of electric vehicles [34]. During the three-month testing in 2020, data on 330 charging iterations from various types of EVs were obtained. The charging data were obtained from seven charging points, each equipped with a 22 kW DC power charger. At four of these seven charging points, it was also possible to use a 50 kW DC charger [34]. A more in-depth analysis of the data provided information about the time spent in the car park and the time required for sufficient charging. The analysis shows that users left the EVs parked and connected to the charger for a longer time than required to recharge the battery. This can be demonstrated by the fact that it is possible to recharge the battery of a EV more with a lower-power charger over the same parking time. Figure 6 shows that users park for longer than is needed to recharge the battery to full capacity, thus unnecessarily blocking the possibility of recharging other electric cars.

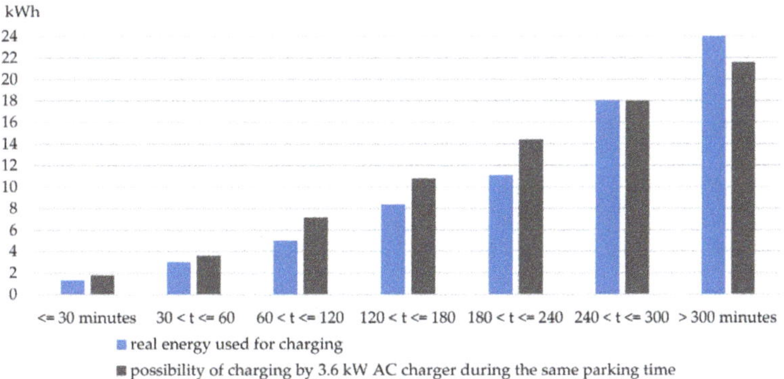

Figure 6. ŠKO-ENERGO charging stations ([34], calculated by authors).

Almost 90% of the car users analysed would have done with a shorter charging time to recharge the required energy or using a 3.6 kW AC charger to achieve the same recharging level in the same amount of parking time (Figure 6).

The EV charging sample also offers a valuable outline of the demand for electricity recharging. The data obtained from 330 charging iterations were sorted based on the energy consumed and subsequently used to calculate the cumulative distribution function (CDF) (Figure 7). The most frequent group (16%) are EVs with an 11 kWh charging requirement. The charging requirements show that 90% of arriving EVs require less than 18 kWh to recharge, which corresponds to approximately 5 h of recharging with a 3.6 kWh charger (Figure 7).

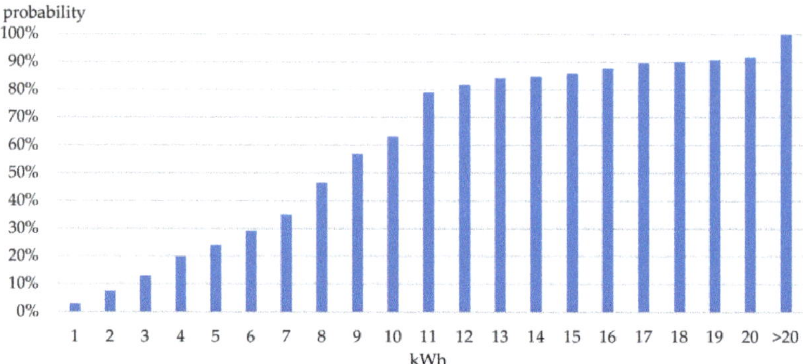

Figure 7. Cumulative distribution function of charging requirements [kWh] ([34], calculated by authors).

4.2. Users Input Data Analysis

The case study considers integration of charging infrastructure into an existing P + R car park with 650 parking places (Q_{pl}) in Prague Letňany. The real data analysis revealed that the average weekend occupancy of the car park (Q) is 100 parking places. Therefore, 550 parking places (Q_r) out of the total parking capacity of 650 (Q_{pl}) can be reserved for residents. A more detailed analysis of the input data describing the car park occupancy set the residents' interval from $X = 7$ am to $Y = 7$ pm. The input data vector describing the one-day car park occupancy is composed of 288 five-minute intervals. The number of predicted nonresidents who arrived in a given five-minute period ($A(t)$) is estimated using the normal distribution function. The mean value and standard deviation needed to calculate the normal distribution are achieved from the five-minute intervals, which correspond with both the day of the week and the placement of the five-minute period on a given day. (For example, the calculation concerning all the first five-minute periods obtained from each Monday.) Similarly, the time spent by a nonresident in the parking place is estimated using the normal distribution function, where the mean value equals to 4 h and the standard deviation is 1 h.

The Poisson distribution was used to predict the demand for charging nonresidents ($E_n(x)$), where λ corresponds to the mean value of the data obtained from the charging stations operated by ŠKO-ENERGO.

The requirements for charging are described as follows:

$$\hat{A}(t) \sim N(\mu_A(t), \sigma_A(t)) \tag{14}$$

$$L(x) \sim N(\mu_L, \sigma_L); \mu_L = 4\,\text{h}, \sigma_L = 1\,\text{h} \tag{15}$$

$$E_n(x) \sim \text{Poisson}(\lambda); \lambda = 8\,\text{kWh} \tag{16}$$

where $\hat{A}$ denotes a vector with the distribution of arrivals of nonresidents in five-minute intervals during the day expressed by the average $\mu_A(t)$, and the standard deviation $\sigma_A(t)$. The variable L is the length of nonresidential stay with a normal distribution with the average μ_L and the standard deviation σ_L. The last variable denotes a recharging requirement approximation by the Poisson distribution.

The case study for the resident group models the charging demand only. The charging demand is based on the study [35] focusing on the PDF of daily distance driven by car (Figure 8). The assumed demand for charging characterizes the residents' behaviour. They primarily charge their EV once a day after returning home, similar to most users with their own charger and a garage. The arrival curve and the time spent by the residents in the car park have not been accounted for. In the case study, we assume that a sufficient number of residents will arrive before the set time (T_l) required for a guaranteed charge.

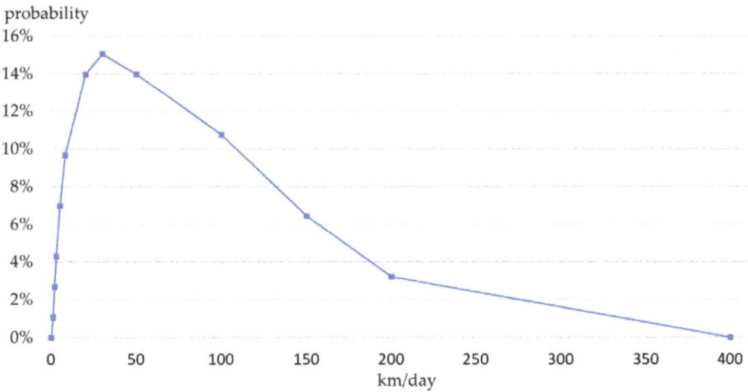

Figure 8. Probability density function of daily distance driven by car [35].

By dividing the daily distance driven by car and the average EV consumption by 100 km (19 kWh), we obtain information about the required energy charge. In order to charge energy, it is also necessary to take into account the compensation for electrical losses that occur during charging. A test of power consumption in electric cars [36] shows that EVs consume between 10 and 25 per cent more electricity than their onboard computers show. However, it can be assumed that charging with a low-power 3.6 kW AC charger results in the losses being closer to the 10% value. By dividing the charging power of the charger and the required charge energy ($E_r(x)$), including losses during charging, we obtain information about the required charging time ($T_r(x)$) (Figure 9).

$$T_r(x) = \frac{E_r(x)}{3.6\text{kW}} \tag{17}$$

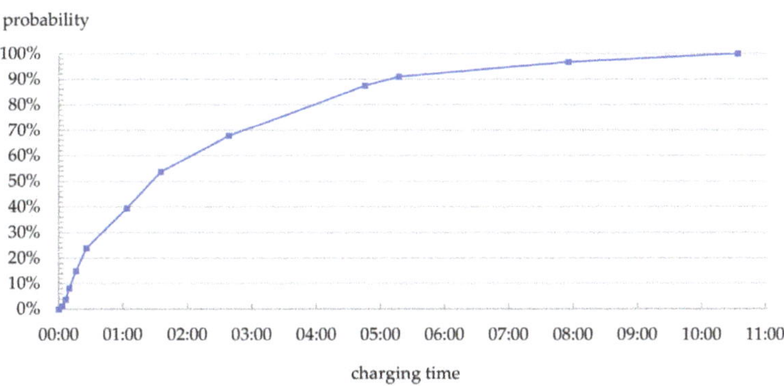

Figure 9. Cumulative distribution function of the resident's EV full charge.

Both the number of guaranteed charging hours (T_g) and the limit that residents must meet to obtain guaranteed charging hours (T_l) were determined based on the time needed to charge by 95% of the residents.

$$AF\big(T_g\big) = p \; ; \; p = 95\% \tag{18}$$

$$T_g = F - 1(95) \cong 6\,\text{h} \tag{19}$$

$$T_l \leq X - T_g; \; X = 7\,\text{h}, \; T_g = 6\,\text{h} \tag{20}$$

$$T_l \leq 1:00\,\text{AM} \tag{21}$$

4.3. Technical Aspects of System Dimensioning

Local transformer station capacity will be first dimensioned to meet the residents' requirements, even though we assume higher utilization of the car park and the entire charging infrastructure by the nonresident group. From the existing series of transformers, it is possible to choose any combination of transformers with different power ratings to ensure the required power output. However, a transformer with an apparent power of 630 kVA is selected for installation in the case study.

The total requirement of recharging 550 resident EVs (Q_r) obtained from the PDF of the daily distance driven by car and average EV consumption corresponds to 5303 kWh (E_r). Within the 12 h reserved for residents ($Y - X$), the assumed local transformer with an apparent power of 630 kVA can supply (E_{TR_XY}) 7182 kWh, and thus fully meets the residents' charging requirements.

$$P_n = S_n \cdot \cos\varphi \tag{22}$$

$$P_n = 630 \cdot 0.95 = 598.5 \tag{23}$$

$$E_{TR\_YX} = P_n \cdot (Y - X)[\text{kWh}] \tag{24}$$

$$E_{TR\_YX} = 598.5 \cdot (19 - 7) = 7182 \, [\text{kWh}] \tag{25}$$

$$E_{TR_{YX}} \geq E_r \tag{26}$$

$$E_{TR_{YX}} \geq Q_r \cdot P_{CH} \cdot T_g \tag{27}$$

$$3.6 \geq T_g \tag{28}$$

$$ch_{par} \leq \frac{P}{P_{CH}} \tag{29}$$

$$ch_{par} \leq \frac{598.5}{3.6} \approx 160 \tag{30}$$

Based on the assumed model parameters, each parking place is equipped with a 3.6 kW AC single-phase charger. From the load optimization point of view, the car park is divided into three parts. Each is connected to one of the three phases of the transformer. The first and second phases of the transformer are connected with 217 chargers. The third phase then provides power to 216 charging stations. Due to the maximum load of one transformer, it will never charge more than 160 EVs at the same time (ch_{par}). Even though one transformer fully satisfies the requirements for recharging 550 residents, they cannot be guaranteed the required 6 h of charging. With one transformer, it is possible to guarantee each resident 3.6 h of charging. However, 3.6 h of guaranteed charging led to a full charge for only 75% of residents (Figure 9).

Besides, the car park control system, which directs arriving users to the parking places to ensure optimal loading of the entire charging system, helps to optimize even loading of both the transformer stations and their phases.

In the following part of the case study, we will focus on the car park operation in the interval reserved for nonresidents. Their behaviour characterized by charging requirements, arrival times, and parking times is described in the last part of the case study focusing on user' characteristics.

In the case study, we will assume that all nonresidents require both parking and charging of electric cars. Several restrictions are associated with the power of the local transformer station (P), determined based on residents' charging requirements. One of them is the maximum number of chargers used simultaneously (ch_{par}). This is also the reason why it is necessary to postpone some of the nonresidents' charging in very busy intervals (Figure 10).

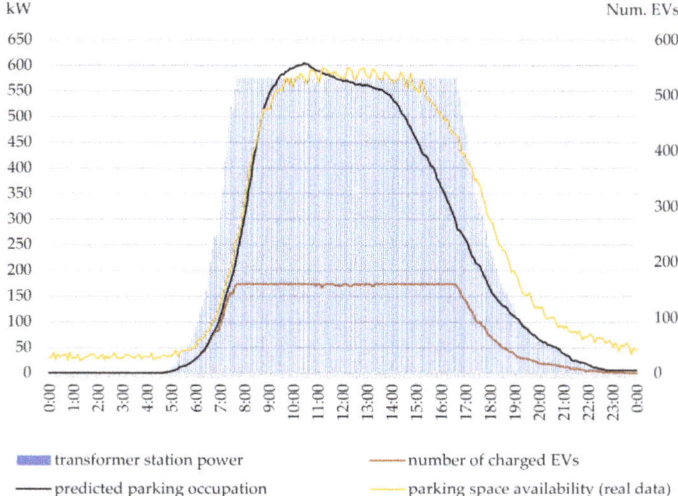

Figure 10. Modelled occupancy in the case of using one transformer.

Nonresidents who have been refused charging wait until one of the previously connected nonresident EVs is fully charged or has finished charging and leaves. The FIFO principle is used for the choice of the next EV for charging from the group of waiting nonresidents, where the longest-waiting nonresident is chosen for charging preferentially (Figure 2).

From the charging point of view, we divide nonresidents into three parts. The first part is composed of nonresidents who were able to recharge throughout the parking time (fully charged). The second part comprises nonresidents whose recharging was shortened due to a shift in the start of charging (partially charged). The third part includes nonresidents whose charging did not start due to a shift in the start of charging (refused).

In the case study, 20 simulations were performed. The number of nonresidents did not change between individual simulations, but their behaviour (charging requirements, arrival times and parking times) did. The results of the simulations showed that on average, 25% of nonresidents were refused, 15% were partially charged, and 60% were fully charged (Figure 11). The course of one of the simulations with one transformer is shown in Figure 10, presenting the load of one transformer and the occupancy of the car park compared to the measured real occupancy.

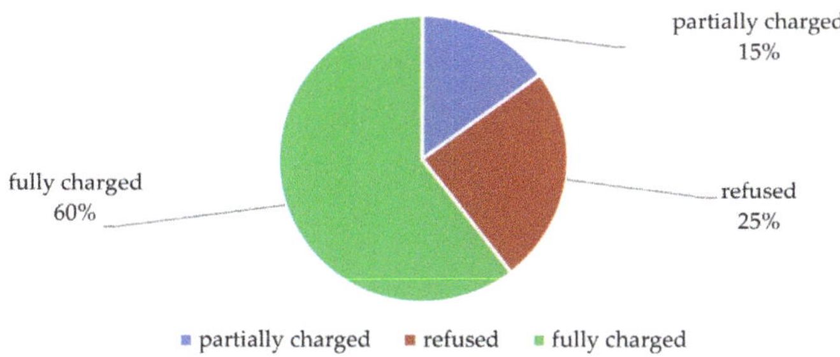

Figure 11. The success of recharging nonresidents in case of using one transformer.

Subsequently, the same set of 20 simulations was performed, but with a difference in the local transformer station power. The existing transformer was supplemented with another one with identical parameters and apparent power in these simulations. This pair of local transformers had a combined power of 1197 kW, enabling the simultaneous charging of 320 EVs. The part of nonresidents with refused charging was eliminated due to the increased local transformer station power (Figure 12).

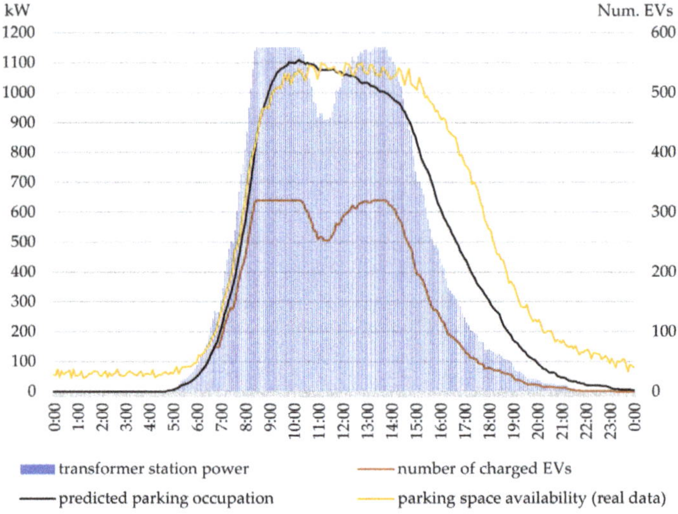

Figure 12. Modelled occupancy in case of using two transformers.

The pair of local transformers reduced the percentage of partially charged nonresidents to 4% and eliminated the part of nonresidents with refused charging (Figure 13). The course of one of the simulations with two transformers is shown in Figure 12, which presents the load of two transformers and the occupancy of the car park compared to the measured real occupancy.

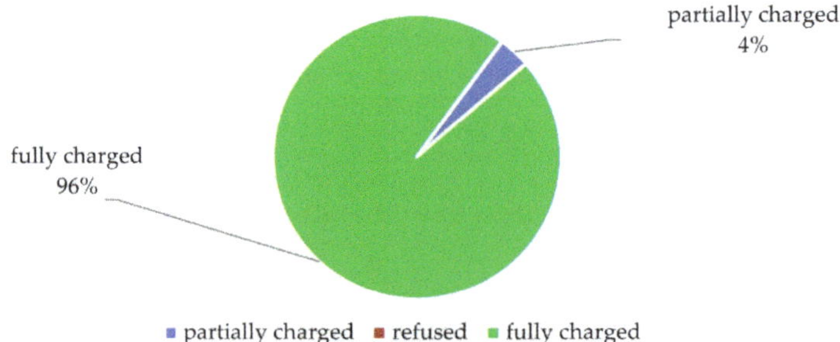

Figure 13. Success of recharging nonresidents in the case of using two transformers.

For the resident group, increasing the output of the local transformer station has the benefit of guaranteeing 6 h of charging (T_g) and a 95% success rate of full charging for residents.

4.4. Economic Analysis of System Dimensioning

From an economic point of view, it is essential to set the business model parameters so that the costs associated with parking and charging can be competitive with home charging. The key issue is to correctly set the ratio of the CAPEX division between nonresidents and residents. The case study presents the range of recharging costs depending on how CAPEX are distributed among users (Table 1).

Table 1. Investment costs.

	TLT	q	Unit Price	Total Costs of 1 TS	Costs of Additional TS	Annuity for 1 TS	Annuity for Additional TS
	Year	-	EUR	EUR	EUR	EUR	EUR
TS-22kV/0.4kV	35	1	13,208	13,208	13,208	3303	3303
Disconnector and isolator	35	1	566	566	566	142	142
Circuit breaker BH400	35	1	453	453	453	113	113
Fuses PM45 16A	35	3	74	221	221	55	55
Cabling [m]	50	2150	9	19,877	0	4969	0
Heat exchanger (cooler)	35	1	102	102	102	25	25
Pipes for cooling	35	1	302	302	302	75	75
Installation material	35	1	377	377	377	94	94
Charging control system	35	1	37,736	37,736	0	9438	0
Charging station	10	650	755	490,566	0	137,394	0
Connection fee	35	1	6038	6038	6038	1510	1510
Labour	35	1	45,283	45,283	45,283	11,325	11,325

Exchange rate: 1 EUR = 26.5 CZK.

The costs associated with electricity consumption (OPEX) are slightly different across countries. Since the car park is located in the Czech Republic, the price of electricity consists of a fixed and a variable part. The fixed part includes the fee for reserved power, renewable energy sources (RES) and the market operator, while the variable part, depending on electricity consumed, includes the price of energy, distribution services, system services and the electricity tax. The modelled annual electricity consumption and actual prices are shown in Table 2.

Table 2. Annual electricity bill [37,38].

	Unit	Unit Costs (Incl. VAT)	Total Annual Costs	Additional TS Annuity Costs
Energy	EUR/MWh	75.34	169,944	-
Distribution services	EUR/MWh	3.37	7604	-
Reserved power	EUR/MW/month	3856	27,762	27,762
System services	EUR/MWh	3.52	7943	-
RES fee	EUR/MW/month	2633	18,961	18,961
Electricity tax	EUR/MWh	1.29	2915	-
Market operator fee	EUR/month	0.23	0.23	-
Additional sale 2nd TS	EUR/MWh	83.52	-	54,800

TS = transformer station; VAT = value added tax 21%; exchange rate: 1 EUR = 26.5 CZK.

Table 3 shows the fixed operating costs related to transformer maintenance and overhauls and the annual energy losses incurred.

Table 3. Fixed operating costs.

	Costs Incl. VAT [EUR]	Annualized Costs [EUR]	Additional TS Annual Costs [EUR]
Annual inspection	457	457	457
Major inspection every five years	2283	849	849
Annual energy losses in TS (WL)	732	732	732
Annual calibration of chargers	4566	4566	-

TS = transformer station; VAT = value added tax 21%; exchange rate: 1 EUR = 26.5 CZK.

Since the transformer station causes most losses, the empirical formula given by CSN 341610 standard is used for the estimation of annual losses in the transformer station:

$$T_L = \left[0.2\frac{T_m}{T_0} + 0.8\left(\frac{T_m}{T_0}\right)^2\right] \cdot T_0 = \left[0.2\frac{1958}{8760} + 0.8\left(\frac{1958}{8760}\right)^2\right] \cdot 8760 = 741 \text{ [hours]} \tag{31}$$

$$W_L = P_{nl}T_0 + P_{kn}\frac{S_m^2}{S_n^2}T_L = 0.6 \cdot 8760 + 6.5 \cdot \frac{606^2}{630^2} \cdot 741 = 9721 \text{ [kWh]} \tag{32}$$

Although the technical part shows that one transformer provides sufficient power for residents but only with a 75% success rate of full charging for residents, the percentage of refused nonresidents creates significant NCFC. Therefore, the previous tables also contain additional annualized costs of deploying one more transformer. The sum of these annualized costs indicates whether it is advantageous to deploy a second transformer. The economic part aims to determine the levelized costs of electricity (LCOE) for both the nonresident and resident groups.

$$LCOE = \frac{N_i + \sum_{t=1}^{n}(M_t + N_{Et})(1+r)^{-t}}{\sum_{t=1}^{n} E_t(1+r)^{-t}} \approx \frac{C_A + M + F_t}{E_A} \tag{33}$$

Here, the variable N_i denotes the initial investment in the equipment; M denotes the annual maintenance expenditures, and N_E is the electricity bill in the year t. Providing that the values are not changed year-on-year, the variables C_A and F_t can be set. C_A indicates the annualized costs for the corresponding discount rate r, and F_t denotes the annual electricity bill. E represents annual electricity sold (kWh).

5. Results

The following chapter presents the results of the model applied to real data obtained from the Letňany car park, Prague, and the users' charging needs from the ŠKO-ENERGO charging station data. The values of levelized costs of electricity showing the competitiveness of charging points are presented.

5.1. LCOE of Charging Tariff Rate

As the LCOE are mainly dependent on the discount rate and the fixed operating cost ratio, the following results show minimum user tariff rate dependence. Furthermore, the tables contain additional thresholds for the investor's more qualified decision.

The initial decision regarding construction of a charging point is to plan the installed capacity of the transformer station. Figure 14 shows the share of uncovered costs caused by the nonresidents with respect to the annualized costs of an additional transformer.

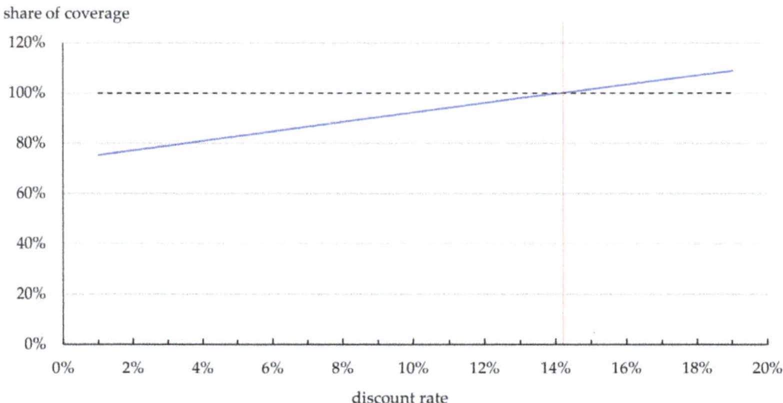

Figure 14. Share of coverage of annualized costs of additional TS.

The first threshold is marked in Figure 14. The share of coverage of annualized costs of additional TS is marked with the red line and is the boundary for the decision to increase the installed capacity of transformers. It follows from the results of the case study that an additional transformer for a discount rate higher than 15% should be installed. This boundary is also shown in Tables 4 and 5.

Table 4. Dependence of residential LCOE on discount rate and on share of inclusion of fixed costs in residential price in case two TS are installed.

		Discount Rate													
	%	5	7	9	11	13	15	17	19	21	23	25	27	29	31
	0	0.08 [1]	0.08 [1]	0.08 [1]	0.08 [1]	0.08 [1]	0.08	0.08	0.08	0.08	0.08	0.08	0.08	0.08	0.08
	10	0.09 [1]	0.09 [1]	0.09 [1]	0.09 [1]	0.09 [1]	0.10	0.10	0.10	0.10	0.10	0.10	0.10	0.10	0.10
	20	0.10 [1]	0.10 [1]	0.10 [1]	0.10 [1]	0.11 [1]	0.11	0.11	0.11	0.11	0.11	0.11	0.11	0.11	0.12
	30	0.11 [1]	0.11 [1]	0.11 [1]	0.11 [1]	0.12 [1]	0.12	0.12	0.12	0.12	0.12	0.13	0.13	0.13	0.13
Share of inclusion of fixed costs in residential price	40	0.12 [1]	0.12 [1]	0.12 [1]	0.13 [1]	0.13 [1]	0.13	0.13	0.13	0.14	0.14	0.14	0.14	0.15	0.15
	50	0.13 [1]	0.13 [1]	0.13 [1]	0.14 [1]	0.14 [1]	0.14	0.14	0.15	0.15	0.15	0.16	0.16	0.16	0.16
	60	0.14 [1]	0.14 [1]	0.14 [1]	0.15 [1]	0.15 [1]	0.15	0.16	0.16	0.16	0.17	0.17	0.17	0.18	0.18
	70	0.15 [1]	0.15 [1]	0.15 [1]	0.16 [1]	0.16 [1]	0.16	0.17	0.17	0.18	0.18	0.18	0.19 [2]	0.19 [2]	0.20 [2]
	80	0.16 [1]	0.16 [1]	0.16 [1]	0.17 [1]	0.17 [1]	0.18	0.18	0.18	0.19 [2]	0.19 [2]	0.20 [2]	0.20 [2]	0.21 [2]	0.21 [3]
	90	0.16 [1]	0.17 [1]	0.17 [1]	0.18 [1]	0.18 [1]	0.19 [2]	0.19 [2]	0.20 [2]	0.20 [2]	0.21 [2]	0.21 [2]	0.22 [3]	0.22 [3]	0.23 [3]
	100	0.17 [1]	0.18 [1]	0.18 [1]	0.19 [1]	0.19 [1]	0.20 [2]	0.20 [2]	0.21 [2]	0.22 [2]	0.22 [3]	0.23 [3]	0.23 [3]	0.24 [3]	0.25 [3]

[1] One TS is sufficient; [2] two TS, price exceeds 0.19 EUR/kWh; [3] one TS, price exceeds 0.19 EUR/kWh.

Table 5. Dependence of nonresidential LCOE on discount rate and on share of inclusion of fixed costs in residential price in case two TS are installed.

		Discount Rate													
	%	5	7	9	11	13	15	17	19	21	23	25	27	29	31
	0	0.14 [1]	0.14 [1]	0.15 [1]	0.15 [1]	0.15 [1]	0.16	0.16	0.17	0.17	0.17	0.18	0.18	0.19 [3]	0.19 [3]
	10	0.14 [1]	0.14 [1]	0.14 [1]	0.14 [1]	0.15 [1]	0.15	0.15	0.16	0.16	0.16	0.17	0.17	0.18	0.18
	20	0.13 [1]	0.13 [1]	0.14 [1]	0.14 [1]	0.14 [1]	0.14	0.15	0.15	0.15	0.16	0.16	0.16	0.17	0.17
	30	0.12 [1]	0.13 [1]	0.13 [1]	0.13 [1]	0.13 [1]	0.14	0.14	0.14	0.14	0.15	0.15	0.15	0.16	0.16
Share of inclusion of fixed costs in residential price	40	0.12 [1]	0.12 [1]	0.12 [1]	0.12 [1]	0.13 [1]	0.13	0.13	0.13	0.14	0.14	0.14	0.14	0.14	0.15
	50	0.11 [1]	0.11 [1]	0.12 [1]	0.12 [1]	0.12 [1]	0.12	0.12	0.12	0.13	0.13	0.13	0.13	0.13	0.14
	60	0.11 [1]	0.11 [1]	0.11 [1]	0.11 [1]	0.11 [1]	0.11	0.11	0.12	0.12	0.12	0.12	0.12	0.12	0.13
	70	0.10 [1]	0.10 [1]	0.10 [1]	0.10 [1]	0.10 [1]	0.11	0.11	0.11	0.11	0.11	0.11	0.11	0.11	0.12
	80	0.10 [1]	0.10 [1]	0.10 [1]	0.10 [1]	0.10 [1]	0.10	0.10	0.10	0.10	0.10	0.10	0.10	0.10	0.10
	90	0.09 [1]	0.09 [1]	0.09 [1]	0.09 [1]	0.09 [1]	0.09	0.09	0.09	0.09	0.09	0.09	0.09	0.09	0.09
	100	0.08 [1]	0.08 [1]	0.08 [1]	0.08 [1]	0.08 [1]	0.08	0.08	0.08	0.08	0.08	0.08	0.08	0.08	0.08

[1] One TS is sufficient; [3] one TS, price exceeds 0.19 EUR/kWh.

Both tables show the LCOE value according to Equation (33) for the case of installation of two transformers. For a discount rate lower than 15%, it is profitable to install one transformer only. For the case of one transformer, the values highlighted in yellow are replaced with the LCOE values.

The table also shows the monetary limits per kWh depending on the number of installed transformer stations. The rate of 0.19 EUR/kWh was chosen as a reference value, which corresponds to the price of home charging. The red area indicates the monetary limits in the case of installation of one transformer. On the contrary, the green area indicates the same limit for installation of two.

5.2. Setting of Minimum Blocking Tariff Rate

If we accept that users in one group can exceed the parking time and thus block others in the other group, then the size of the NCFC is based on the LCOE of the opposite group of users. The following tables (Tables 6 and 7) show the amount of NCFC depending on the discount rate and the share of fixed costs included in the price according to Equations (2) and (4).

Table 6. Nonresident blocking tariff rate in EUR/night.

		Discount Rate													
	%	5	7	9	11	13	15	17	19	21	23	25	27	29	31
Share of fixed costs included in residential price	0	0.0	0.0	0.0	0.0	0.0	0.0	0.0	0.0	0.0	0.0	0.0	0.0	0.0	0.0
	10	0.4	0.4	0.4	0.4	0.4	0.5	0.5	0.5	0.5	0.5	0.6	0.6	0.6	0.6
	20	0.7	0.7	0.8	0.8	0.9	0.9	1.0	1.0	1.0	1.1	1.1	1.2	1.2	1.3
	30	1.1	1.1	1.2	1.2	1.3	1.4	1.4	1.5	1.6	1.6	1.7	1.8	1.9	1.9
	40	1.4	1.5	1.6	1.7	1.7	1.8	1.9	2.0	2.1	2.2	2.3	2.4	2.5	2.6
	50	1.8	1.9	2.0	2.1	2.2	2.3	2.4	2.5	2.6	2.7	2.8	3.0	3.1	3.2
	60	2.1	2.2	2.4	2.5	2.6	2.7	2.9	3.0	3.1	3.3	3.4	3.6	3.7	3.9
	70	2.5	2.6	2.7	2.9	3.0	3.2	3.3	3.5	3.7	3.8	4.0	4.2	4.3	4.5
	80	2.8	3.0	3.1	3.3	3.5	3.6	3.8	4.0	4.2	4.4	4.6	4.7	4.9	5.1
	90	3.2	3.4	3.5	3.7	3.9	4.1	4.3	4.5	4.7	4.9	5.1	5.3	5.6	5.8
	100	3.5	3.7	3.9	4.1	4.3	4.6	4.8	5.0	5.2	5.5	5.7	5.9	6.2	6.4

Table 7. Resident blocking tariff rate in EUR/day.

		Discount Rate													
	%	5	7	9	11	13	15	17	19	21	23	25	27	29	31
Share of fixed costs included in residential price	0	0.9	1.0	1.0	1.1	1.1	1.2	1.3	1.3	1.4	1.4	1.5	1.6	1.6	1.7
	10	0.8	0.9	0.9	1.0	1.0	1.1	1.1	1.2	1.2	1.3	1.4	1.4	1.5	1.5
	20	0.7	0.8	0.8	0.9	0.9	1.0	1.0	1.1	1.1	1.2	1.2	1.3	1.3	1.4
	30	0.7	0.7	0.7	0.8	0.8	0.8	0.9	0.9	1.0	1.0	1.1	1.1	1.1	1.2
	40	0.6	0.6	0.6	0.7	0.7	0.7	0.8	0.8	0.8	0.9	0.9	0.9	1.0	1.0
	50	0.5	0.5	0.5	0.5	0.6	0.6	0.6	0.7	0.7	0.7	0.8	0.8	0.8	0.9
	60	0.4	0.4	0.4	0.4	0.5	0.5	0.5	0.5	0.6	0.6	0.6	0.6	0.7	0.7
	70	0.3	0.3	0.3	0.3	0.3	0.4	0.4	0.4	0.4	0.4	0.5	0.5	0.5	0.5
	80	0.2	0.2	0.2	0.2	0.2	0.2	0.3	0.3	0.3	0.3	0.3	0.3	0.3	0.3
	90	0.1	0.1	0.1	0.1	0.1	0.1	0.1	0.1	0.1	0.1	0.2	0.2	0.2	0.2
	100	0.0	0.0	0.0	0.0	0.0	0.0	0.0	0.0	0.0	0.0	0.0	0.0	0.0	0.0

6. Discussion

The data analysis obtained from real chargers shows some important information: Users leave their EVs parked and connected to the charger for a longer time than is necessary for recharging the battery. A shorter charging time would be sufficient for almost 90% of the 330 EV user sample to charge fully. Users could even use a 3.6 kW AC charger to achieve the same recharging level in the same amount of parking time.

These results also correlate with the studies [4,6] focusing on the average daily distance users travel by car. The average daily distance travelled by a user varies between 48 km and 50 km, and this range also corresponds to the charging requirements in our analysis.

Therefore, in terms of the development of the charging station installation, a sufficient number of chargers is a priority in urban and suburban areas rather than the construction of fewer chargers with higher charging power. The same consideration must be applied when charging infrastructure is built in a P + R car park. Due to the long parking in the daytime, this type of car park is a suitable candidate for constructing a sufficient number of low-power chargers in an urban area.

Our model shows that equipping a car park with chargers is a simple way to increase parking place efficiency. Due to the high occupancy during the daytime and the low occupancy overnight, P + R car parks have a great potential for increasing by installing chargers and introducing a night reservation system. In addition, the evening/night reservation ensures a sufficient number of parking places for EV owners living close to the P + R car park but having no parking place or garage.

The reservation provides user comfort and ensures charging of the EV for the next day comparable to owners of private garages or parking places. By comparison, home charging in the Czech Republic ensures EV owners to use the special electricity tariff D27d [39]. This tariff provides eight-hour charging in the interval from 6 pm to 8 am with a price per kWh approaching 0.08 EUR.

As shown in the economic part of the case study, our solution offers a twelve-hour charging span for a constant price between 0.09 and 0.21 EUR per kWh. This price depends on the discount rate and percentage distribution of non-covered costs between the resident and nonresident groups.

The proposed solution increases the efficiency of existing parking places situated in urban areas, where car parks are usually insufficient. Furthermore, the proposed solution does not dissuade potential new EV owners having no private parking from the purchase of a new EV. Thus, the development of electromobility is significantly supported.

Based on the case study results, the dimensioning of local transformer station power is more in accordance with the EV charging requirements of users arriving during the daytime (nonresidents). Insufficiently dimensioned local transformer station power and the need for simultaneous use of a smaller number of chargers result in financial losses.

The threshold for investors to increase the power of the local transformer station is set by comparing the annualized costs of an additional transformer and total non-covered costs from unsold electricity. Under the conditions described in the case study for a discount rate higher than 15%, the non-covered costs exceed the annualized costs of increasing the local transformer station power. In such cases, the investment in the additional transformer is profitable.

Increasing the power of the local transformer station is beneficial not only for nonresidents, whose percentage of refusal is reduced, but also for the residents, whose guaranteed charging hours may be prolonged. Therefore, doubled installed power allows simultaneous charging of a larger number of EVs, and guarantees up to twice the number of charging hours (Tg).

The model considers installation of oil-immersed transformers, which, unlike the dry type, require a higher degree of fire safety and measures against possible oil leakage. However, if we take into account the same price category, their advantages prevail. Compared to the dry type, the oil-immersed transformer achieves both lower short-circuiting, lower no-load loss and lower noise, guarantees thermal stability during operation at higher loads, and offers the possibility of continuous operation at the rated load [40]. We assume that installing this type of transformer near the P + R car park will meet all the requirements defined by the fire safety standard.

It is evident that an analysis of the behaviour and requirements of specific EV users both living near the site and often using the daytime parking services will be required to achieve a better setting of the parameters of the whole model. A limiting factor of the input data is that they describe only the car park occupancy, losing the information about arrivals and departures within the measured five-minute interval. However, the occupancy data used have been adjusted for public holidays and vacations. Still, it is unknown whether these values may have been affected by, e.g., cultural events which tend to be organized near the car park during the year.

7. Conclusions

The paper analyses the technical and economic potential of installing low-power chargers in an existing car park. The assumed model of a car park equipped with chargers offers a solution to one of the most common obstacles of potential EV users when they decide whether to buy an electric car: the absence of a private charging place. Furthermore, the model solves the problem of the lack of chargers in densely populated districts by installing the charging stations in an existing car park.

Generally, the model defines two groups of users. The first group (nonresidents) are users arriving randomly in the daytime who commute to the site for work or other purposes. Nonresidents usually leave cars parked there for a longer time. The second group (residents) live close to the car park and rent parking places primarily for overnight charging.

However, the major difference between the resident and nonresident groups is that nonresident users' priority is parking rather than charging. The nonresidents' EVs are charged if the current load of the transformer station makes this possible. Contrarily, the resident users' priority is the guaranteed charging. The model is based on real user parking behaviour, which then complements the requirements for recharging. The modelling results are the percentage of satisfaction of individual groups (especially nonresidents) under specifically selected charging infrastructure parameters.

The case study in the paper uses recharging data from the company ŠKO-ENERGO and parking place occupancy data for a P + R car park in Prague Letňany. The charging data contain three months of testing data from the year 2020, 330 charging iterations in total from seven charging points. All charging points were equipped with 22 kW DC power chargers, four of which could also charge with 50 kW DC. The parking occupancy data describe the car park, with a capacity of 650 parking places, in five-minute intervals from the 1st quarter of 2013 to the 4th quarter of 2016. The charging analysis shows that users left the EVs parked and connected to the charger for a longer time than required to recharge the battery. Comparing charging and time spent at the charger, almost 90% of the tested EV users would have done with a shorter charging time to recharge the required energy. They could even use a 3.6 kW AC charger to achieve the same recharging level in the same amount of parking time.

Knowing the above information, the model designs the optimal size of charging equipment. By creating a reservation system for residents, the LCOE of nonresident charging at a discount rate of 15% will drop from around 0.15 EUR to 0.11 EUR, which competes with the price of other types of public charging. The simulation results show that although the model uses low-power chargers with an output of 3.6 kW, the daytime parking time in the P + R car park studied is sufficient to fully charge to 96% for nonresidents. However, in order to guarantee 6 h of resident charging and thus achieve a 95% success rate of full charging, it is necessary to add another transformer to the local transformer station.

Another benefit of the presented solution, in addition to more efficient use of parking places, is ensuring the stability of the electric grid during the charging of a large number of EVs in one place by dividing the power between them. Dividing the power between randomly placed charging stations of different owners within the power grid would not be as easy as dividing the power within the charging stations in one car park. In the case of demand for charging a group of electric vehicles concentrated in one place, the local power grid would be significantly overloaded. Then, it is possible to support the overall stability of the local power grid with battery storage, which would smooth out peaks in the load.

Author Contributions: All authors contributed to the research in the paper. L.D. and M.H. conceived and designed the model; L.D., M.H., M.V. and J.K. provided the data; L.D., M.H. and J.K. analysed the data; L.D. wrote the paper. All authors have read and agreed to the published version of the manuscript.

Funding: This work is supported by the Student Grant Competition of CTU (Grant No. SGS20/125/OHK5/2T/13 and Grant No. SGS20/126/OHK5/2T/13).

Acknowledgments: Above all, we would like to thank the reviewers and ŠKO-ENERGO. We thank the anonymous reviewers for their careful work and ŠKO-ENERGO for providing anonymized data from the operation of their charging stations.

Conflicts of Interest: The authors declare no conflict of interest.

Abbreviations

CAPEX	Capital expenditures
CDF	Cumulative distribution function
DSM	Demand side management
EV	Electric vehicle
HV	High voltage
LCOE	Levelized costs of electricity
LV	Low voltage
NCFC	Non-covered fixed costs
OPEX	Operating expenditures
P2P	Peer-to-peer
PDF	Probability density function
P + R	Park and ride car park
RES	Renewable energy sources
TS	Transformer station

Variables

$\hat{A}$	Vector with distribution of arrivals of nonresidents in five minutes during the day
E	Electricity sold per year [kWh]
E_r	Total resident recharging requirement [kWh]
$E_r(x)$	Recharging requirement of resident x [kWh]
E_{TR_YX}	Energy that can be provided in interval from Y to X by local transformer [kWh]
ch_{par}	Maximum number of chargers used simultaneously
L	Length of nonresidential stay [h]
$LCOE$	Levelized costs of electricity
L_{p_kWh}	Non-covered fixed costs for unsold kilowatt hours [EUR]
M	Annual maintenance expenditures [EUR]
N_E	Electricity bill in year t [EUR]
N_i	Initial investment in equipment [EUR]
N_{var}	Unit price per kWh [EUR]
N_w	Number of EVs waiting to start recharging
P	Transformer station output power [kW]
P_{CH}	Power of one charger [kW]
P_{ch_par}	Power consumption of all chargers used simultaneously ch_{par}. [kW]
P_{kn}	Short-circuit transformer losses [kW]
P_n	TS output power [kW]
P_{nl}	Power transformer no-load loss [kW]
q	Quantity
Q	Number of parking places occupied at weekends
Q_{pl}	Number of parking places in car park
Q_r	Parking places reserved for residents
R	Tariff rate for exceeding the parking time [EUR]
r	Discount rate [%]
S_m	Maximum apparent-power load of transformer [kVA]
S_n	Apparent power of transformer station [kVA]
t	Time step
T_0	Annual operating time of transformer station [h]

T_g	Guaranteed charging time for residents [h]
T_l	Last morning hour of guaranteed h
T_L	Time equivalent of losses at maximum transformer load per year [h]
T_{LT}	Lifetime [years]
T_m	Utility factor [h]
T_o	Average daily use of one charging station [h]
$T_r(x)$	Charging time of x-th resident [h]
$T_w(x)$	Waiting time in queue
W_L	Annual energy losses in transformer
X	End of interval reserved for residents
Y	Beginning of interval reserved for residents

References

1. Xiao, H.; Xu, M.; Yang, H. Pricing strategies for shared parking management with double auction approach: Differential price vs. uniform price. *Transp. Res. Part. E Logist. Transp. Rev.* **2020**, *136*, 101899. [CrossRef]
2. Plenter, F.; Chasin, F.; Von Hoffen, M.; Betzing, J.H.; Betzing, J.H.; Becker, J. Assessment of peer-provider potentials to share private electric vehicle charging stations. *Transp. Res. Part. D Transp. Environ.* **2018**, *64*, 178–191. [CrossRef]
3. Kester, J.; Noel, L.; De Rubens, G.Z.; Sovacool, B.K. Policy mechanisms to accelerate electric vehicle adoption: A qualitative review from the Nordic region. *Renew. Sustain. Energy Rev.* **2018**, *94*, 719–731. [CrossRef]
4. Zhao, Z.; Zhang, L.; Yang, M.; Chai, J.; Li, S. Pricing for private charging pile sharing considering EV consumers based on non-cooperative game model. *J. Clean. Prod.* **2020**, *254*, 120039. [CrossRef]
5. Iversen, E.B.; Moller, J.K.; Morales, J.M.; Madsen, H. Inhomogeneous Markov Models for Describing Driving Patterns. *IEEE Trans. Smart Grid* **2017**, *8*, 581–588. [CrossRef]
6. Xu, Y.; Çolak, S.; Kara, E.C.; Moura, S.J.; González, M.C. Planning for electric vehicle needs by coupling charging profiles with urban mobility. *Nat. Energy* **2018**, *3*, 484–493. [CrossRef]
7. Zhang, Y.; Zhang, Q.; Farnoosh, A.; Chen, S.; Li, Y. GIS-Based Multi-Objective Particle Swarm Optimization of charging stations for electric vehicles. *Energy* **2019**, *169*, 844–853. [CrossRef]
8. Huang, X.; Long, X.; Wang, J.; He, L. Research on parking sharing strategies considering user overtime parking. *PLoS ONE* **2020**, *15*, e0233772. [CrossRef]
9. Cai, Y.; Cheng, J.; Zhang, C.; Wang, B. A parking space allocation method to make a shared parking strategy for appertaining parking lots of public buildings. *Sustainability* **2018**, *11*, 120. [CrossRef]
10. Krause, J.; Ladwig, S.; Schwalm, M. Statistical assessment of EV usage potential from user's perspective considering rapid-charging technology. *Transp. Res. Part. D Transp. Environ.* **2018**, *64*, 150–157. [CrossRef]
11. Švédsko: Základní Charakteristika Teritoria, Ekonomický Přehled. BusinessInfo.cz. Available online: https://www.businessinfo.cz/navody/svedsko-zakladni-charakteristika-teritoria-ekonomicky-prehled/ (accessed on 5 August 2020).
12. Useable Battery Capacity of full Electric Vehicles Cheatsheet—EV Database. Available online: https://ev-database.uk/cheatsheet/useable-battery-capacity-electric-car (accessed on 5 August 2020).
13. Energy Consumption of Full Electric Vehicles Cheatsheet—EV Database. Available online: https://ev-database.uk/cheatsheet/energy-consumption-electric-car (accessed on 5 August 2020).
14. Range of Full Electric Vehicles Cheatsheet—EV Database. Available online: https://ev-database.uk/cheatsheet/range-electric-car (accessed on 5 August 2020).
15. Price of Full Electric Vehicles Cheatsheet—EV Database. Available online: https://ev-database.uk/cheatsheet/price-electric-car (accessed on 5 August 2020).
16. Patt, A.; Aplyn, D.; Weyrich, P.; Van Vliet, O. Availability of private charging infrastructure influences readiness to buy electric cars. *Transp. Res. Part. A: Policy Pr.* **2019**, *125*, 1–7. [CrossRef]
17. Wang, H.; Chen, S.; Yan, Z.; Ping, J. Blockchain-enabled Charging Right Trading among EV Charging Stations: Mechanism, Model, and Method. *Proc. CSEE* **2019**, *258*, 8013. (In Chinese) [CrossRef]
18. Gong, D.; Tang, M.; Buchmeister, B.; Zhang, H. Solving Location Problem for Electric Vehicle Charging Stations—A Sharing Charging Model. *IEEE Access* **2019**, *7*, 138391–138402. [CrossRef]

19. Taljegard, M.; Göransson, L.; Odenberger, M.; Johnsson, F. Impacts of electric vehicles on the electricity generation portfolio—A Scandinavian-German case study. *Appl. Energy* **2019**, *235*, 1637–1650. [CrossRef]

20. Pan, L.; Yao, E.; MacKenzie, D. Modeling EV charging choice considering risk attitudes and attribute non-attendance. *Transp. Res. Part. C: Emerg. Technol.* **2019**, *102*, 60–72. [CrossRef]

21. Pan, L.; Yao, E.; Yang, Y.; Zhang, R. A location model for electric vehicle (EV) public charging stations based on drivers' existing activities. *Sustain. Cities Soc.* **2020**, *59*, 102192. [CrossRef]

22. Sun, S.; Yang, Q.; Yan, W. A Novel Markov-Based Temporal-SoC Analysis for Characterizing PEV Charging Demand. *IEEE Trans. Ind. Inform.* **2018**, *14*, 156–166. [CrossRef]

23. Xu, F.; Chen, X.; Zhang, M.; Zhou, Y.; Cai, Y.; Zhou, Y.; Tang, R.; Wang, Y. A sharing economy market system for private EV parking with consideration of demand side management. *Energy* **2020**, *190*, 116321. [CrossRef]

24. Lopez-Behar, D.; Tran, M.; Froese, T.M.; Mayaud, J.R.; Herrera, O.E.; Merida, W. Charging infrastructure for electric vehicles in Multi-Unit Residential Buildings: Mapping feedbacks and policy recommendations. *Energy Policy* **2019**, *126*, 444–451. [CrossRef]

25. Cost of Charging an Electric Car|Pod Point. Available online: https://pod-point.com/guides/driver/cost-of-charging-electric-car (accessed on 5 August 2020).

26. Wei, Z.; Li, Y.; Zhang, Y.; Cai, L. Intelligent Parking Garage EV Charging Scheduling Considering Battery Charging Characteristic. *IEEE Trans. Ind. Electron.* **2018**, *65*, 2806–2816. [CrossRef]

27. Zhang, L.; Yang, M.; Zhao, Z. Game analysis of charging service fee based on benefit of multi-party participants: A case study analysis in China. *Sustain. Cities Soc.* **2019**, *48*, 101528. [CrossRef]

28. Wu, P.; Chu, F.; Saidani, N.; Chen, H.; Zhou, W. IoT-based location and quality decision-making in emerging shared parking facilities with competition. *Decis. Support. Syst.* **2020**, *134*, 113301. [CrossRef]

29. Country Detail Vehicles and Fleet Compare|EAFO. Available online: https://www.eafo.eu/countries/european-union/23640/vehicles-and-fleet/compare (accessed on 12 August 2020).

30. Friesen, M.; Mingardo, G. Is Parking in Europe Ready for Dynamic Pricing? A Reality Check for the Private Sector. *Sustainability* **2020**, *12*, 2732. [CrossRef]

31. Lei, C.; Ouyang, Y. Dynamic pricing and reservation for intelligent urban parking management. *Transp. Res. Part. C Emerg. Technol.* **2017**, *77*, 226–244. [CrossRef]

32. Electric Vehicles|Mobility and Transport. Available online: https://ec.europa.eu/transport/themes/urban/vehicles/road/electric_en (accessed on 8 August 2020).

33. Oblast Severovýchod—Srovnání let a Dnů v Týdnu—Jana Roulichová. Tableau Public. Available online: https://public.tableau.com/profile/jana.roulichova#!/vizhome/Srovnvacanalza1-PRoblastseverovchod/DBseverovchod-rokyaweekdays (accessed on 12 August 2020).

34. Emobilita. ŠKO-ENERGO—Hřeje Vás Čistá Energie. Available online: https://www.sko-energo.cz/content/upload/file/predpokladana_infrastruktura_dobijecich_stanic_pro_mladou_boleslav.pdf (accessed on 16 November 2020).

35. Pareschi, G.; Küng, L.; Georges, G.; Boulouchos, K. Are travel surveys a good basis for EV models? Validation of simulated charging profiles against empirical data. *Appl. Energy* **2020**, *275*, 115318. [CrossRef]

36. Test by Germany's Motoring Asscciation Reveals E-Cars Use more Power than Dashboards Show|Clean Energy Wire. Available online: https://www.cleanenergywire.org/news/test-germanys-motoring-asscociation-reveals-e-cars-use-more-power-dashboards-show (accessed on 5 October 2020).

37. WHO. Cenové Rozhodnutí Energetického Regulačního Úřadu č. 5/2019 ze dne 26. Listopadu 2019. No. 2019/1. 2019; pp. 1–20. Available online: http://www.eru.cz/documents/10540/6182278/CR_ERU_2020_VVN_VN_zmeny.pdf/79babf8e-3c25-4e5c-b4e3-1f74a254d731 (accessed on 16 November 2020).

38. WHO. Ceník Elektřiny Pro Podnikatele. No. 2019/1. Prague, 2019; pp. 1–20. Available online: moz-extension://ec5f8c9c-ff7c-0c4e-855f-cea9ed1ceed3/enhanced-reader.html?openApp&pdf=https%3A%2F%2Fwww.pre.cz%2FFiles%2Ffirmy%2Felektrina%2Fseznam-produktu%2Faktiv-favorit%2Fcenik-aktiv-favorit-3-predi%2F (accessed on 19 November 2020).

39. Distribuční Sazba D27d: Ideální Při Nabíjení Elektromobilu|Elektřina.cz. Available online: https://www.elektrina.cz/distribucni-sazba-d27d-idealni-pri-nabijeni-elektromobilu (accessed on 6 November 2020).

40. Proč Olejové Transformátory SGB?|Olejové Transformátory|Elpro-Energo. Available online: https://www.elpro-energo.cz/olejove-transformatory/proc-olejove-transformatory-sgb/ (accessed on 15 September 2020).

Publisher's Note: MDPI stays neutral with regard to jurisdictional claims in published maps and institutional affiliations.

Article

Environmental Impacts of Charging Concepts for Battery Electric Vehicles: A Comparison of On-Board and Off-Board Charging Systems Based on a Life Cycle Assessment

Mona Kabus [1], **Lars Nolting [2]**, **Benedict J. Mortimer [3]**, **Jan C. Koj [4,5]**, **Wilhelm Kuckshinrichs [4,5,*]**, **Rik W. De Doncker [3,6,*]** and **Aaron Praktiknjo [2,6,*]**

[1] Chair of Technology and Innovation Management, University of Bayreuth, 95440 Bayreuth, Germany; mona.kabus@uni-bayreuth.de

[2] Institute for Future Energy Consumer Needs and Behavior (FCN), RWTH Aachen University, 52074 Aachen, Germany; lnolting@eonerc.rwth-aachen.de

[3] Institute for Power Generation and Storage Systems (PGS), RWTH Aachen University, 52074 Aachen, Germany; bmortimer@eonerc.rwth-aachen.de

[4] Forschungszentrum Jülich, Institute of Energy and Climate Research–Systems Analysis and Technology Evaluation (IEK-STE), 52425 Jülich, Germany; j.koj@fz-juelich.de

[5] JARA-ENERGY, 52425 Jülich, Germany

[6] JARA-ENERGY, 52074 Aachen, Germany

* Correspondence: w.kuckshinrichs@fz-juelich.de (W.K.); post_erc@eonerc.rwth-aachen.de (R.W.D.D.); apraktiknjo@eonerc.rwth-aachen.de (A.P.)

Received: 31 August 2020; Accepted: 2 December 2020; Published: 9 December 2020

Abstract: We investigate the environmental impacts of on-board (based on alternating current, AC) and off-board (based on direct current, DC) charging concepts for electric vehicles using Life Cycle Assessment and considering a maximum charging power of 22 kW (AC) and 50 kW (DC). Our results show that the manufacturing of chargers provokes the highest contribution to environmental impacts of the production phase. Within the chargers, the filters could be identified as main polluters for all power levels. When comparing the results on a system level, the DC system causes less environmental impact than the AC system in all impact categories. In our diffusion scenarios for electric vehicles, annual emission reductions of up to 35 million kg CO_2-eq. could be achieved when the DC system is used instead of the AC system. In addition to the environmental assessment, we examine economic effects. Here, we find annual savings of up to 8.5 million euros, when the DC system is used instead of the AC system.

Keywords: charging infrastructure; electric vehicle; life cycle assessment; AC charging; DC charging; economic assessment

---

1. Introduction

As the average temperature on earth is increasing [1–3], both environmental and economic consequences are to be expected [4,5]. Hence, there are international efforts to reduce this rise to below 2 °C in the long-term and binding targets have been agreed in the Kyoto Protocol and the Paris Agreement, among others [5,6].

Being one of the largest global economies, Germany emitted ≈858 mn. tons of CO_2-equivalent (CO_2-eq.) in 2018, 19% of which originated in the transport sector [7]. Within the transport sector, passenger cars are a major source of emissions [8]. Modern, non-fossil fuel-based propulsion systems offer opportunities to reduce the emissions and thereby help to mitigate climate change and to reduce dependence on oil imports. In recent decades, electro mobility has proven to be a competitive alternative to existing mobility concepts [9–11]. The continuing decline in battery prices [12] and the falling prices for electric vehicles (EVs) indicate good chances for further market growth [11,13].

An important factor for the diffusion of EVs is the provision of an appropriate charging infrastructure. In the period from 2017 to 2020, 750 mn. euros were being spent for the expansion of the charging infrastructure in Germany [14].

To charge the batteries of EVs, usually, alternating current (AC) from public grids is used. As the battery represents a DC (direct current) power source, a conversion from AC to DC is necessary for proper charging. This conversion is done by power electronics. Using an AC-DC converter, alternating sinusoidal voltage is converted to DC, which is then in turn adapted to the charging requirements of the EV and regulated accordingly (DC-DC converter) [15].

While AC charging requires on-board power electronics, DC also allows for off-board charging. On-board chargers (OBCs) are constrained in their size, and therefore their charging capacity, due to weight, cost, and space restrictions. Off-board Chargers (OfBCs) are less limited in terms of size and weight and thus allow for higher charging capacities. OfBCs are currently the standard for all charging capacities greater than 22 kW.

Life Cycle Assessments (LCAs) have been intensively used in the literature to estimate the environmental impact of electric vehicles compared to fuel-based vehicles (e.g., [10,16–18]). However, only a few works deal with the different charging systems of electric vehicles and their environmental impacts [19–22]. None of the papers compares in detail mutually exclusive AC and DC charging systems in order to generate findings as to whether one of the two systems should be used preferentially based on potential ecological and/or economic advantages. We contribute to this research gap by investigating the potential of DC-based off-board charging technology to reduce the environmental impact and costs of charging technologies for EVs. We conduct a comparative LCA for both charging technologies: AC-based on-board charging and DC-based off-board charging. Further, we assess the systemic impact when upscaling the systems (i.e., larger vehicle stock, more charging infrastructure) as well as the potential cost savings that come with OfBCs within different scenarios.

Our scenarios as well as some assumptions made for modeling are based on the German market; however, most of our results possess a high generalizability. Although the electrical charging infrastructure differs among countries (e.g., different voltage levels) and therefore other components are used, our relative analysis between the systems will remain valid. Accordingly, it can be assumed that the basic results of the paper, namely the result of the system comparison, have a high generalizability.

The remainder of our work is structured as follows: In Section 2, we provide an overview of the materials and methods used and present our systems, our scenarios, and the components modeled for the environmental assessment. In Section 3, we present the results of our environmental assessment on a component and system level. Furthermore, we carry out an economic assessment of the systems on a national level and present the results. Section 4 gives a summary of our paper and states the limitations as well as an outlook for future research.

2. Materials and Methods

In the following paragraphs, we (1) define the technological systems under investigation and (2) the scenarios used for their assessment. Then, we (3) provide the goal definition and scoping as well as the Life Cycle Inventory (LCI) and the Life Cycle Impact Assessment (LCIA) for our LCA.

2.1. Technological Systems

To assess the potential environmental and economic benefits of DC-based charging infrastructure for battery electric vehicles (BEVs), we define two technological systems:

1. In our **AC system**, all electric vehicles have an OBC to charge the battery with a power up to 22 kW. Any publicly accessible charging point provides a charging capacity of 22 kW. For charging

at home, a 3.7 kW charging point (AC) is available. Charging with 3.7 kW requires no additional power electronics.

2. In our **DC system**, each electric vehicle is equipped with an OBC with 3.7 kW charging power that allows for charging at home (AC). Additionally, all electric vehicles can be charged with an OfBC with 50 kW charging power at publicly accessible charging points (DC).

Please note that the introduced names of our systems are determined by the publicly accessible charging infrastructure. In the AC system, charging at public charging infrastructure is only possible with AC; in the DC system, charging at public charging infrastructure is only possible with DC. The AC OBC with 3.7 kW in the DC system is used for a more realistic representation of the systems, but it does not constitute the name of the system.

Furthermore, we differentiate between *chargers* and *charging infrastructure (CIS)*. When talking about CIS, we consider the connection with the grid, the charging cable, and the housing of the charging station. The chargers in our study consist of the necessary power electronics and a housing for the power electronics. The power electronics essentially include the required converters, a filter, and additional electronic components such as printed circuit boards (PCBs) and busbars. The converters itself are composed of different electronic components such as diodes, MOSFETs, coils, capacitors, and transformers.

As indicated in Table 1, the number of components either scale with the number of charging points, the number of charging stations (which can consist of more than one charging point), or the number of BEVs.

Table 1. General scaling of components.

	AC System		DC System		
	22 kW OBC	22 kW CIS	50 kW OfBC	50 kW CIS	3.7 kW OBC
Number of BEVs	x	–	–	–	x
Amount of charging infrastructure [1]	–	x	x	x	–

[1] charging infrastructure can rather be a charging station or a charging point.

2.2. Scenarios

To assess potential reductions of the environmental impact of charging infrastructure on a large scale, we introduce three scenarios regarding the diffusion of BEVs in Germany. Using the Bass diffusion model [23], we differentiate the following potential diffusion curves:

- In the first scenario (**S**), the innovation and imitation coefficients of the Bass model are adopted from [24]. Although [24] data are based on data of global sales of hybrid Toyota vehicles, they are considered to be sufficiently good estimates. The values for the vehicle stock, the number of new registrations, and the number of retired passenger cars are taken from [25]. For a near future distribution, a vehicle stock of 170,000 vehicles is estimated, which seems to be reasonable (136,617 electric vehicles on 1 January 2020 in Germany [26]).

- The second scenario (**M**) is based on data by [27,28], who state the objective that road traffic in Germany should be climate-neutral by 2050. One possibility of how this aim might be reached is to regulate the registration of new cars so that solely BEVs can be registered. For our estimation, we assume that as of 2040, only BEVs can be newly registered, which leads to a vehicle stock of about 220,000 BEVs in the near future. Comparable values can be estimated when calculating innovation and imitation coefficients based on the vehicle stock [25] and the new vehicle registrations of electric vehicles [29].

- The third scenario (**L**) is based on the aim of one million electric vehicles in 2020 [30]. According to [31] and the actual vehicle stock [26], this target is not realistically achievable for 2020, but experts assume that the million level could be reached in 2022 [31]. With the assumptions made for the market volume, we consider this limit as an intermediate target. We further assume that the vehicle stock is increasing exponentially up to that limit, resulting in a vehicle stock of 310,000 BEVs in the near future. Table 2 shows the vehicle stocks in the three scenarios presented.

Table 2. Number of charging points and charging stations in the defined scenarios.

		Scenario S	Scenario M	Scenario L
Vehicle stock		170,000	220,000	310,000
Number of public charging points	AC system	14,579	18,867	25,827
	DC system	6415	8302	11,364
Number of public charging stations	AC system	7290	9434	12,914
	DC system	3207	4151	5682

We base the demand for publicly accessible CIS on the electricity demand of the vehicle stock in each scenario. We calculate this electricity demand considering the number of vehicles in each scenario derived from the bass diffusion models, the load requirements per BEV, and the share of at-home-charging (i.e., non-public charging).

To calculate the load requirements for BEVs, we consider the mileage as well as the average efficiency. To estimate the efficiency of a diverse vehicle stock, we assign the vehicles to three different classes: small car, compact car, and luxury car (sport cars are also regarded as *luxury cars* in our study). We calculate a ratio of the segments on the total stock of 0.28, 0.57, and 0.15 respectively (based on [32]) and also calculate the vehicle stock within each class. Furthermore, we derive average efficiencies for each class (13 kWh/(100 km), 14 kWh/(100 km), and 21 kWh/(100 km) respectively) based on current market data [33–41]. We assume these ratios to be constant over time. For the annual mileage, we assume a constant value of 13,922 km, regardless of the vehicles' class [42].

We estimate the share of public charging based on various studies [43–47]. Hence, we consider a share of the total electricity provided by the public charging infrastructure to be 30% of the total electricity demand for charging. The total energy required by BEVs can either be charged at home or at a publicly accessible charging infrastructure. In our study, we assume that for both systems, home charging is possible (with 3.7 kW AC, see Section 2.1). Since a system comparison is intended, the charging infrastructure required for home charging is not considered. The energy requirements of home charging are considered only to the extent that they influence the required number of publicly accessible charging infrastructure. In the following, we focus on the charging infrastructure in the publicly accessible areas if not stated otherwise.

To derive the number of publicly accessible charging points, the rate of utilization for a charging point, the efficiency of the charging point, and the number of charging points within a charging station need to be estimated.

We define the rate of utilization as the ratio of the time a charging point is used to the time a charging point theoretically can be used. For the theoretically usable time, we assume a theoretically possible utilization of 12 h a day and 302 days a year. In addition, we assume that each charging point is directly available to other users as soon as the previous car is fully charged (i.e., no additional time gaps). In our study, the rate of utilization is 10%, which implies that there are about 12 BEVs per charging point for the AC system and 27 BEVs for the DC system. This is within the range of the number of electric vehicles per charging point stated in other studies [48–50].

Data available from manufacturers [51–54] indicate an efficiency for the charging process from 90% to 95% for their charging stations. Using a conservative estimate, we determine an efficiency of 90%.

With the given data, we calculate the energy provided by each charging point and combine it with the already determined energy demand. For the amount of charging stations, an average of two charging points per charging station is assumed [55]. The resulting number of charging points and charging stations for the scenarios can be seen in Table 2. The formula for the underlying calculations can be found in Appendix A.

2.3. Life Cycle Assessment (LCA): Scope and Goal Definition, Life Cycle Inventory (LCI), and Life Cycle Impact Assessment (LCIA)

Having described the technological systems under investigation and having introduced our scenarios, we will now summarize relevant materials and methods of the LCA that we have conducted to assess the environmental impacts of different charging infrastructures. LCA is a commonly used environmental assessment concept [56]. The structure and methodological aspects of LCA are described in the ISO 14040 and ISO 14044 standards [57,58]. Each LCA can be sub-divided into four phases: the goal and scope definition, Life Cycle Inventory analysis (LCI), Life Cycle Impact Assessment (LCIA), and interpretation.

For our assessment, we use *Ecoinvent 3.4* [59] as a database and apply the so-called cut-off method. We incorporate the database into the software *openLCA 1.7.2* [60].

2.3.1. Goal and Scope

The goal of this LCA is to determine and compare the systemic environmental impact of the previously described AC and DC-based charging systems. We distinguish between *production*, *transport*, and *use phases*. Due to a lack of data on the regarding processes, recycling is not considered. The functional unit is a kilowatt hour (kWh) that is facilitated by the publicly accessible charging infrastructure. The components examined can be seen in Figure 1. The assumed lifetime for all components is 10 years [49,61,62]. Furthermore, we assume that this lifetime applies to all components so that neither the chargers nor the batteries in the BEVs need to be replaced during the lifetime. In accordance with the common LCA method, we conduct the assessment with scaling to a functional unit (i.e., the provision of one kWh charging energy).

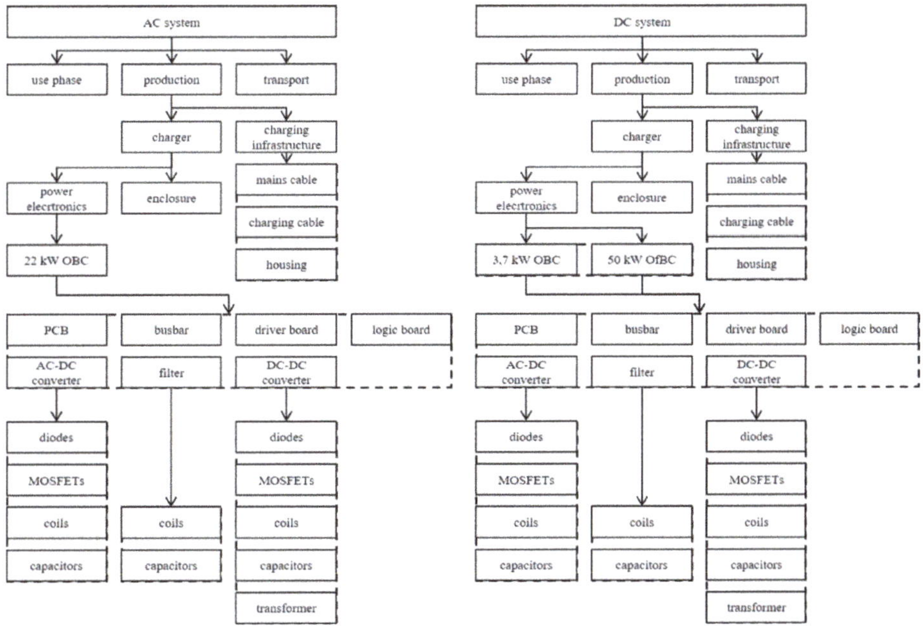

Figure 1. Overview of the modeled systems. Abbreviations: OBC—On-board charger; OfBC—Off-board charger; PCB—printed circuit board.

We neglect the additional consumption of fuel due to the weight of the chargers (for an estimation of weight-related effects, see e.g., [63]) and estimate the life cycle inventories to the extend necessary due to the non-availability of data. Furthermore, the assembly of the chargers is not considered in the production phase. The geographical scope of our analysis is Germany; however, there are few indications that the findings significantly differ in other countries as discussed in the introduction section. Further, it is assumed that the rate of utilization is constant over time.

Modular approaches for chargers are often chosen to provide high flexibility and worldwide network connectivity [54,64–66]. On the one hand, this allows for a better use of economies of scale for power electronic components; on the other hand, it reduces the maintenance cost and the total failure times of the charging stations. In addition, the charging infrastructure can be gradually adapted to changing charging requirements. Moreover, the overall efficiency can be enhanced by modularity. Therefore, all systems are based on an interconnection of 3.7 kW chargers.

The topologies used for the chargers are mainly based on the work of [67,68]. Therefore, the main reasons are the possibility of modular connection and the well-documented use of components used. The chargers are unidirectional.

2.3.2. Life Cycle Inventory (LCI)

In the following, we briefly describe the LCI, starting with the production phase of the chargers' components. Eventually, we summarize our inventory assumptions for the transport and use phase. For reasons of brevity, additional assumptions about the components for the LCI can be found in Appendix B (see Tables A1 and A2).

Production of Components

The components used for the chargers are based on [68,69]. Components for which no sufficient data are available are replaced by appropriate components for which the corresponding data are accessible. The main distinction between the 3.7 kW OBC, the 22 kW OBC, and the 50 kW OfBC is the number of the required components. A general overview of modeled systems in this study can be found in Figure 1. In the following, we present each component separately.

- Filter

Low-pass filters are a frequently used option for limiting high-frequency oscillations and noise emissions. Since *LCL filters* are often used for chargers with a high charging capacity [70–72], we use an LCL filter in our study. The dimensioning of the filter and its components depends on the associated rectifier and hence on all downstream components [70,73,74]. In our paper, we base the dimensioning of the LCL filter on [72,75], respectively. The components used in our study can be found in Table 3. The number of components is determined by the performance of the chargers, their electrical connection, and the specifications of the components used. To estimate the environmental impacts of the coils, the *Ecoinvent* [59] dataset for inductors is used as a fairly good estimate and scaled with the associated mass. The production process of the capacitor is modeled using the *Ecoinvent* [59] dataset "capacitor, film type, for through-hole mounting".

Table 3. Components used to model filters.

Component	Manufacturer	Notation	Ref.	Number of Components		
				3.7 kW	22 kW [1]	50 kW [1]
Coil (mains side)	Fastron Group	TLC/10A-100M-00	[76]	2	4	8
Coil (inverter side)	Fastron Group	TLC/10A-471M-00	[76]	2	4	–
Coil (inverter side)	Fastron Group	TLC/10A-331M-00	[76]	–	–	8
Capacitor	ICAR	MKP-B1X-8-48	[77]	2	4	6

[1] The number of components stated for the 22 kW and 50 kW charger is for each phase.

- AC-DC converter

As for AC-DC converters, we use rectifiers with a downstream power factor correction stage (PFC-converters). The rectifier bridge is combined with a step-up converter. The components used for modeling the AC-DC converter can be found in Table 4. MOSFETs and diodes are modeled with the material data given by the manufacturers and the *Ecoinvent* production processes for transistors and diodes. The ceramic capacitor used in [67] is modeled with the mass of the capacitor and the dataset for the production of capacitors for surface-mounting in the *Ecoinvent* database.

Table 4. Components used for the modeled AC-DC converter.

Component	Manufacturer	Notation	Ref.	Number of Components		
				3.7 kW	22 kW [1]	50 kW [1]
MOSFET	Infineon	IPW60R045CP	[78]	2	2	4
Diode	Infineon	IDW40G65C5B	[79]	6	6	12
Capacitor	Murata	KC355WD72J474MH01#	[80]	2	2	4
Coil	Epcos	EELP 58 Core	[81]	2	2	4

[1] The number of components stated for the 22 kW and 50 kW charger is for each phase. AC: alternating current, DC: direct current.

- DC-DC converter

As for DC-DC converters, a common topology for chargers is the isolated full bridge with phase shifted operation. By adjusting the voltage of the active side, the voltage can be adapted to the state of charge of the battery [82,83]. The components used for the DC-DC converter can be found in Table 5. The MOSFETs are taken from [67]. The material data are given by the manufacturer. For the production process, we use the *Ecoinvent* dataset for transistors. For the diodes as well as for the capacitor, the stated components are used instead of the ones from [67] due to the non-availability of data. The material data for the diodes are given by the manufacturer; for the production process, the *Ecoinvent* dataset for diodes is used. The capacitor is modeled as stated before for the AC-DC converter. The production process is modeled with the *Ecoinvent* dataset for inductors.

Table 5. Components used for the modeled DC-DC converter.

Component	Manufacturer	Notation	Ref.	Number of Components		
				3.7 kW	22 kW [1]	50 kW [1]
MOSFET	Infineon	IPW65R045C7	[84]	4	8	20
Diode	Infineon	IDW40G65C5B	[79]	4	8	20
Capacitor	Murata	KC355WD72J474MH01#	[80]	2	4	10
Coil	Epcos	EELP 64 Core	[81]	1	2	5

[1] The number of components for the 22 kW and 50 kW charger is for each phase.

The dimensioning of the transformer for galvanic isolation is based on [85]. Elgström and Nordgren [85] dimensioned the transformer for a load of 5.5 kW, which is a sufficient estimate for our study. The stated transformer can be used for modular connection as well. Based on the information on the core [85,86] and the turns ratio, we can calculate the amount of ferrite and copper. For the production process, the *Ecoinvent* [59] dataset for wire drawing is used.

- Printed circuit board (PCB), driver board, logic board and busbars

To mount the electronic components, a PCB is used. We estimate the size based on the outer dimension of the chargers. Details of the calculations can be found in the appendix. For the mounting of the PCB, we use input and output data from [87].

We adopt the driver board from [87]. According to Nordelöf and Alatalo [87], the structure of driver boards for three-phase inverters for the automotive sector vary negligibly with the power in a range from 20 to 200 kW. We use one of the stated driver boards for each 3.7 kW charging unit. The design of the logic board is also based on [87], and it is likewise assumed that one logic board is required for each 3.7 kW charging unit. We base the material flows and the production processes on [87].

We base the dimensioning of the busbar on [87,88]. The production process is modeled based on [87].

- Enclosure of electronic components and heatsink.

Based on [67,87,89], we assume that all electronic components share the same enclosure. According to [90], the majority of enclosures is made of aluminum. The material flows and processes used for the production phase of the enclosure with an untreated surface are based on [87].

According to Nordelöf and Alatalo [87], the inverter unit considered can be air-cooled for charging capacities up to 50 kW, whereas higher power needs to be liquid-cooled. According to [91], air cooling can also be used for charging stations with significantly higher charging capacity. Therefore, we model an air-cooled heatsink; the material used is aluminum [87]. Process data are taken from [87].

- Charging infrastructure

The modeled CIS consists of a mains cable for each charging station, a charging cable for each charging point, and a housing for each charging station.

We assume that charging stations with up to 100 kW output can be attached to the low-voltage grid [66]. A five-core cable is used as a mains cable with the diameter depending on the power supply. Details of the modeling can be found in the appendix. The length of the mains cable is set to 15 m.

According to [92,93], for a charging power up to 22 kW, a charging cable of the type $5G6 + 1 \times 0.5$ can be used. The total weight as well as the weight of copper is given in the dataset of [93]. For DC charging, a charging cable with diameter 3×16 mm^2 and $3 \times 2 \times 0.75$ mm^2 is used [94]. To estimate the amount of copper, a cable from [95] with $3 \times 16 + 3G2.5$ is selected. The length of the charging cable is set to 4 m [92].

For 22 kW, a stand-alone housing made of stainless steel is modeled [96–103]. For the dimension of the charging stations with two 22 kW charging points, a rounded average of the manufacturers sizes is used [96–110].

The dimension of the charging station with two 50 kW charging points is taken from [111]. As material for the 50 kW charging station, stainless steel or a material mix with stainless steel is used [112,113]. For simplification, we model a housing out of stainless steel. Both dimensions are shown in Table 6.

Table 6. Dimensioning charging stations.

		Unit	Charging Power in Kw	
			22	**50**
	Height	m	1.5	2
Dimensions	Width	m	0.4	0.85
	Depth	m	0.24	1

Transport and Use Phase

For transportation, the *Ecoinvent* dataset for the transportation of electronic products is used [114]. The total mass is composed of the mass of the aforementioned components. Since the sites of the charging stations are not further specified, the transport of the enclosure, the mains cable, the charging cable, and the housing are not considered in our assessment.

The use phase considers the provision of one kWh at the publicly accessible charging point. To estimate the environmental impacts, we consider the low-voltage system from *Ecoinvent*. The energy mix can be found in Table A1 in Appendix B [115].

2.3.3. Life Cycle Impact Assessment (LCIA)

Various characterization models exist for impact assessment. According to [116], *CML*, *ReCiPe*, and *TRACI* represent examples of selectable models. Deviations between the underlying properties of these models can affect the LCIA results [117]. Consequently, information about the used LCIA methodology should be stated. We describe selected methodological aspects subsequently.

In recent years, the damage-based assessment method *ReCiPe 2016* [118] has been increasingly used and is therefore selected for our study. Furthermore, for these LCIA methods, the midpoint and endpoint level of the environmental impacts can be assessed [119]. Endpoint impact indicators describe aggregated impacts on "areas of protection" (e.g., human health and the natural environment) [119,120]. Midpoint impact categories are located at an intermediate position of environmental impact cause–effect chains and consequently before the endpoint categories [119,120]. An advantage of midpoint indicators in comparison to endpoint indicators is their scientific robustness [121]. We use the *Hierarchist perspective* and carry out all evaluations on the midpoint level. With the choice of method, the impact categories, impact indicators, and characterization models defined in the method are simultaneously chosen. For the impact categories, we use the abbreviations shown in Table 7.

Table 7. Abbreviations and units for impact categories used.

Impact Category	Abbreviation	Unit
Water depletion	WD	m^3
Urban land occupation	ULO	m^2a
Terrestrial ecotoxicity	TET	kg 1,4-DCB-eq.
Terrestrial acidification	TA	kg SO_2-eq.
Photochemical oxidant formation	POF	kg NMVOC
Particulate matter formation	PMF	kg PM_{10}-eq.
Ozone depletion	OD	kg CFC-11-eq.
Natural land transformation	NLT	m^2
Mineral resource depletion	MRD	kg Fe-eq.
Marine eutrophication	ME	kg N-eq.
Marine ecotoxicity	MET	kg 1,4-DCB-eq.
Ionizing radiation	IR	kg U^{235}-eq.
Human toxicity	HT	kg 1,4-DCB-eq.
Freshwater eutrophication	FE	kg P-eq.
Freshwater ecotoxicity	FET	kg 1,4-DCB-eq.
Fossil resource depletion	FD	kg oil-eq.
Climate change	CC	kg CO_2-eq.
Agricultural land occupation	ALO	m^2a

3. Results

Having described relevant materials and methods, we now show our results of the LCA and the economic analysis.

3.1. Results and Interpretation of the Life Cycle Impact Assessment (LCIA)

We show our results of the LCA first on a component level and then aggregate to a system level. Then, we scale our results according to our diffusion scenarios. Subsequently, we conduct sensitivity analyses to account for uncertainties in modeling.

3.1.1. Component-Based Results

In the following, the component level is evaluated to identify individual components with high optimization potential within the power electronics and the charging infrastructure. This implies that the environmental impacts specified below apply to one provided kWh by one charger or one public charging station or one publicly accessible charging point, respectively.

- Power electronics

We identify the filter to be a main polluter for 3.7 kW chargers as well as for 22 kW chargers. For 50 kW chargers, the filter as well as the DC-DC converter are main polluters. For all filters, the inverter-side coils are the main contributors to pollution. In the DC-DC converter, the coils and transformers are the components with a major influence in all impact categories. The transformers outvalue the coils when it comes to mineral resource depletion, which can be explained with the ferrite used for the core. For the AC-DC converter, the coils are main polluters as well, followed by diodes, MOSFETs, and capacitors. The percentage share of the components can be found in Figure 2, which is exemplary for the 22 kW charger. See Table A3 in Appendix C for more detailed results.

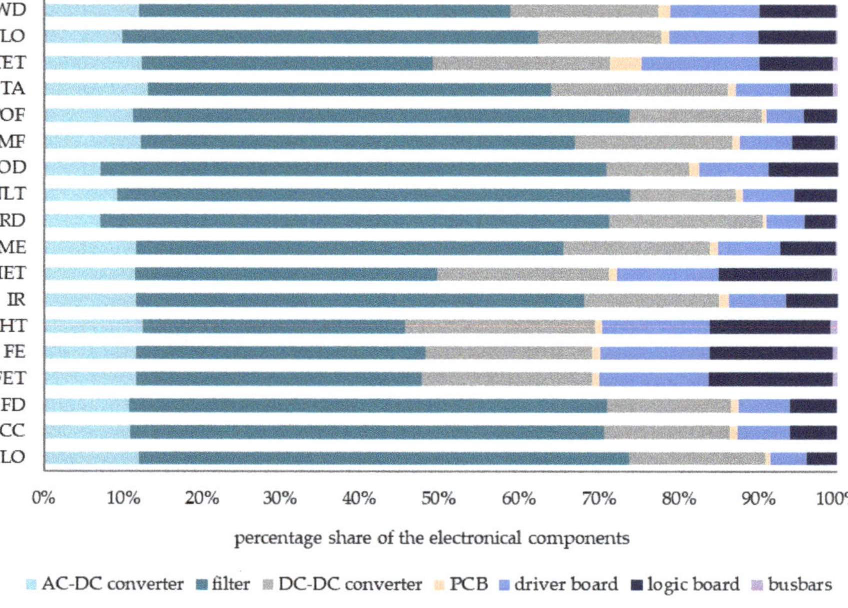

Figure 2. Percentage share of the electronic components for the modeled 22 kW charger per kWh.

When considering the impact of power electronics and the enclosure, it turns out that for the 22 kW charger, the power electronics account for more than 50% of the environmental impacts for all impact categories. The enclosure has the main impact for all chargers when it comes to climate change and the least when considering mineral resource depletion.

- Charging infrastructure (CIS)

As for the CIS, we consider the mains cable, the charging cable, and the housing. For the 22 kW CIS, the mains cable is the main emitter in all impact categories. The mains cable has the highest impact in 13 out of 18 impact categories (except agricultural land occupation, climate change, fossil resource depletion, ionizing radiation, and ozone depletion) for the 50 kW charging station. When comparing the mains cable of the 22 kW CIS and the 50 kW CIS, the mains cable for the 50 kW charging station has between 72% (fossil resource depletion) and 90% (mineral resource depletion) higher impacts than the mains cable for the 22 kW charging station. Considering the charging cable, the charging cable for the 50 kW charging station causes between 39% (fossil resource depletion) and 146% (mineral resource depletion) higher impacts than the charging cable for the 22 kW charging station. For the 22 kW charging station, the charging cable causes less than 3.2% in all impact categories. For the 50 kW charging station, the impact of the charging cable is below 4% in all impact categories. The housing of the 50 kW charging station causes about three times higher impacts than the 22 kW charging station for all impact categories. For both charging stations, the housing has the highest impact in the impact category climate change.

3.1.2. System-Based Results

The following results refer to a system level. The scaling of the components is shown in Table 1.

- Chargers

For the DC system, the total amount of 3.7 kW OBCs (scales with the number of vehicles) leads to higher environmental impacts (more than 70% for all impact categories) than the sum of the 50 kW chargers (scales with the number of charging points). The relative impact of the 50 kW chargers is highest for human toxicity (about 28%). Concerning global warming, the 3.7 kW chargers account for about 80% of the emitted CO_2-eq. (0.07 kg CO_2-eq./kWh).

- Charging infrastructure (CIS)

When comparing the environmental impacts of the CIS on a system base, the AC system has less environmental impacts in 10 out of 18 impact categories when compared to the CIS of the DC system. We find the most significant difference (percentage wise) to be present in climate change with additional emissions of $0.7 \cdot 10^{-3}$ kg CO_2-eq. per kWh when the DC system is used. The environmental impacts of the charging infrastructure for the AC system and the DC system can be found in Table A4 in Appendix C.

When comparing the environmental impacts of the total number of chargers and the CIS, chargers account for over 90% of the environmental impacts for all impact categories in the AC system. For the DC system, the CIS has a more significant environmental impact (percentage wise) than in the AC-based system. The most significant impact of the CIS in the DC system can be seen in human toxicity (about 25% or 0.05 kg 1,4-DCB-eq./kWh). The transport only accounts for less than 0.3% of the emissions in all impact categories; therefore, it is neglected in further examinations.

- Charging infrastructure, chargers, transport, and use phase

When summing up the environmental impacts of the production (chargers as well as charging infrastructure) and comparing them with the environmental impacts of the use phase (energy provision), we find that the use phase is responsible for more than 50% of the environmental impacts in 12 impact categories for the AC system. Only *mineral resource depletion, photochemical oxidant formation, natural land transformation, particulate matter formation, terrestrial ecotoxicity,* and *urban land occupation* predominate in the production phase. When considering the DC system, solely *mineral resource depletion* is dominant during production, while all other environmental impacts are mainly emitted during the use phase. Referring to global change in the AC system, only about 30% of CO_2-eq. are emitted during production; while in the DC system, it is less than 12%. Due to the scaling to one kWh (i.e., the functional unit) and the same charging efficiency for all chargers, the same total amount of emissions is emitted during the use phase in both systems, which also means that changes in the energy mix have the same impact on the use phase for both systems.

When comparing the overall environmental impacts of the two systems (production, transport, and use), a reduction of environmental impacts in all 18 impact categories is achieved by the DC system. Figure 3 shows the results—with the environmental impacts of the AC system scaling to 100%. We observe the most significant differences in percentage terms in *mineral resource depletion* (68.11%), *photochemical oxidant formation* (48.10%), and *natural land transformation* (44.77%). The most significant absolute changes can be observed in the impact categories *human toxicity* (0.36 kg 1,4-DCB-eq./kWh), *mineral resource depletion* (0.19 kg Fe/kWh), and *climate change* (0.19 kg CO_2-eq./kWh).

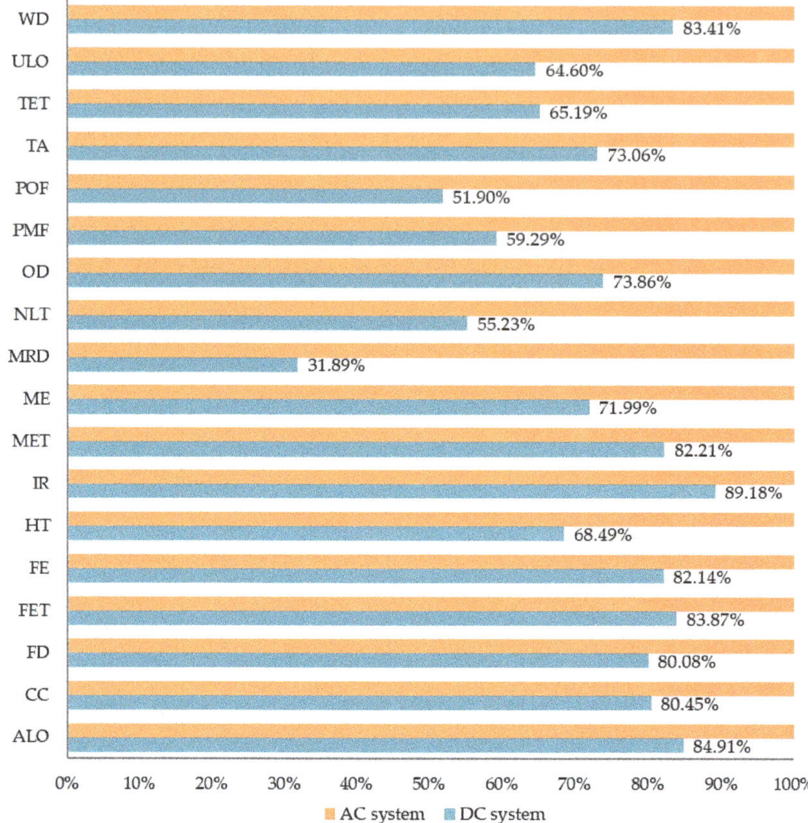

Figure 3. Comparison of the environmental impacts of the AC system and DC system. The environmental impacts of the AC system are scaled to 100%.

For *mineral resource depletion* (see Figure 4a), power electronics have the most significant impact (≈68% and ≈90%, respectively). It was shown before that mainly the DC-DC converter and the filter contribute to that impact. Within the converter and the filter, the coils constitute the main demand for mineral resources. For natural land transformation (see Figure 4b) in the AC-based system, the production of power electronics (mainly the filter) could be identified as main contributors (≈55%) while for the DC system, the use phase is the main contributor. Considering *photochemical oxidant formation* (see Figure 4c), in the AC system, power electronics (filter and DC-DC converter) are the main contributors, while it is the use phase in the DC system.

The most significant absolute changes were observed in the impact categories *human toxicity* (0.36 kg 1,4-DCB-eq./kWh), *mineral resource depletion* (0.19 kg Fe/kWh), and *climate change* (0.19 kg CO_2-eq./kWh). For *human toxicity* (see Figure 5a), the use phase has the highest impact followed by power electronics. Within the power electronics for the 22 kW charger, the main contributor is the DC-DC converter followed by the filter and the logic board. For *climate change* (see Figure 5b), the main source of environmental impacts is the use phase. Within the production of the power electronics, it is mainly the filter that contributes to climate change for both the 22 kW and the 50 kW charger.

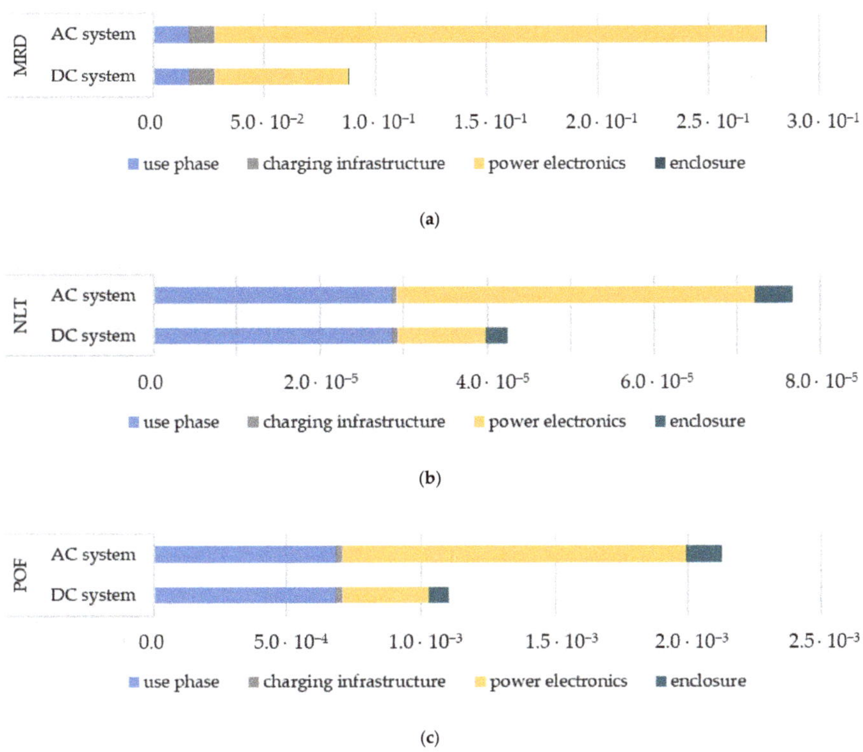

(a)

(b)

(c)

Figure 4. Mineral resource depletion (**a**), natural land transformation (**b**), and photochemical oxidant formation (**c**) for AC system and DC system per kWh.

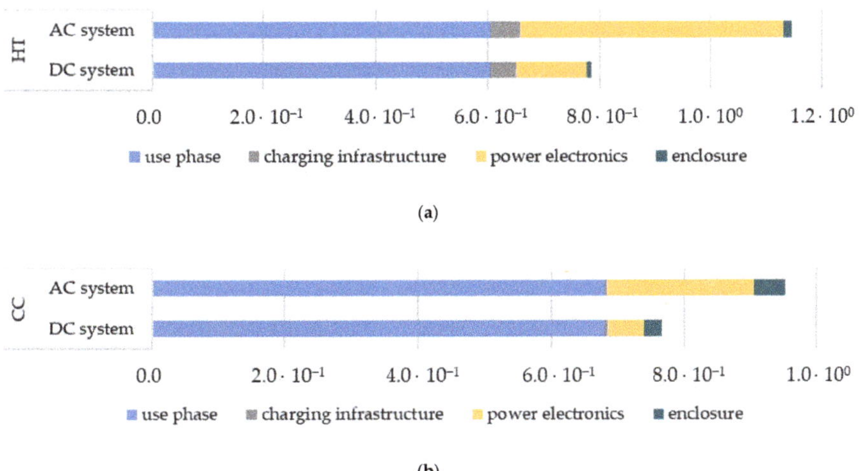

(a)

(b)

Figure 5. Human toxicity (**a**) and climate change (**b**) for AC system and DC system per kWh.

3.1.3. Scenario-Based Results

In a next step, we apply our results to the diffusion scenarios that we have introduced in Section 2.2. Therefore, the analysis is no longer carried out for one kWh of charging electricity (i.e., the functional unit), but rather based on the stock of BEVs and the regarding charging requirements per scenario. The absolute environmental impacts for the scenarios can be found in Table A2 in Appendix B. Table 8 describes the absolute reductions of environmental impacts in selected impact categories that could be achieved by implementing the DC system instead of the AC system with considering a lifetime of 10 years for all components.

Table 8. Absolute reductions for selected environmental impact categories when comparing DC system and AC system for the modeled scenarios considering a lifetime of 10 years for all components.

Impact Category	Absolute Reductions		
	Scenario S	Scenario M	Scenario L
CC	$1.95 \cdot 10^8$	$2.52 \cdot 10^8$	$3.45 \cdot 10^8$
HT	$3.78 \cdot 10^8$	$4.89 \cdot 10^8$	$6.70 \cdot 10^8$
MRD	$1.97 \cdot 10^8$	$2.55 \cdot 10^8$	$3.49 \cdot 10^8$
NLT	$3.60 \cdot 10^4$	$4.66 \cdot 10^4$	$6.38 \cdot 10^4$
POF	$1.07 \cdot 10^6$	$1.39 \cdot 10^6$	$1.90 \cdot 10^6$

In the impact category, climate change of about 20, 25, and 35 mn. kg CO_2-eq. yearly can be saved when implementing the DC system instead of the AC system. That would represent approximately 0.012%, 0.016%, and 0.022% of current annual greenhouse gas emissions in the transport sector in Germany [7]. The analysis also indicates that the potential for reductions increases as the number of BEVs increases.

3.1.4. Sensitivity Analyses

To account for data uncertainties in modeling, we conduct sensitivity analyses. Therefore, we vary the emission intensity of the production of the power electronics. Additionally, we vary the average utilization of the charging points as well as the efficiency of the charging points and the ratio of home charging (i.e., non-public charging, see modeling in Section 2.2). We provide an overview of our parameter variations in Table 9. The investigated parameters represent those with the highest effect on our results. We conduct the analyses for all 18 impact categories and exemplify it for significant impact categories.

Table 9. Parameter variation for sensitivity analysis.

Parameter	Original Value	Interval	Iterations	Impact On
Emission intensity of the production of power electronics	100%	[80%;120%]	20	Production
Degree of utilization	10%	[5%;25%]	20	Production
Efficiency	90%	[80%;100%]	20	Production and use phase
Ratio of charging at home/non-public charging	70%	[60%;80%]	20	Production and use phase

The variation in the environmental impact of power electronics production results in greater environmental impacts in the AC system even for the most unfavorable combination for the DC system (DC system: 120%; AC system: 80%) for all impact categories. As the power electronics have the main impact in the overall systems in *mineral resource depletion* (90% in the AC system and 68% in the DC system), *natural land transformation* (56% in the AC system and 25% in the DC system), and *photochemical oxidant building* (60% in the AC system and 29% in the DC system), the result of the sensitivity analyses is shown to be exemplary for those impact categories (Figure 6a–c). Additionally, the result for climate change is shown in Figure 6d. This suggests that even with greater uncertainties in the environmental impact of power electronics, the DC system has lower environmental impacts than the AC system in our model.

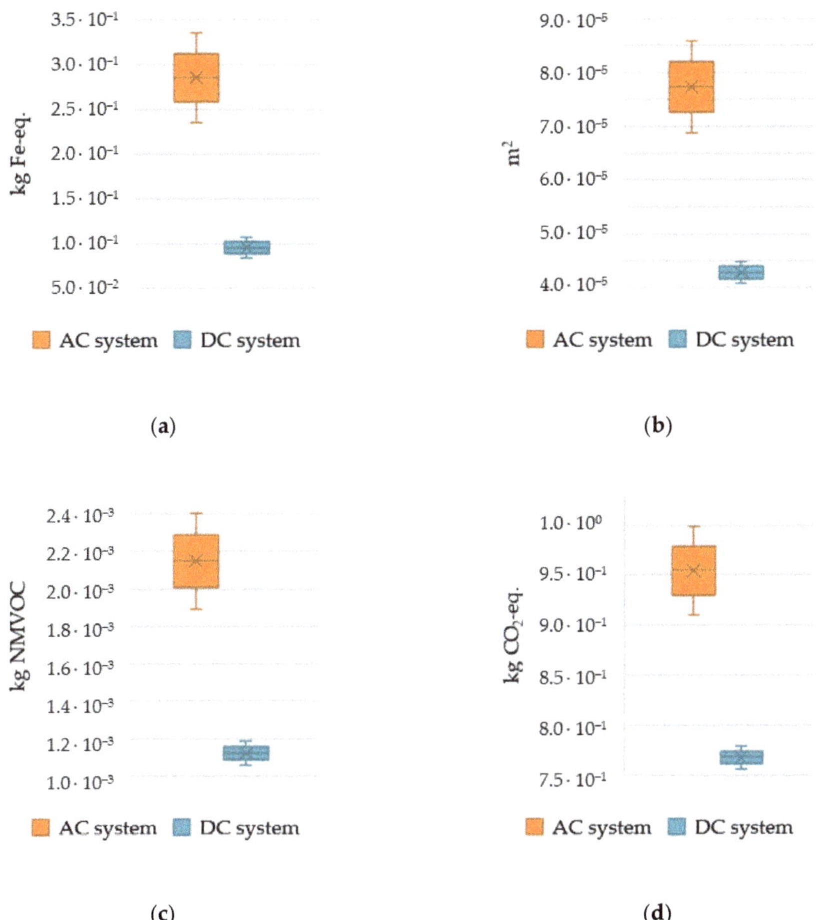

Figure 6. Results per kWh for sensitivity analyses of the emission intensity of the production of power electronics exemplary for (**a**) mineral resource depletion; (**b**) natural land transformation; (**c**) photochemical oxidant building; and (**d**) climate change.

When varying the degree of utilization, the most considerable changes on a percentage basis are in *human toxicity*, *mineral resource depletion*, and *terrestrial ecotoxicity* (see Figure 7a–c). Again, Figure 7d shows the result for climate change.

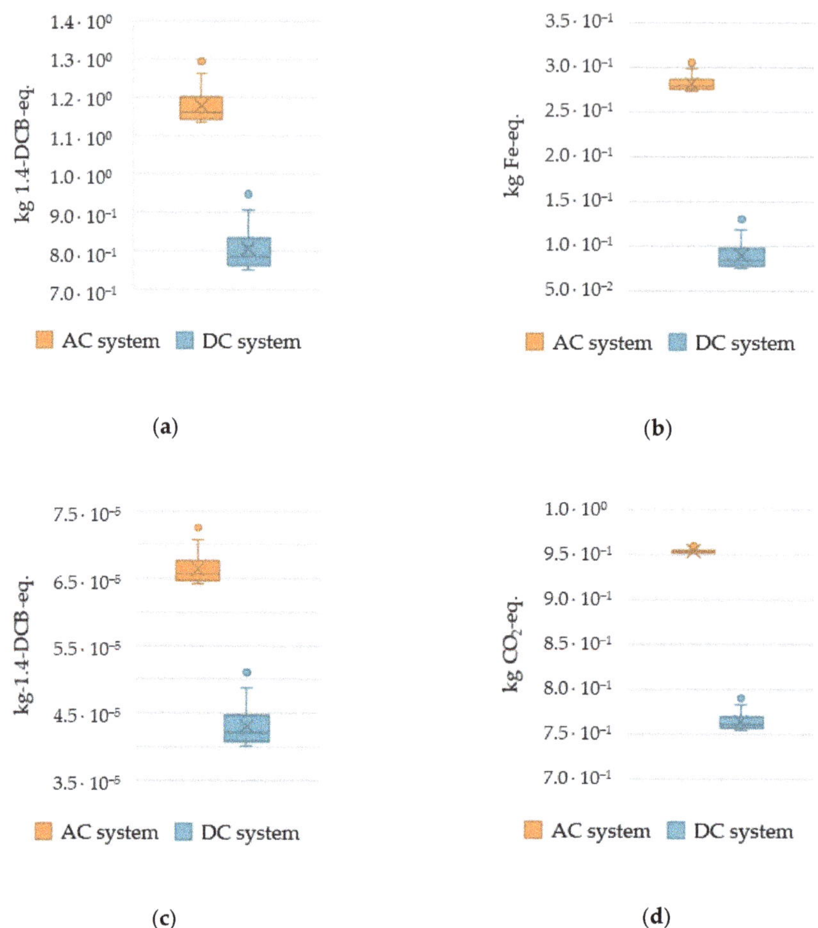

(a)

(b)

(c)

(d)

Figure 7. Results per kWh for sensitivity analyses of the degree of utilization (**a**) human toxicity; (**b**) mineral resource depletion; (**c**) terrestrial ecotoxicity; and (**d**) climate change.

The efficiency of the charging process can vary considerably, although charging stations with higher charging powers usually have a higher efficiency. When comparing the most unfavorable combination for the DC system (DC system: 80%; AC system: 100%), the AC system has less environmental impacts in the impact categories *agricultural land occupation, freshwater ecotoxicity, ionizing radiation,* and *water depletion.* The resulting sensitivities for the impact categories are shown in Figure 8a–d. From our results, we can conclude that although there are cases where the AC system causes less emissions than the DC system in the above-mentioned impact categories, the mean and median of the DC system are significantly below the mean and median of the AC system in all critical impact categories.

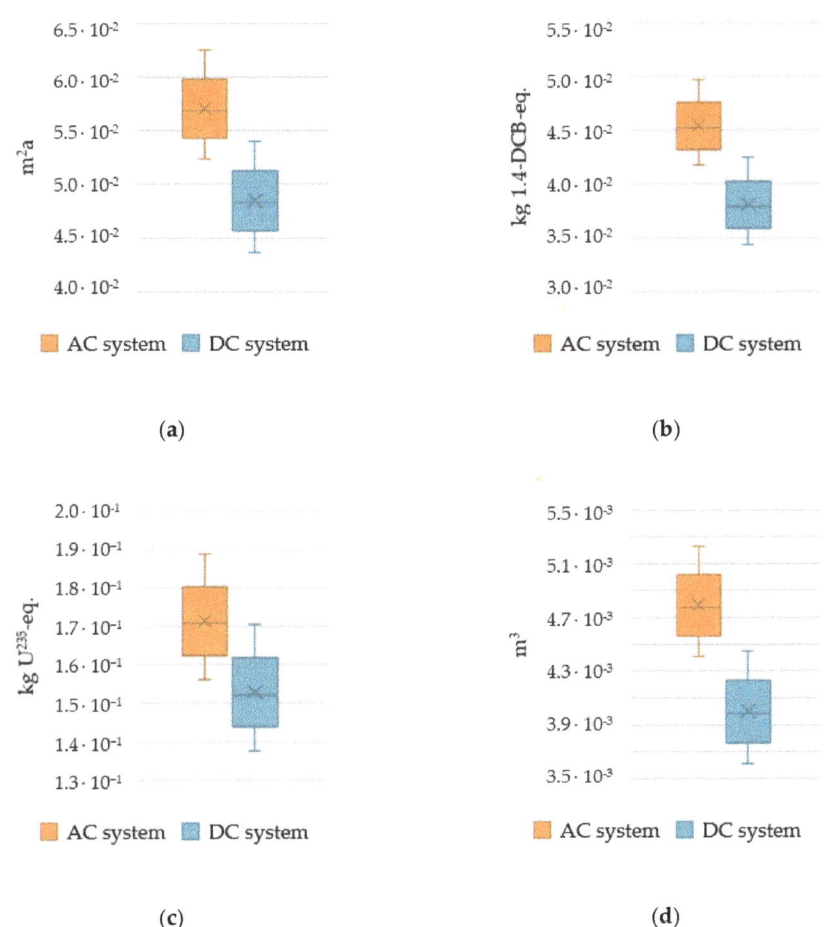

Figure 8. Results per kWh for sensitivity analyses of the efficiency for (**a**) agricultural land occupation; (**b**) freshwater ecotoxicity; (**c**) ionizing radiation; and (**d**) water depletion.

When varying the ratio of home charging (i.e., non-public charging and therefore the number of publicly accessible charging stations, see Section 2.1) and evaluating the DC system's most unfavorable combination (DC system: 80%; AC system: 60%), the DC-based system has less environmental impacts in all impact categories, even though the DC system has more charging points than the AC system in this constellation. The results are shown in Figure 9, which are exemplary for the most significant changes (on a percentage basis) (Figure 9a–c) as well as for *climate change* (Figure 8d). However, it should be noted that the total energy demand does not change if the proportion of home charging is increased. Since the energy used for home charging is not included in the modeling, there are large variations in the considered impact categories. This effect would decrease if the energy required for home charging was included.

Summing up, the sensitivity analyses showed some combinations for which the DC system has more environmental impacts in some impact categories (i.e., *agricultural land occupation, freshwater ecotoxicity, ionizing radiation,* and *water depletion;* see Figure 8). Nevertheless, the overall finding, that the DC system has less environmental impacts than the AC system in the modeled system, seems to be consistent, as this constellation only represents rather unlikely parameter combinations.

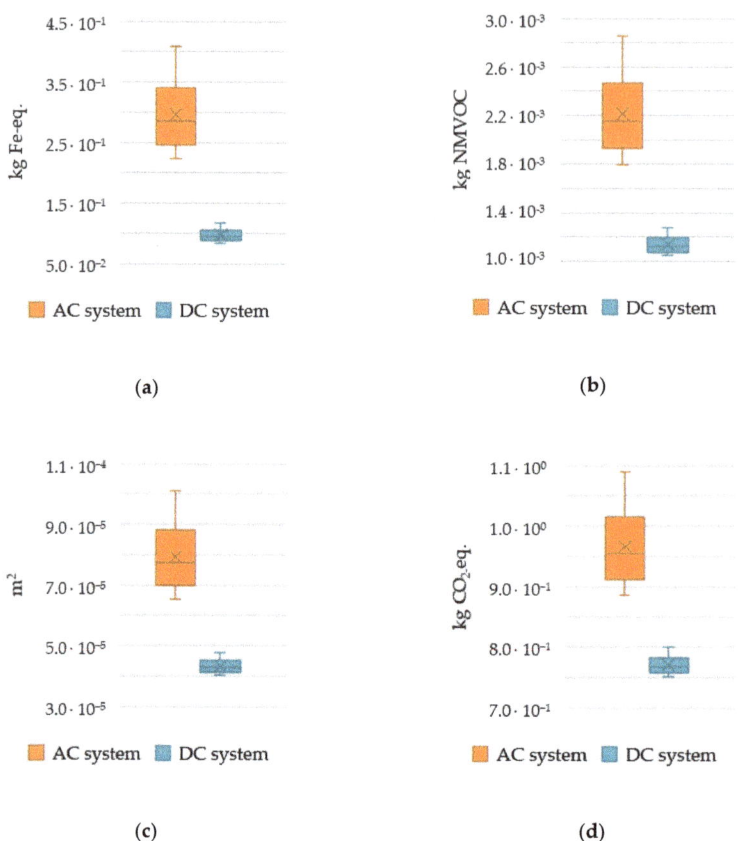

Figure 9. Results per kWh for sensitivity analyses of the ratio of home charging for (**a**) mineral resource depletion; (**b**) photochemical oxidant formatting; (**c**) natural land transformation; (**d**) climate change.

3.2. Economic Analysis

We conduct the economic analysis and distinguish between CIS and charging devices. The considered cost components are shown in Table 10. All estimates are based on [61,122,123]. The costs for the CIS and the chargers are annualized.

Table 10. Cost listing of the charging stations.

Parameter	Unit	System	
Charging power	kW	22	50
Charging points (for each charging station)	Pieces	2	2
Hardware	EUR	2500	3000
Grid connection cost	EUR	2000	10,000
Permission and planning	EUR	1000	3000
Montage, construction cost and signage	EUR	2000	7000
Total investment cost	EUR	7500	50,000
Current cost	EUR/a	750	3000
Interest rate	%	5	5
Life span	a	10	10

Reference: based on [61,122,123].

We calculate the cost for the CIS for all three diffusion scenarios. Figure 10 shows our results. Despite the fewer charging stations in the DC system, the higher costs of the charging stations cannot be compensated. The system costs for the CIS in the DC system are about 180% higher than the system costs in the AC system.

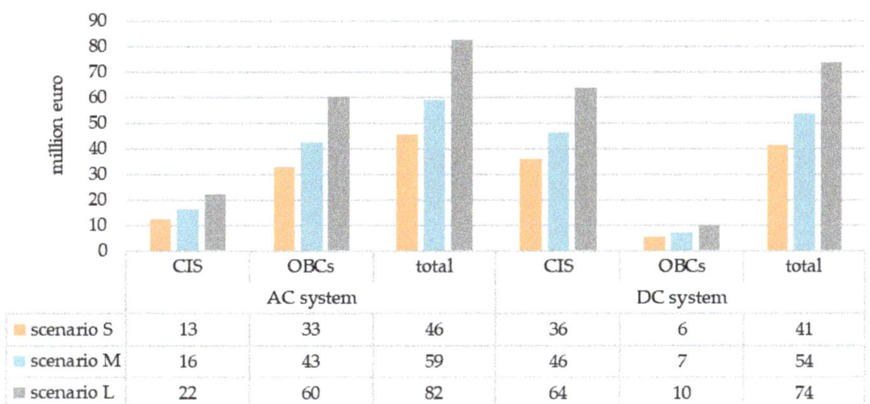

	CIS	OBCs	total	CIS	OBCs	total
		AC system			DC system	
scenario S	13	33	46	36	6	41
scenario M	16	43	59	46	7	54
scenario L	22	60	82	64	10	74

Figure 10. Cost estimation AC system and DC system.

For the OBC, the cost estimation is based on Mathieu [124], who states about 250 EUR for a 3.7 kW charger. Due to the modular design in our paper, the cost for the 22 kW chargers results in 1500 EUR. The costs for the OBCs in the scenarios are shown in Figure 10.

The OBCs have a significant impact for the AC system. About 70% of the total cost for the AC system arise from the OBCs, while in the DC system, the OBCs are only about 13% of the total cost. Thus, in total, the AC system becomes more expensive than the DC system. From a system point of view, in our scenarios, 4.2 mn., 5 mn. and 8.5 mn. euros can be saved annually with the DC-based charging infrastructure compared to the AC-based system. The analysis also indicates that the potential for savings increases with the number of BEVs.

4. Conclusions and Outlook

In our study, we analyzed and compared the environmental impacts of different charging concepts for electric vehicles. Two systems were considered, which essentially differ in the position of the chargers. When charging with alternating current (AC), an on-board charger is used. In our AC system, a charging power of 22 kW is considered for the publicly accessible charging infrastructure, and home charging with 3.7 kW can be conducted without any additional components. Charging at higher power levels requires an off-board charger. In the DC system, a charging power of 50 kW is assumed. Additionally, a 3.7 kW on-board charger is designed to allow for AC charging at home (3.7 kW). In our modeled systems, publicly accessible infrastructure rather solely supports AC charging (22 kW) or DC charging (50 kW). The calculation of the required charging infrastructure is based on the total energy demand for charging in the publicly accessible area. We defined three diffusion scenarios to model the near-future stock of BEVs in Germany and the required amount of charging infrastructure.

According to the goal of our study, we evaluated the environmental impacts of the systems by conducting an LCA. Our results show that the DC charging system causes less environmental impacts than the AC system in all impact categories. Even though the use phase has a large impact on the environmental analysis, it has the same impact for both systems when scaling to 1 kWh due to the model parameters. For the production, we showed that—on a component level—chargers are the main contributor in both systems. In the DC system, the sum of 3.7 kW chargers has a greater impact than the sum of the 50 kW chargers. Within the chargers, we identified the filters as main polluters for the 3.7 kW and the 22 kW chargers. For 50 kW chargers, the filter and the DC-DC converter cause the most environmental impacts. For all filters, the inverter-side coils are the main polluters. In the DC-DC converter, the coils and transformers are the components with a major influence in all impact categories. When assigning the results to our scenarios, for *global warming*, annual emission reductions of 20 mn., 25 mn., and 35 mn. kg CO_2-eq. could be achieved when the DC system is used instead of the AC system, which represent approximately 0.012%, 0.016%, and 0.022% of current annual greenhouse gas emissions in the German transport sector [7].

The main reason for the differences in the two systems is the scaling. While in the AC system each vehicle is equipped with a 22 kW charger (the number of chargers and the corresponding power electronics scales with the number of vehicles (see Table 1), in the DC system, the users of electric vehicles share the chargers and the corresponding power electronics in publicly accessible areas (the 50 kW chargers scale with the number of charging stations (see Table 1)).

Our sensitivity analyses confirmed the advantages of the DC system and showed the robustness of the results. Even in cases for the most unfavorable combination for the DC system, it causes less environmental impacts than the AC system.

For our economic analysis, we calculated the cost on a German-national level. Even though the DC charging infrastructure causes significantly higher cost than the AC charging infrastructure, this becomes less important when the costs of the OBCs are also considered. In our scenarios, annual savings of 4.2 mn., 5.4 mn., and 8.5 mn. euros are possible when the DC system is used instead of the AC system. The reductions of environmental impacts and the cost savings both scale with the stock of BEVs in our model.

All findings gained should be considered within the model boundaries. First, we want to mention that even though the *Ecoinvent* database is commonly used, some generic processes may have changed over time, so there may be some shifts of the environmental impacts. In addition, the database is often based on processes from the European area, which should be taken into account when transferring the results to other countries, although it cannot necessarily be assumed that the processes and environmental impacts differ significantly. Furthermore, our study is only valid for the modeled components. If different or additional components are used, e.g., if the charging power is changed, the resulting environmental impacts of these components must be considered separately. In addition, the density of the publicly available charging infrastructure is not explicitly investigated. Although the sensitivity analysis of the utilization shows that even when the DC system has more charging points than the AC system it causes less environmental impacts than the AC system, this aspect would need to be investigated further. Another assumption made is that the energy mix stays constant during the day. As the CO_2 intensity of the energy supply differs intraday, considering the charging behavior of users within the systems is an interesting supplement as well. Nevertheless, the study provides a systems comparison of AC and DC-based charging and gives an overview of the components that have a relevant environmental impact and that may need to be modeled in more detail in future considerations.

Based on environmental as well as economic results, our study shows that DC charging offers both economic saving and environmental reduction potentials. The advantages of the DC system scale with the number of electric vehicles, so it can be assumed that the findings will become more relevant as the number of BEVs increases. With the current distribution of charging solutions in Germany, on-board charging dominates, while off-board charging accounts for about 7% of all public charging points. The on-board charging capacity of the newly offered vehicles seems to settle at a level of 11 kW. In order to achieve higher charging performance, off-board charging will be increasingly used.

Author Contributions: Conceptualization, M.K., L.N., B.J.M., J.C.K., W.K., R.W.D.D. and A.P.; Data curation, M.K.; Formal analysis, M.K.; Methodology, M.K., L.N., B.J.M., J.C.K., W.K., R.W.D.D. and A.P.; Resources, L.N., B.J.M., J.C.K., W.K., R.W.D.D. and A.P.; Software, M.K.; Supervision, L.N., B.J.M., J.C.K., W.K., R.W.D.D. and A.P.; Validation, L.N., B.J.M., J.C.K., W.K., R.W.D.D. and A.P.; Visualization, M.K.; Writing—original draft, M.K.; Writing—review and editing, M.K., L.N., B.J.M., J.C.K. and A.P. All authors have read and agreed to the published version of the manuscript.

Funding: L.N., B.J.M., R.W.D.D. and A.P. received funding by the Federal Ministry of Education and Research (BMBF) and the Ministry of Culture and Science of the German State of North Rhine-Westphalia (MKW) under the Excellence Strategy of the Federal Government and the Länder (grant ID: (DE-82)EXS-SF-OPSF560). J.C.K. and W.K. gratefully acknowledge funding of the center of excellence "Virtual Institute—Power to Gas and Heat" (EFRE-0400151) by the "Operational Program for the promotion of investments in growth and employment for North Rhine-Westphalia from the European fund for regional development" (OP EFRE NRW) through the Ministry of Economic Affairs, Innovation, Digitalization and Energy of the State of North Rhine-Westphalia.

Conflicts of Interest: The authors declare no conflict of interest.

Appendix A. Background Information to Section 2.2

The total number of charging stations is calculated as follows:

$$n_{CS} = \frac{\sum_{classes\ S,M,L} \left(\alpha_{PC} \cdot am_{car} \cdot n_{BEV,\ class} \cdot c_{BEV,class} \right)}{n_{CP,CS} \cdot UT_{CP} \cdot P_{CP} \cdot \eta_{CP} \cdot t_{PC}} \tag{A1}$$

with

α_{PC} : share public charging [/]

am_{car}: annual mileage of a car [km]

$n_{BEV,\ class}$: number of BEVs in the class [/]

$c_{BEV,class}$: consumption of BEVs in the class [kWh/km]

$n_{CP,CS}$: number of charging points per charging station [/]

UT_{CP}: utilization charging point [/]

P_{CP}: power charging point [kW]

η_{CP}: efficiency charging point [/]

t_{PC}: time for public charging [h]

Appendix B. Background Information for the Life Cycle Inventory (LCI)

In the following, we provide a more detailed description of the components modeled in Section 2.3.2. The structure of the background information is similar to the structure in in the Sub-Section *Production of Components*.

- Filter

For the calculations, we use a maximum current ripple of 15% [70]. For the design of the filter capacity, we assume that the maximum change in the power factor perceived by the grid is 5%. The desired attenuation is set to 20%, the switching frequency of 200 kHz and the DC voltage result from [68,69], respectively. Based on the specified parameters, we select electronic components that sufficiently fulfill the characteristics, are available on the market, and for which material data are available. The components used in our study can be found in Table 3.

- AC-DC Converter

Even though the capacitors for surface-mounting in the *Ecoinvent* database are not specified in [125], the dataset shows a ratio of 40% ceramics and is therefore considered as the best available dataset for ceramic capacitors. As an estimate for the mass of the capacitor, a dataset with mass specification from Murata is used [80]. The coil *EELP 58* is used instead of *EELP 54* [81]. The mass of the core is given in the manufacturers' dataset. For the windings, we assume that 60% of the cores' air space is filled with wire. With the density of copper (8920 kg/m^3), the mass of copper for the windings is estimated. The production process is modeled with the *Ecoinvent* dataset for inductors.

- DC-DC Converter

For the coil core, *EELP64* is used. The amount of copper is calculated as shown for the AC-DC converter.

- Printed circuit board (PCB), driver board, logic board, and busbars

For the size estimation of the size of the PCB, we estimate the outer dimension of the chargers. Schmenger et al. [68] state the volumes of the modular chargers for different charging powers. Based on the assumption that the 50 kW charger has the same power density as the 22 kW charger, we estimate the volume for the 50 kW charger. The ratio of width and height is based on [89]. We assume that the ratio remains constant over power. Based on the model from [89], we assume that the area of the PCB is 30% of the area of the charger.

As stated above, we base the dimensioning of the busbar on [87,88]. As stated by Mersen [88], an ampacity of 5 A/mm^2 is considered and a 5% security surcharge is obeyed for each additional conductor. The minimum cross-sectional area is based on [87]. The maximum current considered originates from [67,68]. We estimate the maximum current based on the modularity of the charging unit. For the AC busbars, a quantity of three trace pitches can be assumed [87]. For the DC busbars, two trace pitches are assumed. We estimate the length of the busbars based on the outer dimensions of the charger [67]. The production process is modeled based on [87]. The surface of the busbars is galvanically nickel-plated. With the assumption of a thickness of 1 mm for the bus bars, a thickness of 20 μM for nickel-plating [87] and the density of 8900 kg/m^3 for nickel the material flow and process data are based on [87].

- Enclosure of electronic components and heatsink

To calculate the material flows of the enclosure for electronic components, the dimensions of the heatsink need to be estimated in a first step. The heatsink for the electronic devices shown in [89] is liquid-cooled. Based on [87], a thickness of 2.5 cm can be assumed. Under the assumption that the remaining space is used by electronic devices, we calculate the height of the enclosure for the electronic components from the dimensions of the charging unit and the height of the heat sink. The heat sink composes the base plate of the charging unit and is modeled separately. The material flows and processes used for the production phase of the enclosure with an untreated surface are based on [87].

The size estimations for our air-cooled heatsink are based on the assumption of a forced cooling with an airflow of at least 7 *m/s* [87]. We do not model the fan. We determine the mass of the air-cooled heatsink on the basis of a standard heatsink [126]. In our estimations, the size of the heat sink matches the base area of the charging unit. With the given ratio of weight and length of the heatsink [126], we estimate the mass of aluminum. For surface refinement, the surface area of the heatsink is anodized [87]. We estimate the surface to be anodized by taking the number of lamellas from [126] without considering the changing number of lamellas with the changing width. Therefore, each lamella has two sides to be anodized. The side of the heatsink that forms the bottom surface of the housing is not treated. With the anodizing surface calculated, accordingly, the further process data is taken from [87].

- Charging infrastructure (CIS)

Based on [127] for a charging station with two charging points at 22 kW, we select the mains cable by [128] (Type NYY-J 5 × 50 SW) and assume that the difference of the total weight to the weight of copper is caused by insulation. Due to the fact that the material for the insulation is not differentiated any further in the dataset of the manufacturer, we base our estimation of the composition of the insulation and the processes needed for production on the *Ecoinvent* dataset *"cable connector for computer, without plugs"* [125]. For 50 kW charging power, a cross-section of the mains cable of 75 mm^2 or 90 mm^2 is used [129]. Due to the non-availability of data, we use a 95 mm^2 cable for modeling [128] (Type NYY-J 5 × 95 SW). We calculate the quantity of insulation as described above.

For the charging cable, the insulation is estimated as mentioned above as the difference between the weight of copper and the total weight. The materials used for insulation as well as the production processes are based on the aforementioned dataset in *Ecoinvent*.

Regarding the housing, the most common stainless steel alloy (steel of grade *304* and *304L* which consists of 18% chrome and 8% nickel and has a density of 8000 kg/m^3 [130]) is used. The wall thickness is set to 2 mm [131]. For its production, the *Ecoinvent* dataset for sheet rolling is used. According to [96], the stainless steel is powder-coated. The related process is taken from *Ecoinvent*.

Table A1. Energy mix considered for modeling.

Energy Source	Percentage Share [%]
Coal	43.51
Nuclear	25.05
Natural gas	9.32
Wind power	4.03
Hydropower	3.32
oil	1.47
Hydropower, pumped storage	1.06
biomass	0.57
biogas	0.49
photovoltaic	0.09
Other	11.08

Reference: for detailed information, please refer to [115].

Table A2. Environmental impacts of the AC system and the DC system for modeled scenarios considering a lifetime of 10 years.

Impact Category	AC System			DC System		
	Scenario S	Scenario M	Scenario L	Scenario S	Scenario M	Scenario L
WD	$4.97 \cdot 10^6$	$6.43 \cdot 10^6$	$8.80 \cdot 10^6$	$4.14 \cdot 10^6$	$5.36 \cdot 10^6$	$7.34 \cdot 10^6$
ULO	$8.44 \cdot 10^6$	$1.09 \cdot 10^7$	$1.49 \cdot 10^7$	$5.45 \cdot 10^6$	$7.05 \cdot 10^6$	$9.65 \cdot 10^6$
TET	$6.79 \cdot 10^4$	$8.79 \cdot 10^4$	$1.20 \cdot 10^5$	$4.43 \cdot 10^4$	$5.73 \cdot 10^4$	$7.84 \cdot 10^4$
TA	$4.52 \cdot 10^6$	$5.86 \cdot 10^6$	$8.02 \cdot 10^6$	$3.31 \cdot 10^6$	$4.28 \cdot 10^6$	$5.86 \cdot 10^6$
POF	$2.23 \cdot 10^6$	$2.88 \cdot 10^6$	$3.94 \cdot 10^6$	$1.16 \cdot 10^6$	$1.50 \cdot 10^6$	$2.05 \cdot 10^6$
PMF	$1.46 \cdot 10^6$	$1.89 \cdot 10^6$	$2.59 \cdot 10^6$	$8.65 \cdot 10^5$	$1.12 \cdot 10^6$	$1.53 \cdot 10^6$
OD	$5.20 \cdot 10^1$	$6.73 \cdot 10^1$	$9.22 \cdot 10^1$	$3.84 \cdot 10^1$	$4.97 \cdot 10^1$	$6.81 \cdot 10^1$
NLT	$8.04 \cdot 10^4$	$1.04 \cdot 10^5$	$1.43 \cdot 10^5$	$4.44 \cdot 10^4$	$5.75 \cdot 10^4$	$7.87 \cdot 10^4$
MRD	$2.89 \cdot 10^8$	$3.74 \cdot 10^8$	$5.12 \cdot 10^8$	$9.22 \cdot 10^7$	$1.19 \cdot 10^8$	$1.63 \cdot 10^8$
ME	$8.42 \cdot 10^5$	$1.09 \cdot 10^6$	$1.49 \cdot 10^6$	$6.06 \cdot 10^5$	$7.84 \cdot 10^5$	$1.07 \cdot 10^6$
MET	$4.36 \cdot 10^7$	$5.64 \cdot 10^7$	$7.73 \cdot 10^7$	$3.59 \cdot 10^7$	$4.64 \cdot 10^7$	$6.35 \cdot 10^7$
IR	$1.78 \cdot 10^8$	$2.31 \cdot 10^8$	$3.16 \cdot 10^8$	$1.59 \cdot 10^8$	$2.06 \cdot 10^8$	$2.82 \cdot 10^8$
HT	$1.20 \cdot 10^9$	$1.55 \cdot 10^9$	$2.13 \cdot 10^9$	$8.22 \cdot 10^8$	$1.06 \cdot 10^9$	$1.46 \cdot 10^9$
FE	$1.31 \cdot 10^6$	$1.70 \cdot 10^6$	$2.33 \cdot 10^6$	$1.08 \cdot 10^6$	$1.40 \cdot 10^6$	$1.91 \cdot 10^6$
FET	$4.66 \cdot 10^7$	$6.03 \cdot 10^7$	$8.25 \cdot 10^7$	$3.91 \cdot 10^7$	$5.05 \cdot 10^7$	$6.92 \cdot 10^7$
FD	$2.67 \cdot 10^8$	$3.46 \cdot 10^8$	$4.74 \cdot 10^8$	$2.14 \cdot 10^8$	$2.77 \cdot 10^8$	$3.79 \cdot 10^8$
CC	$9.97 \cdot 10^8$	$1.29 \cdot 10^9$	$1.77 \cdot 10^9$	$8.02 \cdot 10^8$	$1.04 \cdot 10^9$	$1.42 \cdot 10^9$
ALO	$5.93 \cdot 10^7$	$7.67 \cdot 10^7$	$1.05 \cdot 10^8$	$5.03 \cdot 10^7$	$6.52 \cdot 10^7$	$8.92 \cdot 10^7$

Appendix C. Detailed Results on Component Level

Table A3. Absolute environmental impacts of the electronic components for the modeled 22 kW charger per kWh.

Impact Category	AC-DC Converter	Filter	DC-DC Converter	PCB	Driver Board	Logic Board	Busbars
WD	$6.92 \cdot 10^{-10}$	$2.71 \cdot 10^{-9}$	$1.07 \cdot 10^{-9}$	$9.21 \cdot 10^{-11}$	$6.48 \cdot 10^{-10}$	$5.58 \cdot 10^{-10}$	$1.59 \cdot 10^{-11}$
ULO	$2.12 \cdot 10^{-9}$	$1.12 \cdot 10^{-8}$	$3.29 \cdot 10^{-9}$	$2.18 \cdot 10^{-10}$	$2.40 \cdot 10^{-9}$	$2.08 \cdot 10^{-9}$	$4.44 \cdot 10^{-11}$
TET	$2.20 \cdot 10^{-11}$	$6.51 \cdot 10^{-11}$	$3.95 \cdot 10^{-11}$	$6.83 \cdot 10^{-12}$	$2.63 \cdot 10^{-11}$	$1.63 \cdot 10^{-11}$	$1.19 \cdot 10^{-12}$
TA	$1.10 \cdot 10^{-9}$	$4.26 \cdot 10^{-9}$	$1.86 \cdot 10^{-9}$	$8.32 \cdot 10^{-11}$	$5.77 \cdot 10^{-10}$	$4.46 \cdot 10^{-10}$	$5.21 \cdot 10^{-11}$
POF	$8.54 \cdot 10^{-10}$	$4.72 \cdot 10^{-9}$	$1.25 \cdot 10^{-9}$	$5.18 \cdot 10^{-11}$	$3.59 \cdot 10^{-10}$	$3.11 \cdot 10^{-10}$	$1.07 \cdot 10^{-11}$
PMF	$5.00 \cdot 10^{-10}$	$2.24 \cdot 10^{-9}$	$8.09 \cdot 10^{-10}$	$3.98 \cdot 10^{-11}$	$2.75 \cdot 10^{-10}$	$2.19 \cdot 10^{-10}$	$1.52 \cdot 10^{-11}$
OD	$6.89 \cdot 10^{-15}$	$6.12 \cdot 10^{-14}$	$1.00 \cdot 10^{-14}$	$1.19 \cdot 10^{-15}$	$8.52 \cdot 10^{-15}$	$8.35 \cdot 10^{-15}$	$4.34 \cdot 10^{-17}$
NLT	$2.34 \cdot 10^{-11}$	$1.63 \cdot 10^{-10}$	$3.36 \cdot 10^{-11}$	$2.20 \cdot 10^{-12}$	$1.64 \cdot 10^{-11}$	$1.36 \cdot 10^{-11}$	$2.05 \cdot 10^{-13}$
MRD	$1.04 \cdot 10^{-7}$	$9.36 \cdot 10^{-7}$	$2.81 \cdot 10^{-7}$	$7.64 \cdot 10^{-9}$	$7.19 \cdot 10^{-8}$	$5.54 \cdot 10^{-8}$	$4.36 \cdot 10^{-9}$
ME	$1.89 \cdot 10^{-10}$	$8.76 \cdot 10^{-10}$	$3.00 \cdot 10^{-10}$	$1.71 \cdot 10^{-11}$	$1.27 \cdot 10^{-10}$	$1.14 \cdot 10^{-10}$	$4.54 \cdot 10^{-12}$
MET	$6.58 \cdot 10^{-9}$	$2.18 \cdot 10^{-8}$	$1.22 \cdot 10^{-8}$	$5.29 \cdot 10^{-10}$	$7.33 \cdot 10^{-9}$	$8.14 \cdot 10^{-9}$	$3.88 \cdot 10^{-10}$
IR	$1.65 \cdot 10^{-8}$	$7.99 \cdot 10^{-8}$	$2.40 \cdot 10^{-8}$	$1.72 \cdot 10^{-9}$	$1.03 \cdot 10^{-8}$	$9.12 \cdot 10^{-9}$	$6.74 \cdot 10^{-11}$
HT	$3.50 \cdot 10^{-7}$	$9.26 \cdot 10^{-7}$	$6.61 \cdot 10^{-7}$	$2.50 \cdot 10^{-8}$	$3.79 \cdot 10^{-7}$	$4.25 \cdot 10^{-7}$	$2.33 \cdot 10^{-8}$
FE	$1.99 \cdot 10^{-10}$	$6.30 \cdot 10^{-10}$	$3.56 \cdot 10^{-10}$	$1.65 \cdot 10^{-11}$	$2.35 \cdot 10^{-10}$	$2.67 \cdot 10^{-10}$	$1.05 \cdot 10^{-11}$
FET	$6.40 \cdot 10^{-9}$	$1.99 \cdot 10^{-8}$	$1.17 \cdot 10^{-8}$	$4.80 \cdot 10^{-10}$	$7.52 \cdot 10^{-9}$	$8.62 \cdot 10^{-9}$	$3.63 \cdot 10^{-10}$
FD	$3.92 \cdot 10^{-8}$	$2.18 \cdot 10^{-7}$	$5.62 \cdot 10^{-8}$	$3.73 \cdot 10^{-9}$	$2.38 \cdot 10^{-8}$	$2.13 \cdot 10^{-8}$	$1.99 \cdot 10^{-10}$
CC	$1.43 \cdot 10^{-7}$	$7.77 \cdot 10^{-7}$	$2.06 \cdot 10^{-7}$	$1.37 \cdot 10^{-8}$	$8.65 \cdot 10^{-8}$	$7.75 \cdot 10^{-8}$	$7.88 \cdot 10^{-10}$
ALO	$7.80 \cdot 10^{-9}$	$4.03 \cdot 10^{-8}$	$1.11 \cdot 10^{-8}$	$4.44 \cdot 10^{-10}$	$3.02 \cdot 10^{-9}$	$2.42 \cdot 10^{-9}$	$7.86 \cdot 10^{-11}$

Table A4. Comparison of total environmental impacts and changes in environmental impacts per kWh when considering the charging infrastructure of the AC system and the DC system.

Impact Category	AC System	DC System	Δ Absolute [1]	Δ Relative [1]
WD	$3.25 \cdot 10^{-5}$	$3.13 \cdot 10^{-5}$	$1.14 \cdot 10^{-6}$	3.50%
ULO	$1.13 \cdot 10^{-4}$	$1.17 \cdot 10^{-4}$	$-3.86 \cdot 10^{-6}$	−3.41%
TET	$2.74 \cdot 10^{-6}$	$2.51 \cdot 10^{-6}$	$2.34 \cdot 10^{-7}$	8.54%
TA	$9.74 \cdot 10^{-5}$	$8.87 \cdot 10^{-5}$	$8.69 \cdot 10^{-6}$	8.92%
POF	$2.70 \cdot 10^{-5}$	$2.73 \cdot 10^{-5}$	$-2.79 \cdot 10^{-7}$	−1.03%
PMF	$3.47 \cdot 10^{-5}$	$3.55 \cdot 10^{-5}$	$-7.55 \cdot 10^{-7}$	−2.18%
OD	$1.74 \cdot 10^{-10}$	$2.13 \cdot 10^{-10}$	$-3.92 \cdot 10^{-11}$	−22.54%
NLT	$5.53 \cdot 10^{-7}$	$5.78 \cdot 10^{-7}$	$-2.55 \cdot 10^{-8}$	−4.61%
MRD	$1.14 \cdot 10^{-2}$	$1.17 \cdot 10^{-2}$	$-3.27 \cdot 10^{-4}$	−2.86%
ME	$1.12 \cdot 10^{-5}$	$1.08 \cdot 10^{-5}$	$4.04 \cdot 10^{-7}$	3.61%
MET	$9.32 \cdot 10^{-4}$	$8.85 \cdot 10^{-4}$	$4.67 \cdot 10^{-5}$	5.01%
IR	$2.25 \cdot 10^{-4}$	$2.67 \cdot 10^{-4}$	$-4.26 \cdot 10^{-5}$	−18.99%
HT	$5.17 \cdot 10^{-2}$	$4.48 \cdot 10^{-2}$	$7.00 \cdot 10^{-3}$	13.52%
FE	$2.35 \cdot 10^{-5}$	$2.05 \cdot 10^{-5}$	$3.01 \cdot 10^{-6}$	12.81%
FET	$8.73 \cdot 10^{-4}$	$8.33 \cdot 10^{-4}$	$3.98 \cdot 10^{-5}$	4.56%
FD	$8.99 \cdot 10^{-4}$	$1.06 \cdot 10^{-3}$	$-1.59 \cdot 10^{-4}$	−17.66%
CC	$3.05 \cdot 10^{-3}$	$3.79 \cdot 10^{-3}$	$-7.39 \cdot 10^{-4}$	−24.22%
ALO	$2.51 \cdot 10^{-4}$	$2.77 \cdot 10^{-4}$	$-2.55 \cdot 10^{-5}$	−10.13%

[1] AC system—DC system.

References

1. Yilmaz, M.; Krein, P.T. Review of Battery Charger Topologies, Charging Power Levels, and Infrastructure for Plug-In Electric and Hybrid Vehicles. *IEEE Trans. Power Electron.* **2013**, *28*, 2151–2169. [CrossRef]
2. Zeit Online. Forscher Warnen Vor Einer Heißzeit. 2018. Available online: https://www.zeit.de/wissen/umwelt/2018-08/klimaerwaermung-heisszeit-kippelemente-studie-klimaforschung (accessed on 10 August 2020).
3. Umweltbundesamt. Beobachteter Klimawandel. 2013. Available online: https://www.umweltbundesamt.de/themen/klima-energie/klimawandel/beobachteter-klimawandel (accessed on 10 August 2020).

4. Stern, N. *The Economics of Climate Change: The Stern Review*; Cambridge University Press: Cambridge, UK, 2007. [CrossRef]

5. Breidenich, C.; Magraw, D.; Rowley, A.; Rubin, J.W. The Kyoto protocol to the United Nations framework convention on climate change. *Am. J. Int. Law* **1998**, *92*, 315–331. [CrossRef]

6. Agreement, P. The Paris Agreement. In Proceedings of the Parties to the United Nations Framework Convention on Climate Change, Paris, France, 30 November–12 December 2015.

7. Umweltbundesamt. Trendtabellen Treibhausgase 1990–2018 (Stand: EU-Submission). 2020. Available online: https://www.umweltbundesamt.de/dokument/trendtabellen-treibhausgase-1990-2018-stand-eu (accessed on 5 December 2020).

8. Umweltbundesamt. Emissionsdaten. 2020. Available online: https://www.umweltbundesamt.de/themen/verkehr-laerm/emissionsdaten#verkehrsmittelvergleich_personenverkehr (accessed on 10 August 2020).

9. Eisenkopf, A.; Fricke, H.; Friedrich, M.; Haasis, H.-D.; Knieps, G.; Knorr, A. Nach der Klimakonferenz in Paris: Wird eine neue Klimastrategie für den Verkehr benötigt? *Z. Verk.* **2016**, *87*.

10. Hawkins, T.R.; Singh, B.; Majeau-Bettez, G.; Strømman, A.H. Comparative Environmental Life Cycle Assessment of Conventional and Electric Vehicles. *J. Ind. Ecol.* **2012**, *17*, 53–64. [CrossRef]

11. International Energy Agency. *Global EV Outlook 2017*; International Energy Agency: Paris, France, 2017. [CrossRef]

12. Nykvist, B.; Nilsson, M. Rapidly falling costs of battery packs for electric vehicles. *Nat. Clim. Chang.* **2015**, *5*, 329–332. [CrossRef]

13. Faria, R.; Moura, P.; Delgado, J.; De Almeida, A.T. A sustainability assessment of electric vehicles as a personal mobility system. *Energy Convers. Manag.* **2012**, *61*, 19–30. [CrossRef]

14. Elektromobilität, N.P. *Fortschrittsbericht 2018—Markthochlaufphase*; Nationale Plattform Elektromobilität (NPE): Berlin, Germany, 2018.

15. Musavi, F.; Edington, M.; Eberle, W.; Dunford, W.G. Evaluation and Efficiency Comparison of Front End AC-DC Plug-in Hybrid Charger Topologies. *IEEE Trans. Smart Grid* **2012**, *3*, 413–421. [CrossRef]

16. Nordelöf, A.; Messagie, M.; Tillman, A.-M.; Söderman, M.L.; Van Mierlo, J. Environmental impacts of hybrid, plug-in hybrid, and battery electric vehicles—What can we learn from life cycle assessment? *Int. J. Life Cycle Assess.* **2014**, *19*, 1866–1890. [CrossRef]

17. Tintelecan, A.; Dobra, A.C.; Marțiș, C. Literature Review—Electric Vehicles Life Cycle Assessment. In Proceedings of the 2020 ELEKTRO, Taormina, Italy, 23 March 2020; pp. 1–6.

18. Faria, R.; Marques, P.; Moura, P.; Freire, F.; Delgado, J.; De Almeida, A.T. Impact of the electricity mix and use profile in the life-cycle assessment of electric vehicles. *Renew. Sustain. Energy Rev.* **2013**, *24*, 271–287. [CrossRef]

19. Zhang, Z.; Sun, X.; Ding, N.; Yang, J. Life cycle environmental assessment of charging infrastructure for electric vehicles in China. *J. Clean. Prod.* **2019**, *227*, 932–941. [CrossRef]

20. Lucas, A.; Silva, C.A.; Neto, R.C. Life cycle analysis of energy supply infrastructure for conventional and electric vehicles. *Energy Policy* **2012**, *41*, 537–547. [CrossRef]

21. Nansai, K.; Tohno, S.; Kono, M.; Kasahara, M.; Moriguchi, Y. Life-cycle analysis of charging infrastructure for electric vehicles. *Appl. Energy* **2001**, *70*, 251–265. [CrossRef]

22. Traut, E.; Hendrickson, C.; Klampfl, E.; Liu, Y.; Michalek, J.J. Optimal design and allocation of electrified vehicles and dedicated charging infrastructure for minimum life cycle greenhouse gas emissions and cost. *Energy Policy* **2012**, *51*, 524–534. [CrossRef]

23. Bass, F.M. A New Product Growth for Model Consumer Durables. *Manag. Sci.* **1969**, *15*, 215–227. [CrossRef]

24. Massiani, J.; Gohs, A. The choice of Bass model coefficients to forecast diffusion for innovative products: An empirical investigation for new automotive technologies. *Res. Transp. Econ.* **2015**, *50*, 17–28. [CrossRef]

25. Kraftfahrt-Bundesamt (KBA)–Federal Motor Transport Authority. Bestand an Pkw in den Jahren 2009 bis 2018 nach Ausgewählten Kraftstoffarten. 2018. Available online: www.kba.de/DE/Statistik/Fahrzeuge/Bestand/Umwelt/b_umwelt_z.html?nn=663524 (accessed on 15 December 2018).

26. Kraftfahrt-Bundesamt (KBA)–Federal Motor Transport Authority. Bestand an Personenkraftwagen am 1. Januar 2020 nach Bundesländern und Ausgewählten Kraftstoffarten Absolut. 2020. Available online: https://www.kba.de/DE/Statistik/Fahrzeuge/Bestand/Umwelt/fz_b_umwelt_archiv/2020/2020_b_umwelt_dusl.html?nn=2601598 (accessed on 6 August 2020).

27. Bergk, F.; Biemann, K.; Heidt, C.; Knörr, W.; Lambrecht, U.; Schmidt, T. Klimaschutzbeitrag des Verkehrs Bis 2050. Available online: https://www.umweltbundesamt.de/sites/default/files/medien/1410/publikationen/texte_56_2016_klimaschutzbeitrag_des_verkehrs_2050_getagged.pdf (accessed on 5 December 2020).

28. Umweltbundesamt. Klimabilanz 2017: Emissionen Gehen Leicht Zurück. 2018. Available online: https://www.umweltbundesamt.de/presse/pressemitteilungen/klimabilanz-2017-emissionen-gehen-leicht-zurueck (accessed on 6 August 2020).

29. Kraftfahrt-Bundesamt (KBA)–Federal Motor Transport Authority. Neuzulassungen von Pkw in den Jahren 2008 bis 2017 nach Ausgewählten Kraftstoffarten. 2018. Available online: https://www.kba.de/DE/Statistik/Fahrzeuge/Neuzulassungen/Umwelt/n_umwelt_z.htmlnn=652326 (accessed on 15 December 2018).

30. Bundesregierung. *Regierungsprogramm Elektromobilität*; Publ. Bundesregierung: Berlin, Germany, 2011.

31. Nationale Plattform Elektromobilität (NPE). Die Ziele. 2020. Available online: http://nationale-plattform-elektromobilitaet.de/hintergrund/die-ziele/ (accessed on 6 August 2020).

32. Kraftfahrt-Bundesamt (KBA)–Federal Motor Transport Authority. Bestand am 01. Januar 2018 nach Segmenten. 2018. Available online: https://www.kba.de/DE/Statistik/Fahrzeuge/Bestand/Segmente/segmente_node.html (accessed on 15 May 2018).

33. Kompetenzzentrum Elektromobilität NRW. Data Sheet: Hyundai Kona Elektro (100 kW). 2018. Available online: https://www.elektromobilitaet.nrw.de/fileadmin/Daten/Download_Dokumente/Fahrzeuge/Hyundai_Kona_Elektro__100_kW_.pdf (accessed on 10 January 2020).

34. Kompetenzzentrum Elektromobilität NRW. Data Sheet: Jaguar I-PACE. 2018. Available online: https://www.elektromobilitaet.nrw.de/fileadmin/Daten/Download_Dokumente/Fahrzeuge/Jaguar_I-PACE.pdf (accessed on 10 January 2020).

35. Kompetenzzentrum Elektromobilität NRW. Data Sheet: Nissan Leaf. 2018. Available online: https://www.elektromobilitaet.nrw.de/fileadmin/Daten/Download_Dokumente/Fahrzeuge/Nissan_Leaf.pdf (accessed on 10 January 2020).

36. Kompetenzzentrum Elektromobilität NRW. Data Sheet: Hyundai Kona Elektro (150 kW). 2018. Available online: https://www.elektromobilitaet.nrw.de/fileadmin/Daten/Download_Dokumente/Fahrzeuge/Hyundai_Kona_Elektro__150_kW_.pdf (accessed on 10 January 2020).

37. Kompetenzzentrum Elektromobilität NRW. Data Sheet: VW e-Golf. 2018. Available online: https://www.elektromobilitaet.nrw.de/fileadmin/Daten/Download_Dokumente/Fahrzeuge//VW_e-Golf.pdf (accessed on 10 January 2020).

38. Kompetenzzentrum Elektromobilität NRW. Data Sheet: Opel Ampera-e. 2018. Available online: https://www.elektromobilitaet.nrw.de/fileadmin/Daten/Download_Dokumente/Fahrzeuge/Opel_Ampera-e.pdf (accessed on 10 October 2018).

39. Kompetenzzentrum Elektromobilität NRW. Data Sheet: Renault Zoe, Z.E. 2018. Available online: https://www.elektromobilitaet.nrw.de/fileadmin/Daten/Download_Dokumente/Fahrzeuge/Renault_Zoe_Z.E..pdf (accessed on 10 October 2018).

40. Kompetenzzentrum Elektromobilität NRW. Data Sheet: Tesla Model S 100D. 2018. Available online: https://www.elektromobilitaet.nrw.de/fileadmin/Daten/Download_Dokumente/Fahrzeuge/Tesla_Model_S_100D.pdf (accessed on 10 October 2018).

41. Kompetenzzentrum Elektromobilität NRW. Data Sheet: Tesla Model X 100D. 2018. Available online: https://www.elektromobilitaet.nrw.de/fileadmin/Daten/Download_Dokumente/Fahrzeuge/Tesla_Model_X_100D.pdf (accessed on 10 January 2020).

42. Kraftfahrt-Bundesamt (KBA)–Federal Motor Transport Authority. Erneut mehr Gesamtkilometer bei Geringerer Jahresfahrleistung je Fahrzeug. 2018. Available online: https://www.kba.de/DE/Statistik/Kraftverkehr/VerkehrKilometer/2017_vk_kurzbericht_pdf.pdf?__blob=publicationFile&v=14 (accessed on 3 June 2018).

43. Kang, J.E.; Recker, W. An activity-based assessment of the potential impacts of plug-in hybrid electric vehicles on energy and emissions using 1-day travel data. *Transp. Res. Part D Transp. Environ.* **2009**, *14*, 541–556. [CrossRef]

44. Robinson, A.; Blythe, P.; Bell, M.; Hübner, Y.; Hill, G. Analysis of electric vehicle driver recharging demand profiles and subsequent impacts on the carbon content of electric vehicle trips. *Energy Policy* **2013**, *61*, 337–348. [CrossRef]

45. Siegrist, A.; Schnabl, P.; Burkhart, S.; de Haan, P.; Bianchetti, R. Elektromobilität: Studie Ladeinfrastruktur Region Basel. Available online: https://docplayer.org/8358125-Elektromobilitaet-studie-ladeinfrastruktur-region-basel.html (accessed on 5 December 2020).

46. Nationale Plattform Elektromobilität. Vision und Roadmap der Nationalen Plattform Elektromobilität. *Gem. Geschäftsst. Elektromobilität Bundesreg* **2013**, *27*. Available online: http://nationale-plattform-elektromobilitaet. de/fileadmin/user_upload/Redaktion/1379668151_de_2122747422.pdf (accessed on 5 December 2020).

47. Christensen, L.; Nørrelund, A.V.; Olsen, A. Travel behaviour of potential Electric Vehicle drivers. The need for changing A contribution to the Edison project. In Proceedings of the European Transport Conference 2010, Glasgow, UK, 11–13 October 2010.

48. EAFO. Country Detail Electricity. 2019. Available online: https://www.eafo.eu/electric-vehicle-charging-infrastructure (accessed on 6 August 2020).

49. Lucas, A.; Prettico, G.; Flammini, M.G.; Kotsakis, E.; Fulli, G.; Masera, M. Indicator-Based Methodology for Assessing EV Charging Infrastructure Using Exploratory Data Analysis. *Energies* **2018**, *11*, 1869. [CrossRef]

50. Ladesäulen-Bestand in Deutschland für Dreimal so Viele Elektroautos Ausreichend. Available online: https://www.emobilserver.de/nachrichten/energie-ladetechnik/1131-ladesäulen-bestand-in-deutschland-für-dreimal-so-viele-elektroautos-ausreichend.html (accessed on 6 August 2020).

51. ABB. Terra 184 Infographic. 2020. Available online: https://search.abb.com/library/Download.aspx?DocumentID=9AKK107680A8808&LanguageCode=en&DocumentPartId=&Action=Launch (accessed on 7 August 2020).

52. Sheet NLG664—On Board Fast Charger. Available online: https://www.brusa.biz/fileadmin/template/Support-Center/Datenbl%C3%A4tter/BRUSA_DB_EN_NLG664.pdf (accessed on 5 December 2020).

53. EVTEC AG. Data: Move&Charge 3in1. 2020. Available online: https://www.evtec.ch/en/products/move_and_charge/ (accessed on 7 August 2020).

54. Stercom Power Solutions GmbH: Feel the Power. Available online: https://stercom.de/_media//PDFs/Deutsch/OBC_D1.7.pdf (accessed on 5 December 2020).

55. Bundesnetzagentur. Ladesäulenkarte. 2020. Available online: https://www.bundesnetzagentur.de/DE/Sachgebiete/ElektrizitaetundGas/Unternehmen_Institutionen/HandelundVertrieb/Ladesaeulenkarte/Ladesaeulenkarte_node.html (accessed on 5 August 2020).

56. McManus, M.C.; Taylor, C.M. The changing nature of life cycle assessment. *Biomass Bioenergy* **2015**, *82*, 13–26. [CrossRef]

57. International Organization for Standardization. *ISO 14040: Environmental Management—Life Cycle Assessment; Principles and Framework*; ISO: Geneva, Switzerland, 2006.

58. International Organization for Standardization. *ISO 14044: Environmental Management—Life Cycle Assessment; Requirements and Guidelines*; ISO: Geneva, Switzerland, 2006.

59. Wernet, G.; Bauer, C.; Steubing, B.; Reinhard, J.; Moreno-Ruiz, E.; Weidema, B.P. The ecoinvent database version 3 (part I): Overview and methodology. *Int. J. Life Cycle Assess.* **2016**, *21*, 1218–1230. [CrossRef]

60. Green Delta. 2020. OpenLCA. Available online: http://www.openlca.org/ (accessed on 5 December 2020).

61. Schroeder, A.; Traber, T. The economics of fast charging infrastructure for electric vehicles. *Energy Policy* **2012**, *43*, 136–144. [CrossRef]

62. Stella, K.; Wollersheim, O.; Fichtner, W.; Jochem, P.; Schücking, M.; Nastold, M. *Über 300.000 Kilometer unter Strom: Physikalisch-Technische, Ökonomische, Ökologische und Sozialwissenschaftliche Begleitforschung des Schaufensterprojektes RheinMobil: Grenzüberschreitende, Perspektivisch Wirtschaftliche Elektrische Pendler-und Dienstwagenverkehre im Deutsch-Französischen Kontext: Schaufenster Elektromobilität-eine Initiative der Bundesregierung: LivingLab BWe Mobil*; Karlsruher Institut für Technologie (KIT): Karlsruhe, Germany, 2015.

63. Shiau, C.S.N.; Samaras, C.; Hauffe, R.; Michalek, J.J. Impact of battery weight and charging patterns on the economic and environmental benefits of plug-in hybrid vehicles. *Energy Policy* **2009**. [CrossRef]

64. ABB. Global Product Portfolio-Smarter Mobility. 2019. Available online: https://new.abb.com/about/our-businesses/electrification/e-mobilitysolutions (accessed on 5 December 2020).

65. Evtec. Data Sheet: Espresso&Charge. 2018. Available online: https://www.evtec.ch/en/products/espressoandcharge-usp/ (accessed on 7 August 2020).

66. Mortimer, B.; Olk, C.; Roy, G.K.; Tarnate, W.R.; Doncker, R.W.; Monti, A. *Fast-Charging Technologies, Topologies and Standards 2.0*; E.ON Energy Research Center Series: Aachen, Germany, 2019; Volume 11.

67. Schmenger, J.; Zeltner, S.; Kramer, R.; Endres, S.; Marz, M. A 3.7 kW on-board charger based on modular circuit design. In Proceedings of the IECON 2015—41st Annual Conference of the IEEE Industrial Electronics Society, Yokohama, Japan, 9–12 November 2015; pp. 001382–001387.

68. Schmenger, J.; Endres, S.; Zeltner, S.; März, M. A 22 kW on-board charger for automotive applications based on a modular design. In Proceedings of the 2014 IEEE Conference on Energy Conversion (CENCON), Johor Bahru, Malaysia, 13–14 October 2014; pp. 1–6.

69. Schmenger, J.; Kramer, R.; Marz, M. Active hybrid common mode filter for a highly integrated on-board charger for automotive applications. In Proceedings of the 2015 IEEE 13th Brazilian Power Electronics Conference and 1st Southern Power Electronics Conference (COBEP/SPEC), Santos, Brazil, 29 November–2 December 2015; pp. 1–7.

70. Hultin, A.; Wolgers, J. Integration of Fast Charger Stations. Master's Thesis, Chalmers University of Technology, Göteborg, Sweden, 2017.

71. Liserre, M.; Blaabjerg, F.; Hansen, S. Design and Control of an LCL-Filter-Based Three-Phase Active Rectifier. *IEEE Trans. Ind. Appl.* **2005**, *41*, 1281–1291. [CrossRef]

72. Reznik, A.; Simoes, M.G.; Al-Durra, A.; Muyeen, S.M. LCL Filter design and performance analysis for grid-interconnected systems. *IEEE Trans. Ind. Appl.* **2014**. [CrossRef]

73. Beres, R.N.; Wang, X.; Liserre, M.; Blaabjerg, F.; Bak, C.L. A Review of Passive Power Filters for Three-Phase Grid-Connected Voltage-Source Converters. *IEEE J. Emerg. Sel. Top. Power Electron.* **2016**, *4*, 54–69. [CrossRef]

74. Plaßmann, W.; Schulz, D. *Handbuch Elektrotechnik*, 6th ed.; Springer: Berlin/Heidelberg, Germany, 2013.

75. Reznik, A.; Simões, M.G.; Al-Durra, A.; Muyeen, S.M. LCL filter design and performance analysis for small wind turbine systems. In Proceedings of the 2012 IEEE Power Electronics and Machines in Wind Applications, Denver, CO, USA, 16–18 July 2012; pp. 1–7.

76. Fastron Group. Data Sheet: TLC/10A. 2016. Available online: http://fastrongroup.com/toroid-line-chokes/tlc10a (accessed on 5 December 2020).

77. ICAR. Data Sheet: Power Electronics Capacitors: MKP Series. 2011. Available online: https://www.icar.com/wp-content/uploads/2018/05/MKP-Icar-Capacitors.pdf (accessed on 5 December 2020).

78. Infineon. Data Sheet IPW60R045CP. 2013. Available online: https://www.mouser.es/datasheet/2/196/Infineon-IPW60R045CP-DS-v02_02-en-1227462.pdf (accessed on 5 December 2020).

79. Infineon. Data Sheet IDW40G65C5B. 2015. Available online: https://www.infineon.com/dgdl/Infineon-IDW40G65C5B-DS-v02_00-EN.pdf?fileId=5546d4624e24005f014e451799b149e1 (accessed on 5 December 2020).

80. Murata Manufacturing Co. Ltd. Data KC355WD72J474MH01#. 2020. Available online: https://www.murata.com/en-eu/products/productdetail?partno=KC355WD72J474MH01%23 (accessed on 7 August 2020).

81. TDK. Epcos Data Book 2013: Ferrites and Accessories. 2013. Available online: https://www.tdk-electronics.tdk.com/download/519704/069c210d0363d7b4682d9ff22c2ba503/ferrites-and-accessories-db-130501.pdf (accessed on 5 December 2020).

82. Chae, H.J.; Kim, W.Y.; Yun, S.Y.; Jeong, Y.S.; Lee, J.Y.; Moon, H.T. 3.3kW on board charger for electric vehicle. In Proceedings of the 8th International Conference on Power Electronics—ECCE, Jeju, Korea, 30 May–3 June 2011.

83. Lai, J.S.; Miwa, H.; Lai, W.H.; Tseng, N.H.; Lee, C.S.; Lin, C.H. A high-efficiency on-board charger utilitzing a hybrid LLC and phase-shift DC-DC converter. In Proceedings of the 2014 International Conference on Intelligent Green Building and Smart Grid, IGBSG, Taipei, Taiwan, 23–25 April 2014.

84. Infineon. Data Sheet IPW65R045C7. 2013. Available online: https://www.infineon.com/dgdl/Infineon-IPW65R045C7-DS-v02_01 (accessed on 5 December 2020).

85. Elgström, T.; Nordgren, L. Evaluation and Design of High Frequency Transformers for on Board Charging Applications. Master's Thesis, Chalmers University of Technology, Gothenburg, Sweden, 2016.

86. TDK Electronics AG. Data Sheet: Ferrite Cores for Switching Power Supplies: Large PQ Series. 2016. Available online: https://product.tdk.com/info/en/catalog/datasheets/ferrite_mz_sw_large_pq_en.pdf (accessed on 5 December 2020).

87. Nordelöf, A.; Alatalo, M. A scalable life cycle inventory of an automotive power electronic inverter unit—Part I: Design and composition. *Int. J. Life Cycle Assess.* **2019**, *24*, 78–92. [CrossRef]

88. Laminated Bus Bar Solutions—Mersen. Available online: https://ep-us.mersen.com/sites/mersen_us/files/2018-12/Laminated-Busbar-Solutions.pdf (accessed on 5 December 2020).

89. Fraunhofer Institute for Integrated Systems and Device Technology IISB. 3.7 kW Battery Charger for Automotive Applications. 2014. Available online: https://www.iisb.fraunhofer.de/content/dam/iisb2014/en/Documents/Research-Areas/vehicle_electronics/FraunhoferIISB_ProductSheet_FE_3k7W-Onboard-Battery-Charger_www.pdf (accessed on 5 December 2020).

90. Conrad Electronics International GmbH & CoKG. Data: Universal Enclosures. 2020. Available online: https://www.conrad.com/o/universal-enclosures-0203042 (accessed on 7 August 2020).

91. Star Charge. 180 kW Super DC Charger_DC Series. 2018. Available online: http://en.starcharge.com/show.asp?id=37 (accessed on 7 August 2020).

92. ESL. Ladekabel Elektroauto. 2020. Available online: https://esl-emobility.com/de/ladekabel-elektroauto.html (accessed on 20 November 2018).

93. Lapp Group. Data Sheet: ÖLFLEX® CHARGE. 2020. Available online: https://lappaustria.lappgroup.com/produkte/katalog-e-shop/anschluss-und-steuerleitungen/besondere-anwendungen/emobility/oelflex-charge.html (accessed on 5 December 2020).

94. Phoenix Contact. Data Sheet: DC Charging Cable–EV-T2M4CC-DC60A-2,5M16ESBK00-1623043. 2020. Available online: https://www.phoenixcontact.com/online/portal/pi?uri=pxc-oc-itemdetail:pid=1623043&library=pien&tab=1 (accessed on 5 December 2020).

95. Lapp Group. Data Sheet: ÖLFLEX® SERVO 2YSLCY-JB. 2020. Available online: https://products.lappgroup.com/online-catalogue/power-and-control-cables/servo-applications/pvc-sheath/oelflex-servo-2yslcy-jb.html (accessed on 5 December 2020).

96. Walther-Werke Ferdinand Walther GmbH. Data Sheet: E41X01A249B0. 2017. Available online: https://www.walther-werke.de/pdf/datasheets/PD_E41X01A249B0_DE.pdf (accessed on 5 December 2020).

97. Schrack Technik GmbH. Data: I-CHARGE Public. 2020. Available online: https://www.schrack.at/know-how/alternativenergie/elektromobilitaet/i-charge-stromtankstellen/i-charge-public/ (accessed on 7 August 2020).

98. Elektrohandel Wandelt GmbH. Data: Mennekes 1311510SW Ladestation Basic 22. 2020. Available online: https://www.elektro-wandelt.de/Mennekes-1311510SW-Ladestation-Basic-22.html (accessed on 7 August 2020).

99. Elektrohandel Wandelt GmbH. Data: Mennekes 1319610SW Ladestation Smart 22 SW. 2020. Available online: https://www.elektro-wandelt.de/Mennekes-1319610SW-Ladestation-Smart-22-SW.html (accessed on 7 August 2020).

100. Circocontrol, S.A. Data Sheet: Serie eStreet (AC). Available online: https://circontrol.de/wp-content/uploads/saule-estreet-datasheet.pdf (accessed on 5 December 2020).

101. Data: Alfen Twin. 2020. Available online: https://alfen.com/de/icu-twin (accessed on 7 August 2020).

102. ABL SURSUM. Data Sheet: Ladesäule eMC2. 2018. Available online: https://www.ablmobility.de/global/downloads/datenblaetter/emc2/eMC2_DE/2P4445.pdf?m=1597417459& (accessed on 5 December 2020).

103. ABL SURSUM. Data Sheet: LADESÄULE eMC3. 2019. Available online: https://www.ablmobility.de/global/downloads/anleitungen/emc3/eMC3_UM_IM_DE.pdf?m=1602138773& (accessed on 5 December 2020).

104. Compleo Charging Solutions GmbH. Data Sheet: Compleo Advanced. Available online: https://www.compleo-cs.com/fileadmin/user_upload/ladestationen/datenblaetter_advanced_digital.pdf (accessed on 5 December 2020).

105. EBG Group. Data Sheet: CITO BM2 500. Available online: https://pub-mediabox-storage.rxweb-prd.com/exhibitor/document/e2c22e9f-8306-4880-836f-6e42a889b7ae (accessed on 5 December 2020).

106. Charge-ON GmbH. Data Sheet: Ladestation Pro. Available online: https://www.eon-drive.de/content/dam/eon-emobility-de/Playground/PDFs/Datasheets/Pro_Datenblatt.pdf (accessed on 5 December 2020).

107. Walther-Werke Ferdinand Walther GmbH. Data: Charging Station Evolution 350 Key with Two Charging Points Type 2 up to 22kW and Basis Monitoring. Available online: https://www.walther-werke.de/nc/products-page/filter/ctrl/product/action/show/partno/e21901ae0910-14/?tx_waltherwerkeproducts_productfilter%5Bebene%5D=2&tx_waltherwerkeproducts_productfilter%5Bebene2%5D=13&tx_waltherwerkeproducts_productfilter%5Bebene3%5D (accessed on 7 August 2020).

108. PC Electric GmbH. Data Sheet: Ladesäule LS4. Available online: https://oxomi.com/p/3000237/catalog/10163042 (accessed on 5 December 2020).

109. Castellan, A.G. Data: Empfang-Ladestation. Available online: https://www.castellan-ag.de/details/empfang-ladestation.html (accessed on 7 August 2020).

110. Swarco Traffic Systems GmbH. Data Sheet: Evolt Public 7". 2017. Available online: https://www.swarco.com/sites/default/files/public/downloads/2018-10/01_2017%20SWARCO%20EVOLT%20Public%207%20Final.pdf (accessed on 5 December 2020).

111. Delta Electronics (Netherlands) B.V. Data Sheet: 150 kW Ultra-Fast Charger. 2019. Available online: https://emobility.delta-emea.com/downloads/Brochure%20-%20UFC%20Ultra%20Fast%20Charger%20V7%20EN%202019-10-07.pdf.pdf (accessed on 5 December 2020).

112. ABB Inc. Data Sheet: Terra 53 Multi-Standard DC Charging Station. 2014. Available online: https://new.abb.com/ev-charging/de/produkte/ladestationen-pkws/terra-schnelllades%c3%a4ulen/terra-54-cjg (accessed on 5 December 2020).

113. Petrotec Group. Data Sheet: PFAST MULTI. 2014. Available online: https://www.petrotec.com/media/33069/modibp-fast-multi-05_2014.pdf (accessed on 5 December 2020).

114. Borken-Kleefeld, J.; Weidema, B.P. Global default data for freight transport per product group. *Manuscr. Spec. Ecoinvent* **2013**, *3*.

115. Frischknecht, R.; Tuchschmid, M.; Emmenegger, M.F.; Bauer, C.; Dones, R. *Strommix. Sachbilanzen von Energiesystemen: Grundlagen Für Den Ökologischen Vergleich von Energiesystemen und den Einbezug von Energiesystemen in Ökobilanzen für Die Schweiz*; Dones, R., Ed.; Paul Scherrer Institut, Villigen & Swiss Centre for Life Cycle Inventories: Dübendorf, Switzerland, 2007; ISBN 3-905594-38-2.

116. Koj, J.C.; Wulf, C.; Zapp, P. Environmental impacts of power-to-X systems-A review of technological and methodological choices in Life Cycle Assessments. *Renew. Sustain. Energy Rev.* **2019**, *112*, 865–879. [CrossRef]

117. Rosenbaum, R.K. Selection of Impact Categories, Category Indicators and Characterization Models in Goal and Scope Definition. In *Goal and Scope Definition in Life Cycle Assessment*; Springer: Berlin/Heidelberg, Germany, 2017; pp. 63–122.

118. Huijbregts, M.; Steinmann, Z.J.N.; Elshout, P.M.F.M.; Stam, G.; Verones, F.; Vieira, M.D.M. ReCiPe 2016. *Natl. Inst. Public Health Environ.* **2016**, *194*. [CrossRef]

119. Hauschild, M.Z.; Rosenbaum, R.K.; Olsen, S. *Life Cycle Assessment*; Springer: Berlin/Heidelberg, Germany, 2018.

120. Jolliet, O.; Müller-Wenk, R.; Bare, J.; Brent, A.; Goedkoop, M.; Heijungs, R. The LCIA midpoint-damage framework of the UNEP/SETAC life cycle initiative. *Int. J. Life Cycle Assess.* **2004**, *9*, 394. [CrossRef]

121. Bare, J.C.; Hofstetter, P.; Pennington, D.W.; De Haes, H.A.U. Midpoints versus endpoints: The sacrifices and benefits. *Int. J. Life Cycle Assess.* **2000**, *5*, 319. [CrossRef]

122. Elektromobilität, N.P. *Ladeinfrastruktur für Elektrofahrzeuge in Deutschland: Statusbericht und Handlungsempfehlungen*; Nationale Plattform Elektromobilität (NPE): Berlin, Germany, 2015.

123. Funke, S.Á. *Techno-Ökonomische Gesamtbewertung Heterogener Maßnahmen zur Verlängerung der Tagesreichweite von Batterieelektrischen Fahrzeugen*. Ph.D. Thesis, Universität Kassel, Kassel, Germany, 2018.

124. Mathieu, G. Design of an On-Board Charger for Plug-In Hybrid Electrical Vehicle (PHEV). Master's Thesis, Chalmers University of Technology, Gothenburg, Sweden, 2009.

125. Hischier, R.; Classen, M.; Lehmann, M.; Scharnhorst, W. *Life Cycle Inventories of Electric and Electronic Equipment: Production, Use and Disposal*; Final Report Ecoinvent Data v2.0; Swiss Centre for Life Cycle Inventories: Dübendorf, Switzerland, 2007. [CrossRef]

126. Austerlitz Electronic GmbH. Data Sheet: Heatsinks. 2018. Available online: http://austerlitz-electronic.de/katalog/Austerlitz_Katalog-13_DE-EN.pdf (accessed on 5 December 2020).

127. MENNEKES Elektrotechnik GmbH & Co. KG. Data Sheet: Basic 3,7, Basic 11, Basic 22, Basic S 22. 2013. Available online: https://datatransfer.chargeupyourday.de/Graphics/XPic9/00039836_0.pdf (accessed on 5 December 2020).

128. Klaus Faber, A.G. Data Sheet: Power Cable NYY-J/-O. 2020. Available online: https://www.faberkabel.de/upload/files/downloads/204/2017_faber_katalog.pdf (accessed on 5 December 2020).

129. Charge-ON GmbH. Data Sheet: E.ON Ladestation Fast. Available online: https://www.eon-drive.de/content/dam/eon-emobility-de/Playground/PDFs/Datasheets/Fast_Datenblatt.pdf (accessed on 5 December 2020).

130. Fine Tubes Ltd. Data Sheet: Stainless Steel Alloys 304/304L. 2018. Available online: https://www.finetubes.co.uk/-/media/ametekfinetubes/files/downloads/alloy%20data%20sheets/stainless%20steel%20-%20alloys%20304%20and%20304l.pdf?utm_source=&utm_medium=undefined (accessed on 5 December 2020).

131. Classen, M.; Althaus, H.-J.; Blaser, S.; Tuchschmid, M.; Jungbluth, N.; Doka, G. *Life Cycle Inventories of Metals*; Final Report Ecoinvent Data v2.1 No.10; Swiss Centre for Life Cycle Inventories: Dübendorf, Switzerland, 2009.

Publisher's Note: MDPI stays neutral with regard to jurisdictional claims in published maps and institutional affiliations.

Article

Impact of Different Charging Strategies for Electric Vehicles in an Austrian Office Site

Carlo Corinaldesi *, Georg Lettner, Daniel Schwabeneder, Amela Ajanovic and Hans Auer

Energy Economics Group (EEG), Vienna University of Technology, Gußhausstraße 25-29, E370-3, 1040 Vienna, Austria; lettner@eeg.tuwien.ac.at (G.L.); schwabeneder@eeg.tuwien.ac.at (D.S.); ajanovic@eeg.tuwien.ac.at (A.A.); auer@eeg.tuwien.ac.at (H.A.)
* Correspondence: corinaldesi@eeg.tuwien.ac.at; Tel.: +43-(0)1-58801-370370

Received: 1 October 2020; Accepted: 3 November 2020; Published: 10 November 2020

Abstract: Electric vehicles represent a necessary alternative for wheeled transportation to meet the global and national targets specified in the Paris Agreement of 2016. However, the high concentration of electric vehicles exposes their harmful effects on the power grid. This reflects negatively on electricity market prices, making the charging of electric vehicles less cost-effective. This study investigates the economic potential of different charging strategies for an existing office site in Austria with multiple charging infrastructures. For this purpose, a proper mathematical representation of the investigated case study is needed in order to define multiple optimization problems that are able to determine the financial potential of different charging strategies. This paper presents a method to implement electric vehicles and stationary battery storage in optimization problems with the exclusive use of linear relationships and applies it to a real-life use case with measured data to prove its effectiveness. Multiple aspects of four charging strategies are investigated, and sensitivity analyses are performed. The results show that the management of the electric vehicles charging processes leads to overall costs reduction of more than 30% and an increase in specific power-related grid prices makes the charging processes management more convenient.

Keywords: electric mobility; charging strategies; economics; promotion policies; mixed-integer optimization; flexible systems

---

1. Introduction

The global warming and the increasing GHG emission challenges in recent years are expected to be an accelerator for the deployment of electric vehicles (EVs) as a more sustainable alternative for wheeled transportation around the world [1–4]. The European Environment Agency recognizes transport as a critical source of environmental pressures in the European Union, and it has a decisive influence on climate change and air pollution. The transport sector consumes one-third of the final energy in the European Union and is responsible for a large share of the European's greenhouse gas emissions. This makes transport one of the major contributors to climate change [5]. The Intergovernmental Panel on Climate Change (IPCC) affirms that wheeled transportation produces more than 70% of the overall greenhouse gas emissions from transport [6]. Because of their significant environmental advantages, numerous countries are working on different strategies to make EVs monetarily more convenient than traditional wheeled vehicles [7].

However, as the number of EVs increases, the negative effects of their charging on the power grid become more evident, especially at the low-voltage level. In fact, high concentrations of EVs have various harmful effects due to the overlap between EV charging, residential peak loads and renewable generation [8–10]. For these reasons, the growing number of EVs, in parallel with the growing penetration of renewable energy sources, leads the power grid facing a challenging future [11,12].

Charging management systems and different charging strategies of EVs have been extensively studied in order to reduce the greenhouse gas emissions, to improve the power grid operation and to reduce the electricity costs for end-users [13–16]. In the last decade, several unidirectional and bidirectional charging strategies of EVs have been investigated in different contributions [17–20]. Furthermore, the possibility of combining stationary battery storages (SBSs) with the charging infrastructures has also been studied in-depth, in order to further reduce the peak load power and the electricity costs and allow the fast charging of EVs even with low grid connection power [21,22].

The core objective of this study is to investigate the economic potential of different charging strategies of EVs for an existing office site in Austria with multiple charging infrastructures. The economic potential is given by trading electricity in the Day-Ahead (DA) spot market considering the overall power and energy procurement costs of the office site's charging infrastructures. Profit opportunities could incentivize the end-users to apply flexible electricity consumption patterns to their EVs charging schedules, and they could also incentivize the office site to install further SBSs.

In this paper, multiple optimization problems are defined in order to determine the monetary potential of different operating strategies for managing the EVs' charging processes. The optimization approach aims to define the power flows between the EVs, the SBSs and the power grid, in order to minimize the electricity costs and to best allocate the flexibility of the Austrian office site. However, a detailed description of the technical operation of the above-mentioned components is needed in order to implement them in a mathematical model. In this work, the components of the Austrian office site are described with the exclusive use of linear relationships, which can be implemented in mixed-integer optimization problems. The optimization problems are modeled using the *Python* toolbox *Pyomo* [23] and solved with the *Gurobi* solver [24].

The paper is organized as follows. Section 2 provides an overview of the state of the art in the scientific literature. Next, Section 3 describes the mathematical representation of the Austrian office site, including the EVs and the SBSs in the optimization problem and its objective function, which aims to minimize the overall costs. Section 4 illustrates the description of the investigated real-life use case in Austria with measured data used to simulate the different charging strategies. Section 5 presents the comprehensive results and sensitivity analyses of the case study. Finally, Section 6 concludes the paper and discusses possible directions for future research.

2. State of the Art

The growing share of renewable energy sources, such as wind and solar photovoltaic, increases the volatility of electricity generation in the power grid [25–27]. At the same time, the increasing integration of high-power consumption loads, such as EVs, are setting new challenges for the distribution system operators [28,29]. Hence, the active participation of the demand-side can play a crucial role in the European transition to a carbon-free energy sector [30–32]. For this reason, nowadays, one of the key challenges is to enhance the use of the potential flexible demand in the power grid. An EV represents a flexible load type, and the growing number of grid-connected EVs accords to them a growing flexibility potential for the power grid [33]. Flexibility is the electrical components capability to alter their scheduled consumption in reaction to external signals, for example, spot market prices or grid costs [34]. However, an exhaustive mathematical description of the flexible components, such as EVs and SBSs, is needed in order to efficiently coordinate and aggregate multiple flexible load types.

In several studies, for example [33,35,36], the flexibility of EVs is utilized in order to support the variable renewable energy injection and minimize the overall system costs. Weis et al. [37] quantify the benefits of managed charging of EVs achieving 1.5–2.3% cost savings in the simulations. Sheikhi et al. [38] introduce a charging management strategy for EVs aimed to reduce peak loads. The benefits and the drawbacks of bidirectional charging of EVs are thoroughly investigated in [39–41]. In several simulations, the electricity cost reductions achieved through the vehicle-to-grid charging are overcompensated by higher battery degradation costs. If the higher battery degradation costs are not considered, the overall costs reduction could reach 13.6% [39]. However, managed charging and vehicle-to-grid charging of EVs are able to alleviate congestion in the power grid [42]. In the last decade, different methods were developed in order to mathematically represent flexible loads in optimization frameworks. In Hao et al. [43], a method to describe the flexibilities of different technologies as virtual batteries is presented. In this work, flexibilities are represented in a mixed-integer optimization problem with the exclusive use of linear relationships, which makes the model scalable and able to handle a growing amount of components.

This paper presents the mathematical implementation of EVs and SBSs in optimization problems, which aim to minimize the overall costs of their operation, applying peak shaving and load shifting to aggregated charging infrastructures. Furthermore, in this work different charging strategies are defined and compared in order to determine the optimal charging strategy for an office site of an electric utility company in Austria with multiple charging infrastructures. The simulated period covers the entire year 2019. The main contributions of this paper beyond the-state-of-the-art are as follows.

- Development of a method to implement EVs and SBSs in optimization problems with the exclusive use of linear relationships,
- Costs comparison of four different charging strategies of EVs,
- Investigation of the potential of coupling charging infrastructures with SBSs,
- Application of the methods to a real-life use case in Austria with measured data.

3. Methods

The real operation of EVs and SBSs is characterized by non-linear relationships, which lead to non scalability of the calculations. In Section 3.1, a simplified method to implement EVs and SBSs in optimization problems as linear systems is developed. Moreover, the optimization problem and its cost function are presented in Section 3.2. Lastly, in Section 3.3, the investigated EV charging strategies are defined.

3.1. Components

A mathematical description of SBSs and EVs is needed in order to efficiently implement them in a mixed-integer optimization problem and define the optimal allocation of their power flows using a linear optimization model. In the following Sections 3.1.1 and 3.1.2, the mathematical representations of SBSs and EVs are presented. In this work, the time is considered as discrete, and the optimized time range $\mathcal{T}$ is divided into a number of constant time intervals Δt.

3.1.1. Stationary Battery Storage (SBS)

The operation of a SBS is bounded to its physical limits such as power and capacity limits. The input and output power ($p_t^{SBS,in}$ and $p_t^{SBS,out}$) are confined between 0 and the maximum input and output power ($p_{max}^{SBS,in}$ and $p_{max}^{SBS,out}$) are specified as follows.

$$0 \leq p_t^{SBS,in} \leq p_{max}^{SBS,in} \qquad \forall\, t \in \mathcal{T} \qquad (1)$$

$$0 \leq p_t^{SBS,out} \leq p_{max}^{SBS,out} \qquad \forall\, t \in \mathcal{T} \qquad (2)$$

The capacity limits bound the state of charge of the SBS (soc_t^{SBS}) between 0 and its nominal capacity (E^{SBS}) as mathematically described below.

$$0 \leq soc_t^{SBS} \leq E^{SBS} \qquad \forall t \in \mathcal{T} \qquad (3)$$

If the charging and discharging efficiency ($\eta^{SBS,in}$ and $\eta^{SBS,out}$) and the standby losses percentage ($E_{Loss\%}^{SBS}$) are taken into account, the energy balance of an SBS can be defined in each time period in $\mathcal{T}$ as follows.

$$soc_t^{SBS} = soc_{t-1}^{SBS} \cdot \left(1 - E_{Loss\%}^{SBS}\right) + \left(p_t^{SBS,in} \cdot \eta^{SBS,in} - \frac{p_t^{SBS,out}}{\eta^{SBS,out}}\right) \cdot \Delta t \qquad \forall\, t \in \mathcal{T} \qquad (4)$$

Furthermore, the costs of the operation of the SBS can be implemented in the optimization problem using the levelized cost of storage (LCOS) method [44]. LCOS can be described as the cost per unit of charged electricity for a specific storage technology. It can be formally defined as follows.

$$LCOS = \frac{C_{Inv}^{SBS}}{n \cdot E^{SBS}} \qquad (5)$$

where C_{Inv}^{SBS} indicates the investment costs of the SBS and n the number of charging cycles that the SBS is able to support before its capacity falls under 80% of its original capacity [45]. Hence, the total costs of storage C_{LCOS}^{SBS} within the optimized time range $\mathcal{T}$ can be described as the LCOS multiplied by the SBS's cumulative delivered electricity and can be expressed as below.

$$C_{LCOS}^{SBS} = LCOS \cdot \sum_{t=1}^{T} \left(p_t^{SBS,in}\right) \qquad (6)$$

According to the physical limits and the total costs of storage, the optimization algorithm determines the optimal input and output power ($p_t^{SBS,in}$ and $p_t^{SBS,out}$) and the state of charge (soc_t^{SBS}) of the SBS, hence defining its optimal operation.

3.1.2. Electric Vehicle (EV)

In this paper, we consider different possibilities of charging EVs: managed charging (MC) and vehicle-to-grid (V2G). The flexibility available from a charging cycle can vary in terms of duration and amount of energy. In fact, EV batteries are only available for a limited period of time, or rather only when they are connected to the charging infrastructure. Hence, the operation of a charging cycle is bounded to its physical limits such as power, capacity and time limits. The time limits are given by the connection time (S^{EV}) and the disconnection time (D^{EV}) of the EV at the charging infrastructure. We formally consider the initial state of charge ($soc_{S^{EV}}^{EV}$) equal to 0, since most of today's charging infrastructures do not allow to know the state of charge of an EV when it is connected to one of them. Formally, the state of charge of an EV is confined as follows.

$$soc_{S^{EV}}^{EV} = 0 \qquad (7)$$
$$soc_{D^{EV}}^{EV} = E^{EV} \qquad (8)$$
$$0 \leq soc_t^{EV} \leq E^{EV} \qquad \forall t \in (S^{EV}, D^{EV}) \qquad (9)$$

The input and output power ($p_t^{EV,in}$ and $p_t^{EV,out}$) are confined between 0 and the maximum input and output power ($p_{max}^{EV,in}$ and $p_{max}^{EV,out}$) as formally described below.

$$0 \leq p_t^{EV,in} \leq p_{max}^{EV,in} \qquad \forall t \in (S^{EV}, D^{EV}) \qquad (10)$$
$$0 \leq p_t^{EV,out} \leq p_{max}^{EV,out} \qquad \forall t \in (S^{EV}, D^{EV}) \qquad (11)$$

If V2G is not considered, the maximum output power ($p_{max}^{EV,out}$) is formally set to 0. Moreover, when the EV is not connected to the charging station, the input and output power is set to 0 as specified below.

$$p_t^{EV,in} = 0 \qquad\qquad \forall t \notin (S^{EV}, D^{EV}) \qquad (12)$$

$$p_t^{EV,out} = 0 \qquad\qquad \forall t \notin (S^{EV}, D^{EV}) \qquad (13)$$

If the charging and discharging efficiency ($\eta^{EV,in}$ and $\eta^{EV,out}$) and the standby losses percentage ($E_{Loss\%}^{EV}$) are considered, the energy balance of an EV could be formally defined in the optimization problem in each time period in $\mathcal{T}$ as follows.

$$soc_t^{EV} = soc_{t-1}^{EV} \cdot \left(1 - E_{Loss\%}^{EV}\right) + \left(p_t^{EV,in} \cdot \eta^{EV,in} - \frac{p_t^{EV,out}}{\eta^{EV,out}}\right) \cdot \Delta t \qquad \forall t \in (S^{EV}, D^{EV}) \qquad (14)$$

where the AC-DC and the DC-AC conversion losses are included in the charging and discharging efficiency factors ($\eta^{EV,in}$ and $\eta^{EV,out}$). A graphical representation and the associated power flows of the flexibility of a charging process of an EV are shown in Figure 1.

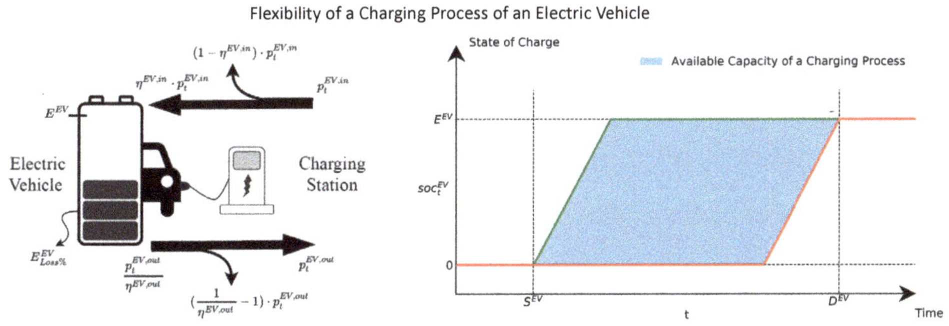

Figure 1. Power flows and available capacity of an electric vehicle during a charging process.

The graphic in Figure 1 shows the upper (green) and lower (red) capacity bounds of a charging process of an EV. The blue painted area represents the available capacity of a charging process, where the flexibility can eventually be activated.

Conforming to these formal bounds given by the physical limits of a charging process, the optimization algorithm determines the optimal operation of a charging process defining the optimal input and output power ($p_t^{EV,in}$ and $p_t^{EV,out}$) and the state of charge (soc_t^{EV}) of the EV.

3.2. Optimization Framework

The optimization framework simulates the DA electricity spot market auctions and coordinates the power flows of the flexible components in order to minimize the overall costs of the Austrian office site. The flexible components are connected to a common grid connection point (GCP). The power passing through the grid connection point ($p_t^{GCP,Load}$ and $p_t^{GCP,Feed-in}$) is directly traded in the DA spot market. A graphical representation of the flexible components and the associated power flows is shown in Figure 2.

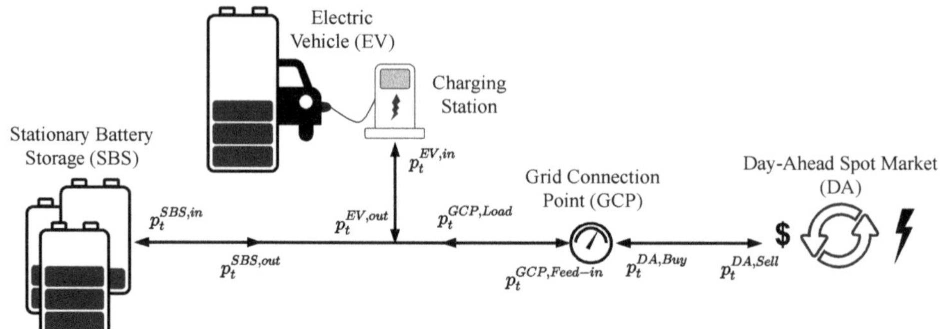

Figure 2. Flexible components and associated power flows.

The power balance equations of the interconnected flexible components can be formally defined as below.

$$p_t^{EV,in} - p_t^{EV,out} + p_t^{SBS,in} - p_t^{SBS,out} = p_t^{GCP,Load} - p_t^{GCP,Feed-in} \qquad \forall\, t \in \mathcal{T} \qquad (15)$$

$$p_t^{GCP,Load} - p_t^{GCP,Feed-in} = p_t^{DA,Buy} - p_t^{DA,Sell} \qquad \forall\, t \in \mathcal{T} \qquad (16)$$

Furthermore, the electricity procurement costs are given by the DA spot market costs (C^{DA}) and the grid costs (C^{Grid}). The DA spot market prices are implemented in the optimization problem as an exogenous time-series for each time period in $\mathcal{T}$ as follows.

$$P_t^{DA} = (P_1^{DA}, P_2^{DA} \dots, P_T^{DA}) \qquad (17)$$

Hence, the overall DA spot market costs in $\mathcal{T}$ are given by

$$C^{DA} = \sum_{t=1}^{T} \left(P_t^{DA} \cdot \left(p_t^{GCP,Load} - p_t^{GCP,Feed-in} \right) \right) \qquad \forall t \in \mathcal{T} \qquad (18)$$

The grid costs (C^{Grid}) are given by three main components that vary according to the grid level, where the flexible components are connected. The three components are

- A fixed flat rate ($C^{Grid,FR}$),
- An energy-related component ($C^{Grid,E}$),
- And a power-related component ($C^{Grid,P}$).

The fixed flat rate ($C^{Grid,FR}$) depends on the time of use of the grid connection, while the energy-related component ($C^{Grid,E}$) depends on the total amount of electricity passing through the grid connection point. Moreover, the power-related component ($C^{Grid,P}$) depends on the power peak flowing through the grid connection point during the investigated year. The components that form the grid costs are formally defined below.

$$C^{Grid,E} = \sum_{t=1}^{T} \left(p^{Grid,E} \cdot p_t^{GCP,Load} \cdot \Delta t \right) \qquad \forall t \in \mathcal{T} \qquad (19)$$

$$C^{Grid,P} = p^{Power} \cdot p_{max}^{GCP,Load} \qquad (20)$$

$$C^{Grid} = C^{Grid,E} + C^{Grid,P} + C^{Grid,FR} \qquad (21)$$

p^{Power} and $p^{Grid,E}$ represent the specific power-related-and energy-related grid costs, which depend on the grid level, where the flexible components are connected.

Hence, the overall costs are given by the electricity trades in the DA spot market, the grid costs and the total costs of storage within the time periods in $\mathcal{T}$ and are mathematically defined as follows.

$$C^{\text{Overall}} = C^{\text{DA}} + C^{\text{Grid}} + C^{\text{SBS}}_{LCOS} \tag{22}$$

Lastly, the objective function of the mixed-integer optimization problem is the minimization of the overall costs and is defined as follows.

$$\min_{\mathcal{T}} \quad C^{\text{Overall}} \tag{23}$$

3.3. Use Cases

In this study, different EV charging strategies are analyzed and compared with traditional EV charging. In fact, nowadays, EVs are charged as soon as they are connected to the charging infrastructure. If the maximum capacity of the EV is reached before the disconnection time, then it remains connected to the charging infrastructure until the disconnection time occurs. At an office site, EV drivers usually plug in their vehicles at the same time when they start working. Consequently, unmanaged charging leads to significant peak loads, which have a large potential negative impact on the grid from a technical, financial and environmental point of view. In the case of MC, the EV is charged when it is economically more convenient taking into account the grid costs, the electricity spot market costs and the losses of the EV. Furthermore, the V2G charging strategy allows the EV to reinject electricity into the power grid at times of high electricity prices in order to achieve lower overall costs. Moreover, these two different charging strategies are investigated with the further coordination of the charging infrastructures with stationary battery storage. The investigated use cases in this work are listed below.

- Baseline (BL): This use case coincides with the unmanaged charging. The EVs are charged as soon as they are connected to the charging infrastructure and remain connected to the charging infrastructure until the disconnection time occurs.
- Managed Charging (MC): In the MC use case, the timing of EVs charging is optimized in order to minimize the overall costs.
- Vehicle-to-Grid (V2G): In this use case, the electricity of the EVs battery can be reinjected to the power grid. The EVs are charged and discharged in order to minimize the overall costs.
- Coordinated Managed Charging (CMC): In the CMC use case, the charging of EVs is managed and coordinated with the operation of a SBS.
- Coordinated Vehicle-to-Grid (CV2G): In the CV2G use case, the charging and discharging of EVs is enabled and coordinated with the operation of a SBS.

In the interests of clarity, charging processes of an EV with MC and V2G are shown and compared with the BL in Figure 3.

In Figure 3, three different EV charging strategies are observable for the same charging process. The upper graphic shows the available capacity (blue painted area), the upper (green) and lower (red) capacity bounds as already shown in Figure 1. The lower graphic in Figure 3 shows a synthetic time-series of DA spot market prices. Furthermore, the state of charge of an EV applying three different charging strategies are shown: BL in orange, MC in blue and V2G in black. As illustrated in Figure 3, when the BL strategy is applied, the EV is charged as soon as it is connected to the charging infrastructure without taking into account the DA spot market prices (purple). When the MC strategy is applied, the EV is charged when the electricity prices at the DA spot market are low in order to minimize the overall costs, despite the standby losses. The V2G strategy enables the reinjection of electricity into the power grid. As shown in Figure 3, in fact, the electricity is bought at times when prices at the DA spot market are low and then resold when the prices are high in order to minimize the overall costs. Furthermore, charging processes of an EV with coordinated MC and coordinated V2G are shown and compared with the BL in Figure 4.

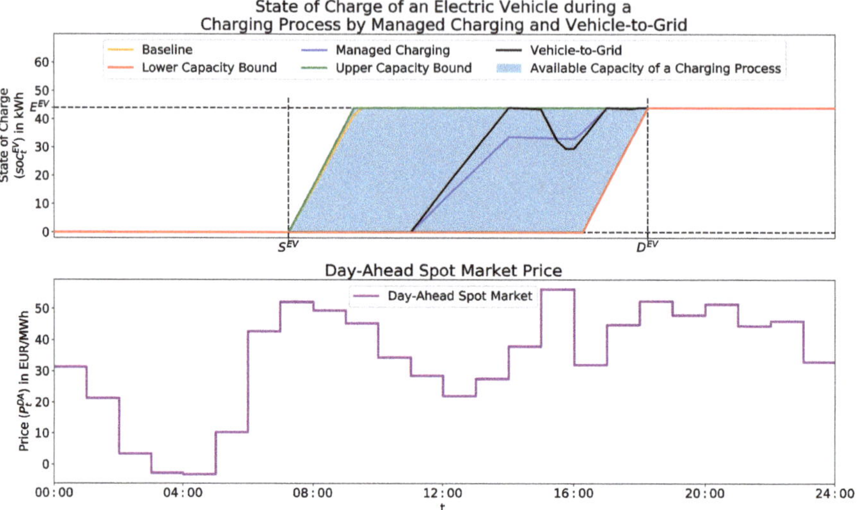

Figure 3. Charging processes of an electric vehicle by managed charging and vehicle-to-grid compared with the baseline.

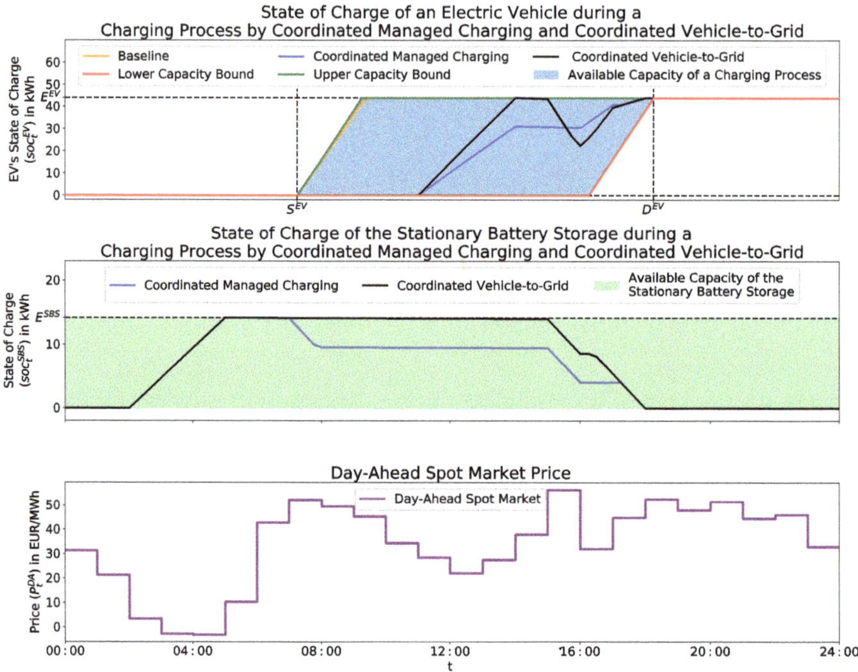

Figure 4. Charging processes of an electric vehicle with coordinated managed charging and coordinated vehicle-to-grid compared with the baseline.

Figure 4 shows the state of charge of the EVs and the SBSs during a charging process by coordinated MC and coordinated V2G. As in Figure 3, the upper graphic shows the available capacity (blue painted area), the upper (green) and lower (red) capacity bounds. Moreover, the state of charge of the EV and the SBS by CMC (blue) and CV2G (black) are illustrated. In these cases, electricity is bought before the charging process begins at a low price. The electricity bought at a low price at the DA spot market can be stored in the SBS and further sold at a high price or used to recharge the EV. The coordination between EVs and SBSs allows the system to use more price signals in order to further reduce the overall costs.

4. Description of the Case Study

In this work, the parking lot of an office site of an electric utility company in Austria is investigated (More specifically WEB, Windenergie AG, Pfaffenschlag, Austria with the coordinates N 48.843594, E 15.200681). The simulated period covers the year 2019. Furthermore, the considered grid tariffs are those applied in Austria (according to the Austrian electricity regulatory office [46]). The investigated parking lot has 30 charging infrastructures with different nominal powers; 4 charging stations with a nominal power of 22 kW, 10 with a nominal power of 11 kW and 16 with a nominal power of 3.7 kW. This study aims to understand which is the most advantageous charging strategy and how convenient it is to install an 80 kWh-capacity stationary battery storage. The considered stationary battery storage consists in a battery pack of lithium-iron-phosphate batteries (model *"BYD BATTERY-BOX HV"* with an external inverter). Over the period of one year, 3739 charging processes of EVs take place with a total of 54.927 MWh. The consumption data of the EVs are actually measured data of the investigated office site. The European Power Exchange (EPEX) DA spot market prices are considered, available on the ENTSO-E transparency platform [47]. Table 1 shows the grid costs and the LCOS that are assumed in the simulations.

Table 1. Assumed grid costs and LCOS in the simulations.

Parameter	Value	Unit	Description
$C^{\text{Grid,FR}}$	1492 [46]	$\frac{\text{€}}{Year}$	Fixed flat rate of the power grid
$p^{\text{Grid,E}}$	33.1 [46]	$\frac{\text{€}}{MWh}$	Specific energy-related grid price
p^{Power}	46.725 [46]	$\frac{\text{€}}{kW_p \cdot Year}$	Specific power-related grid price
$LCOS$	20 [48]	$\frac{\text{€ct}}{kWh}$	Levelized cost of storage

The objective of this work is to compare the economic potential of four charging strategies (MC, V2G, CMC, CV2G) with the unmanaged charging of EVs for an existing office site in Austria with multiple charging infrastructures. It is assumed that the operations of the managed components are optimized through automated technologies. Furthermore, no communication problems of any kind between the operating components are considered. Moreover, since this work aims to understand the potential of the different charging strategies for EVs, the optimization problems are solved under the assumption of perfect load forecast.

5. Results

In this section, the results of the simulations are presented, and the economic potential of the four investigated EVs charging strategies of (MC, V2G, CMC, CV2G) are compared with the unmanaged charging (BL).

The results of the simulations are described and discussed, focusing on quantifying the value of different charging managements strategies of EVs. Figure 5 shows the results of the investigated case study and raises the monetary potential of four charging strategies (MC, V2G, CMC, CV2G) compared with the unmanaged charging of EVs (BL) for the studied office site in Austria.

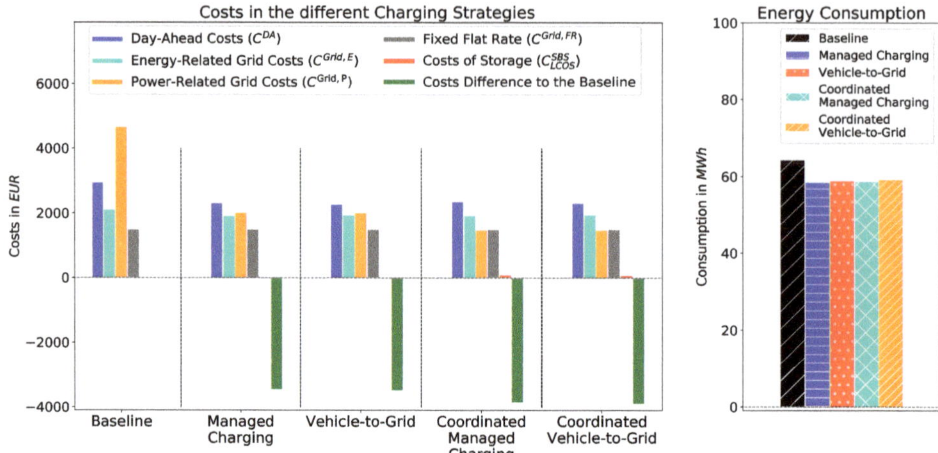

Figure 5. Costs and energy consumption in the different charging strategies.

In Figure 5, it is observable that in the BL, the overall costs (in the graph on the left) and the electricity consumption (in the graph on the right) are higher compared to the remaining four investigated charging strategies. In the BL, the electricity consumption is higher because the EVs are charged as soon as they are connected to the charging infrastructure and remain connected to the charging infrastructure until the disconnection time occurs. This leads to high standby losses. Consequently, more electricity is needed to reach the required state of charge of the EV's batteries when the disconnection time occurs. Therefore, the energy-related grid costs (light blue) are higher in the BL compared to the other four investigated charging strategies. The costs at the DA spot market (blue) are mainly higher for two reasons. Firstly, as already mentioned, more electricity is required and therefore it is necessary to buy more electricity in the DA spot market. The second reason is that the capacity of a charging process is not utilized (as already shown in Figure 3). In the BL, in fact, electricity is bought regardless of the DA spot market prices as soon as an EV connects to the charging infrastructure without applying any load shifting. Furthermore, in the BL, no peak shaving is operated for the aggregated charging infrastructures. This leads to high power-related grid costs (orange). The fixed flat rate is the same for all investigated use cases since it depends exclusively on the time of use of the grid connection, which in this study amounts to one year.

The MC achieves a reduction in overall costs of 30.7% compared to the BL. This is due to the fact that in this case, the load shifting and the peak shaving are applied with the possibility of charging EVs when the DA spot market prices are low and avoiding congestion of EVs charging processes. Congestion of EVs causes the increase in peak power, and consequently of the power-related grid costs. In fact, the power-related grid costs in the MC use case are reduced by 56.6% compared to the BL. Moreover, in this case, the electricity is bought when it is economically more convenient taking into account the grid costs, DA spot market costs and losses of the EV. This leads to a reduction in DA spot market costs of 21.2% and in the energy-related grid costs of 9% as observable in Figure 5.

Furthermore, the V2G strategy achieves overall costs reduction of 31% compared to the BL. In this case, the reinjection of electricity from the EVs into the power grid is enabled together with the possibility to operate load shifting and peak shaving, when it is economically convenient. The power-related grid costs, in this case, are reduced by 56.6%, the DA spot market costs of 22.6% and the energy-related grid costs of 8.2% compared to the BL. The possibility to reinject the electricity into the power grid, in fact, allows the resale of electricity purchased at a low price for a higher price. Hence, in this case, an increased amount of electricity is consumed, but the costs at the DA spot market are lower compared to those in the MC use case.

Moreover, in the CMC and CV2G strategies, the use of the SBS allows achieving further costs reductions compared to the other investigated use cases. As we can see in Figure 5, the overall cost reduction in both cases is more than 34%. The DA spot market costs are reduced by 19.7% applying the CMC and 21.3% in the CV2G use case. The energy-related grid costs are reduced by 8.5% in the CMC use case and 7.7% applying the CV2G strategy compared to the BL. The operation of the SBS aims mainly to reduce the power peak in order to reduce the power-related grid costs in both cases, achieving a reduction of 68.3% compared to the BL.

The high LCOS makes the SBS use very rare for load shifting. For this reason, the SBS is used more in the CMC than in the CV2G use case, since the possibility to inject the electricity directly from the EVs to the power grid makes the SBS less useful. In fact, applying the CV2G strategy, the SBS is charged with an amount of electricity equal to 489.1 kWh, while in the CMC use case the SBS is charged with 519.7 kWh. The amount of electricity flown in the SBS, as already mentioned, is very low due to the high LCOS. Section 5.1 presents a sensitivity analysis aimed to determine the effect of variations in LCOS on the overall costs and battery usage.

5.1. Sensitivity Analysis: Levelized Costs of Storage vs. Overall Costs and Stationary Battery Storage Usage

In an EVs parking lot, the SBS can contribute to the reduction of the overall costs in three ways.

- Through the buying and selling of electricity in the DA spot market, so as to be able to take advantage of price volatilities,
- applying a further load shifting by buying the electricity needed to charge the EVs when the EVs are not yet connected to the charging infrastructures,
- by doing peak shaving, thus reducing power-related grid costs. In all three cases, the LCOS costs play a key role because they determine the monetary convenience of operating the SBS.

In all three cases, the LCOS costs play a key role because they determine the monetary convenience of operating the SBS. Figure 6 presents a sensitivity analysis, aimed to determine the effect of variations in LCOS on the overall costs and the SBS usage in the CMC and CV2G use cases.

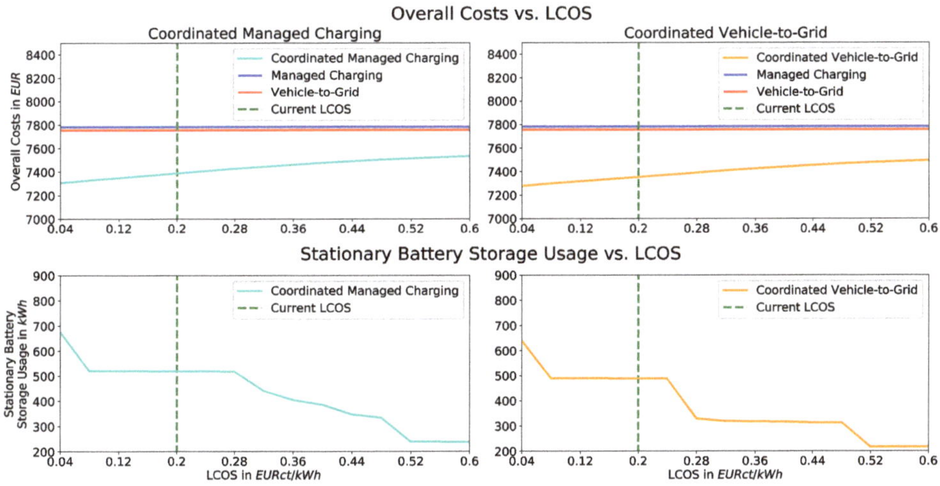

Figure 6. Levelized costs of storage vs. overall costs and stationary battery storage usage.

As shown in Figure 6, the overall costs of the CMC (light blue) and CV2G (orange) use cases decrease with the lowering of the LCOS. As the LCOS increases, the overall costs of the CMC and CV2G use cases tend to the overall costs of the MC (blue) and V2G (red) use cases, as the SBS usage decreases. The SBS in the CMC use case is utilized more than in the CV2G use case since, with high

LCOS, it is more convenient to inject and withdraw electricity from the EVs when their capacity is available. In the CV2G use case, in fact, the SBS is used mostly for peak shaving and less for electricity marketing and load shifting as we can see from the lower graphs of Figure 6. The use of the SBS, in fact, decreases with the increase of the LCOS, since it becomes no longer economically convenient to cover a high number of power peaks and that is why it is preferable not to operate peak shaving and pay a higher power-related grid rate. Moreover, in Section 5.2, a sensitivity analysis aimed to determine the effect of variations in specific energy-related grid price on the overall costs is presented.

5.2. Sensitivity Analysis: Specific Energy-Related Grid Price vs. Overall Costs

In this section, it is investigated whether increasing or reducing the specific energy-related grid price ($p^{\text{Grid,E}}$) can, in some way, incentivize the investigated Austrian office site to apply one of the studied charging strategies. Figure 7 consists of 5 graphs showing the overall costs as a function of the specific energy-related grid price for the four investigated use cases and the BL.

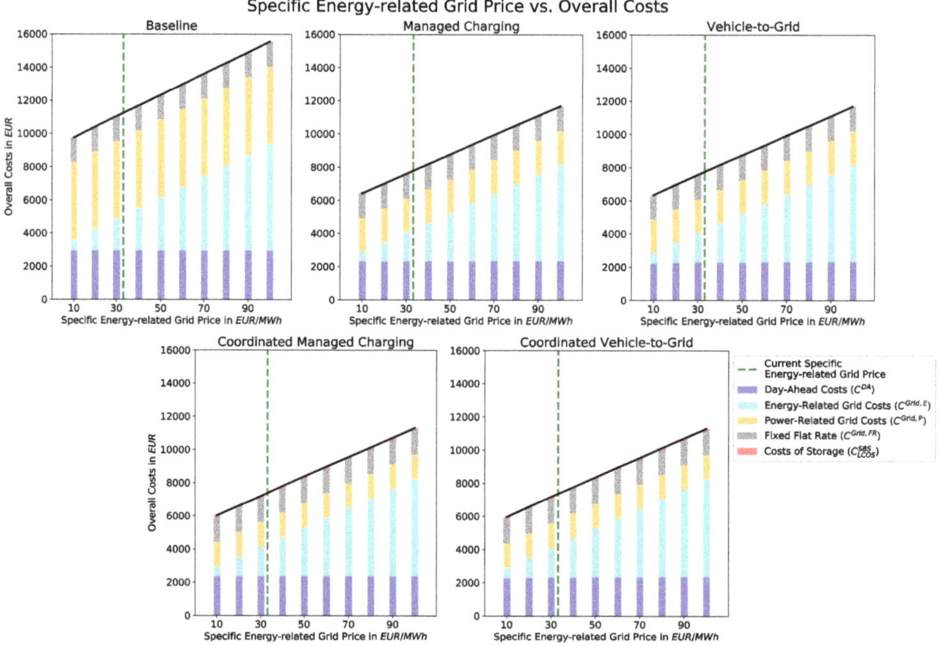

Figure 7. Specific energy-related grid price vs. overall costs.

Figure 7 shows that the changes in specific energy-related grid price have a similar influence on all investigated use cases. This is due to the fact that in all use cases studied, EVs must be charged with the same amount of electricity. This means that at high specific energy-related grid price, the optimal solution leads to minimizing the losses, so as to have the least amount of electricity flowing through the grid connection point. With the increase of the specific energy-related grid price, it does not become more economically convenient to operate electricity marketing on the DA spot market since the storage of electricity causes efficiency and standby losses, and also because the electricity purchase price increases linearly with the specific energy-related grid price. An increase in specific energy-related grid price, therefore does not make any of the four investigated charging strategies more advantageous over the other. Even at a very low specific energy-related grid price, the SBS is mostly operated for peak shaving, as current LCOSs often do not make it convenient to operate electricity marketing in the

DA spot market. Therefore, in Section 5.3 a sensitivity analysis is presented in order to determine the effect of variations in specific power-related grid price on the overall costs.

5.3. Sensitivity Analysis: Specific Power-Related Grid Price vs. Overall Costs

As already seen in Figure 5, the power-related grid costs constitute the most expensive cost component in the BL. Therefore, the four investigated charging strategies aim mainly to reduce the power-related grid costs lowering the peak power. Figure 8 presents a sensitivity analysis aimed to determine the effect of variations in the specific power-related grid price on the overall costs.

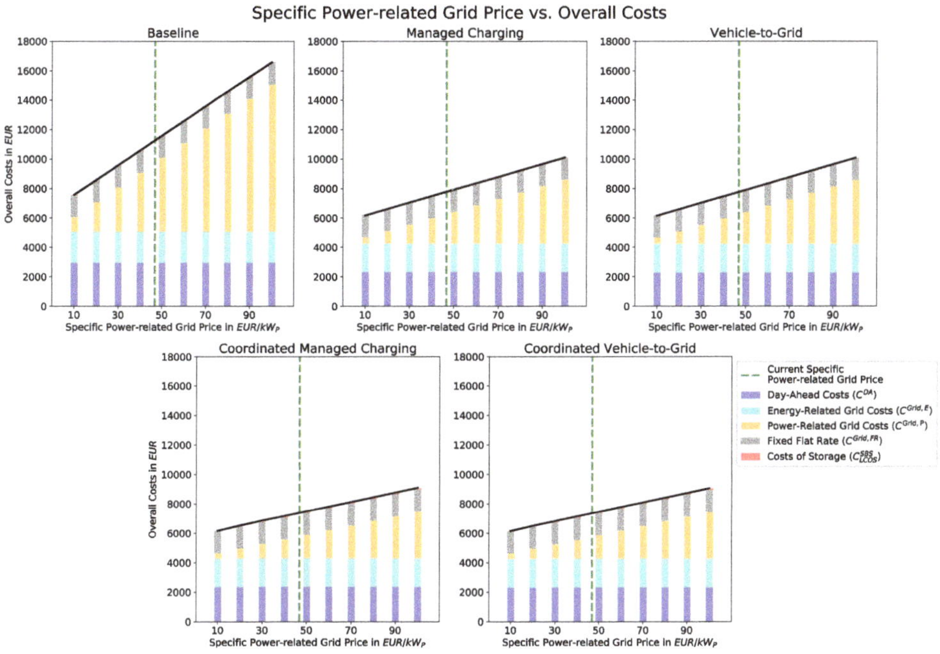

Figure 8. Specific power-related grid price vs. overall costs.

In Figure 8, it is observable that the overall costs in all investigated use cases increase almost linearly with the increase of the specific power-related grid price. However, it is evident how the slope of the total costs as a function of the specific power-related grid price changes in the studied use cases. The BL has the most significant slope, while the cases where the EVs are coordinated with the SBS the slope is minor. This is due to the fact that in the CMC and CV2G use cases, there is more capacity available to perform peak shaving. Again, the difference in overall costs between the MC and the V2G use cases and between the CMC and the CV2G use cases is minimal, confirming the fact that injecting electricity into the grid from an EV is rarely cost-effective. In conclusion, it can be demonstrated to the hand of Figure 8 that an increase in specific power-related grid price would further incentivize the investigated Austrian office to implement a smart charging strategy and invest in the SBS.

6. Conclusions and Outlook

This paper presents the monetary potential of different EVs charging strategies for an existing Austrian office site with multiple charging infrastructures. Hence, multiple optimization problems are defined in order to determine the operation of the investigated components in the different EVs charging strategies. Various diversities and potentials of different EVs charging strategies are identified comparing four use cases for the same case study. Furthermore, the method to implement EVs and SBSs in optimization problems with the exclusive use of linear relationships is presented.

The simulations have shown that unmanaged charging of EVs leads to high power peaks, which result in high overall costs. The implementation of a managed charging strategy can lead the overall costs of the Austrian office site's parking lot to a reduction of more than 34%. In particular, the power-related grid costs can decrease by 68% if the charging of EVs is managed and coordinated with the operation of a SBS. Moreover, the simulations have shown that the charging processes management leads to less electricity consumption because of the lower standby losses. This results in reduced energy-related grid costs, which can decrease by 9%. The results highlight the fact that for an office there is no big difference in overall costs if the V2G is operated. In fact the overall costs between the V2G and MC use case and between the CV2G and CMC use cases differ only by 0.3%. This is mainly due to the fact that the connection time of EVs to the charging infrastructures at an office is not long enough to allow the capture of market price signals and charging the EV at the same time. In fact, the average connection time of EVs to the charging infrastructures during the year 2019 was 98 min. However, it should be noted that this cannot be ruled out for private end-users, where the average connection time of EVs is certainly longer and it is possible that the V2G charging is monetary convenient.

The performed sensitivity analyses show that the SBS at current LCOS operates mostly for peak shaving and not for marketing electricity in the DA spot market. Marketing electricity in the DA spot market is economically convenient for the SBS when the LCOS are lower than 0.08 $\frac{\text{\euro}ct}{kWh}$. It has also been shown that variations in specific energy-related grid price are unlikely to incentivize the Austrian office site to implement managed charging in their parking lot. Instead, an increase in specific power-related grid price makes EV's charging management more convenient and therefore incentivize the Austrian office site to implement it and also to invest in a SBS. Therefore, one of the future challenge is to investigate to what extent the specific power-related grid price will increase. Nowadays, the growing share of renewables in the power grid raised the specific power-related grid prices in different European countries. Therefore, it is reasonable to assume that in the future the charging process management of EVs can be further incentivized.

Author Contributions: Conceptualization, C.C., D.S., G.L.; methodology, C.C., D.S.; software, C.C.; validation, C.C., G.L., A.A., H.A.; formal analysis, C.C.; investigation, C.C., D.S., A.A.; resources, G.L.; data curation, C.C.; writing–original draft preparation, C.C.; writing—review and editing, C.C., G.L., A.A., H.A.; visualization, C.C., A.A.; supervision, H.A.; project administration, C.C., G.L.; funding acquisition, G.L. All authors have read and agreed to the published version of the manuscript.

Funding: This research was funded by the Flex+ project. The Flex+ project (No 864996) is being funded under the 4th call of the energy research program of the Austrian Research Promotion Agency (FFG) and the Climate Energy Fund. The authors are particularly grateful for the metered data provided by F. Mader of the WEB, Windenergie AG, Pfaffenschlag, Austria.

Acknowledgments: The authors acknowledge TU Wien Bibliothek for financial support through its Open Access Funding Program.

Abbreviations

The following abbreviations are used in this manuscript:

BL	Baseline
CMC	Coordinated Managed Charging
CV2G	Coordinated Vehicle-to-Grid
DA	Day-Ahead
EPEX	European Power Exchange
EV	Electric Vehicle
GCP	Grid Connection Point
IPCC	Intergovernmental Panel on Climate Change

MC   Managed Charging
SBS   Stationary Battery Storage
V2G   Vehicle-to-Grid

References

1. Hannappel, R. The Impact of Global Warming on the Automotive Industry. *AIP Conf. Proc.* **2017**, *1871*, 060001. [CrossRef]
2. Jacobson, M.Z. Review of Solutions to Global Warming, Air Pollution, and Energy Security. *Energy Environ. Sci.* **2009**, *2*, 148–173. [CrossRef]
3. Kebriaei, M.; Niasar, A.H.; Asaei, B. Hybrid Electric Vehicles: An Overview. In Proceedings of the 2015 International Conference on Connected Vehicles and Expo (ICCVE), Shenzhen, China, 19–23 October 2015; pp. 299–305. [CrossRef]
4. Hawkins, T.R.; Singh, B.; Majeau-Bettez, G.; Strømman, A.H. Comparative Environmental Life Cycle Assessment of Conventional and Electric Vehicles. *J. Ind. Ecol.* **2013**, *17*, 53–64. [CrossRef]
5. Transport—European Environment Agency. Available online: https://www.eea.europa.eu/themes/transport/intro (accessed on 13 August 2020).
6. Transport—IPCC. Available online: https://www.ipcc.ch/report/ar5/wg3/transport/ (accessed on 13 August 2020).
7. Foley, A.; Winning, I.; Ó Gallachóir, B.P.Ó. State-of-the-Art in Electric Vehicle Charging Infrastructure. In Proceedings of the 2010 IEEE Vehicle Power and Propulsion Conference, Lille, France, 1–3 September 2010; pp. 1–6. [CrossRef]
8. Knezović, K.; Marinelli, M.; Zecchino, A.; Andersen, P.B.; Traeholt, C. Supporting Involvement of Electric Vehicles in Distribution Grids: Lowering the Barriers for a Proactive Integration. *Energy* **2017**, *134*, 458–468. [CrossRef]
9. Prajapati, V.K.; Mahajan, V. Congestion Management of Power System with Uncertain Renewable Resources and Plug-in Electrical Vehicle. *IET Gener. Transm. Distrib.* **2019**, *13*, 927–938. [CrossRef]
10. Wei, W.; Liu, F.; Mei, S. Charging Strategies of EV Aggregator Under Renewable Generation and Congestion: A Normalized Nash Equilibrium Approach. *IEEE Trans. Smart Grid* **2016**, *7*, 1630–1641. [CrossRef]
11. Dow, L.; Marshall, M.; Xu, L.; Romero Agüero, J.; Willis, H.L. A Novel Approach for Evaluating the Impact of Electric Vehicles on the Power Distribution System. In Proceedings of the IEEE PES General Meeting, Providence, RI, USA, 25–29 July 2010; pp. 1–6. [CrossRef]
12. Shao, S.; Pipattanasomporn, M.; Rahman, S. Challenges of PHEV Penetration to the Residential Distribution Network. In Proceedings of the 2009 IEEE Power Energy Society General Meeting, Calgary, AB, Canada, 26–30 July 2009; pp. 1–8. [CrossRef]
13. Zhang, L.; Li, Y. Optimal Management for Parking-Lot Electric Vehicle Charging by Two-Stage Approximate Dynamic Programming. *IEEE Trans. Smart Grid* **2017**, *8*, 1722–1730. [CrossRef]
14. Venayagamoorthy, G.K. Dynamic, Stochastic, Computational, and Scalable Technologies for Smart Grids. *IEEE Comput. Intell. Mag.* **2011**, *6*, 22–35. [CrossRef]
15. Habib, S.; Kamran, M.; Rashid, U. Impact Analysis of Vehicle-to-Grid Technology and Charging Strategies of Electric Vehicles on Distribution Networks—A Review. *J. Power Sources* **2015**, *277*, 205–214. [CrossRef]
16. Lopes, J.P.; Almeida, P.M.R.; Silva, A.M.; Soares, F.J. Smart Charging Strategies for Electric Vehicles: Enhancing Grid Performance and Maximizing the Use of Variable Renewable Energy Resources. In *CPES—Articles in International Conferences*; EVS24 International Battery, Hybrid and Fuel Cell Electric Vehicle Symposium: Stavanger, Norway, 2019.
17. Moghaddam, Z.; Ahmad, I.; Habibi, D.; Phung, Q.V. Smart Charging Strategy for Electric Vehicle Charging Stations. *IEEE Trans. Transp. Electrif.* **2018**, *4*, 76–88. [CrossRef]
18. Cui, H.; Li, F.; Fang, X.; Long, R. Distribution Network Reconfiguration with Aggregated Electric Vehicle Charging Strategy. In Proceedings of the 2015 IEEE Power Energy Society General Meeting, Denver, CO, USA, 26–30 July 2015; pp. 1–5. [CrossRef]
19. Clairand, J.M.; Rodríguez-García, J.; Álvarez-Bel, C. Electric Vehicle Charging Strategy for Isolated Systems with High Penetration of Renewable Generation. *Energies* **2018**, *11*, 3188. [CrossRef]

20. Turker, H.; Bacha, S. Optimal Minimization of Plug-In Electric Vehicle Charging Cost With Vehicle-to-Home and Vehicle-to-Grid Concepts. *IEEE Trans. Veh. Technol.* **2018**, *67*, 10281–10292. [CrossRef]

21. Aziz, M.; Oda, T.; Ito, M. Battery-Assisted Charging System for Simultaneous Charging of Electric Vehicles. *Energy* **2016**, *100*, 82–90. [CrossRef]

22. Bryden, T.S.; Hilton, G.; Dimitrov, B.; Ponce de León, C.; Cruden, A. Rating a Stationary Energy Storage System Within a Fast Electric Vehicle Charging Station Considering User Waiting Times. *IEEE Trans. Transp. Electrif.* **2019**, *5*, 879–889. [CrossRef]

23. Pyomo. Available online: http://www.pyomo.org (accessed on 13 August 2020).

24. Gurobi—The Fastest Solver—Gurobi. Available online: https://www.gurobi.com/ (accessed on 13 August 2020).

25. Ballester, C.; Furió, D. Effects of Renewables on the Stylized Facts of Electricity Prices. *Renew. Sustain. Energy Rev.* **2015**, *52*, 1596–1609. [CrossRef]

26. Hammons, T.J. Integrating Renewable Energy Sources into European Grids. *Int. J. Electr. Power Energy Syst.* **2008**, *30*, 462–475. [CrossRef]

27. Jones, L.E. *Renewable Energy Integration: Practical Management of Variability, Uncertainty, and Flexibility in Power Grids*; Academic Press: Cambridge, MA, USA, 2017.

28. Knezović, K.; Marinelli, M.; Codani, P.; Perez, Y. Distribution Grid Services and Flexibility Provision by Electric Vehicles: A Review of Options. In Proceedings of the 2015 50th International Universities Power Engineering Conference (UPEC), Staffordshire University, Staffordshire, UK, 1–4 September 2015; pp. 1–6. [CrossRef]

29. Shao, S.; Pipattanasomporn, M.; Rahman, S. Grid Integration of Electric Vehicles and Demand Response With Customer Choice. *IEEE Trans. Smart Grid* **2012**, *3*, 543–550. [CrossRef]

30. Corinaldesi, C.; Fleischhacker, A.; Lang, L.; Radl, J.; Schwabeneder, D.; Lettner, G. European Case Studies for Impact of Market-Driven Flexibility Management in Distribution Systems. In Proceedings of the 2019 IEEE International Conference on Communications, Control, and Computing Technologies for Smart Grids (SmartGridComm), Beijing, China, 21–23 October 2019; pp. 1–6. [CrossRef]

31. Voerman, M.; de Bont, K.F.M.; Zeiler, W.; Department of the Built Environment, Building Physics and Building Services. Active Participation of Buildings as a Source of Energy Flexibility. In Proceedings of the ASHRAE Annual Conference, Houston, TX, USA, 23–27 June 2018.

32. Aghajani, G.R.; Shayanfar, H.A.; Shayeghi, H. Demand Side Management in a Smart Micro-Grid in the Presence of Renewable Generation and Demand Response. *Energy* **2017**, *126*, 622–637. [CrossRef]

33. Schuller, A.; Flath, C.M.; Gottwalt, S. Quantifying Load Flexibility of Electric Vehicles for Renewable Energy Integration. *Appl. Energy* **2015**, *151*, 335–344. [CrossRef]

34. Huo, Y.; Bouffard, F.; Joós, G. An Energy Management Approach for Electric Vehicle Fast Charging Station. In Proceedings of the 2017 IEEE Electrical Power and Energy Conference (EPEC), Saskatoon, SK, Canada, 22–25 October 2017; pp. 1–6. [CrossRef]

35. Corinaldesi, C.; Schwabeneder, D.; Lettner, G.; Auer, H. A Rolling Horizon Approach for Real-Time Trading and Portfolio Optimization of End-User Flexibilities. *Sustain. Energy Grids Netw.* **2020**, *24*, 100392. [CrossRef]

36. Sadeghianpourhamami, N.; Refa, N.; Strobbe, M.; Develder, C. Quantive Analysis of Electric Vehicle Flexibility: A Data-Driven Approach. *Int. J. Electr. Power Energy Syst.* **2018**, *95*, 451–462. [CrossRef]

37. Weis, A.; Jaramillo, P.; Michalek, J. Estimating the Potential of Controlled Plug-in Hybrid Electric Vehicle Charging to Reduce Operational and Capacity Expansion Costs for Electric Power Systems with High Wind Penetration. *Appl. Energy* **2014**, *115*, 190–204. [CrossRef]

38. Sheikhi, A.; Bahrami, S.; Ranjbar, A.M.; Oraee, H. Strategic Charging Method for Plugged in Hybrid Electric Vehicles in Smart Grids; a Game Theoretic Approach. *Int. J. Electr. Power Energy Syst.* **2013**, *53*, 499–506. [CrossRef]

39. Kiaee, M.; Cruden, A.; Sharkh, S. Estimation of Cost Savings from Participation of Electric Vehicles in Vehicle to Grid (V2G) Schemes. *J. Mod. Power Syst. Clean Energy* **2015**, *3*, 249–258. [CrossRef]

40. Schuller, A.; Dietz, B.; Flath, C.M.; Weinhardt, C. Charging Strategies for Battery Electric Vehicles: Economic Benchmark and V2G Potential. *IEEE Trans. Power Syst.* **2014**, *29*, 2014–2022. [CrossRef]

41. Noel, L.; Zarazua de Rubens, G.; Kester, J.; Sovacool, B.K. Beyond Emissions and Economics: Rethinking the Co-Benefits of Electric Vehicles (EVs) and Vehicle-to-Grid (V2G). *Transp. Policy* **2018**, *71*, 130–137. [CrossRef]

42. López, M.A.; Martín, S.; Aguado, J.A.; de la Torre, S. V2G Strategies for Congestion Management in Microgrids with High Penetration of Electric Vehicles. *Electr. Power Syst. Res.* **2013**, *104*, 28–34. [CrossRef]
43. Hao, H.; Somani, A.; Lian, J.; Carroll, T.E. Generalized Aggregation and Coordination of Residential Loads in a Smart Community. In Proceedings of the 2015 IEEE International Conference on Smart Grid Communications (SmartGridComm), Miami, FL, USA, 2–5 November 2015; pp. 67–72. [CrossRef]
44. Lai, C.S.; Jia, Y.; Xu, Z.; Lai, L.L.; Li, X.; Cao, J.; McCulloch, M.D. Levelized Cost of Electricity for Photovoltaic/Biogas Power Plant Hybrid System with Electrical Energy Storage Degradation Costs. *Energy Convers. Manag.* **2017**, *153*, 34–47. [CrossRef]
45. Schmidt, O.; Melchior, S.; Hawkes, A.; Staffell, I. Projecting the Future Levelized Cost of Electricity Storage Technologies. *Joule* **2019**, *3*, 81–100. [CrossRef]
46. Tariff Model in Austria. Available online: https://www.apg.at/en/markt/strommarkt/tarife (accessed on 13 August 2020).
47. ENTSO-E Transparency Platform. Available online: https://transparency.entsoe.eu/ (accessed on 13 August 2020).
48. Jülch, V. Comparison of Electricity Storage Options Using Levelized Cost of Storage (LCOS) Method. *Appl. Energy* **2016**, *183*, 1594–1606. [CrossRef]

Publisher's Note: MDPI stays neutral with regard to jurisdictional claims in published maps and institutional affiliations.

Article

Efficient Load Management for BEV Charging Infrastructure in Multi-Apartment Buildings

Jasmine Ramsebner *[ID], **Albert Hiesl and Reinhard Haas**

Energy Economics Group (EEG), Technische Universität Wien, Gusshausstraße 25-29, E370-3, 1040 Vienna, Austria; hiesl@eeg.tuwien.ac.at (A.H.); haas@eeg.tuwien.ac.at (R.H.)
* Correspondence: ramsebner@eeg.tuwien.ac.at

Received: 27 August 2020; Accepted: 10 November 2020; Published: 13 November 2020

Abstract: Interest in and demand for battery electric vehicles (BEVs) is growing strongly due to the increasing awareness of climate change and specific decarbonization goals. One of the largest challenges remains the provision of large-scale, efficient charging infrastructure in multi-apartment buildings. Successful load management (LM) for BEV charging directly influences the technical requirements and the economic and environmental aspects of charging infrastructure and can prevent costly distribution grid expansion. The main objective of this paper is to evaluate potential LM approaches in multi-apartment buildings to avoid an increase in existing electricity demand peaks with BEV diffusion. Using our model parameters, off-peak charging achieved a 40% reduction in the building's demand peak at 100% BEV diffusion compared to uncontrolled charging and reduced the correlation between BEV charging and the national share of thermal power generation. The most efficient charging capacity in the private network was achieved at 0.44 kW/BEV. A verification of the model results with the demonstration phase of the "Urcharge" project supports our overall findings. Our results outline the advantages of LM across a large-scale BEV charging network to control the impact on the electricity system along with the diffusion of e-mobility.

Keywords: electric mobility; charging infrastructure; load management; battery electric vehicle; urban area; multi-apartment building; zero emission mobility; private charging

---

1. Introduction

Interest in and demand for battery electric vehicles (BEVs) is growing strongly. Reasons for this growth include increasing awareness of climate change, as well as the obligation of car manufacturers to reduce the CO2 emissions of their fleets substantially by 2030 [1]. One of the largest keys to success associated with this development is the provision of appropriate charging infrastructure (IS) [2]. Since BEV charging currently remains largely connected to single homes with private charging stations or charging at public stations, large-scale load management (LM) applications are still not commonly applicable. Major questions include not only where charging can take place (e.g., private and public charging, charging at work, etc.) and at which charging speed and capacity but also how the potential temporal distribution of this load behaves. The successful management of BEV charging demand is directly related to the economic aspects of charging IS installations and distribution grid performance and expansion requirements. This work is a result of the Austrian research project "Urcharge", which aims at an extension of currently available LM functionalities for large-scale private charging IS at the technical, economic, and customer levels [3,4]. From a technical perspective, Urcharge aims at substantial enhancements of LM functionalities from a static management across a maximum of 15 sub-stations towards the dynamic management of more than 150 charging points, including improved data gathering and analysis, appropriate billing functionalities, and efficiency improvements. This is one of the few projects globally currently investigating private charging IS on such a large scale. Three projects in the US–in San Diego [5], Los Angeles [6] and Columbus–Ohio [7]–and one in Germany [8] aim at higher availability of BEV charging points in multi-unit dwellings.

For public fast charging, single-family buildings, and small company applications, many projects have already been conducted, with Hall et al. [9] providing a global best practice overview, IEA [10] describing the Nordic EV outlook, two projects for BEV charging IS in Denmark [11,12], and Project Better Energy in the UK [13], as well as one in Germany [14]. Nevertheless, our literature review reveals that research on large-scale charging IS with optimal load management (LM) in residential buildings remains scarce. Efficient LM is also referred to as smart charging in the literature, defined as an adaption of the BEV charging cycle to both the restrictions of the distribution grid and the needs of BEV users [15]. International Renewable Energy Agency (IRENA) [15] claims that smart BEV charging enables peak shaving, network congestion management, and reduced grid capacity investments. Various studies cover research on the electricity demand profiles of BEVs and potential peak increases. Van Vliet et al. [16] discovered the negative effects of uncontrolled BEV charging on the distribution grid and found that with a BEV diffusion of 30%, the Dutch peak electricity demand would grow by 7%, and the household (HH) or overall building peak demand would grow by 54%. Several studies observed the benefit of off-peak charging, in which the base load is increased while the national peak load remains untouched. While van Vliet et al. [16] highlighted the benefit of off-peak charging for the energy use cost and CO_2 emissions of BEVs, Bitar and Xu [17] suggested a price that decreases with the deadline that is granted by the user for the completion of the charging process. Limmer and Rodemann [18] highlighted the cost aspects of peak charging at public stations. They argued that although the majority of public charging stations for BEVs are presently uncontrolled, controlled charging should offer a price that decreases with the latest deadline the user allows for charging to increase the availability of the BEV at the station. From our perspective, however, this pricing-scheme requires an appropriate interface for the user and would lead to parking lots being blocked for a considerable amount of time. This scheme, therefore, is designated for applications with an assigned parking lot.

Several studies on LM for BEV charging widely address pricing mechanisms to limit negative effects on the distribution grid. Some suggest a time-of-use (ToU) pricing mechanism that incentivizes off-peak charging with a lower charging price [19,20]. New peaks, however, may still arise if too much charging power is pushed into off-peak times. Therefore, so-called dynamic load-based pricing is a potential solution in which charging power decreases with the amount of BEVs charged at a time [17,19]. Dynamic pricing while the BEV is already plugged in, however, represents high price uncertainty and would need to be based on a forecast of charging demand. It has to be kept in mind that such an LM or restrictions on charging power can also be implemented top-down from the control station across the charging points to guarantee distribution grid performance. On the other hand, Yan et al. [21] found that traffic jam and weather forecasting can enhance the operation of LM. Temperatures have an impact on battery performance and heating and cooling demands, while the traffic conditions influence the electricity consumption of driving.

Recently, an increasing number of studies has been published concerning the impact of e-mobility on the distribution grid, power quality, and power generation capacities. Crozier et al. [22] observed that controlled charging has a significant benefit for Great Britain's electricity network. It can reduce expansion requirements for electricity generation capacities, as well as investment into the distribution network. Das et al. [23] focused on the grid integration of BEV charging IS, whereas Khalid et al. [24] emphasized the power quality aspects in the utility grid and Brinkel et al. [25] analyzed the need for grid reinforcement because of growing e-mobility. The benefits of controlled charging have also been discussed within the context of smart grids and smart cities, in which an optimal charging scheduling is carried out to improve grid system operation, voltage, and efficiency [26,27]. Another paper developed an optimized charging framework using knowledge of the upcoming trip schedule of the BEVs [28]. However, such detailed information is difficult to obtain.

Using the battery of the BEV to switch flexibly between charging and providing energy to the power grid has been analyzed from various perspectives throughout the last decade. Some focus on the potential of vehicle-to-grid (V2G) for BEV integration into smart grids [29,30] and its impact on the grid in general [31]. From a user perspective, V2G sometimes is regarded as the potential missing link for the acceptance of BEVs as a cost-effective support policy. Chen et al. [32], for example, found that in Nordic countries, V2G capability—apart from charging time—is a technical aspect that improves BEV adoption. Sortomme and El-Sharkawi [33] studied optimal charging strategies for V2G application. Various studies, however, analyze the techno-economic feasibility of V2G [34–36] and find that, economically, this strategy may only be feasible in specific scenarios which highly depend on the battery degradation costs related to V2G cycling. Noel et al. [37] and Parsons et al. [38] investigated the willingness to pay for V2G applications and concluded that V2G is only relevant in countries with higher overall education or knowledge of the technology. Furthermore, the concept needs to provide a financial incentive for the user to be successful. While Habib et al. [39] studied the impact of V2G technology and charging strategies on the distribution network, the parker project represented a field test of V2G infrastructure and found that V2G can be commercialized by providing frequency containment reserves [40]. Rezania [41] even concluded in a study on Austria that, from an electricity system point of view, the participation of BEVs can hardly be competitive in the frequency reserve market. The competition from other providers with technological and cost advantages, such as heat pumps and pumped hydro, is too strong and the battery degradation cost limits the potential economic benefit.

Some recent studies also investigate in the ability of LM to reduce greenhouse gas (GHG) emissions and promote the integration of variable renewable energy (VRE). Apart from the cost and emission optimization of BEV charging by Brinkel et al. [25], Tu et al. [42] optimized BEV charging to minimize the emissions from electricity consumption. Additionally, Dixon et al. [43] modelled a BEV charging schedule that minimizes carbon intensity and seeks to absorb otherwise curtailed renewable electricity. Their method achieved an average 25% reduction of the CO_2/km compared to uncontrolled charging from Great Britain's grid.

Nevertheless, despite this technique's great potential, limited work is available that focuses specifically on private BEV charging IS and the potential of large-scale LM in multi-apartment buildings. Lopez-Behar et al. [44] claim that in many cities, most of the residents live in multi-apartment buildings not yet equipped with proper BEV charging IS. Simulation results based on real-world driving data by Wang et al. [45] show that home charging is able to meet the energy demands of the majority of BEVs under average conditions. Kim et al. [46] analyzed residential BEV charging behavior. Additionally, Yi et al. [20] optimized the charging of BEVs via charging behavior models based on a real-world data set of a Nissan Leaf during weekdays, measured from thousands of charging points over several years. The authors found that BEV charging tends to occur together with distribution grid peak loads. Five different scenarios with varying BEV penetrations were simulated to show the difference between uncontrolled and controlled charging (Figure 1). Each scenario includes 100,000 HHs and considers BEV penetrations between 10% and 90%. Jang et al. [47] eventually optimized residential BEV charging to avoid peak loads and minimize charging cost for the user and Khalkhali et al. [48] investigated a residential smart parking lot to control grid performance.

The main objective of this paper is to evaluate the potential LM approaches for BEV diffusion to avoid an increase in existing HH electricity demand peaks in multi-apartment buildings and thus substantial long-term, system-wide capacity investments and operational costs. While the 6-month demonstration phase within Urcharge tested LM for 51 BEVs representing 50% E-mobility in this area, our model provides a framework to analyze the impact on electricity demand profiles up to 100% E-Mobility. We compare the situation of uncontrolled charging to the results with LM approaches and address the environmental effects of a substitution of conventional cars with BEVs and their electricity consumption. Another important goal within this analysis is to define the minimum charging capacity required for each BEV (kW/BEV) as an indicator for cost-efficient sizing. Once the power connection assigned to the area is exceeded by the HH plus BEV electricity demand, substantial costs occur to reinforce the power cables or even buy additional power capacity from the grid operator. Our contribution mainly benefits from being embedded in a holistic research project for Austria, and our model results are verified by experience from the project's demonstration phase in Section 4.2.

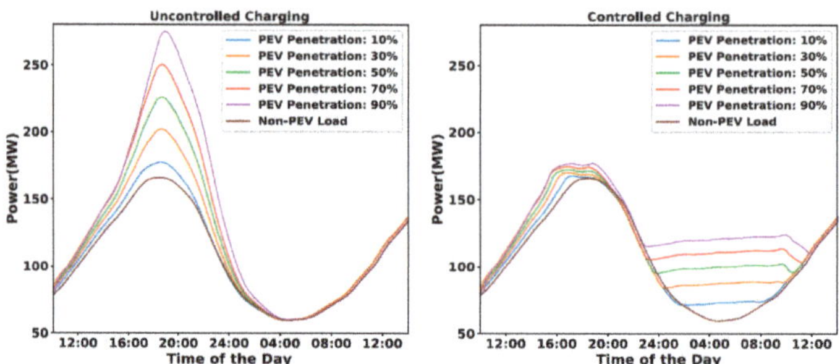

Figure 1. Residential power load for uncontrolled and controlled charging under different plug-in electric vehicle (PEV) penetration levels (PEVs in the study are all battery electric vehicles (BEVs)) [21].

This paper begins with a description of our methodology, model parameters, and assumptions and the considered charging strategies. In Section 3, we present the results of this research. We also describe the existing HH electricity demand, and the impact of E-mobility with uncontrolled BEV charging, compared with the changes through LM and the minimum required charging capacity for each BEV. Finally, in Section 4, we carry out a detailed discussion of our modelling results with a verification using the monitoring data from the project demonstration phase in Section 4.2. The environmental impact of E-mobility is evaluated in Section 4.3, and Section 5 provides the comprehensive conclusions of our work.

2. Materials and Methods

To analyze the impact of different LM approaches on the building's electricity demand, we first define a yearly HH load profile for a residential area with 106 households. The BEV charging demands and availability times at home or public charging points, depending on the trip's start time and length, are taken from a modeling paper by Hiesl et al. [49] based on an Austrian traffic survey. The approach is briefly outlined in Section 2.2. Our model coordinates individual passenger BEV charging in the home network according to the constraints for each charging scenario (Section 2.3.1). The results allow assessing the impact of LM on large-scale BEV charging on electricity demand peaks and valleys. The model parameters largely reflect the common technical standards on battery capacity, charging speed, and efficiency for private stations.

On the one hand, our model analysis is subject to simplifications from more detailed, real-world technical functionalities to enable more generally applicable conclusions. On the other hand, it helps us to carry out additional analyses that were not tested throughout the Urcharge demonstration phase and draw conclusions on the impact of E-mobility toward 100% BEV diffusion. Simplifications to the model include the operation of the LM optimization model under full information of BEV electricity consumption and the trip start times across one year, as well as the assumption of one uniform battery size and uniform technical characteristics for all BEVs. Furthermore, our model operates under knowledge of the battery's state of charge, which is currently not the case in a real setting. A detailed description of the differences between the modeled and real environment in the project, as well as a verification of the model results, are provided in Section 4.2. The model does not cover any load flows of the distribution grid or V2G approaches but instead focuses on an analysis of potential LM approaches to reduce the impact of e-mobility on the overall building load.

2.1. Household Electricity Demand

For our analysis, the electricity demand profile throughout the day is of greatest importance. Of course, demand may vary with parameters such as the HH size and the habitants' ages and types of profession. Furthermore, the HH electricity demand profile is subject to changes due to increasing electrification. However, we only considered the relative impact of e-mobility on existing building electricity demand and did not model any future scenarios.

To determine the aggregate HH electricity demand that largely resembles the demonstration site's 106 HHs, two weeks of data were measured on-site at the respective building's power connection. To achieve a realistic yearly profile for our model, an alternative yearly dataset from a similar project was used. We extrapolated this alternative yearly dataset to the demand level during the measured two weeks at the Urcharge demo site such that the average electricity demand during the same two-week time periods matched. Eventually, the patterns were similar, and the adapted yearly data were considered feasible for our study. The two weeks of measured Urcharge HH electricity demands and the aligned yearly data are shown in Figure 2.

For our further analysis, a threshold to differentiate the HH load valleys and peaks using the 70th percentile (P70) was defined, which is also described in Figure 2. This parameter represents the threshold that 70% of all HH load values lie below, which is 64.5 kW. Then, the LM approaches mostly aim at minimizing the impact on the peaks or shifting BEV charging demands in the HH load valleys. The detailed approaches for LM are described in Section 2.3.1.

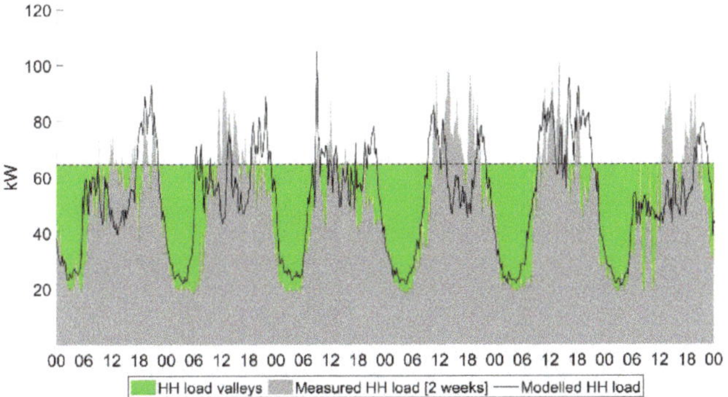

Figure 2. Measured household load in the residential area of the Urcharge project (2 weeks data) and the aligned common yearly household (HH) load profiles with peaks and valleys.

The impact of e-mobility on the existing building electricity demand and the effectiveness of LM approaches was evaluated according to the specific indicators determined to be appropriate, as described in Table 1.

Table 1. Important result parameters (see Figure 2).

Parameter	Description
Maximum demand (kW)	Maximum electricity demand discovered in the year
Minimum demand (kW)	Minimum electricity demand discovered in the year
Peak demand volume (MWh)	Electricity demand volume exceeding the P70 threshold
Base demand volume (MWh)	Electricity demand volume within the HH load valleys beneath P70

2.2. BEV Driving Profiles

To determine common BEV charging demands, the driving profiles according to Hiesl et al. [49] are applied, which were derived from an Austrian traffic survey [50], as a static input for our optimization model. Eight typical driving purposes with the relevant distribution of common distances, related electricity consumption, and potential start times for urban Austrian car users were applied to determine when and after which level of energy consumption each BEV is parked and ready to charge within the private IS [50]. The modeled energy consumption aims at meeting the traffic survey results for Austrian cities (excluding Vienna) with an average yearly distance traveled by each driver of 12.237 km/a.

BEV consumption is modeled as an interpolation between 15 kWh/100 km in summer and 17 kWh/100 km in winter due to the additional power consumption for heating and the temperature sensitivity of the batteries [51]. Due to computing performance, the model calculates the distribution of BEV charging availability for 10 different BEV profiles. Assuming one parking lot for each HH, we extrapolated these charging profiles to the number of BEVs for specific percentages of BEV diffusion according to Table 2.

Table 2. Number of BEVs in the modeled multi-apartment building for BEV diffusion.

No. of BEVs	BEV Diffusion
11	10%
32	30%
53	50%
80	75%
106	100%

2.3. Optimization Model

The BEV charging demand is controlled within a charging network including one central control or master station, thereby coordinating charging for each private sub-station in the area (Figure 3).

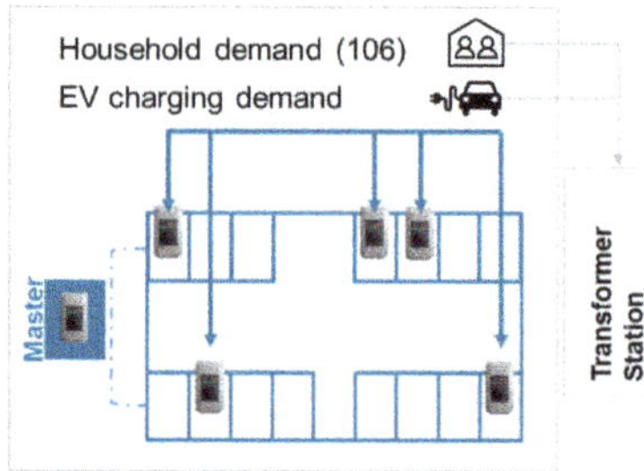

Figure 3. Private charging network: The master station controls all sub-stations.

The maximum charging power of private stations is defined as 3.7 kW, whereas public charging stations operate at 22 kW. 3-phase charging is not examined, with which a vehicle can charge at a speed of 11 kW (3×3.7) at home, but only 1-phase charging. The battery capacity of the individual passenger BEVs is set to 40 kWh. We do not aim at modeling a specific vehicle model. The current battery state of charge is determined by the state at the end of the former time step and the energy consumed and charged in the current time step.

Our model optimizes private charging in 15 min time-steps based on the full information of the yearly BEV power consumption, potential charging times, and HH load and is established as a linear optimization model. The objective function aims at minimizing the costs of BEV charging (c^{ch}) depending on the cost (c) and consumed charging power (p) at the home (h) and public (p) charging point in each time step t (see Equation (1)). The BEV demand is a static input to this LM model from Hiesl et al. [49] and depends on the distance traveled for each driving purpose and the respective electricity consumption (see Section 2.2). Overall, charging should primarily be carried out within the private, controlled network, while public charging should only take place at times of urgent need.

$$\min_{\substack{\{p_t^p,\ p_t^h\} \\ \{c_t^p,\ c_t^h\}}} c^{ch}(p,\ c) \tag{1}$$

This objective function derives the optimal allocation of charging power across all time steps to fulfill the BEVs demands within the constraints for each BEV charging scenario described in Section 2.3.1. This function aims at recharging the BEV before the next journey starts. Obviously, reality this far does not provide any information on the upcoming consumption of the BEV. Nevertheless, modeling with full yearly information offers valuable insights into the benefits of exact customer information.

2.3.1. Charging Strategies

The BEV charging strategies represent different approaches for the control of charging power by the master station top–down across all charging points (Figure 3). We use an uncontrolled charging scenario, without any control measure, as a reference. This represents the start of charging once the BEV is plugged in at the home station. Then, these results are compared to the expected improvements with two LM approaches aimed at a specific distribution of BEV electricity demand. For all charging approaches, it is assumed that the BEVs are plugged in and available for charging while parked at home.

The following BEV charging approaches are being analyzed:

1. Uncontrolled charging (UC)
   As a reference scenario, there are no LM measures applied.
2. Low charging capacity (LCC) LM approach
   The LCC approach refers to slow charging at very low charging power for each BEV to avoid excessive peaks in charging demand. Charging is operated a high simultaneity rate.
3. Time-of-use (ToU) LM approach
   With the ToU approach, the master station shifts BEV charging power from the HH load peaks into the valleys, as described in Figure 2. The HH load peaks are defined as any value exceeding the threshold P70 (see Section 2.1).

As important indicators of the impact of different charging strategies on the distribution of charging power and the length of charging processes, we analyze the simultaneity factor, the average charging power applied, and the charging time ratio. These factors are also compared to the results of the project demonstration phase in Section 4.2.

1. Simultaneity factor
   The simultaneity factor describes the average amount of vehicles charging at the same time across the whole year. Therefore, LM approaches that promote longer charging time periods result in a higher factor of simultaneously charging BEVs.
2. Charging time ratio
   This indicator relates the time a vehicle is charged to the time it is parked. This factor is expected to be much higher for the LM scenarios ToU and LCC than for the uncontrolled approach, which leads to shorter charging time periods at higher charging power.
3. Average and maximum charging power
   The average charging power refers to the average power the BEV is charged at during active charging. We also analyze the maximum charging power to reveal the charging load peak, which is an indicator for the speed at which charging is processed. It is assumed that uncontrolled charging, that does not promote the management of charging power, will lead to a higher average charging power than the ToU or LCC approach.

2.4. Minimum Charging Capacity

Determining the charging capacity required for each BEV involves an investigation independent of the optimization described in Section 2.3. This analysis can reveal the most efficient sizing for charging capacity down to a level at which there is no flexibility left for load shifting. This means an operation at the highest simultaneity of charging processes at the slowest possible speed resulting in the highest distribution of charging power across time.

The cost for charging capacity for the whole area is subject to step fixed costs, which increase substantially once the regional power connection capacity needs to be expanded. This analysis uses the objective function described in Equation (2) to minimize the average charging power (p^h) consumed across the number of all charging points (n) in each time step (t). The minimum charging capacity that needs to be installed for the charging network is derived from the highest charging power consumed by the sum of all home charging points across a year (see Equation (3)). To eliminate outliers in the BEV electricity demand from rare very high BEV electricity consumption due to few long-distance trips, the highest 5% of charging power consumed are removed. Then, this aggregated capacity is divided by the amount of charging points to arrive at the desired indicator kW/BEV. This indicator represents the minimum charging power for each BEV required to fulfill the demand throughout the year.

Two modeling approaches representing different levels of information are applied: yearly optimization with full information and daily rolling optimization. For the daily approach, a higher required minimum capacity is expected due to the inability to balance charging over more than one day.

$$\min_{\{p^h_{n,\,t}\}} \sum_n p^h_t \tag{2}$$

$$cap^h = max \left(\sum_n p^h_t \right) \tag{3}$$

2.5. Environmental Impact of e-Mobility

Considering the results of the former analyses, the environmental impact of e-mobility in the considered residential area are evaluated based on the following aspects:

1. Savings in GHG emissions from the substitution of fossil-fuel-powered vehicles with BEVs

The savings in CO_2-eq, particulate matter (PM), and nitrogen oxide (NOx) are analyzed through the continuous substitution of conventional vehicles using diesel- or petrol-powered internal combustion engines with BEVs across the whole life cycle from vehicle and battery construction to energy provision and driving. This is based on the average yearly km traveled by each car from our model results and the Austrian data on specific emissions for each type of fuel [52]. Scenarios on the future development of individual transport, such as a shift to public transport or bicycles along with the energy transition or an increase in BEV use resulting from a rebound effect due to driving without regret of air pollution are not included. The emissions are based on current measures and will likely decrease further in the future for conventional and battery-powered vehicles. The CO_2-eq emissions are 178 g/km for diesel and 225 g/km for petrol-powered vehicles. However, diesel includes much higher amounts of NOx, with 0.385 g/km compared to petrol with 0.162 g/km. PM emissions amount to 0.021 g/km and 0.024 g/km for diesel and petrol, respectively. The Austrian shares of diesel and petrol vehicles from the stock of 2019 are applied, whose diesel share of 56% is much higher than the European average [53].

According to the Austrian climate goals, by 2030, national electricity generation must be 100% renewable ([54,55]). For our scenario of 10% BEV diffusion, which represents the status in 2020, we, therefore, use the specific emissions of the current Austrian electricity mix represented in the UBA scenario "BEV(Aut-Mix)" [52]. Between 30% and 100% BEV diffusion, representing the situation from 2030 onwards, the electricity consumed is assumed to be almost fully renewable with specific emissions, as in the UBA "BEV(UZ-46-Mix)" [52] scenario.

2.   Impact of LM on the environmental aspect of the electricity mix consumed for charging

Based on the correlation between the Austrian fossil electricity generation (p_{foss}) from coal and gas [56] and the BEV charging power consumption (p_{ch}) from the model, we evaluate the potential effect on the environmental aspect of electricity consumption:

$$r_{foss} = \frac{Cov_{p_{ch},\ p_{foss}}}{\sigma_{p_{ch}},\ \sigma_{p_{foss}}}. \tag{4}$$

3.   Environmental impact of E-mobility and LM on potential grid expansion

From the resulting impact of uncontrolled and controlled charging on the maximum and peak demand, as well as the importance of the minimum charging capacity, conclusions on the consequences for grid expansion and its environmental effects are derived.

3. Results

The modeling results in this section are structured as follows.

Section 3.1 describes the existing HH electricity demand as a basis to evaluate the impact of e-mobility on the building load. Sections 3.2 and 3.3 show the effects of BEV charging with uncontrolled charging and by applying the LM approaches (LCC and ToU). Section 3.4 addresses the minimum charging capacity that must be available for each BEV. In general, we observe that throughout the diffusion of e-mobility up to 100%, the model mainly uses the private charging network, while public charging only fulfills up to 3% of the yearly demand. Therefore, we conclude that the sufficient availability of efficient private charging IS reduces the need for fast charging options, at least for common daily business.

3.1. Household Electricity Demand Characteristics

The HH electricity demand is the basis for our analysis of the impact of increasing BEV diffusion and has a duration curve, as shown in Figure 4. The yearly total demand accounts for 487 MWh, and the profile shows the characteristics outlined in Table 3.

While the LM approaches aim at minimizing an increase in the building's maximum and peak volume demand through e-mobility, a successful shift into the HH load valley will result in an increase of the minimum and base volume demand (see Figure 2). This will result in a more balanced daily building demand profile with less variability. On the one hand, this can mean greater predictability and stability in the building load profile for the distribution grid. On the other hand, the increasing electrification of a growing number of applications could result in additional or new peaks in the future. However, the HH load valleys largely coincide with the overall national electricity demand valleys. Therefore, a BEV load shift into these times may avoid the expansion requirements for the distribution grid and electricity generation capacities for a reasonable amount of time. The material production, infrastructure implementation, land-use associated with the installation of power plants and cables, and disposal obviously entail substantial environmental and economic impacts.

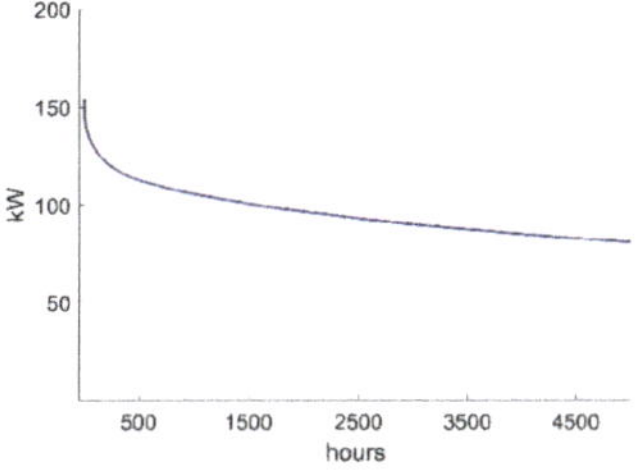

Figure 4. Duration curve for household electricity demands.

Table 3. Household load characteristics.

Parameter	
Maximum demand (kW)	153.8
Minimum demand (kW)	19.2
Peak demand volume (MWh)	49.2
Base demand volume (MWh)	437.6

3.2. Uncontrolled Charging

As explained in the introduction, charging without any load control mechanism increases electricity demand peaks, showing a strong correlation between the electricity demand for BEV charging and that for the HHs. Without any LM approach, most BEV users return home during the evenings and immediately plug in their vehicle for charging. Figure 5 shows the impact of BEV charging on electricity demands without control for an exemplary day along with the diffusion of e-mobility. In this uncontrolled scenario, only 50% of all vehicles charge at the same time (simultaneity factor) and on average, the vehicles only charge 9% of the time they are parked (charging time ratio). The latter proves the short time period during which charging is completed, which implies high charging power in the uncontrolled scenario. The average charging power consumed per BEV is 0.40 kW, and the maximum charging power consumed accounts for 1.80 kW.

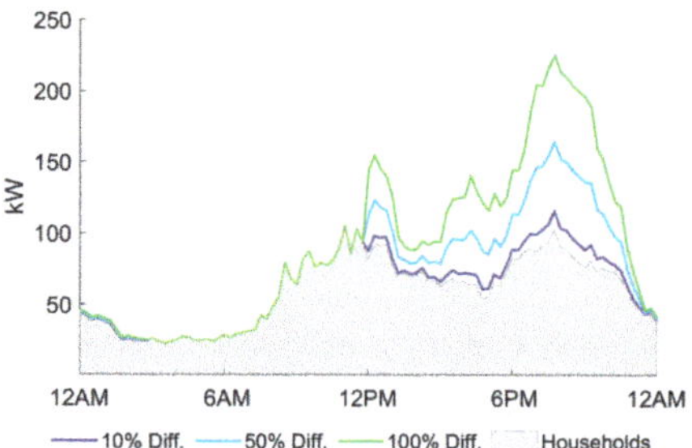

Figure 5. Charging profiles and household electricity demands without control (UC).

The impact of uncontrolled BEV charging on the electricity demand peaks becomes even more visible in the duration curve shown in Figure 6, where the total building electricity demand (BEV charging plus HHs) is ranked from the highest to the lowest for the highest 4500 h in a year. This curve reveals a remarkable increase in the building's maximum and peak volume demand along with growing e-mobility. Toward 100% BEV diffusion, the maximum electricity demand increases by 72% from 153.8 kW without e-mobility to 265.2 kW, and the peak volume demand grows almost four-fold. This increase in peak electricity demand may have severe consequences for the distribution grid and capacity investment in the area. The minimum electricity demand remains untouched, since charging usually coincides with higher HH electricity demand. The base volume demand also only increases by 20% up to 100% BEV diffusion. Therefore, the next section outlines the advantages of the defined LM approaches.

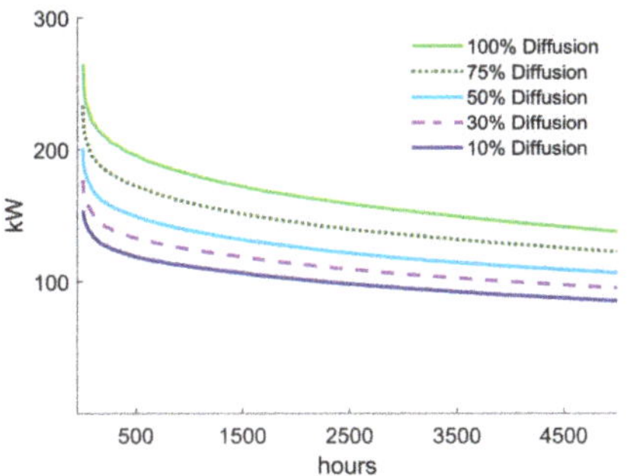

Figure 6. Duration curve with uncontrolled charging (UC).

3.3. Load Management Scenarios

LM can have a significant impact on the characteristics and distribution of the BEV electricity demand profile. If the BEVs are available for charging for a sufficient amount of time, the charging demand can successfully be shifted away from the HH load peaks into the valleys, as indicated in Figure 2. The results of the two LM approaches are also compared to uncontrolled charging. A more detailed quantitative discussion of all charging strategies is provided in Section 4.1.

Figure 7 shows the results for the LCC approach whose slow charging speed leads to a greater BEV load distribution over time compared to uncontrolled charging. In this scenario, on average, 86% of all vehicles charge at the same time, and the vehicles charge 16% of the time they are parked. This represents very long charging times and great simultaneity, though at substantially reduced charging power. The charging times still correlate with the evening HH load peaks. Nevertheless, the extreme peaks in the evening times experienced with uncontrolled charging are reduced substantially with the limitations on overall charging capacity. The duration curve of the building electricity demand in Figure 8 reveals a much smaller peak increase throughout BEV diffusion with a steadier distribution of BEV charging power compared to uncontrolled charging.

The ToU approach achieves the best results in shifting the charging demand away from the HH load peaks (see Figure 9). On average, 81% of all vehicles charge at the same time, which is a bit less than with the LCC approach, and again the vehicles charge 15% of the time they are parked. For both LM approaches, the average charging power consumed per BEV is cut down to 0.24 kW; this is only half of that with uncontrolled charging, and the maximum charging power consumed only accounts for 0.67 kW. This clearly proves the greater distribution of BEV charging power with the LM approaches. The duration curve in Figure 10 proves the capability of this approach to minimize the impact of e-mobility on electricity demand peaks in the building. Here, we observe an increase of only 4% in the yearly maximum demand under the full diffusion of BEVs.

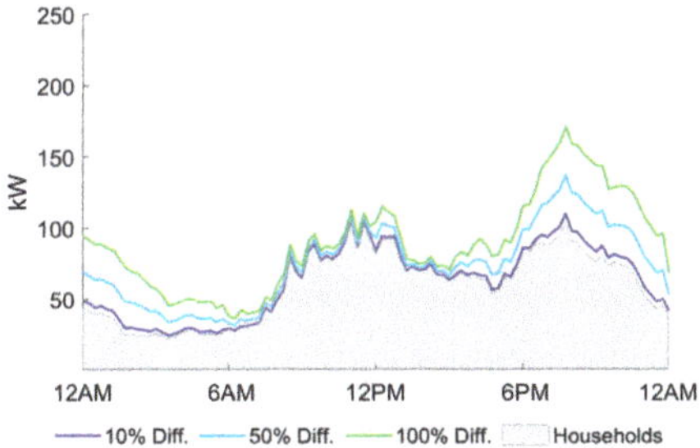

Figure 7. Household load and charging profile applying a low charging capacity (LCC).

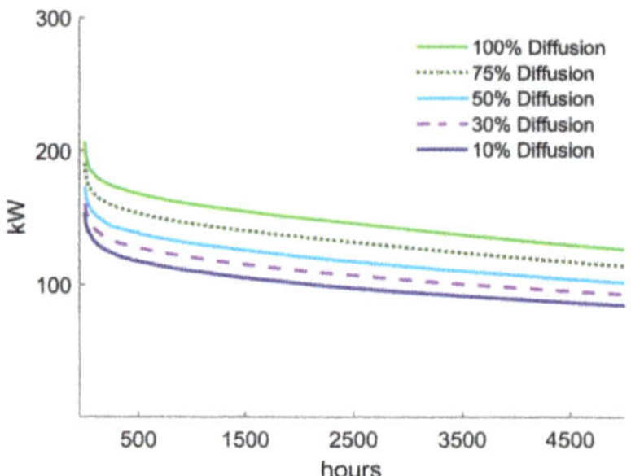

Figure 8. Duration curve applying a low charging capacity (LCC).

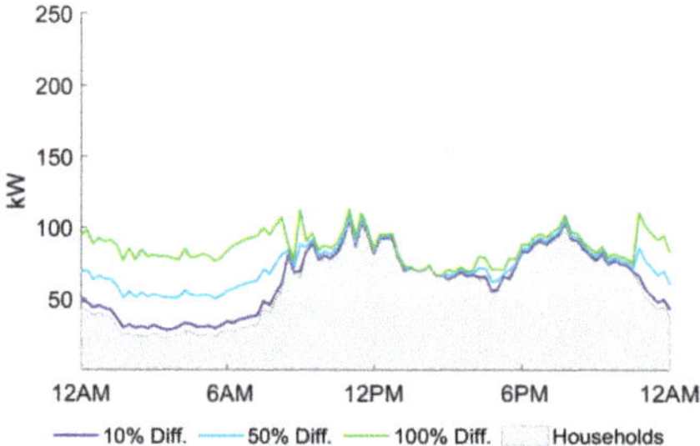

Figure 9. Household load and charging profile applying the time-of-use (ToU) approach.

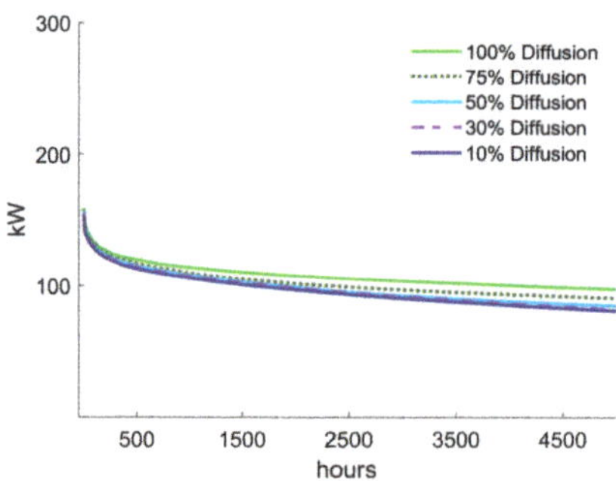

Figure 10. Duration curve applying the ToU approach.

3.4. Minimum Charging Capacity

Using our approach described in Section 2.4, first the minimum charging capacity for each BEV is calculated applying yearly optimization with full information. The yearly information of charging demand allows for a very efficient dimensioning of charging capacity. The minimum charging capacity that needs to be available accounts for 0.44 kW/BEV. However, charging mostly takes place at home, and only 4% of the BEV power demand is covered by public stations. Figure 11 depicts the resulting distribution of BEV charging demand, which largely relies on the availability of BEVs over time.

If now daily rolling optimization is applied, charging can only be carried out based on the information available for one day, without any knowledge of later demand. In our model, less information leads to a substantially higher minimum charging capacity of 1.3 kW/BEV. This capacity can result in higher costs for IS implementation. Hence, even foresight into the next day could help provide more flexible charging and integrate important aspects into the LM process, such as a positive weather forecast leading to longer weekend trips or an alignment of charging with renewable power availability. The efficient sizing of charging capacity reduces the risks and costs associated with exceeding the available regional power connection and thus guarantees a fair price to the final user for private charging.

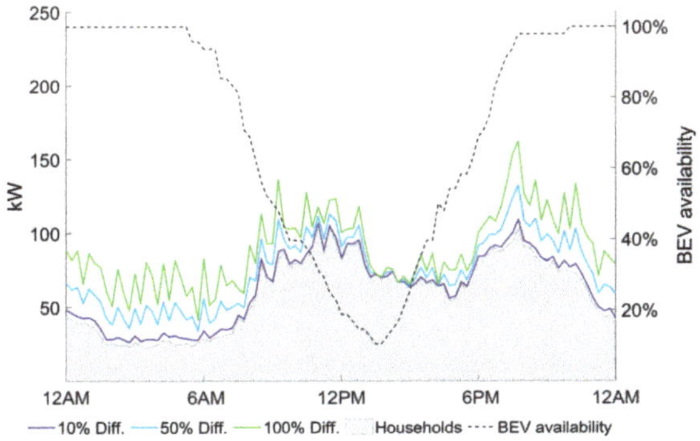

Figure 11. BEV charging demand at the home station under minimum charging capacity.

4. Discussion

In this section, the modeling results are discussed in detail and the environmental implications of e-mobility and the advantages of LM are evaluated in this context.

4.1. Impact of LM on BEV Charging Profiles

Here, the development of each parameter defined in Table 1 for BEV diffusion under the different charging strategies (UC, LCC, and ToU) is discussed and the results are compard to the situation without any e-mobility (see the HH load characteristics in Section 3.1). In general, we find that the simultaneity of charging, which indicates how many BEVs charge at the same time, increases from 9% under uncontrolled charging to 16% under both LM approaches. The charging time ratio increases from 50% to 86% between uncontrolled charging and the LCC approach. Both factors indicate the ability of LM to achieve longer charging periods at lower charging power.

Figure 12 shows that while uncontrolled charging (UC) leads to a rise in the detected yearly maximum electricity demand up to 100% BEV diffusion of 72%, reaching 265 kW, a restriction in the charging capacity with the LCC approach already achieves some reductions. With the LCC approach, the maximum demand reaches 207 kW with the full diffusion of BEVs—which is an increase of 35%—but it hardly increases at all until 50% BEV diffusion. The ToU approach almost manages to keep the electricity demand below the existing HH maximum demand, which only increases by 4% with full BEV diffusion. This way, e-mobility has hardly any impact on the yearly maximum demand.

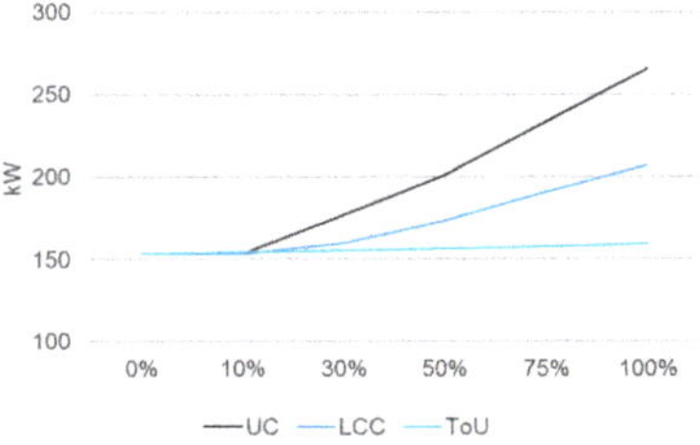

Figure 12. Development of the building's maximum demand.

The peak volume demand, which represents the sum of the demand above the P70 threshold, rises exponentially for all charging strategies between 50% and 100% BEV diffusion (see Figure 13). The uncontrolled scenario leads to a yearly peak volume of 236 MWh up to full BEV diffusion—a fourfold rise compared to no e-mobility. While the LCC still results in a threefold increase up to 193 kWh, the ToU approach achieves a successful shift in BEV charging demand away from the HH load peaks, resulting in a twofold increase in peak volume demand.

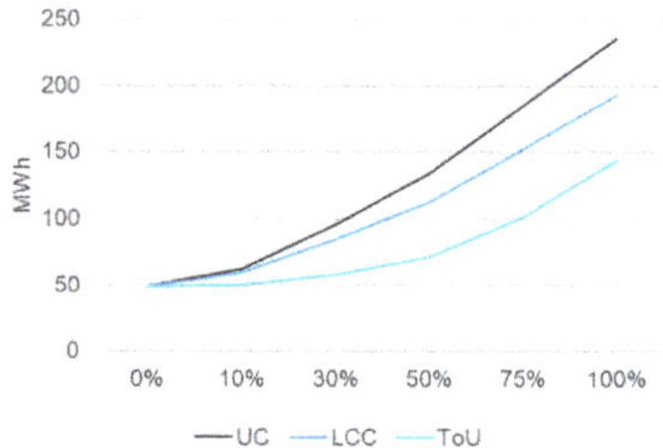

Figure 13. Development of the building's peak volume demand.

Compared to uncontrolled charging, where BEV charging coincides with a higher HH electricity demand, the LCC approach operates at a slower charging speed, leading to an extension of charging times into the load valleys. The most successful load shift into the valleys is obviously achieved by the ToU approach, which is based around this exact objective (see Figure 14). While the demand minimum does not change through e-mobility in the uncontrolled charging scenario, it increases by 7.5% with the ToU approach. Similar developments can be observed for the base volume demand below the P70 threshold (see Figure 15). With uncontrolled charging increasing the base volume demand by 20% toward full BEV diffusion, the LCC approach already leads to an increase of 42%. The aim of the ToU approach to avoid HH load peaks and shift the demand into the base volume represents the most successful load shifting measure. As a result, the base volume demand increases substantially under this LM strategy up to 551 kWh—a 74% increase compared to no e-mobility.

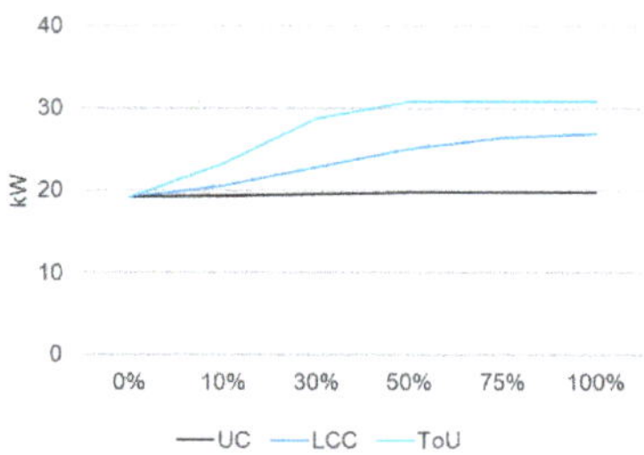

Figure 14. Development of the building's minimum demand.

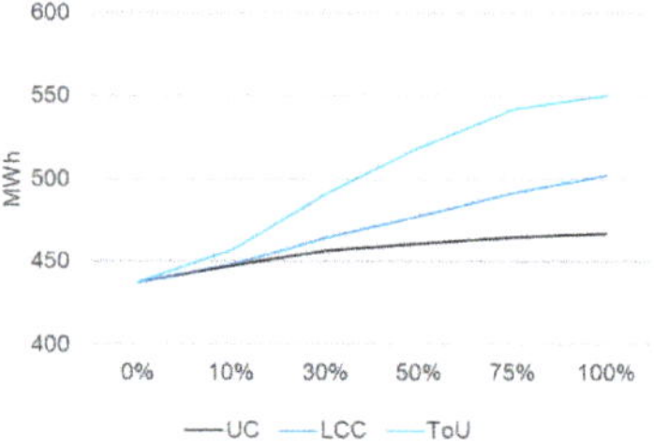

Figure 15. Development of the building's base volume demand.

If the value of a ToU approach is compared to the situation under uncontrolled charging between 10% and 100% BEV diffusion, we discover a 10–40% reduction in the maximum demand and a 17–39% reduction in the peak volume demand. Toward full BEV diffusion, the minimum demand rises by 56% with the ToU approach, and the base volume demand increases by 18% compared to uncontrolled charging. This is the result of effective load shifting away from existing HH peaks into valleys. These results represent a substantial reduction in the maximum and peak volume demands caused by E-mobility and, therefore, they could have significant value related to the sufficiency of the local power connection, distribution grid performance, and the prevention of IS expansion requirements.

This highlights the importance of LM for the further development of charging solutions. At the same time, fears of the impact of E-mobility on these aspects or skepticism for the sufficient availability of charging points in nearby public or, ideally, private environments is often a reason not to switch from conventional cars to this more environmentally friendly alternative. Therefore, we claim that proving the ability to control BEV electricity consumption may help legitimate such solutions in urban, multi-apartment buildings—a decision that needs to be made by the facility owner—and also promote the acceptance of e-mobility among potential users. Thus, overall, the availability of solutions for large-scale charging IS may contribute to the goal of zero-emission mobility.

4.2. Verification with Results from Project Demonstration Phase

The demonstration phase of the Urcharge project tested extensions of the LM algorithm over 6 months in a residential area with 50% e-mobility, where 27 BEVs were controlled with LM and 24 were charging uncontrolled. The project mostly included BEVs of the type Renault Zoe, a few Nissan Leaf, and two Tesla models. Mode 3 charging points with Type 2 chargers were used. Additionally, the customer perspective was analyzed with surveys and interviews among the participants. In this real environment, the basic LM approach, which is enhanced by specific technical functionalities, involves control over the charging capacity across all or single charging stations to avoid exceeding the building's available power connection. Therefore, the charging processes of single vehicles are only interrupted or postponed when the current defined capacity is reached. This largely represents the LCC approach of our model (see Section 2.3.1). However, the modeled environment cannot fully depict the reality of the relevant technical functionalities. The main differences are described in Table 4.

Extrapolated to one year, the average distance traveled during the demonstration phase by the project participants would be 11.100 km, which is slightly below the average of big cities in Austria (except Vienna) of 12.237 km/a [50]. The analysis of customer perspective within the project, carried out through surveys, revealed that this result could be due to the Covid-19 restrictions on daily driving for purposes such as work, visits, and leisure. On the other hand, few have increased their driving to avoid crowds in public transport. The monitoring shows that with an increasing experience, the user reduces the frequency of charging at the private station due to increasing knowledge of the BEV's charging requirements. The average user plugs in every fourth day for about 14 h. The overall charging time ratio related to the total plug-in time in the demonstration phase is 50%, whereas the plug in to parked time ratio is about 45%. This results in a charging time ratio related to the parking time of 22.5%. Our model has an average charging to plug-in time ratio of 9% with uncontrolled and 15% with ToU charging, which can be explained by the fact that the BEVs are always assumed to be plugged in while parked (100%). In our model, sometimes, all BEVs are charging at the same time. However, on average, the uncontrolled scenario leads to 50% of simultaneous charging and the LCC approach leads to 86%. The project's demonstration phase showed significantly lower simultaneity and charging time. At maximum, 12 BEVs were plugged in at the same time, which represents 44% of all BEVs, and a maximum of 11 BEVs were charging at the same time (41%). The average results were 24% and only 12%, respectively.

Of course, this is a major difference to our model parameters in which all BEVs are assumed to be available and plugged in at the charging point during any time while parked at home. The average amount of energy consumed during each charging session and BEV is 22 kWh. The plug-in times are mostly focused between 5 and 7 pm, whereas plug-out times typically range between 6 and 7 am. This fact supports the assumption of high flexibility due to long charging periods in private networks. Public charging is used very rarely.

Table 4. Differences between the modeled and the real environment.

No.	Topic	Model	Demo Phase
1	Information on energy consumption and BEV charging availability	Full information for LM optimization (Section 3.3) Comparison between full and daily information for minimum charging capacity (Section 3.4)	No information
2	Battery state of charge	Known	Unknown
3	Ratio plugged in/time parked	100%, always plugged in when parked	45% of the time plugged in when parked
4	Ratio plugged in/charging	Charging takes place during 30% of the plug-in time for uncontrolled and 47% for ToU charging	Charging takes place during 50% of the plug-in time
5	BEV share analyzed	0–100%	50%
6	LM	Constraints according to charging strategies in Section 2.3.1. ToU based on HH electricity demand	Charging as a system totally separate from the HH load. However, the HH load should be guaranteed. If BEV charging exceeds the building's power connection, charging is stopped first.
7	Phase charging	Each charging point provides 3.7 kW for charging.	1 or 3 phase charging: 1 phase vehicles currently always block all 3 phases with 11 kW but only consume 3.7 (11/3) kW. The LM algorithm was further developed in this project to release the available 2 phases to other 1- or 2-phase BEVs. In the future, this will release a substantial amount of capacity.

Despite the assumptions in the modeled environment, the conclusions from the project's real environment support the findings in this study. To test the capabilities of the LM functions and the minimum required charging capacity to guarantee the fulfillment of the BEV charging demand, during the demonstration phase, the available charging capacity was reduced several times from 35 kW for 27 BEVs (1.3 kW/BEV) down to 25 kW (0.9 kW/BEV). Therefore, the required charging capacity across all charging points could be reduced below the expected level of about 1 kW/BEV, and critical situations with respect to the power connection capacity would not occur. This matches our analysis provided in Section 3.4 of the minimum charging capacity, which can have a significant impact on the cost of charging infrastructure.

Figure 16 shows the BEV charging power with an overall initial charging capacity of 35 kW together with the number of cars currently plugged in. In this exemplary period, 31 kW was reached once during the evening, resulting in a significant impact of e-mobility on the overall building load. On the other hand, Figure 17 shows the situation with a reduced overall charging capacity down to 25 kW, which results in longer charging periods and significantly limits the electricity demand peaks. Furthermore, the dependence on the availability of BEVs plugged in at the charging station, which only accounted for 45% of the parking time in this project, becomes more important under this situation, since charging operates much slower. The project experience also revealed that a higher share of plug-in times would improve LM efficiency. In the first week with 1.3 kW/BEV of charging capacity, the building's electricity demand maximum (HH + BEV) increased almost twofold due to e-mobility. Later, with a reduction down to 0.9 kW/BEV, the LM successfully operated below this threshold, which reduced this surge.

The users recognized little reduction in charging performance under a reduced capacity. From the demonstration phase findings, it can be expected that LM will enable the coordination of 100% e-mobility in the residential area with sufficient comfort for the users and without any impact on the distribution grid capacity. Our model, as calculated in Section 3.4, determined a minimum charging capacity of 0.44 kW/BEV under a scenario with full information and 1.3 kW/BEV with a daily rolling approach, roughly matching the threshold tested in the demonstration phase. With the phase-dependent charging currently being developed and tested, as explained in line 6 of Table 4, the capacity gains will lead to more efficient LM in this respect.

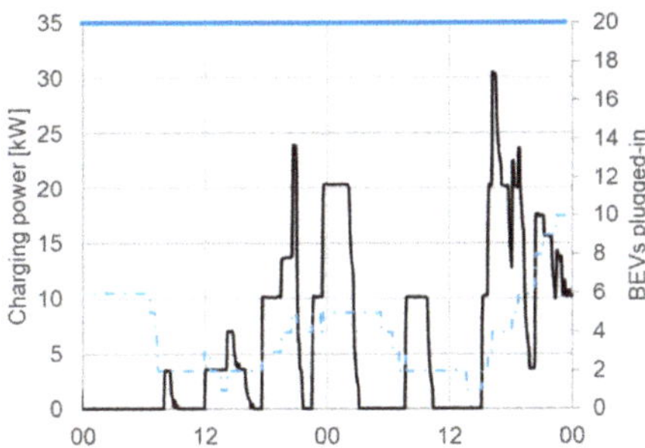

Figure 16. Load management with a maximum charging capacity of 35 kW (1.3 kW/BEV).

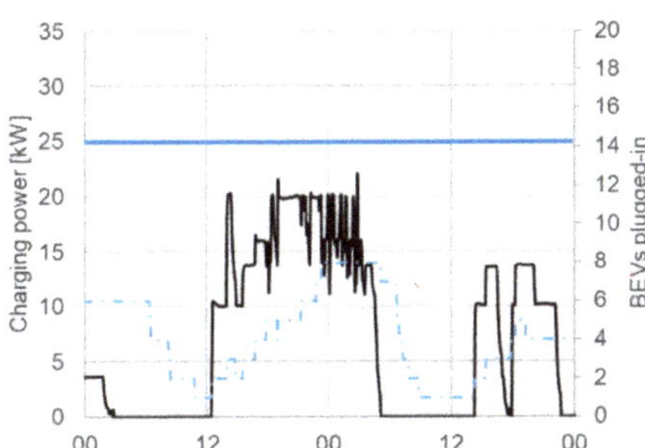

Figure 17. Load management with a maximum charging capacity of 25 kW (0.9 kW/BEV).

4.3. Environmental Aspects of E-Mobility and Advantages of LM

This section discusses the results for the environmental effects of e-mobility, as well as the advantages through LM and the efficient sizing of charging capacity. Of course, all the considerations for minimization of the capacity requirements have not only environmental but also economic aspects.

1. Savings in GHG emissions from the substitution of conventional vehicles with BEVs

In the considered residential area, e-mobility accounts for an increase in the yearly electricity consumption of 13% with a 30% BEV share and 43% with full BEV diffusion, as shown in Figure 18. According to our methodology, the development of the yearly CO_2-eq and NOx emissions, as well as particulate matter (PM), under the substitution of conventional cars with BEVs is shown in Figure 19.

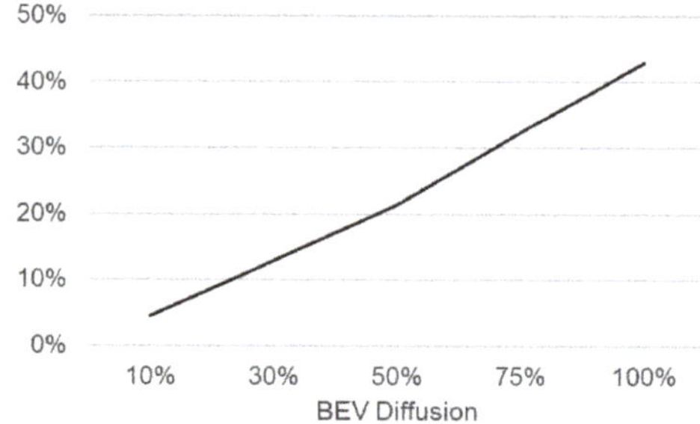

Figure 18. Increase in yearly building electricity demand through e-mobility.

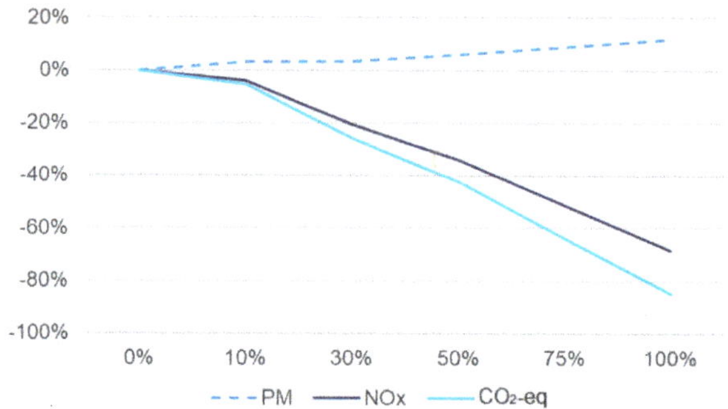

Figure 19. Expected yearly emission development through e-mobility.

All conventional cars in the residential area, considering an Austrian mix of diesel- and petrol-powered engines, emit 268 tCO_2-eq/a. By 2030, Austria aims at a BEV share of 30% representing a substitution of 32 conventional cars with BEVs in the residential area. This would result in 26% savings in yearly CO_2-eq emissions, accounting for 12 tCO_2-eq and 21% NOx savings by 2030. However, according to the UBA [52], during their life cycles, BEVs emit more PM than diesel- or petrol-powered combustion engines due to the emissions caused during vehicle and battery construction. This would lead to a 4% increase in PM emissions for 30% BEV diffusion. The overall 2030 goal in Austria is a 36% reduction of CO_2 emissions in transport [55]. Considering the aim of increasing the use and decarbonization of public transport, this goal may be achievable if the diffusion of BEVs can be realized as planned.

2.   Impact of LM on the environmental aspects of the electricity mix consumed for charging

Our model provides evidence on the negative environmental impact of electricity demand increases with respect to the electricity mix consumed at least from the correlation coefficient between BEV charging and fossil electricity generation (r_{foss}). Whereas uncontrolled charging shows a very weak positive correlation with 0.10 according to Pearson [57], the ToU approach results in a weak negative correlation of −0.24 and therefore seems to increase the use of renewable energy sources.

Furthermore, an analysis of the Austrian national electricity demand and the amount of national fossil power generation (coal and gas) in 2019 revealed a strong positive correlation between the two of 0.65. Additionally, an average HH load profile shows a moderate correlation with the overall national load profile of 0.49. Therefore, we claim that an increase in the building's electricity demand peaks due to e-mobility, such as with uncontrolled charging, leads to a negative environmental impact on the electricity mix consumed. LM can control the distribution of charging power to reduce the consumption of fossil fuel electricity generation during demand peaks.

3. Environmental impact of E-mobility and LM on potential grid expansion

The electricity demand profile of BEVs has a widespread impact on the cost and environmental effects related to the required capacity of power connection, the distribution grid, and even electricity generation capacities.

A reduction in electricity demand peaks through LM can guarantee the sufficiency of existing distribution grid capacities and thereby obviate the need for expansion investments [16]. Additionally, short-term peaks, apart from the described impacts on the electricity mix and use of flexible fossil power plants, would have an effect on the installed electricity generation capacities. A greater distribution of the overall electricity demand would yield a flatter demand profile and avoid the significant investment costs, as well as the associated resource consumption, land-use, and emissions from component production.

5. Conclusions

Firstly, our modeling results prove that uncontrolled charging results in a correlation between BEV electricity demand and HH load peaks. This results in a negative impact of growing e-mobility on distribution grid performance, as well as a negative impact on the environmental aspect of the electricity mix consumed while charging. Ou that LM is an important solution to control BEV charging demands and reduce this impact. Under full information, with between 10% and 100% BEV diffusion, an off-peak ToU approach can achieve a reduction in the building's maximum electricity demand between 10% and 40% compared to uncontrolled charging. The peak volume demand—the sum of the demand exceeding a certain threshold—can be reduced between 17% and 39%. This is partly achieved by a time shift of charging into valleys and by a reduction in the charging power consumed and an associated increase in the charging time periods. As a result, the electricity demand that is shifted from the HH load peaks fills up the load valleys and increases the former demand minimum. This may result in a modification of known building electricity demand profiles. On the one hand, these profiles could become more even due to the valleys being filled. On the other hand, increasing the electrification of a growing amount of applications could result in additional (or new) peaks in the future.

Our study reveals the value of information on user behavior and BEV energy consumption for LM efficiency. In this study, we calculated the minimum size of the charging capacity required to fulfill the BEV electricity demands in a private charging network. In a yearly optimization model with full information and under the assumption that parked BEVs are also plugged in, the minimum charging capacity is 0.44 kW/BEV. Less information in a daily rolling optimization leads to an almost threefold capacity requirement of about 1.3 kW/BEV. The Urcharge demonstration phase revealed that the average user plugs in for charging only every fourth day, reducing the flexibility of LM in a real environment. In reality, future BEV energy consumption and plug-in times are unknown. Nevertheless, the charging capacity in the demonstration phase could be decreased to 0.9 kW/BEV, thereby cutting the demand peaks and reducing the impact on the distribution grid. This study clearly highlights the benefit of using more information to provide longer foresight and flexibility with higher predictability for upcoming BEV charging demands.

One way for an electricity provider to collect users' behavioral information and charging preferences is via a mobile app. Currently, because the battery's state of charge remains unknown to the master station, electricity providers could assign pre-defined charging profiles to single charging points or groups aligned with the user's preferences communicated via an app (charging deadlines, day/night charging, etc.). Efficient LM and charging capacity sizing also have significant positive environmental effects. While 100% E-mobility increases yearly electricity demands in the residential area by 43%, the substitution of fossil-fuel-powered vehicles results in significant reductions in CO_2-eq and NOx emissions over a vehicle's lifecycle. Interestingly, off-peak charging can achieve a negative correlation between fossil fuel electricity generation and BEV electricity consumption, thereby promoting the use of renewable energy sources. Without an increase in peak demand through e-mobility and efficiently sized private charging capacity, not only could the expansion of the distribution grid and power generation capacities be avoided, but the emissions associated with the electricity mix consumed could also be controlled.

In general, our work proves the ability of LM in multi-apartment buildings to solve most of the common challenges associated with distribution grid performance throughout growing BEV diffusion. The implementation of this controlled, large-scale, and private charging IS offers vast economic, technical, and environmental advantages compared to more isolated and uncontrolled solutions. First, a private charging network offers substantial advantages for LM due to the high availability of the BEVs at the charging station and the lower time-criticality of charging. Secondly, LM avoids an increase in electricity demand peaks and thus prevents distribution grid expansion and the use of thermal power generation. As a result, the total cost of BEV charging, as well as the environmental impacts, can be managed. In the future, for the benefit of the whole energy system in terms of costs and environmental effects, regulatory and market frameworks need to be established to promote the implementation of a charging IS that applies controlled, joint solutions rather than isolated, fast charging options. LM could be enhanced by research on forecasting the relevant parameters, such as weather and traffic conditions, and the optimal integration of renewable energy. Eventually, BEV charging needs to be integrated into an overall system perspective. These are the next steps for future research and development.

Author Contributions: Conceptualization, J.R. and A.H.; formal analysis, J.R.; funding acquisition, J.R. and R.H.; methodology, J.R. and A.H.; project administration, J.R.; software, A.H. and J.R.; supervision, R.H.; Visualization, J.R.; writing—original draft, J.R.; writing—review and editing, J.R. and A.H. All authors have read and agreed to the published version of the manuscript.

Funding: This research was carried out within the cooperative R&D project URCHARGE powered by the Austrian Climate and Energy Fund within the program "Zero Emission Mobility" with a focus on IS.

Conflicts of Interest: The authors declare no conflict of interest. The funders had no role in the design of the study; in the collection, analyses, or interpretation of data; in the writing of the manuscript, or in the decision to publish the results.

References

1. European Parliament (EP). *European Commission (EC) REGULATION (EU) 2019/631 of the European Parliament and of the Council—Setting Co2 Emission Performance Standards for New Passenger Cars and for New Light Commercial Vehicles, and Repealing Regulations (Ec) No 443/2009 And (Eu) No 510/2011*; European Commission: Brussels, Belgium, 2019.
2. Spöttle, M.; Jörling, K.; Schimmel, M.; Staats, M.; Grizzel, L.; Jerram, L.; Drier, W.; Gartner, J. *Research For Tran Committee—Charging Infrastructure for Electric Road Vehicles*; European Parliament: Strasbourg, France, 2018.
3. Klima- und Energiefonds URCHARGE. Available online: https://www.klimafonds.gv.at/unsere-themen/mobilitaetswende/leuchttuerme-der-elektromobilitaet/urcharge/ (accessed on 25 September 2020).
4. LINZ AG Urcharge—ein Projekt Für Zukunftsfitte E-Mobilität. Available online: https://www.linzag.at/portal/de/ueber_die_linzag/projekte/urcharge (accessed on 25 September 2020).
5. Netze BW GmbH NETZlabor E-Mobility-Carré—Netze BW GmbH. Available online: https://www.netze-bw.de/unsernetz/netzinnovationen/netzintegration-elektromobilitaet/e-mobility-carre (accessed on 25 September 2020).
6. Sandag. *Multi-Unit Dwelling Electric Vehicle Charging*; Sandag: San Diego, CA, USA, 2019.
7. Smart Columbus Case Study: Increasing EV Charging Access at Multi-Unit Dwellings. Available online: https://smart.columbus.gov/playbook-assets/electric-vehicle-charging/case-study--increasing-ev-charging-access-at-multi-unit-dwellings (accessed on 25 September 2020).
8. Balmin, J.; Bonett, G.; Kirkeby, M. *Increasing Electric Vehicle Charging Access in Multi-Unit Dwellings in Los Angeles*; Luskin Center for Innovation: Los Angeles, CA, USA, 2012.
9. Cazzola, P.; Gorner, M.; Scheffer, S.; Scuitmaker, R.; Tattini, J. *Nordic EV Outlook 2018*; International Energy Agency (IEA): Paris, France, 2018.
10. Hall, D.; Lutsey, N. *Emerging Best Practices for Electric Vehicle Charging Infrastructure*; The International Council on Clean Transportation: Washington, DC, USA, 2017.

11. Resource Efficient Cities Implementing Advanced Smart City Solutions—READY. Available online: http://www.smartcity-ready.eu/about-ready/ (accessed on 25 September 2020).

12. CleanTechnica Crowd Charging—New Pilot Project In Denmark. Available online: https://cleantechnica.com/2018/01/12/crowd-charging-new-pilot-project-denmark/ (accessed on 25 September 2020).

13. Project Better Energy Limited EV Home Chargers | Home Car Charging Points | Project EV. Available online: https://www.projectev.co.uk/ev-home-chargers/ (accessed on 25 September 2020).

14. Netze BW GmbH Elektromobilität hautnah erleben—Netze BW GmbH. Available online: https://www.netze-bw.de/e-mobility-allee (accessed on 25 September 2020).

15. International Renewable Energy Agency (IRENA). Electric-Vehicle Smart Charging—Innovation Landscape Brief. Available online: https://irena.org/-/media/Files/IRENA/Agency/Publication/2019/Sep/IRENA_EV_smart_charging_2019.pdf?la=en&hash=E77FAB7422226D29931E8469698C709EFC13EDB2 (accessed on 25 September 2020).

16. Van Vliet, O.; Brouwer, A.S.; Kuramochi, T.; van den Broek, M.; Faaij, A. Energy use, cost and CO_2 emissions of electric cars. *J. Power Sources* **2011**, *196*, 2298–2310. [CrossRef]

17. Hu, Z.; Zhan, K.; Zhang, H.; Song, Y. Pricing mechanisms design for guiding electric vehicle charging to fill load valley. *Appl. Energy* **2016**, *178*, 155–163. [CrossRef]

18. Limmer, S.; Rodemann, T. Peak load reduction through dynamic pricing for electric vehicle charging. *Int. J. Electr. Power Energy Syst.* **2019**, *113*, 117–128. [CrossRef]

19. Crozier, C.; Morstyn, T.; McCulloch, M. The opportunity for smart charging to mitigate the impact of electric vehicles on transmission and distribution systems. *Appl. Energy* **2020**, *268*, 114973. [CrossRef]

20. Bitar, E.; Xu, Y. Deadline Differentiated Pricing of Deferrable Electric Loads. *IEEE Trans. Smart Grid* **2016**, *8*, 13–25. [CrossRef]

21. Yi, Z.; Scoffield, D.; Smart, J.; Meintz, A.; Jun, M.; Mohanpurkar, M.; Medam, A. A highly efficient control framework for centralized residential charging coordination of large electric vehicle populations. *Int. J. Electr. Power Energy Syst.* **2020**, *117*, 105661. [CrossRef]

22. Yan, J.; Zhang, J.; Liu, Y.; Lv, G.; Han, S.; Alfonzo, I.E.G. EV charging load simulation and forecasting considering traffic jam and weather to support the integration of renewables and EVs. *Renew. Energy* **2020**, *159*, 623–641. [CrossRef]

23. Das, H.S.; Rahman, M.M.; Li, S.; Tan, C.W. Electric vehicles standards, charging infrastructure, and impact on grid integration: A technological review. *Renew. Sustain. Energy Rev.* **2020**, *120*, 109618. [CrossRef]

24. Khalid, M.R.; Alam, M.S.; Sarwar, A.; Jamil Asghar, M.S. A Comprehensive review on electric vehicles charging infrastructures and their impacts on power-quality of the utility grid. *ETransportation* **2019**, *1*, 100006. [CrossRef]

25. Brinkel, N.B.G.; Schram, W.L.; AlSkaif, T.A.; Lampropoulos, I.; van Sark, W.G.J.H.M. Should we reinforce the grid? Cost and emission optimization of electric vehicle charging under different transformer limits. *Appl. Energy* **2020**, *276*, 115285. [CrossRef]

26. Luo, Y.; Zhu, T.; Wan, S.; Zhang, S.; Li, K. Optimal charging scheduling for large-scale EV (electric vehicle) deployment based on the interaction of the smart-grid and intelligent-transport systems. *Energy* **2016**, *97*, 359–368. [CrossRef]

27. Monteiro, V.; Pinto, J.G.; Afonso, J.L. Improved vehicle-for-grid (iV4G) mode: Novel operation mode for EVs battery chargers in smart grids. *Int. J. Electr. Power Energy Syst.* **2019**, *110*, 579–587. [CrossRef]

28. Usman, M.; Knapen, L.; Yasar, A.-U.-H.; Vanrompay, Y.; Bellemans, T.; Janssens, D.; Wets, G. A coordinated Framework for Optimized Charging of EV Fleet in Smart Grid. *Procedia Comput. Sci.* **2016**, *94*, 332–339. [CrossRef]

29. Hariri, A.-M.; Hejazi, M.A.; Hashemi-Dezaki, H. Reliability optimization of smart grid based on optimal allocation of protective devices, distributed energy resources, and electric vehicle/plug-in hybrid electric vehicle charging stations. *J. Power Sources* **2019**, *436*, 226824. [CrossRef]

30. Tan, K.M.; Ramachandaramurthy, V.K.; Yong, J.Y. Integration of electric vehicles in smart grid: A review on vehicle to grid technologies and optimization techniques. *Renew. Sustain. Energy Rev.* **2016**, *53*, 720–732. [CrossRef]

31. Aguilar-Dominguez, D.; Dunbar, A.; Brown, S. The electricity demand of an EV providing power via vehicle-to-home and its potential impact on the grid with different electricity price tariffs. *Energy Rep.* **2020**, *6*, 132–141. [CrossRef]

32. Gough, R.; Dickerson, C.; Rowley, P.; Walsh, C. Vehicle-to-grid feasibility: A techno-economic analysis of EV-based energy storage. *Appl. Energy* **2017**, *192*, 12–23. [CrossRef]

33. Chen, C.; Zarazua de Rubens, G.; Noel, L.; Kester, J.; Sovacool, B.K. Assessing the socio-demographic, technical, economic and behavioral factors of Nordic electric vehicle adoption and the influence of vehicle-to-grid preferences. *Renew. Sustain. Energy Rev.* **2020**, *121*, 109692. [CrossRef]

34. Mullan, J.; Harries, D.; Bräunl, T.; Whitely, S. The technical, economic and commercial viability of the vehicle-to-grid concept. *Energy Policy* **2012**, *48*, 394–406. [CrossRef]

35. Noel, L.; Papu Carrone, A.; Jensen, A.F.; Zarazua de Rubens, G.; Kester, J.; Sovacool, B.K. Willingness to pay for electric vehicles and vehicle-to-grid applications: A Nordic choice experiment. *Energy Econ.* **2019**, *78*, 525–534. [CrossRef]

36. Kempton, W.; Tomić, J. Vehicle-to-grid power fundamentals: Calculating capacity and net revenue. *J. Power Sources* **2005**, *144*, 268–279. [CrossRef]

37. Sortomme, E.; El-Sharkawi, M.A. Optimal Charging Strategies for Unidirectional Vehicle-to-Grid. *IEEE Trans. Smart Grid* **2011**, *2*, 131–138. [CrossRef]

38. Habib, S.; Kamran, M.; Rashid, U. Impact analysis of vehicle-to-grid technology and charging strategies of electric vehicles on distribution networks—A review. *J. Power Sources* **2015**, *277*, 205–214. [CrossRef]

39. Parsons, G.R.; Hidrue, M.K.; Kempton, W.; Gardner, M.P. Willingness to pay for vehicle-to-grid (V2G) electric vehicles and their contract terms. *Energy Econ.* **2014**, *42*, 313–324. [CrossRef]

40. Peter, B.A.; Seyedmostafa, H.T.; Thomas, M.S.; Bjørn, E.C.; Jens Christian, M.L.H.; Antonio, Z. *The Parker Project: Final Report*; Energy Technology Development and Demonstration Program (EUDP): Copenhagen, Denmark, 2019.

41. Rezania, R. Integration of Electric Vehicles in the Austrian Electricity System. Ph.D. Thesis, TU Wien, Vienna, Austria, 2013.

42. Tu, R.; Gai, Y.; Farooq, B.; Posen, D.; Hatzopoulou, M. Electric vehicle charging optimization to minimize marginal greenhouse gas emissions from power generation. *Appl. Energy* **2020**, *277*, 115517. [CrossRef]

43. Dixon, J.; Bukhsh, W.; Edmunds, C.; Bellkeith, K. Scheduling electric vehicle charging to minimise carbon emissions and wind curtailment. *Renew. Energy* **2020**, 1072–1091. [CrossRef]

44. Lopez-Behar, D.; Tran, M.; Froese, T.; Mayaud, J.R.; Herrera, O.E.; Merida, W. Charging infrastructure for electric vehicles in Multi-Unit Residential Buildings: Mapping feedbacks and policy recommendations. *Energy Policy* **2019**, *126*, 444–451. [CrossRef]

45. Jang, H.S.; Bae, K.Y.; Jung, B.C.; Sung, D.K. Apartment-level electric vehicle charging coordination: Peak load reduction and charging payment minimization. *Energy Build.* **2020**, *223*, 110155. [CrossRef]

46. Wang, D.; Gao, J.; Li, P.; Wang, B.; Zhang, C.; Saxena, S. Modeling of plug-in electric vehicle travel patterns and charging load based on trip chain generation. *J. Power Sources* **2017**, *359*, 468–479. [CrossRef]

47. Kim, J.D. Insights into residential EV charging behavior using energy meter data. *Energy Policy* **2019**, *129*, 610–618. [CrossRef]

48. Khalkhali, H.; Hosseinian, S.H. Multi-class EV charging and performance-based regulation service in a residential smart parking lot. *Sustain. Energy Grids Netw.* **2020**, *22*, 100354. [CrossRef]

49. Hiesl, A.; Ramsebner, J.; Haas, R. Modelling stochastic electricity demand of EVs based on traffic surveys—The case of Austria. *Energies* **2020**, *13*. in press.

50. Bundesministerium für Verkehr, Innovation und Technologie (BMVIT). *Österreich Unterwegs 2013/2014*; Federal Ministry for Transport, Innovation and Technology: Vienna, Austria, 2016.

51. American Automobile Association Inc. *Electric Vehicle Range Testing 2019*; American Automobile Association Inc.: Heathrow, FL, USA, 2019.

52. Umweltbundesamt (UBA). Update: Ökobilanz Alternativer Antriebe. Available online: https://www.umweltbundesamt.at/fileadmin/site/publikationen/DP152.pdf (accessed on 13 August 2020).

53. Statistics Austria Stock of Motor Vehicles and Trailers. Available online: https://www.statistik.at/web_en/statistics/EnergyEnvironmentInnovationMobility/transport/road/stock_of_motor_vehicles_and_trailers/index.html (accessed on 13 August 2020).

54. Oesterreichs Energie Daten & Fakten Zur Stromerzeugung. Available online: https://oesterreichsenergie.at/daten-fakten-zur-stromerzeugung.html (accessed on 13 August 2020).

55. Bundesministerium für Nachhaltigkeit und Tourismus (BMNT). *#MISSION2030—Austrian Climate and Energy Strategy*; Bundesministerium für Nachhaltigkeit und Tourismus: Vienna, Austria, 2018.

56. ENTSO-E Transparency Platform. Available online: https://transparency.entsoe.eu (accessed on 1 October 2020).
57. Evans, J.D. *Straightforward Statistics for the Behavioral Sciences*; Duxbury Press: Pacific Grove, CA, USA, 1996; ISBN 978-0-534-23100-2.

Publisher's Note: MDPI stays neutral with regard to jurisdictional claims in published maps and institutional affiliations.

Article

Integrating Bidirectionally Chargeable Electric Vehicles into the Electricity Markets

Timo Kern [1,2,*], Patrick Dossow [1] and Serafin von Roon [1]

[1] Forschungsgesellschaft für Energiewirtschaft mbH (FfE), 80995 Munich, Germany; pdossow@ffe.de (P.D.); sroon@ffe.de (S.v.R.)

[2] Department of Electrical and Computer Engineering, Technical University of Munich (TUM), Arcisstraße 21, 80333 München, Germany

* Correspondence: tkern@ffe.de; Tel.: +49-89-158121-35

Received: 31 August 2020; Accepted: 3 November 2020; Published: 6 November 2020

Abstract: Replacing traditional internal combustion engine vehicles with electric vehicles (EVs) proves to be challenging for the transport sector, particularly due to the higher initial investment. As EVs could be more profitable by participating in the electricity markets, the aim of this paper is to investigate revenue potentials when marketing bidirectionally chargeable electric vehicles in the spot market. To simulate a realistic marketing behavior of electric vehicles, a mixed integer linear, rolling horizon optimization model is formulated considering real trading times in the day-ahead and intraday market. Results suggest that revenue potentials are strongly dependent on the EV pool, the user behavior and the regulatory framework. Modeled potential revenues of EVs of current average size marketed with 2019 German day-ahead prices are found to be at around 200 €/EV/a, which is comparable to other findings in literature, and go up to 500 €/EV/a for consecutive trading in German day-ahead and intraday markets. For future EVs with larger batteries and higher efficiencies, potential revenues for current market prices can reach up to 1300 €/EV/a. This study finds that revenues differ widely for different European countries and future perspectives. The identified revenues give EV owners a clear incentive to participate in vehicle-to-grid use cases, thereby increasing much needed flexibility for the energy system of the future.

Keywords: V2G; bidirectionally chargeable electric vehicles; smart charging; unmanaged charging; spot markets; day-ahead market; intraday auction; continuous intraday trading; mixed integer linear optimization; revenues potentials of EVs

1. Introduction

The energy system transformation entails structural changes in all sectors. While the electricity supply sector in Germany has already been subjected to massive adjustments due to the expansion of renewable energies, the final energy sectors of private households, industry, transport, as well as small and medium enterprises have been very slow to switch to more climate-friendly technologies.

Emissions from the transport sector in Germany even increased since 1990 [1] and will, therefore, be unable to contribute to achieving the climate targets for 2020. Switching to electric vehicles (EVs) is one possible solution for lowering emissions. When comparing the carbon footprint of an EV to an internal combustion engine vehicle (ICEV), both the production of the battery as well as tailpipe emissions of the EV depend on the underlying energy system [2]. Due to a targeted reduction of CO_2-emission of the European energy sector by up to 95% [3], at least operational emissions of EVs resulting mainly from charging electricity will decrease. Consequently, EVs are promising for lowering CO2-emissions.

However, due to higher initial investment costs compared to ICEVs, the integration of EVs proves to be challenging. Investment in EVs could become more profitable by participating in the electricity markets by smart or bidirectional charging. In this respect, the "Bidirectional Charge Management" (BCM) project, which was launched in May 2019, focuses on the analysis of revenue potentials of bidirectionally chargeable EVs in the different electricity markets [4]. The main region of our investigations is Germany since it is the largest electricity market in Europe that is characterized by a heterogenic generation portfolio of renewable energy sources as well as conventional power plants [5]. Due to the German energy transition, volatility in electricity generation will continue to increase thereby enhancing the potential benefits of bidirectional chargeable EVs. German spot markets include the day-ahead market (auction at 12 noon one day before delivery), the intraday auction (at 3 pm one day before delivery) and the continuous intraday trading (starting at 3 pm one day before delivery) [6]. In addition, the results compare revenue potentials of bidirectionally chargeable EVs in 28 European countries, where findings of the parameter analysis can be generalized to other regions.

The BCM project defined 13 different use cases for bidirectional charging management separated into the three revenue creation groups: "Vehicle-to-Grid" (V2G), "Vehicle-to-Business" (V2B) and "Vehicle-to-Home" (V2H) [4]. Two use cases of the V2G use cases group are examined in this paper, which refer to arbitrage trading in the day-ahead market and the intraday market. Since bidirectionally chargeable EVs that participate in the spot markets increase the flexibility of the markets and energy system, these two use cases can have an impact on a cost-effective integration of renewable energies while providing revenues to the EV owners. The aim of this paper is to show these revenue potentials for the marketing of EVs in an aggregated pool in the spot markets at the same time ensuring faster integration of e-mobility into the energy system of the future.

Several scientific publications have discussed the revenue potentials by participation of EVs in the spot markets [7]. Using smart (but unidirectional) charging, an aggregator can considerably reduce the costs of charging electric vehicles [8,9]. Such costs can be further reduced and revenues can be generated by bidirectional charging according to [10]. The studies mentioned above are limited to marketing in the day-ahead and reserve market. Rominger et al. [11] point out revenues for intraday trading, but only restricted on flexibilities with constant availability such as stationery storage. Schmidt et al. [12] consider prices in the day-ahead market and the intraday auction for EV charging, but is limited to a smart charging process without discharging to the grid. We are thus extending existing research with a more detailed representation of bidirectionally chargeable EVs participating in the spot markets by considering day-ahead and intraday trading in a rolling optimization with a limited time horizon. Another limitation of most studies mentioned is the focus on revenues of EVs by a fixed EV parameterization. As revenue potentials of bidirectionally chargeable EVs are strongly dependent on many influencing parameters, this paper points out how a variation of user parameters like the minimum safety state of charge (SoC), minimum SoCs at departure, plug-in probabilities and location of charging infrastructure change revenue potentials.

Peterson et al. deal with profits of bidirectionally chargeable EVs using arbitrage trading in three different US local markets [13]. Pelzer et al. find out that revenue potentials of bidirectionally chargeable EVs in the US and Singapore markets are highly dependent on spatial and temporal resolution of market prices [14]. These studies regard battery degradation costs, but are limited to an EV participation in only one electricity market and modeling without consideration of user behavior parameters.

Furthermore, several studies investigate the effects of V2G on the electricity markets. The participation of bidirectional vehicles in the spot market has a smoothing effect on electricity prices. In times of surplus feed-in by renewable energies, V2G is able to reduce price drops resulting in higher market values of renewable energies [15]. Rodríguez et al. [16] show a flattening impact of V2G applications on the demand curve that results in a higher load factor. These findings showing promising effects on the energy system are expanded by our study to determine whether the vehicle owner can also benefit by participating in the electricity market. Since the regulatory framework is of significant importance for the evaluation of revenues, the influence of additional charges on purchased energy is pointed out.

2. Methods

To determine V2G revenue potentials, we developed an aggregated storage optimization model that covers the use cases of arbitrage trading in the spot markets. The holistic implementation of all V2G use cases facilitates the applicability and avoids building several parallel models.

A simplified representation of the modeling process is displayed in Figure 1, where the model consists of four different parts. Based on a model of the EV's battery, optimized charging strategies can be developed for different scenarios depending on charging and discharging restrictions. An uncoupled aggregated EV pool can be modeled to participate in different markets, where marketing strategies are optimized with a rolling forecast of market prices. The optimized strategy depends on electricity prices of the respective energy markets and is based on historical and simulated future market prices. In addition to bidirectionally chargeable EVs, two reference scenarios are considered that cover smart, unidirectionally chargeable EVs and simple, directly chargeable EVs. The entire model is implemented in Matlab, where a CPLEX solver (optimization software package) is used for optimization.

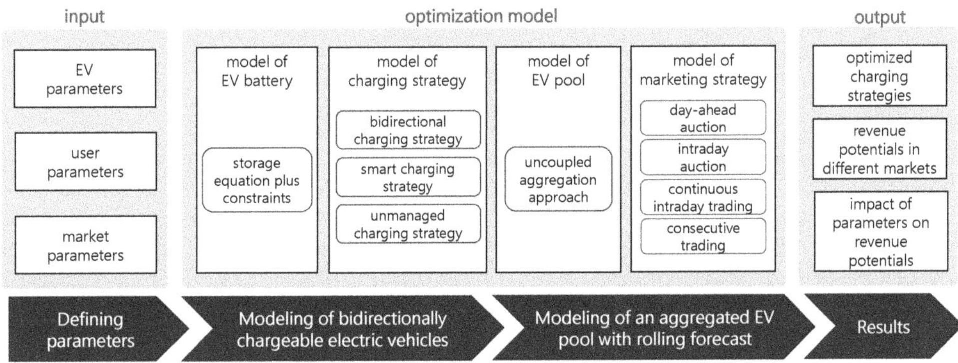

Figure 1. Schematic representation of the developed optimization model for Vehicle-to-Grid (V2G) use cases of bidirectional charging.

2.1. Modeling of Bidirectionally Chargeable Electric Vehicles

Modeling a bidirectionally chargeable EV consists mainly of a model of the electric battery of the vehicle similar to stationary electricity storage. The EV battery is modeled by the storage equation displayed below, which relates the state of charge (SoC) of the battery to the different amounts of electricity charged into or discharged out of the battery:

$$SoC(t) = SoC(t-1) + P_{charge}(t){\cdot}\eta_{charge}{\cdot}\Delta t - \frac{P_{discharge}(t)}{\eta_{discharge}}{\cdot}\Delta t$$

$$+\frac{P_{counter-purchase}(t)}{\eta_{discharge}}{\cdot}\Delta t - P_{counter-sale}(t){\cdot}\eta_{charge}{\cdot}\Delta t \tag{1}$$

$$+P_{schedule}(t){\cdot}\Delta t + P_{fastcharge}(t){\cdot}\Delta t{\cdot}\eta_{charge} - E_{consumption}(t)$$

here, t stands for the modeled point in time, Δt is the difference between two points in time and η represents efficiency, which differs for the charging and discharging process. $P_{charge}(t)$ is the charging power and $P_{discharge}(t)$ the discharging power at the modeled time t. The variables are considered on the alternating current side of the charging station and thus correspond to the amounts of energy traded in the market. $P_{schedule}$ is the sum of purchases and sales already made at the modeled time t. Accordingly,

$P_{counter-purchase}(t)$ and $P_{counter-sale}(t)$ represent the power which can be purchased or sold in the market to counteract transactions that have already taken place (countertrading). For example, purchased energy in the day-ahead market could be sold in the intraday markets resulting in a countertrade $P_{fastcharge}(t)$ is the power, that can be used to rapidly charge the vehicle if necessary and $E_{consumption}(t)$ is the EV's energy consumption by driving at time t. The limit values of all time-dependent variables are explained in the following section. Table A1 in the Appendix A provides an overview of all these variables and their limit values.

2.1.1. State of Charge

$SoC(t-1)$ represents the battery's storage level at the point in time prior to the modeled point in time. The change in capacity of the battery storage corresponds to the difference from $SoC(t)$ to $SoC(t-1)$. Due to the limited storage capacity of the electric vehicle and user requirements for a minimum storage level, the variable $SoC(t)$ can assume a limited range of values.

The maximum storage level of the battery storage is limited by SoC_{max}. If a value of 100% is set for this parameter, the entire storage capacity is available for the charging process. The minimum value SoC_{min} varies depending on the vehicle's location status, which is known for each point in time. Equation (2) summarizes the values that SoC_{min} can assume:

$$SoC_{min}(t) = \begin{cases} SoC_{min,safe} & \text{for status} = \text{connected} \\ SoC_{min,dep} & \text{for status} = \text{departure} \\ SoC_{min,disconnected} & \text{for status} = \text{not connected} \end{cases} \tag{2}$$

If the vehicle is connected to the electric grid, SoC_{min} equals $SoC_{min,safe}$. The storage level must not fall below this value or must load onto it as quickly as possible. If the vehicle is connected to a charging station and is at the point of departure, SoC_{min} assumes the value $SoC_{min,dep}$. Before departure, the charging strategy is thus optimized in a manner such that the SoC at the time of departure at least corresponds to $SoC_{min,dep}$. If the vehicle is not connected to the electric grid, the value $SoC_{min,disconnected}$ results in the minimum SoC.

To ensure that $SoC(t)$ lies between the minimum and maximum possible storage level, Equation (3) is implemented:

$$SoC_{min}(t) \cdot C \leq SoC(t) + P_{supplement}(t) \cdot \Delta t \leq SoC_{max}(t) \cdot C, \tag{3}$$

where C describes the storage capacity of the electric vehicle and $P_{supplement}$ stands for additional, theoretical electric power to be charged. $P_{supplement}(t)$ is incorporated to meet the storage level restriction in Equation (2) at any time. The possibility that the value of the storage level is below SoC_{min} exists even if the EV is connected to the electric grid. This might occur if the storage level is lower than the minimum SoC when the vehicle arrives at a charging station or if it is not possible to charge to the minimum SoC before departure because the vehicle has not been connected for long enough. In the case that the minimum storage level cannot be reached with the charging strategy, $P_{supplement}(t)$ takes on a value greater than 0 to simulate a hypothetical charging process. The variable thus can be interpreted as penalty costs that arise from the driving behavior of a user who disregards the requirements for a minimal SoC. By introducing $P_{supplement}(t)$, the model can optimize all driving profiles regardless of driving behavior or consumption, so that a selection of unsuitable driving profiles does not have to be made in advance. $P_{supplement}(t)$ is not taken into account in the storage equation and does not change the actual storage level.

2.1.2. Charging/ Discharging Power and Already Traded Energy

In the storage Equation (1), $P_{charge}(t)$ and $P_{discharge}(t)$ describe the purchase and sale of power and determine the change in storage capacity for each time step. Due to the limited power of any EV charging station, the variables are limited to the maximum charging and discharging power

$P_{charge,max}(t)$ and $P_{discharge,max}(t)$ and to the minimum charging and discharging power $P_{charge,min}(t)$ and $P_{discharge,min}(t)$. If the vehicle is connected to a charging station, the EV battery can be charged with $P_{charge,max}(t)$ or discharged with $P_{discharge,max}(t)$. If the vehicle is not connected to the electric grid, both maximum and minimum charging and discharging power become 0.

The boolean variables b_{charge} and $b_{discharge}$ describe the state of the battery during charging and discharging processes. If a charging process takes place, b_{charge} is *true*. Analogously, the variable $b_{discharge}$ becomes *true* during discharging. Due to the fact that it is not possible to purchase and feed electricity into the electric grid at the same time, only one of the boolean variables can assume the value 1 (= *true*) at the modeled point in time Equation (4). A simultaneous purchase and sale in the market is, therefore, excluded.

$$b_{charge}(\mathrm{t}) + b_{discharge}(\mathrm{t}) \leq 1 \tag{4}$$

The resulting constraints regarding these variables are shown in Equations (5) and (6):

$$\begin{aligned} P_{charge,min}(t) \cdot b_{charge}(\mathrm{t}) &\leq P_{charge}(\mathrm{t}) \\ &\leq \left(P_{charge,max}(t) - P_{schedule,\,purchase}(t) \right) \cdot b_{charge}(\mathrm{t}), \end{aligned} \tag{5}$$

$$\begin{aligned} P_{discharge,min}(t) \cdot b_{discharge}(\mathrm{t}) &\leq P_{discharge}(\mathrm{t}) \\ &\leq \left(P_{discharge,max}(t) - P_{schedule,\,sale}(\mathrm{t}) \right) \cdot b_{discharge}(\mathrm{t}). \end{aligned} \tag{6}$$

There is a possibility that the storage capacity of the electric vehicle has already been marketed through previous trading on the electricity markets, for example through consecutive trading on different spot markets. Such traded power must be taken into account in subsequent storage optimizations. In Equations (5) and (6), $P_{schedule,\,purchase}$ and $P_{schedule,sale}$ correspond to already made purchases or sales at the modeled time t. These amounts of electricity reduce the maximum charging or discharging power in such a way that only the capacity that has not yet been traded can be marketed. $P_{schedule}$ can be defined as the difference between $P_{schedule,\,purchase}$ and $P_{schedule,sale}$ Equation (7) and is included in the storage Equation (1).

$$P_{schedule}(\mathrm{t}) = P_{schedule,\,purchase}(t) \cdot \eta_{charge} - \frac{P_{schedule,sale}(t)}{\eta_{discharge}} \tag{7}$$

2.1.3. Countertrades

Electricity spot markets in Germany include consecutive day-ahead and intraday trading resulting in the opportunity to countertrade day-ahead purchases or sells in the intraday market. The model not only accounts for already marketed storage capacities, it also includes the possibility of compensation transactions (countertrades) that compensate for the previous trade in the opposite direction. In this regard, the variable $P_{counter-purchase}(t)$ corresponds to a buyback (counter purchase), the variable $P_{counter-sale}(t)$ to a sellback (counter sale) of already traded storage capacities. The volume of a countertrade can at most assume the previously inversely traded amount of energy. Countertrades do not describe a physical loading or unloading process. The resulting constraints are shown in Equations (8) and (9):

$$P_{counter-purchase}(t) \leq P_{schedule,\,sale}(\mathrm{t}) \cdot b_{counter-purchase}, \tag{8}$$

$$P_{counter-sale}(t) \leq P_{schedule,\,purchase}(\mathrm{t}) \cdot b_{counter-sale}. \tag{9}$$

Similar to the charging and discharging processes, the boolean variables $b_{counter-purchase}$ and $b_{counter-sale}$ describe the state of countertrades. If a counter-purchase takes place at time t ($b_{counter-purchase} = 1$), the EV battery cannot be discharged at the same time. Conversely, no charging process can be conducted during a counter-sale ($b_{counter-sale} = 1$). Equations (10) and (11) show these constraints:

$$b_{counter-purchase}(\mathrm{t}) + b_{discharge}(\mathrm{t}) \leq 1, \tag{10}$$

$$b_{counter-sale}(t) + b_{charge}(t) \leq 1.\tag{11}$$

2.1.4. Electricity Consumption and Fast Charging

The battery of a vehicle has an electric energy consumption $E_{consumption}(t)$ at the modeled time t, which is considered in the storage equation. Due to the foresight of the driving profiles (explained in Section 2.4), EV consumption is known at all times. Each driving phase of the vehicle results in a capacity reduction.

Depending on the user's driving behavior, it might occur that the electricity consumption of an EV is so high at one point in time that the current storage level is not sufficient to meet the energy demand, for example if the EV has not been connected to a charging station for too long. In this case, the fast charging power $P_{fastcharge}(t)$ is utilized to comply with the restrictions of $SoC_{min}(t)$ and to avoid a supposed negative storage level. The employment of the fast charging process is accompanied by an increase in storage capacity. $P_{fastcharge}(t)$ represents the charging power at a public charging station rather than at a bidirectional charging station. The vehicle user is thus given the opportunity to charge the vehicle on the road.

2.2. Formulation of Optimization Model

The developed model of a bidirectionally chargeable EVs allows for the implementation of different optimized charging and discharging strategies, which differ in particular in the structure of the objective function. For the assessment of the use cases of arbitrage trading on the day-ahead market as well as on the intraday market, three different charging strategies are implemented: a strategy for bidirectional charging, a strategy for smart charging, and a strategy for unmanaged charging.

First, the bidirectional charging strategy, which allows for charging and discharging of the EV, is restricted to the storage equation and its aforementioned constraints. The objective of this charging strategy is to charge at minimum costs while discharging at maximum revenue. To do so, the objective function of the optimization model aims at minimizing all costs considered:

$$\begin{aligned}
min\Bigg(&\sum_{t=1}^{T} p_{market,buy}(t) \cdot \left[P_{charge}(t) + P_{counter-purchase}(t)\right] \\
&- \sum_{t=1}^{T} p_{market,sell}(t) \cdot \left[P_{discharge}(t) + P_{counter-sale}(t)\right] \\
&+ \sum_{t=1}^{T} p_{fastcharge}(t) \cdot P_{fastcharge}(t) + \sum_{t=1}^{T} p_{supplement}(t) \cdot P_{supplement}(t) \Bigg)
\end{aligned}\tag{12}$$

where T corresponds to the number of time steps of the optimization. Depending on the respective market, traded energy quantities per time step as well as corresponding market prices $p_{market,i}(t)$ are considered. In this regard, $p_{market,buy}$ is the price at which electricity is bought and $p_{market,sell}$ is the price at which electricity is sold, where both prices can include respective transaction costs and possibly additional electricity price components. Charged power corresponds to a purchase transaction and is associated with costs as is each counter purchase of power. In contrast, discharged power and counter sales are traded with corresponding revenues, which is why $P_{discharge}$ and $P_{counter-sale}$ are subtracted.

Fast charging power and supplement power are also included in the objective function. Both fast charging costs $p_{fastcharge}(t)$ and penalty costs $p_{supplement}(t)$ are fixed to be a relatively high value, so that only the minimum necessary power is charged to meet the requirements for minimum storage level. As $P_{fastcharge}(t)$ should only be utilized to the extent that a negative SoC is avoided, $p_{fastcharge}(t)$ should be selected sufficiently larger than $p_{supplement}(t)$. Thus, the following condition must also be fulfilled to guarantee a functioning bidirectional charging strategy:

$$p_{market,i}(t) \ll p_{supplement}(t) \ll p_{fastcharge}(t).\tag{13}$$

Second, the smart charging strategy is implemented as a reference scenario to simulate already existing smart charging stations. The objective is to minimize electricity purchase costs by intelligent charging of the EV. Since discharging the EV battery is impossible in this scenario, Equation (14) is implemented. By eliminating the discharge power, the objective function already defined via Equation (12) for the bidirectional charging strategy can also be used for the smart charging strategy.

$$0 = P_{discharge,min} \leq P_{discharge} \leq P_{discharge,max} = 0 \tag{14}$$

Third, the unmanaged charging strategy accounts for today's most commonly installed simple charging stations as a second reference scenario, where the EV battery is charged as soon as the vehicle is connected without an optimized charging control. As with the smart charging strategy, discharging is not possible Equation (14). The aim of the unmanaged charging strategy is, therefore, to maximize the storage level that is equal to minimizing the negative value of $SoC(t)$ at all times. The battery is accordingly charged at maximum charging power until the battery's storage level corresponds to $SoC_{max}(t)$ or until the EV leaves the location. In addition, as for the previously explained strategies, fast charging costs and penalty costs for an insufficient SoC with regard to the location-based limit values are included. The resulting objective function is expressed as follows:

$$min\left(\sum_{t=1}^{T} p_{fastcharge}(t) \cdot P_{fastcharge}(t) + \sum_{t=1}^{T} p_{supplement}(t) \cdot P_{supplement}(t) - \sum_{t=1}^{T} SoC(t)\right) \tag{15}$$

2.3. Optimization with Limited Forecast in Consecutive Spot Markets

To investigate the influence of different characteristics and requirements of the considered markets on revenue potentials of bidirectional charging, optimized trading strategies based on price forecasts are simulated by a rolling optimization model, where realistic trading behavior results from a limited foresight of market prices.

Acting in the market under uncertainty is modeled in the same manner as described in [17], where each day is divided into 8 time slices of three hours each. The model regards real trading times in the spot markets. Figure 2 illustrates the methodical procedure of consecutive trading in the day-ahead and intraday markets with rolling price forecast horizons, where each horizontal bar displays the prices known in the respective optimization run of three hours. At 12 noon of day d, for instance, a market participant sees averaged continuous intraday prices of the following 12 quarter-hourly products. At the same time, less precise quarter-hourly prices of the continuous intraday are assumed for the interval from 3 pm to midnight. For day $d + 1$, day-ahead market prices are known and for $d + 2$ a forecast of the day-ahead market prices is presented. The participant's trading decision, which is the optimized marketing strategy, is based on this limited foresight.

As described by the example, foresight of market prices varies for the individual markets. Since the auction on the day-ahead market takes place daily at 12 noon for the respective following day $d + 1$, precise prices forecasts for $d + 1$ are known shortly before 12 noon on day d. To prevent unrealistic trading behavior at the end of day $d + 1$, such as discharging all batteries to maximize revenues, estimated prices for day $d + 2$ are included in the forecast horizon, where prices are also presented at 12 noon of day d. The forecast period $d + 2$ can be one or more days representing a worse or better foresight and is evaluated in Appendix B. The length of the optimization time steps for day-ahead trading is 1 h.

For the intraday auction, precise price forecasts of day $d + 1$ are known shortly before 3 pm of day d, since the auction takes place at 3 pm. The length of the optimization time steps is 0.25 h.

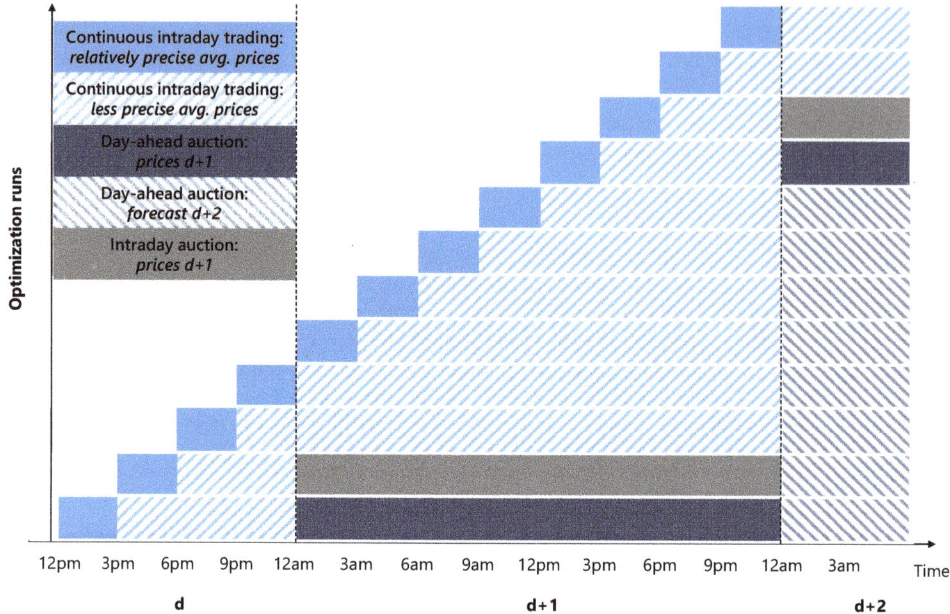

Figure 2. Schematic representation of the optimization steps with limited foresight of market prices.

Following the intraday auction, continuous intraday trading starts at 4 pm with quarter-hourly products, which defines the length of the optimization time steps. Here, a first forecast horizon of relatively precise prices is set to three hours covering the following 12 quarter-hourly products, where prices are based on trading transactions of these three hours. For the period following the three-hour time window, all continuous intraday transactions of this interval are used to calculate a second forecast price, thereby reflecting the uncertainty of market prices.

Hence, optimization runs before noon include market price information of the remaining day *d* and the following day *d* + 1. The optimization runs from 12 noon on the trading day also include day *d* + 2. The total revenue of the marketed EV battery corresponds to the summed costs and revenues of all traded products (filled areas). The cross-hatched areas in Figure 2 are not regarded as revenues, since these are only price forecasts serving as reference points for the trading strategy.

If consecutive trading takes place in several markets, storage capacities already marketed must be taken into account in subsequent optimization runs and can be countertraded as described before. The storage level at the end of real continuous intraday trading (filled blue area) of each optimization run determines the actual charging and discharging behavior of the vehicle. This storage level is applied as the starting value for the subsequent optimization run.

2.4. Input Data and Parameterization of Electric Vehicle (EV) Pool Scenarios

In the model, parameters related to the EV are the battery's storage capacity C, charging and discharging power $P_{charge/discharge}$, and different efficiency parameters. To investigate the range of revenue potentials in detail, three different sets of EV parameters are implemented: First, a currently common-sized EV is modeled (EV1), comparable to a 2018 BMW i3 [18] and a 2018 Renault Zoe [19], using realistic values regarding storage capacity, charging and discharging power and efficiencies. Second, a relatively large EV and a highly efficient charging station are defined representing a future EV (EV2). Third, a set of ambitious, yet plausible future values is selected to model maximum revenue potentials (EV3). These parameter sets were discussed and agreed upon within the research project BCM. Table 1 summarizes the chosen parameter values for the three sets of EV models.

Table 1. List of relevant electric vehicle (EV) parameters and chosen set of parameter values.

Parameter		EV1	EV2	EV 3
Storage capacity	C	38 kWh	100 kWh	100 kWh
Charging power	P_{charge}	11 kW	11 kW	22 kW
Discharging power	$P_{discharge}$	10 kW	11 kW	22 kW
Charging efficiency (AC-DC)	η_{charge}	92.5%	94.5%	95.0%
Discharging efficiency (DC-AC)	$\eta_{discharge}$	92.0%	94.5%	95.0%
Roundtrip efficiency (AC-AC)	$\eta_{roundtrip}$	85.1%	89.3%	90.3%

All losses and efficiencies considered in the model are based on discussions and on the consultation with experts from the BCM project. Other studies assume roundtrip efficiencies that are similar to EV1 [13] or slightly lower [14]. Constant values are set for the efficiencies in order to allow for a linear optimization problem, which results in much faster optimization times and thus enables many more optimization runs, i.e., more results. In real operation, however, efficiencies follow a declining, non-linear course for decreasing charging power. Hence, resulting revenue potentials of the presented model overestimate real revenues of bidirectional charging.

The user parameters result from characteristics, requirements and behavior of the vehicle user. In contrast to a large-scale stationary storage system, the battery of an EV is not continuously connected to the grid. The availability of an EV battery for V2G use cases largely depends on the individual driving profile of the user, the location of an appropriate charging station and the probability that the user has connected the vehicle to this charging station.

As a detailed representation of the driving behavior of the user, vehicle-specific driving profiles describe the EV's whereabouts as well as its energy consumption while driving in a chronological sequence. Based on data regarding household and route information as well as individual user logbooks from the "Mobility in Germany 2017" study [20] and a methodology first developed in the MOS 2030 [21] project, annual driving profiles of various EVs are created that are available as Supplementary Materials (see Section 6). Each profile meets the following standards:

- A change of location is always accompanied by a driving phase.
- During each driving phase, the EV has discrete consumption, which leads to a reduction of the storage level.
- The EV can be located and connected either at the place of residence, the place of work or the public space

The temporal resolution of the driving profiles is quarter-hourly intervals. For each profile, the energy consumption is calculated based on information regarding driving speed, outside temperature and vehicle type.

The basic data is additionally used to cluster these driving profiles into user groups to further analyze the influence of user behavior on revenue potentials resulting in a set of commuter groups which display typical commuter behavior, and a set of non-commuter groups with homogeneous behavior different to commuter behavior. The commuter set consists of 12 commuter groups. These are defined by the time of arrival of the vehicle at the place of work and the distance traveled from the place of residence to the place of work. The non-commuter set is made up of three user groups, which are determined by age and number of persons in a household. The number of created commuter and non-commuter profiles per group reflects the real distribution within the German vehicle fleet [22]. Since revenue potentials are strongly related to driving behavior, these two different pools of driving profiles are defined as input for the model:

- a commuter pool consisting of representatives of all 12 commuter groups;

- a non-commuter pool consisting of representatives of all 3 non-commuter groups.

Table 2 summarizes the characteristics of the two pools of driving profiles including the probability of the EVs' whereabouts, which is the averaged probability of the EV's location at any given point in time. The sum of probability of all three locations apart from the driving phase is 94.5% for the commuter pool and 96.8% for the non-commuter pool, which represents the theoretical availability for bidirectional charging management if an appropriate charging station is installed at their location. To analyze the influence of possible charging station locations on revenue potentials of the discussed V2G use cases, the charging point location parameter can be flexibly selected in the model for each individual EV, where the distinguished three locations can be individually defined as available for bidirectional charging or not available.

Table 2. Characteristics of user pools.

Pools of Driving Profiles	Probability of Whereabouts				Averaged Consumption (kWh/100 km)	Averaged Driving Distance (km/a)
	Place of Residence	Place of Work	Public Space	Driving Phase		
Commuter Pool	68.8%	22.1%	3.6%	5.5%	17.4	13,600
Non-commuter Pool	87.5%	1.4%	7.9%	3.2%	17.4	8300

The probability of each individual EV user to plug the vehicle into an available bidirectional charging station upon arrival determines the plug-in probability, where the expected value of a normally distributed probability is defined as a parameter. A higher plug-in probability results in a greater availability of the EV for V2G use cases. The parameter can be set flexibly to any value between 0% and 100%. As users will most likely be rewarded in some way for plugging in their EV, the plug-in probability is expected to be very high, up to a 100% certainty.

The parameter $SoC_{min,safe}$ states the minimum storage level not to be undercut when the EV is connected to the electric grid, which guarantees a certain safety range in the event of an unscheduled departure. This parameter can be set flexibly to meet the requirements of users. The storage level that must be reached at the time of a scheduled departure, $SoC_{min,dep}$, should be adjustable by the user according to his/her preferences in a real implementation. In the model, the parameter can be set between 0% and 100%.

The charging and discharging behavior model is determined in particular by the time series of market prices. For the use cases of arbitrage trading, actual price time series of day-ahead and intraday markets from 2019 are used to represent price forecasts of a maximum of two and a half days [23,24]. For trading in the day-ahead and the intraday markets, corresponding auction prices of 2019 are used. Regarding the continuous intraday market, real prices are bilaterally determined for each transaction, where buy and sell orders are constantly matched. Thus, two representative forecast prices are determined. For the relatively precise forecast of the next three hours after modeling time t, ID3 is calculated, which is the volume-weighted quarter-hourly price of all transactions in the market for the last three hours, where market liquidity is sufficient to determine a representative market price. For the more uncertain time beyond three hours after modeling time, ID_{Avg} is used, which is the volume-weighted quarter-hourly price of all transactions for this forecasted time horizon.

The regulatory framework for bidirectional charging applications is not yet fully defined to the point that simulated revenue potentials might determine what kind of regulatory incentive or obstacle enables or respectively prevents the considered V2G use cases. A market design with a reduction of different electricity price components such as grid fees and taxes would decrease the marginal costs of the EV accordingly and thus lead to an increased discharging behavior. To incorporate this highly important role of the market design in the model, various values are assigned to the additional charges on purchased energy parameter and resulting differences in revenues are assessed. The applied additional charges on purchased energy range from 0 €/MWh, which corresponds to a complete exemption from all additional electricity price components, to 234 €/MWh, which reflects the amount of all electricity price components for households in Germany in 2019, excluding electricity purchase prices [25].

3. Results

For the following investigations, user and EV parameters that are introduced in Section 2.4 are combined to form six EV pool scenarios (EV1, EV2, EV3 each for commuters and non-commuters). Displayed revenues of bidirectional or smart charging EVs always refer to the difference of these revenues to revenues of the unmanaged charging scenario. As long as there is no presentation explicitly showing single profile revenues, displayed revenues always refer to mean revenues of the considered EV pool scenario. User, modeling and regulatory parameters are set to the values shown in Table 3. The process of determining suitable parameters is further discussed in Section 3.3 as well as Appendices B and C.

Table 3. Fixed user, modeling and regulatory parameters for the investigation of revenue potentials.

Parameter	Value	Type	Further Discussion of Parameters' Influence on Revenue Potentials
Minimum SoC at departure	70%	User	Section 3.3.1
Minimum safety SoC	20% (EV1 and EV2) 30% (EV3)	User	Section 3.3.1
Plug-in probability	100%	User	Section 3.3.1
Charging point location	At place of residence	User	Section 3.3.1
Additional charges of purchased energy	0 €/MWh	Regulatory	Section 3.3.2
Forecast period	1 day	Model	Appendix B
EV pool size	Commuter: 50 Non-commuter: 75	Model	Appendix C

3.1. Revenue Potentials for Vehicle-to-Grid (V2G) Use Cases

All resulting datasets in Section 3.1, individual revenue potentials depending on EV pool scenario, and driving profiles, have been made freely available (see Section 6). The following analysis shows average revenue potentials and thus represents an aggregated extract of the provided result data.

3.1.1. Revenue Potential in the German Spot Market

The German spot market is divided into the day-ahead auction with hourly products, the intraday auction with quarter-hourly products, and the continuous intraday trading offering quarter-hourly and hourly products. Figure 3 shows the revenue potentials for the EV pool scenarios considered in the different markets with a separate participation in the markets compared to consecutive marketing. In comparison, the revenue potential of smart charging is shown in red bars. For commuters, smart charging is more attractive than for non-commuters because of the higher annual driving and thus the higher need for charging. However, for both commuters and non-commuters arbitrage trading with bidirectional charging leads to much higher revenues compared to smart charging. Revenues of non-commuters are slightly higher than revenues of commuters due to the higher availability at the place of residence.

For trading in the day-ahead market, revenues range from almost 200 €/EV/a (EV1) to 600 €/EV/a (EV3). Quarter-hourly intraday trading leads to higher revenues due to the higher volatility in prices. Consecutive trading in the day-ahead market, the intraday auction and continuous intraday trading results in best-case revenues of 400 €/EV/a (EV1) to 1300 €/EV/a (EV3). Comparing the EV1 to EV2 pool, the 2.6 times higher capacity of EV2 implicates higher flexibility for charging and discharging times. Therefore, revenues of EV2 are 200 €/EV/a higher than the revenues of EV1. EV3 almost doubles the revenues of EV2 due to doubled charging and discharging power. The increase of charging and discharging power is, thus, even more relevant for revenue potentials than the increase of battery capacity.

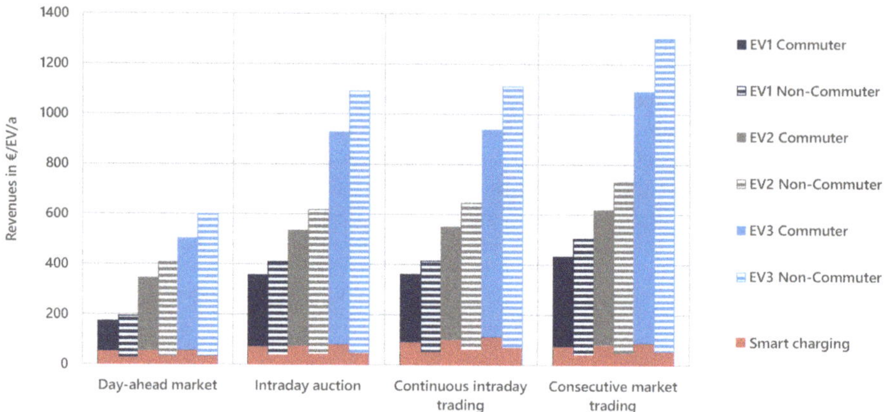

Figure 3. Revenues for bidirectionally chargeable and smart charging EVs participating in different spot markets in Germany.

3.1.2. Revenue Potential in European Markets

Since the energy systems change to a volatile and renewable production in most European countries, flexibility will be needed to cover the demand at any particular time. Therefore, bidirectionally chargeable EVs are a possible flexibility option in all European countries. To quantify revenue potentials in European countries other than Germany, we use entso-e data (European Network of Transmission System Operators for Electricity) of electricity day-ahead prices for 2019 as an input for the developed optimization model [24].

Figure 4 shows the resulting revenues for bidirectionally chargeable EVs compared to unmanaged charging for 28 European countries for commuters and non-commuters. Revenues are the highest in Ireland, Romania, Bulgaria and Hungary. These countries had scarcity prices of more than 100 €/MWh during approximately 200 h to go along with a high standard deviation of electricity prices in 2019, giving bidirectionally chargeable EVs an opportunity to use arbitrage trading more profitably. On the other hand, the revenues are the lowest for Norway and Sweden. High capacities of hydropower and some nuclear power plants in Sweden characterize the energy system in those countries [26] with almost constant marginal costs, resulting in barely volatile electricity prices.

In other countries, like Germany, Austria and France, bidirectionally chargeable EVs generate medium revenues in arbitrage trading. These energy systems are more heterogenic with some volatile renewable production as well as gas-fired, coal-fired or nuclear electricity production. Revenues in this group of countries are still varying. For example, nuclear power plants with almost constant marginal costs dominate electricity production in France [27] resulting in narrow price spreads. As another example, Austria's energy system shows high capacities of pump storage facilities [26] resulting in flattened electricity prices. These structural characteristics lead to slightly lower revenues for bidirectionally chargeable EVs for the year 2019. On the other hand, Germany has a heterogenic production portfolio of volatile wind and solar generation as well as conventional power plants with widely varying marginal costs. Price spreads and the resulting revenues for bidirectionally chargeable EVs are thus higher there.

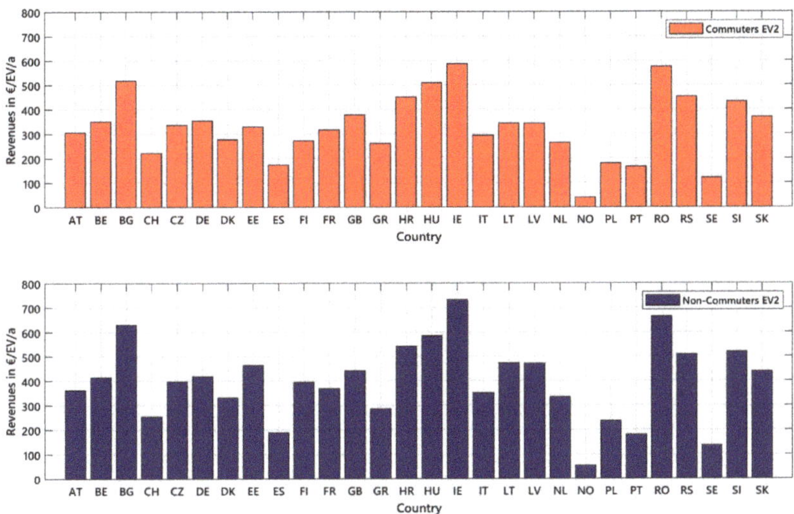

Figure 4. Revenues for bidirectionally chargeable EVs in different European day-ahead markets for market prices of 2019.

The structure of the energy system is consequently crucial for revenue potentials for bidirectionally chargeable EVs. In European countries, structural characteristics of electricity production differ a lot. In regard to energy transition in Europe accompanied by a shift to different volatile renewable production technologies, revenue potentials could vary even more in the future. For this reason, future revenue potentials in Germany are quantified and discussed in the following section.

3.1.3. Revenue Potential for Future Day-Ahead Market Prices

In addition to an assessment of current revenue potentials, future revenue potentials are also important for an investment in a bidirectionally chargeable EV and corresponding EV supply equipment (EVSE). For an estimation of the changed revenues, price time series for future years from the DYNAMIS project are used [28] (see Section 6). The DYNAMIS project performs a dynamic and intersectoral evaluation of measures for the cost-efficient decarbonization of the energy system. The multi-energy system model ISAaR (Integrated simulation model for unit dispatch and expansion with regionalization) determines the design of the future energy system in a model-based way to be able to evaluate the measures [29].

Based on the multi-stage, exploratory assessment of measures and packages of measures, a climate protection scenario has been developed that aims to reduce greenhouse gas emissions by 95% by 2050. This scenario is characterized on the supply side by a cost-optimized provision of energy sources and on the application side takes into account the technology- and sector-specific boundary conditions and restrictions. The expansion of renewable energies is the most important measure. Green fuels (including all solid, liquid and gaseous fuels produced from biomass, renewable electricity or a combination of both) will increasingly be used from 2040 onwards in applications that can only be electrified at considerable expense. Domestic power-to-x technologies increase the available flexibility in the electricity system due to the good storage capacity of green fuels. Bidirectionally chargeable electric vehicles are not yet modeled in this scenario path and, therefore, their effects on the energy system could not be investigated there either.

One output of the model are hourly marginal costs representing day-ahead electricity prices for the years 2020 to 2050. The mean annual day-ahead price and its daily standard deviation are shown in Table 4. It can be seen that the level and in particular the standard deviation of the electricity price increases sharply. This is mainly due to the severely changed energy system that includes high capacities of renewable energies resulting in production surpluses. This results in increasing times with electricity prices of 0 €/MWh. On the other hand, rising fuel and carbon prices many times lead to very high electricity prices due to the unavoidable use of power plants with expensive marginal costs.

Table 4. Modeled mean day-ahead prices and standard deviation of day-ahead prices for the years 2020 to 2050 in Germany [28] compared to empirical prices in 2019 [23].

Year	2020 (Modeled)	2030 (Modeled)	2040 (Modeled)	2050 (Modeled)	2019 (Real Prices)
Mean day-ahead price in €/MWh	46.3	61.2	63.8	80.4	37.7
Daily standard deviation of day-ahead price in €/MWh	5.0	8.7	15.8	25.9	9.0

The hourly price time series are transferred as input into the optimization model for bidirectionally chargeable EVs to estimate future revenue potentials in the day-ahead market. Figure 5 shows the revenues from 2020 to 2050 as a box plot. The black cross shows the mean revenues for the EVs considered. The top and bottom edges of the blue boxes indicate the 25th and 75th percentiles. The whiskers show the lowest and highest revenues, excluding outliers. Outliers that represent values that are 1.5 times bigger than the interquartile range are illustrated as red plus signs. Comparing the revenues of 2020 to those pointed out in Section 3.1.1 for 2019, mean revenues for the modeled prices are much lower than the mean revenues for the empirical data. Modeled electricity prices of energy system models often tend to be less volatile than real prices [30]. Consequently, lower price spreads lead to lower revenues.

Regarding modeled electricity prices for future years, mean revenues of bidirectionally chargeable EVs compared to unmanaged charging EVs increase by a factor of 5 to 6. This is mainly due to the future structural change of production units to renewable volatile production in combination with high carbon and fuel prices leading to many low electricity prices around 0 €/MWh and many high electricity prices. Bidirectionally chargeable EVs can harvest the resulting high price spreads to generate revenues. Another interesting aspect is the range of revenues within a user group that increases considerably in future years, showing higher uncertainty of revenue potentials. For non-commuters in particular, there is a heterogenic distribution of revenues.

The results for future electricity prices show a much higher revenue potential than for current electricity prices. Regarding these results, one has to keep in mind that there are many uncertainties about the design of the future energy system and the resulting electricity prices. Furthermore, bidirectionally chargeable EVs will have a retroactive effect on electricity prices, reducing price spreads and revenue potentials. Nevertheless, the use case of arbitrage trading for bidirectional EVs will most certainly get more attractive in future years.

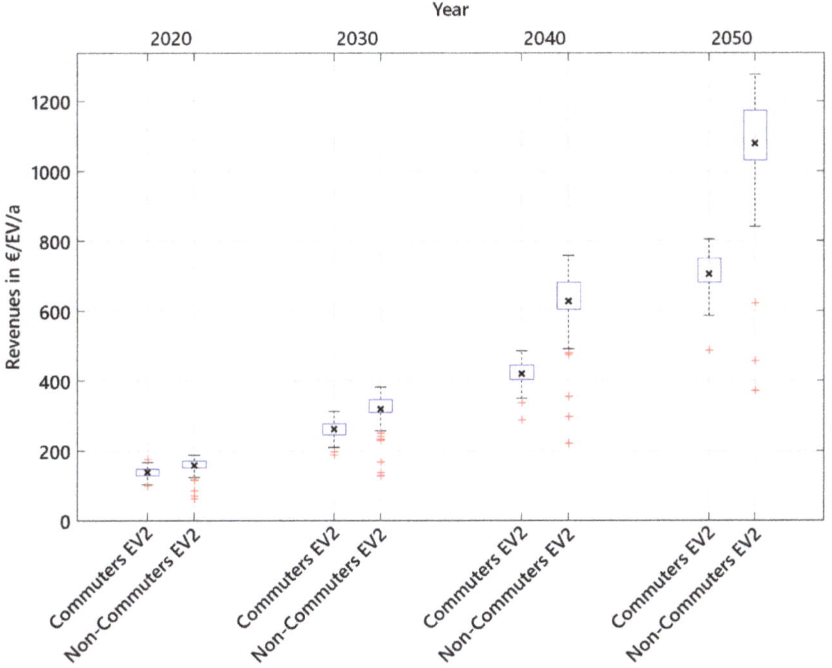

Figure 5. Revenues of bidirectionally chargeable EVs compared in accordance with future hourly day-ahead market prices.

3.2. Effect of V2G Use Cases on Full Cycles and Operating Hours

3.2.1. Effect of Unrestricted Trading in the Electricity Markets

The V2G use case of arbitrage trading leads to higher usage of the battery of the EVs and relevant supplying equipment as well as information and communication technology. Relevant parameters that show the additional charge of EVs are full battery cycles and total operating hours. High yearly full cycles and operating hours in particular mean a faster ageing of the battery. For the revenue modeling in Section 3.1, we deliberately applied no restrictions on cycles or operation hours. In the BCM project, a separate model will be used for evaluation of the impact of V2G use cases on EV components.

Table 5 shows the impact of the EV operation in Section 3.1.1 on the EV parameter full cycles, revenues per full cycle and operating hours. For arbitrage trading, a strong increase in full cycles by 100–500 full cycles/a and in operating hours by 1500–5000 h/a is determined. Revenues per full cycle are around 1 to 3 €/full cycle for arbitrage trading.

If one sets the highlighted parameter values in relation to currently warrantied lifetime values of lithium-ion batteries (e.g., 5000 to 6000 full cycles for residential storage systems [31,32] and a typical 10,000 operating hours in automotive applications [33]), it becomes clear that strong, relevant, additional loads of the battery arise for the use cases of arbitrage trading. These additional loads are significant, yet the use cases can still become economic without overloading the battery. Since large battery systems (as in EV2 and EV3) do not have many full cycles from driving, an alternative usage of the battery is a logical addition.

Table 5. Impact of V2G use cases on EV's full cycles, revenues per full cycle and operating hours.

Market Modeling	Affected EV Parameter	Commuters			Non-Commuters		
		EV1	EV2	EV3	EV1	EV2	EV3
Reference	Full cycles	60	25	25	35	15	15
Unmanaged charge	Operating Hours	400	400	340	250	250	190
Arbitrage: Day-ahead market	Full cycles	230	210	300	320	270	400
	Revenues/Full cycle	0.8	1.7	1.7	0.6	1.5	1.5
	Operating Hours	1860	3900	2920	2710	5070	3930
Arbitrage: Intraday auction	Full cycles	490	270	490	640	340	630
	Revenues/Full cycle	0.7	2.0	1.9	0.7	1.9	1.7
	Operating Hours	3760	4880	4660	4890	6180	5970
Arbitrage: Continuous intraday trading	Full cycles	450	250	470	590	320	600
	Revenues/Full cycle	0.8	2.2	2.0	0.7	2.0	1.9
	Operating Hours	3450	4670	4450	4490	5950	5680
Arbitrage: Consecutive trading	Full cycles	440	240	450	570	300	570
	Revenues/Full cycle	1.0	2.6	2.5	0.9	2.4	2.3
	Operating Hours	3280	4350	4110	4340	5590	5310

3.2.2. Effect of Restricted Trading in the Electricity Markets

Since modeled full cycles and operating hours in the previous section could critically decrease the lifetime of the EV's battery and power electronics, a minimum spread of electricity prices as a limit value for arbitrage trading could lower the EV's operation while still generating high profits. The minimum spread refers to the spread of sold to purchased energy. Consequently, the selling price for these simulations has to be lowered by the minimum, modeled price spread divided by the roundtrip efficiency.

Figure 6 illustrates the effect of a minimum price spread of 0 to 50 €/MWh on full cycles and operating hours compared to revenues of EV2. Revenues refer to the difference of revenues of bidirectionally chargeable EVs to the revenues of smart charging EVs to show only the added benefit by bidirectional charging. The decrease of full cycles and operating hours is displayed in percentage referring to the simulation with no minimum price spread. The maximum decrease of full cycles and operating hours arises when increasing the minimum price spread from 0 to 5 €/MWh, whereas revenues do not decrease significantly for this change. For a minimum price spread of 10 €/MWh, additional full cycles for bidirectional charging of EV2 decrease by more than 50% to 95 full cycles per year for commuters and to 125 full cycles per year for non-commuters. For the same restriction, additional operating hours for bidirectional charging of EV2 also decrease by more than 50% to 1800 operating hours per year for commuters and to 2400 operating hours per year for non-commuters. For a minimum price spread of 10 €/MWh, revenues for both commuters and non-commuters decrease by only 20% to 240 €/EV/a and to 310 €/EV/a, respectively. By applying higher minimum price spreads, full cycles and operating hours further decrease and both the revenue per full cycle rate and revenue per operating hour rate increase. Full data for full cycles and operating hours of all EV scenarios is attached in Appendix D.

As a result, applying a minimum price spread is an effective method of limiting full cycles and operating hours while maintaining adequate profits. Even though revenues are generally decreased, this approach leads to an increase of the revenue per full cycle rate and revenue per operating hour rate and might thus represent a practicable approach for a future operation of bidirectionally chargeable EVs.

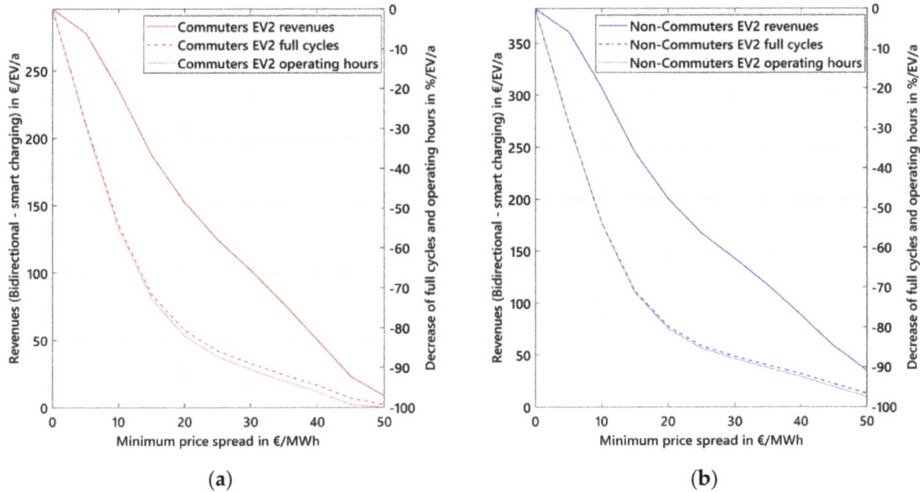

Figure 6. Effect of restricted minimum price spread on revenues, full cycles and operating hours of EV2 for commuters (**a**) and non-commuters (**b**).

3.3. Analysis of User Parameters and Regulatory Framework on Revenue Potentials of V2G Use Cases

The revenues of bidirectional charging are determined by a multitude of input parameters. We use German day-ahead prices in 2019 (Sections 3.3.1 and 3.3.2) as well as German intraday auction prices in 2019 (Section 3.3.2) to show the influence of user and regulatory parameters on the revenue potentials of bidirectionally chargeable EVs.

3.3.1. Influence of User Parameters

Minimum SoC at Departure

The user parameter $SoC_{min,dep}$ describes the minimum battery storage level that has to be reached at a scheduled departure. Considering this restriction, a higher minimum SoC at departure leads to a reduction of flexibility regarding the bidirectional charging strategy due to the smaller range between SoC_{max} and $SoC_{min,dep}$ in which the storage level can vary. This reduces the extent to which profitable price spreads and thus a revenue-maximizing discharge behavior can be used. The effects of different parameterization of $SoC_{min,dep}$ on the revenue potential in the day-ahead market are shown for the six different EV pool scenarios in Figure 7a. Potential revenues with $SoC_{min,dep}$ of 10% to 100% are compared to a reference of no minimum SoC at departure ($SoC_{min,dep} = 0\%$).

All scenarios show an exponential decrease in revenues with an increasing $SoC_{min,dep}$. If flexibility is gradually increased starting from a $SoC_{min,dep}$ of 100% up to a $SoC_{min,dep}$ of 0%, the impact on revenues is highest at the first adjustment from 100% to 90%. This is because with this first increase in flexibility, the highest electricity prices present at the considered time can be used to discharge at great profit. As low prices are used for charging, a great specific profit can be made. As flexibility is further increased due to a lower minimum SoC on departure, lower spot prices are also increasingly used for discharging. The price spread between charging and discharging decreases, which means that specific profits are reduced and revenues change less. Consequently, if an EV user can define the parameter himself, he/she should choose the lowest possible minimum SoC on departure to maximize revenues. In particular, users should avoid selecting an unnecessarily high $SoC_{min,dep}$. For investigations in this paper, a realistic minimum SoC of 70% at departure is assumed after consultation with the project BCM.

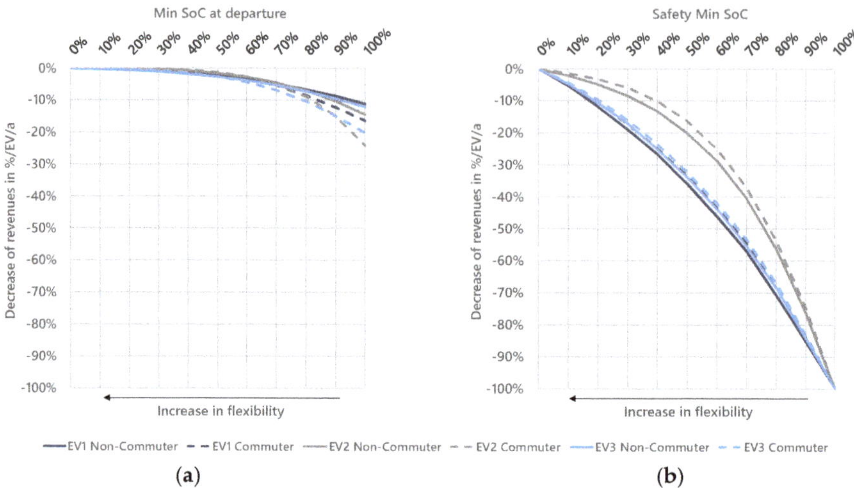

Figure 7. (**a**) Influence of a minimum SoC at departure on revenue potentials of bidirectionally chargeable EVs using arbitrage trading; (**b**) influence of a minimum SoC at the place of residence on revenue potentials of bidirectionally chargeable EVs using arbitrage trading.

Minimum Safety SoC

The user parameter $SoC_{min,safe}$ describes the minimum battery storage level that an EV always should have when connected in order to guarantee a drive to the hospital or other relevant short-distance routes at any time. If an EV arrives at a charging station with a lower SoC than $SoC_{min,safe}$, it will start charging immediately until it reaches the minimum parameterized SoC. A higher minimum safety SoC alike a higher minimum SoC at departure leads to a reduction of flexibility, since the useable capacity for marketing in the spot markets of $SoC_{max} - SoC_{min,safe}$ decreases as the minimum safety SoC is increased. To quantify the impact of a minimum safety SoC, revenues with varied parameterization of $SoC_{min,safe}$ are compared in Figure 7b showing the relative decrease of revenues compared to a reference with no safety SoC.

There is an exponential decrease of revenues for all EV pool scenarios depending on the minimum safety SoC. EV1 and EV3 have similar functions, while EV2 has a much smaller gradient for low safety SoCs and a steeper gradient for higher safety SoCs. This is due to the large battery capacity of EV2 and the fact, that the capacity of the charging/discharging power ratio (E/P) is much higher for EV2 at around 9, compared to the ratios of EV1 and EV3 at 3.5 and 4.5, respectively. A large battery capacity of the EV in combination with a fixed SoC on departure makes it less likely that the vehicle comes back with a low SoC on arrival and thus it is less limited by a low safety SoC than an EV with a low capacity. A low E/P ratio, on the other hand, means that EV batteries can be charged and discharged quickly with a great cycle depth, which is limited even with a low safety *SOC*. EV2 rarely gets these low SoCs, so a low safety SoC has little effect.

Consequently, users with EVs that have a low E/P ratio should care more about a parameterization of a very low safety SoC than users that own an EV with a higher E/P ratio. In the investigations in this paper, a minimum safety SoC of 30% for EV1 representing the status quo for BMW electric vehicles and a lower safety SoC of 20% for EV2 and EV3 representing future electric vehicles are assumed.

Plug-in Probability

An EV user has the possibility to connect his vehicle to a charging station upon arrival at a location where a charging station is available. The probability that the user connects his vehicle to the charging station is called plug-in probability. There are several factors influencing plug-in probability. First, there is the incentive for a user to plug in his/her EV. Most importantly, the user is motivated to charge his EV for the next driving phase, where the desire could be to charge the EV directly or the possibility of smart or bidirectional charging. Since the incentives for EVs using bidirectional charging have hardly been investigated yet, there are no data on the plug-in probability of those EVs. Therefore, Figure 8a shows the influence of a changed plug-in probability on revenue potentials.

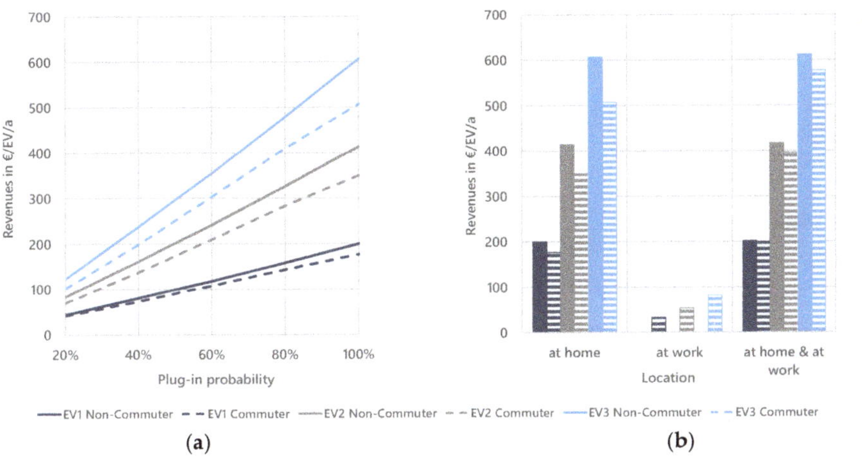

Figure 8. (**a**) Influence of plug-in probability on revenue potentials of bidirectionally chargeable EVs using arbitrage trading; (**b**) influence of charging point location on revenue potentials of bidirectionally chargeable EVs using arbitrage trading.

In all EV pool scenarios, there is a positive, approximately linear relationship between the plug-in probability and the revenues. If the EV is connected to a bidirectional charging station more frequently, a discharge process that maximizes revenues can be carried out more often. The gradient of the curves differs in the various scenarios. The increase in revenues in this investigation varies between 2 € for EV1 and 5 € to 6 € for EV3 with a 1% increase in the plug-in probability. Hence, for bidirectionally chargeable EVs participating in the spot markets, a higher plug-in probability means equally higher revenues. For investigations in this paper, a plug-in probability of 100% is assumed under the assumption that using bidirectional charging for arbitrage trading is profitable and users are incentivized to plug-in their EVs.

Charging Point Location

A bidirectional charging management system can only be operated if a bidirectional charging station is available at the EV's location. In accordance with the driving profiles from Section 2.4, possible locations for the EV are the place of residence, the place of work and the public space. An extensive expansion of bidirectional charging stations in public spaces is unlikely since the main reason for using public charging stations is the fast charging of the EV. Further analysis, therefore, concentrates on charging points at the place of residence and the place of work.

Figure 8b shows the revenue potentials of bidirectionally chargeable EVs depending on the charging point location. Comparing mean revenues at the place of residence and the place of work, there is a much higher incentive even for commuters to use bidirectional charging at the place of residence for the use case of arbitrage trading. This is due to the shorter period of time spent at the place of work while restrictions for minimum safety and departure SoC still have to be regarded. EVs can be discharged profitably less frequently than at the place of residence. In the non-commuter pool, 55% of the driving profiles are never located at a place of work, which is why an evaluation is not appropriate for these vehicle pools. If charging points are located both at the place of residence and at the place of work, revenues are slightly higher than if a charging point is only available at the place of residence. Compared to total revenues, the increase is quite low.

Consequently, the revenue potential of bidirectional charging at the place of work as opposed to the place of residence is low. Bidirectional charging should, therefore, be prioritized for EV users who are ready to install a bidirectional charging station at the place of residence. In the investigations within the framework of this paper, a charging station located at or near the place of residence is assumed.

3.3.2. Impact of Regulatory Framework on Revenue Potentials

Bidirectionally chargeable EVs are a new technology whose regulatory framework conditions have not yet been developed at the European Union (EU) level [34]. An essential question is whether bidirectionally chargeable EVs are classified as storage devices and, consequently, what additional charges they have to pay for charged electricity that is discharged later.

In Germany, the wholesale market price of electricity accounts for just around 15% (around 40 to 50 €/MWh) of the price of electricity for households. The other 85% (around 260 €/MWh) of the price is accounted for by additional charges such as the EEG surcharge (surcharge of electricity for remuneration of renewables) and grid fees as well as distribution [25]. Pumped storage facilities on the other hand are exempted from most of the EEG surcharge, grid fees and other levies, so that only additional charges of around 18 €/MWh have to be paid on electricity purchases [35], which can lead to a profitable arbitrage trading for storage facilities. As most of the exemptions refer to electricity purchases, storage losses are included [36].

For the use case of arbitrage trading, Figure 9 shows the influence of additional charges for purchased energy in the day-ahead or intraday market on revenue potentials of bidirectionally chargeable EVs for commuters and non-commuters. The solid lines represent revenues of bidirectionally chargeable EVs compared to unmanaged charging EVs and the dashed lines show revenues of smart charging EVs compared to the unmanaged charging ones. The diagram makes clear that possible revenues are strongly dependent on the regulatory framework. Starting from mean revenues with no additional charges at around 550 to 600 €/MWh for intraday auction trading, the revenues decrease by over 40% for additional charges of only 10 €/MWh. Non-commuters have an even sharper decrease since the starting revenues are a bit higher and the minimum revenues representing smart charging are lower. Revenue potential of smart charging is higher because of the increased annual driving the associated increase in the annual charging demand. If a bidirectionally chargeable EV is regulatory-equal to a pumped storage facility with additional charges of 18 €/MWh on purchased electricity, mean revenue of non-commuters will decrease by over 60%, and commuters will have less than 50% revenue. For additional charges of more than 50 €/MWh, the added revenue of bidirectionally chargeable EVs compared to smart charging EVs is less than 20 €/EV/a.

Consequently, the future regulatory framework will decide if there is a chance for profitable arbitrage trading for bidirectional EVs. In investigations in this paper, additional costs are set to zero €/MWh to show the potential revenues for bidirectionally chargeable EVs.

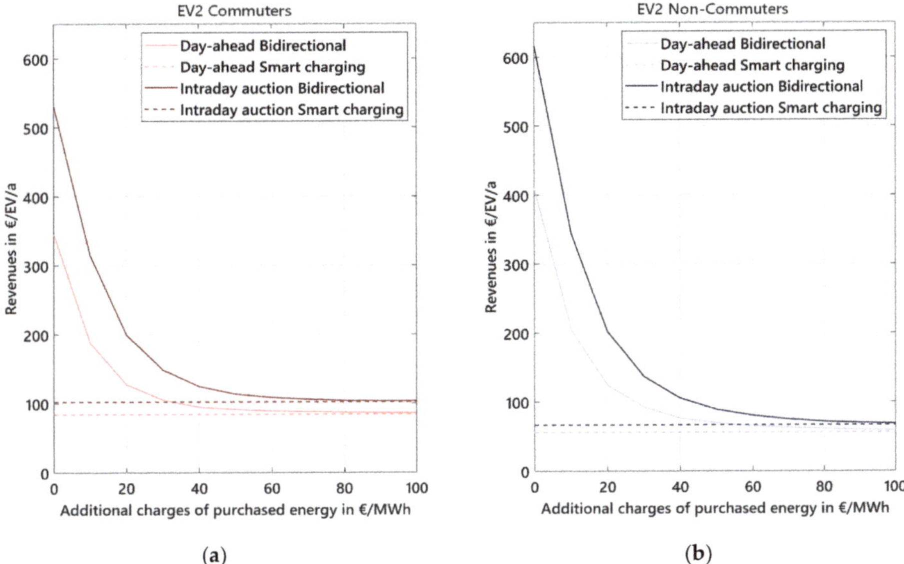

(**a**) (**b**)

Figure 9. Influence of additional charges for purchased energy on revenue potentials of bidirectionally chargeable EVs compared to revenues of smart charging in the day-ahead market and intraday auction for commuters (**a**) and non-commuters (**b**).

4. Discussion

Bidirectionally chargeable EVs can use arbitrage trading to generate revenues for a potential improvement of their economic efficiency. Section 3.1 pointed out that revenues differ widely depending not only on the EV pool scenarios but also on the considered markets and the years under review. For an evaluation based on empirical prices of 2019 in Germany, revenues by bidirectional arbitrage trading range from 200 to 1300 €/EV/a depending on EV pool and market participation. Profits on other European day-ahead markets compared to the German day-ahead market vary between +70% and −90% and are, therefore, highly dependent on the structure of the considered energy system. For future electricity prices, there is high revenue potential for bidirectionally chargeable EVs, which is pointed out by revenues for arbitrage trading in 2050 up to six times as high as the revenue potentials determined for 2020.

Most V2G related studies in literature refer to profits in reserve markets [37–39]. However, some studies deal with V2G profits of arbitrage trading. Peterson et al. point out revenues of 140 to 250 US$/EV/a (120 to 210 €/EV/a) for 16 kWh EV batteries in local markets in three US cities [13]. Pelzer et al. determine 60 to 300 US$/EV/a (50 to 250 €/EV/a) depending on spatial and temporal variation in US and Singapore markets under consideration of battery degradation costs [14]. These revenues have a similar level as our calculated EV1 revenues for day-ahead trading in Germany in 2019. We do not consider battery degradation costs but show the maximization of revenues by consecutive trading in day-ahead and intraday markets and the crucial influence of user parameters and regulatory framework on revenue potentials.

Concerning analyzed user parameters, our results confirm those of Szinai et al., who found that smart charging value at residential locations is much higher than at work or public locations [40]. In addition, we found this to also be true for bidirectionally chargeable EVs. Geske et al. indicate that the minimum range and range anxiety are the most important determinants for users participating in V2G use cases [41]. In this regard, we show the quantitative effect of the parameters 'minimum SOC at departure' and 'safety minimum SOC' on the revenue potentials of bidirectionally chargeable

EVs, thereby addressing the minimum range and range anxiety. Increasing these parameters leads to an exponential decrease of revenues depending on the EV type. Hence, a tradeoff exists between the users' range anxiety and potential revenues.

With regard to the indicated high future revenues for arbitrage trading, one has to consider the retroactive effects that bidirectional EVs participating in the considered markets will have on market prices. As for arbitrage trading, where EVs charge when spot prices are low and discharge when spot prices are high, a flattening impact on spot prices is foreseeable. From a spot market perspective, offers of electric energy increase during times of high spot prices, and the demand for electricity increases during times of low prices, leading to higher prices when prices are low and lower prices when prices are high. These retroactive effects will lower revenue potentials if significant quantities of bidirectionally chargeable EVs participate in the markets. For quantitative evaluation of these retroactive effects, an energy system model is needed that models the supply and demand curves and thus can model price changes attributable to bidirectionally chargeable EVs. This additional research will be addressed by the BCM project in a following publication.

For an assessment of the impact of bidirectionally chargeable EVs in the markets, Table 6 shows the market volumes of considered and relevant markets and fitting EV quantities with 10 kW bidirectional charging station that would cover the market completely. Regarding the table and considering Germany's aim of 10 million EVs by 2030, the probable retroactive effect in the markets can be derived. If a significant quantity of those EVs will have a bidirectional charging station and use V2G, there will be a high impact on prices in the intraday auction, and the quarter-hourly and the hourly continuous intraday trading market. A lower retroactive effect will occur for day-ahead prices in the German spot market.

Table 6. Average market volumes and fitting EV quantities for complete covering of considered and relevant German markets.

Market	Average Market Volume in Germany 2019	EVs with 10 kW Charging Station to Completely Cover the Market
Day-ahead market	26,000 MW (EPEX Spot) [1] 58,000 MW (German demand) [2]	2.6 mil (EPEX Spot) 5.8 mil (German demand)
Quarter hourly intraday auction	800 MW [1]	80,000
Hourly continuous intraday trading	4500 MW [1]	450,000
Quarter hourly continuous intraday trading	800 MW [1]	80,000

[1] Average power trading based on data of EPEX Spot [23]. [2] Average demand calculated by net consumption in 2019 of 512 TWh [42] divided by 8760 h.

Besides restrictions of market volumes, Section 3.2 points out the effect of V2G use cases on full cycles and operating hours of the EV's battery storage. For the use case of arbitrage trading, there is a high increase in battery usage resulting in an accelerated ageing of the battery. Although warranties for the full cycle lifetime of battery systems increase, the additional charge on the battery will reduce the use case's economic efficiency. However, in Section 3.2.2, we point out that both full cycles and operating hours can be decreased significantly, while revenues are still high by implementing a minimum price spread for arbitrage trading. In the BCM project, a detailed battery-ageing model is used for evaluating the impact of V2G on the battery and power electronics of the EV.

Another important restriction is the regulatory framework. Section 3.3.2 has shown the immense effect of additional charges on the profitability of the use case of arbitrage trading. It will be decisive if bidirectionally chargeable EVs are regulatory classified as a storage and if so, which additional charges will arise.

Regarding model limitations, constant values are set for the efficiencies of charging and discharging in order to achieve much faster optimization times. For arbitrage trading, charging and discharging processes usually use the highest possible power as price signals express either purchasing, selling or doing nothing. In a following publication in the BCM project, revenues of vehicle-to-home (V2H) use cases are compared by using constant and non-constant efficiencies, in which it is pointed out that modeling a non-constant efficiency for V2H use cases is necessary.

For the use case of arbitrage trading, we model revenues without perfect foresight for a rolling horizon of two to three days. The implemented market prices are real and not forecasted prices, which are used for real trading. On the other hand, it can be assumed that price forecasts over- and underestimated actual market prices to the same extent and on an average alignment with real market prices resulting in a realistic modeling of revenues for bidirectionally chargeable EVs.

Finally, for the evaluation of the economy of V2G use cases, additional costs have to be considered. The main additional hardware costs result from a bidirectional charging station. Currently, the cost for the only bidirectional charging station soon available in the German market is around 6000 € [43]. Medium-term cost projections in the project BCM for a bidirectional charging station is around 2000 €. In regard to the current stage of the BCM project, cost projections for additional hardware and operating costs are not defined, and the economic efficiency of the use cases can, therefore, not yet be evaluated.

5. Conclusions

Based on the developed aggregated storage optimization model, revenues of bidirectionally chargeable EVs have been calculated for the V2G use case arbitrage trading. As a detailed description of optimization constraints and input data are provided, readers are able to reconstruct the indicated revenues of bidirectionally chargeable EVs. The major findings of this research are:

- We developed a rolling optimization model that regards real trading times of European spot markets and allows countertrading in consecutive traded markets while considering user behavior parameters leading to a realistic representation of revenue potentials of bidirectionally chargeable EVs using arbitrage trading.
- Revenues of bidirectionally chargeable EVs are dependent on user parameters. An increase of the safety minimum SoC at the place of residence or the minimum SoC at departure leads to an exponential decrease of revenues for bidirectionally chargeable EVs.
- For a participation of bidirectionally chargeable EVs in the German spot markets in 2019, potential revenues range from 200 to 1300 €/EV/a depending on the modeled EV pool scenario under the assumption of no additional charges for purchased electricity.
- Revenues of currently available EV models participating in the day-ahead market are comparable to findings of other literature, while our research shows a significant increase in revenues for consecutive trading in all spot markets.
- The regulatory framework concerning additional charges of purchased energy is the most decisive parameter for the potential revenues of bidirectionally chargeable EVs.
- Considering additional charges amounting for example to the payments of a pumped storage facility for bidirectionally chargeable EVs results in a decrease of revenues by 50% to 60%. Thus, if V2G arbitrage trading is supposed to give flexibility to the future energy system, the market regulator will have to exempt bidirectionally chargeable EVs from the major part of additional charges.
- Unrestricted arbitrage trading of bidirectionally chargeable EVs results in a sharp increase of full cycles and operating hours by 200 to 600 full cycles/a, respectively, by 2000 to 6000 h/a resulting in much faster battery degradation. Restricted arbitrage trading with a minimum price spread can lower this additional load for EV and EVSE. For a minimum price spread of 10 €/MWh, operating hours and full cycles decrease by 50% while revenues only decrease by 20%.
- Revenues of bidirectionally chargeable EVs differ widely depending on the electricity production structure of the energy system. European day-ahead market revenues for EV2 in 2019 range from

50 €/EV/a in Norway to 700 €/EV/a in Ireland. Modeled potential future revenues are 2 times higher in 2030 and 5 to 6 times higher in 2050 than modeled revenues in 2020.

In general, potentially high revenue opportunities are identified for bidirectionally chargeable EVs in the electricity markets. Thus, participating in V2G use cases could promote electric mobility and, thereby, provide the flexibility needed for the energy system of the future. For future profitable usage of V2G use cases, the design of the regulatory framework and battery lifetime are decisive. For further investigations, especially retroactive effects of bidirectionally chargeable EVs on market prices and resulting decrease of revenue opportunities is of interest.

6. Data Availability

The driving profile data are available in Supplementary Materials 'input and results data\driving profiles'. The modelled future electricity prices are available in 'https://openenergy-platform.org/dataedit/view/scenario/ffe_dynamis_emission_factors_marginal_cost'.

In Section 3.1 which analyzed revenue potentials of individual EVs for German spot market prices, European day-ahead market prices and future day-ahead market prices are available in Supplementary Materials 'input and results data\revenues_Section_3_1'.

Supplementary Materials: The following are available online at http://www.mdpi.com/1996-1073/13/21/5812/s1, Driving Profiles data: 50 commuter and 75 non-commuter driving profiles including quarter-hourly resolved time series for location and consumption of individual EVs: http://opendata.ffe.de/dynamis-emission-factors, Revenues Section 3.1: Individual user revenues depending on driving profiles as additional data for shown aggregated revenues in Section 3.1.

Author Contributions: Conceptualization, T.K.; Data curation, T.K.; Formal analysis, T.K. and P.D.; Investigation, T.K.; Methodology, T.K. and P.D.; Software, T.K.; Validation, T.K.; Visualization, T.K.; Writing—original draft, T.K. and P.D.; Writing—review & editing, T.K., P.D. and S.v.R. All authors have read and agreed to the published version of the manuscript.

Funding: The described work is conducted within the project BCM by Forschungsgesellschaft für Energiewirtschaft mbH and funded by Bundesminsterium für Wirtschaft und Energie (BMWi) under the funding code 01MV18004C.

Conflicts of Interest: The authors declare no conflict of interest. The funders had no role in the design of the study; in the collection, analyses, or interpretation of data; in the writing of the manuscript; or in the decision to publish the results.

Appendix A

Table A1. List of time-dependent variables of the storage equation and their respective limit values.

Time-Dependent Variables		Minimum Value	Maximum Value
State of charge	SoC	SoC_{min}	SoC_{max}
Charging power	P_{charge}	$P_{charge,min}$	$P_{discharge,max}$
Discharging power	$P_{discharge}$	$P_{discharge,min}$	$P_{discharge,max}$
Discharging boolean	b_{charge}	0	1
Charging boolean	$b_{discharge}$	0	1
Counter purchase power	$P_{counter-purchase}$	0	$P_{schedule,\ sale}$
Counter sale power	$P_{counter-sale}$	0	$P_{schedule,\ purchase}$
Counter purchase boolean	$b_{counter-purchase}$	0	1
Counter sale boolean	$b_{counter-sale}$	0	1
Supplementary power	$P_{supplement}$	0	∞
Fast charging power	$P_{fastcharge}$	0	∞

Appendix B

Influence of Forecast Period

For modeling realistic revenues of bidirectionally chargeable EVs, a rolling, limited time horizon is implemented. The selected limited time horizon is important for two reasons. Firstly, market prices are perfectly forecasted, which leads to a perfect EV charging strategy for the considered time horizon. As an example, if there are very low prices 10 days ahead of the starting point, the EV may shift its charging in the future. In reality, the low prices may depend on renewable energies that cannot be forecast 10 days in advance accurately. Secondly, EV driving behavior is perfectly forecast. In reality, regular driving (e.g., commuting to work) can be predicted most of the time, whereas more spontaneous driving (e.g., for free time activities) is much more uncertain. In this regard, it is hard to declare a realistic horizon for forecasting in the model.

The upper diagram in Figure A1 illustrates the effect of an adapted forecasting period on revenue potentials of bidirectionally chargeable EVs with no additional charges on purchased energy. Starting from a forecast horizon of ten days going down to four days, there is no change in revenues. A shorter forecast period results in slightly decreased revenues, but the impact of the forecast period on revenues is very low. This is mainly due to daily characteristics of the electricity price. Price spreads are used to charge with low prices and discharge with high prices with just a slight dependence on future departures or future electricity prices. Depending on the regulatory framework, additional charges can arise for purchased energy. The bottom diagram shows the effect of a varied forecast horizon on the revenues with additional charges of 100 €/MWh on purchased energy. There is huge decrease of revenues even resulting in negative revenues for short forecast periods, since EVs often do not know a future departure and consequently discharge although the future charging price is much higher. Only for forecast periods of 7 days and longer are revenues relatively stable compared to a horizon of 10 days.

The forecast horizon is mainly applied to prevent unrealistic future trading after the optimized period that is remunerated. For simulations with no additional charges, the rolling optimization is still necessary as it allows consideration of real sequential market trading. Consequently, for investigations in this paper a short forecast period of one day is defined. For an evaluation of the regulatory framework in Section 3.3.2, a longer forecast period of seven days is applied to prevent unrealistic discharging although charging prices are higher.

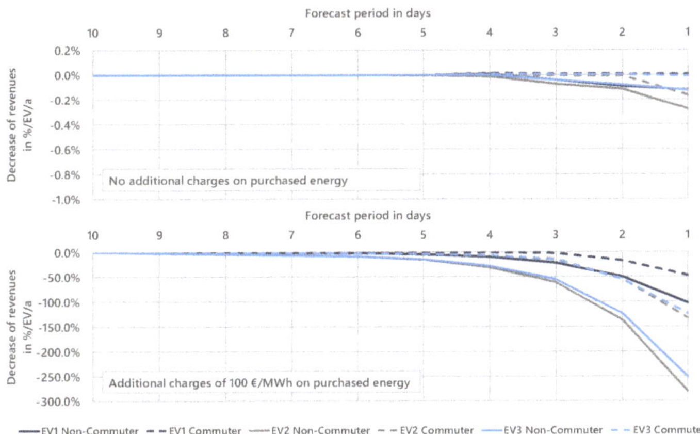

Figure A1. Influence of forecasting horizon on revenue potentials of bidirectionally chargeable EVs for the use cases of arbitrage trading with no additional charges (top) and additional charges of 100 €/MWh (bottom) on purchased energy.

Appendix C

Determination of a Realistic Pool Size

The modeling of a realistic vehicle pool is characterized by the EV pool size. The revenue potential of bidirectional charging of a single vehicle depends strongly on the individual driving profile of the user. In order to show a meaningful average revenue potential for different EV pool groups, a relevant number of vehicles must be determined for the model calculations. As computational time is linearly dependent on the number of profiles considered, a trade-off has to be faced. The aim of the investigations is to identify a pool size at which the addition of further vehicles has a low influence on revenues compared to the higher computational time. For all evaluated EV pool scenarios, the revenue potential on the German day-ahead market in 2019 was investigated for different pool sizes. For 5 to 150 profiles, a random drawing of 50,000 profile groups leads to a statistically significant analysis. Figure A2 shows the maximum revenue deviation for different numbers of profiles compared to the best case of 150 profiles. The maximum number of individual profiles is 150 in order to limit computational time. If all 150 profiles are drawn, there will be no revenue deviation to the best case of 150 profiles.

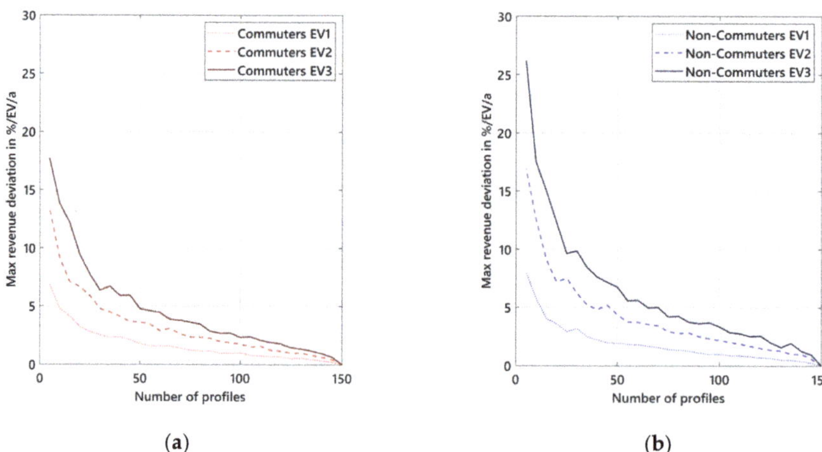

Figure A2. Maximum revenue deviation for a varying number of profiles compared to the best case of 150 profiles for a drawing of 50,000 EV profile groups for commuters (**a**) and non-commuters (**b**).

The more vehicles that are modeled, the less is the maximum revenue deviation. A relevant pool size for the EV pool scenarios can be determined by the revenue difference falling below a defined threshold value. A maximum deviation of less than 5% per vehicle per year is assumed to be sufficiently accurate while limiting computational time. This results in a relevant pool size of 50 vehicles for a commuter EV pool. A representative non-commuter EV pool needs a number of 75 vehicles. Revenues of non-commuters are more heterogenic than revenues of commuters because commuters have a more regular driving profile during weekdays.

Appendix D

Table A2. Revenues, full cycles and operating hours for all EV scenarios (difference of bidirectional charging to smart charging) with restricted minimum price spread.

EV1—Commuter

Minimum Price spread in €/MWh	Revenues in €/EV/a	Full Cycles per Year	Operating Hours per Year	Average Price Spread in €/MWh	Revenue/ Full Cycle in €/ Full Cycle	Revenue/ Operating Hour in €/Operating Hour
0	125.1	231.0	1898	14.2	0.54	0.07
5	117.7	166.7	1363	18.6	0.71	0.09
10	97.8	102.9	841	25.0	0.95	0.12
15	75.5	60.4	494	32.9	1.25	0.15
20	59.7	38.7	319	40.6	1.54	0.19
25	46.2	25.6	210	47.4	1.80	0.22
30	37.1	18.6	153	52.6	2.00	0.24
35	27.7	13.1	109	55.7	2.12	0.25
40	17.2	8.0	67	56.8	2.16	0.26
45	6.6	2.8	24	62.9	2.39	0.27
50	2.7	0.9	7	77.2	2.93	0.35

EV1—Non-Commuter

Minimum Price Spread in €/MWh	Revenues in €/EV/a	Full Cycles per Year	Operating Hours per Year	Average Price Spread in €/MWh	Revenue/ Full Cycle in €/Full Cycle	Revenue/ Operating Hour in €/Operating Hour
0	173.3	324.2	2737	14.1	0.53	0.06
5	162.5	226.7	1911	18.9	0.72	0.08
10	135.6	139.5	1173	25.6	0.97	0.12
15	107.1	85.0	712	33.2	1.26	0.15
20	86.1	56.1	472	40.4	1.54	0.18
25	68.7	38.8	327	46.6	1.77	0.21
30	56.5	29.0	245	51.3	1.95	0.23
35	44.2	22.1	189	52.7	2.00	0.23
40	33.7	16.5	140	53.9	2.05	0.24
45	19.9	10.1	87	51.9	1.97	0.23
50	10.8	5.5	48	51.4	1.95	0.23

EV2—Commuter

Minimum Price Spread in €/MWh	Revenues in €/EV/a	Full Cycles per Year	Operating Hours per Year	Average Price Spread in €/MWh	Revenue/ Full Cycle in €/Full Cycle	Revenue/ Operating Hour in €/Operating Hour
0	296.1	211.4	3963	14.0	1.40	0.07
5	278.2	150.5	2819	18.5	1.85	0.10
10	236.1	96.8	1815	24.4	2.44	0.13
15	187.3	59.9	1123	31.3	3.13	0.17
20	152.5	41.3	766	37.0	3.70	0.20
25	125.0	30.3	561	41.3	4.13	0.22
30	102.1	23.5	434	43.4	4.34	0.24
35	77.2	17.5	322	44.0	4.40	0.24
40	50.3	11.9	216	42.3	4.23	0.23
45	22.9	5.2	93	44.3	4.43	0.25
50	8.6	1.6	30	52.9	5.29	0.29

Table A2. *Cont.*

EV2—Non-Commuter

Minimum Price Spread in €/MWh	Revenues in €/EV/a	Full Cycles per Year	Operating Hours per Year	Average Price Spread in €/MWh	Revenue/ Full Cycle in €/Full Cycle	Revenue/ Operating Hour in €/Operating Hour
0	383.4	270.7	5113	14.2	1.42	0.07
5	361.0	192.5	3645	18.8	1.88	0.10
10	307.0	125.4	2383	24.5	2.45	0.13
15	245.0	78.7	1498	31.1	3.11	0.16
20	200.7	55.2	1045	36.4	3.64	0.19
25	167.5	41.9	795	40.0	4.00	0.21
30	142.8	34.5	654	41.5	4.15	0.22
35	117.1	28.5	541	41.0	4.10	0.22
40	88.6	22.8	429	38.9	3.89	0.21
45	58.5	15.9	299	36.8	3.68	0.20
50	34.7	9.3	174	37.3	3.73	0.20

EV3—Commuter

Minimum Price Spread in €/MWh	Revenues in €/EV/a	Full Cycles per Year	Operating Hours per Year	Average Price Spread in €/MWh	Revenue/ Full Cycle in €/Full Cycle	Revenue/ Operating Hour in €/Operating Hour
0	451.2	301.4	2994	15.0	1.50	0.15
5	430.6	227.5	2243	18.9	1.89	0.19
10	369.2	151.5	1487	24.4	2.44	0.25
15	290.4	92.4	898	31.4	3.14	0.32
20	231.8	60.6	582	38.3	3.83	0.40
25	187.6	42.6	402	44.0	4.40	0.47
30	152.0	31.9	297	47.7	4.77	0.51
35	123.6	24.5	227	50.4	5.04	0.54
40	85.1	17.1	156	49.8	4.98	0.55
45	41.7	8.5	75	49.2	4.92	0.55
50	16.7	2.8	25	59.1	5.91	0.66

EV3—Non-Commuter

Minimum Price Spread in €/MWh	Revenues in €/EV/a	Full Cycles per Year	Operating Hours per Year	Average Price Spread in €/MWh	Revenue/ Full Cycle in €/Full Cycle	Revenue/ Operating Hour in €/Operating Hour
0	574.6	397.5	3979	14.5	1.45	0.14
5	545.8	291.1	2913	18.8	1.88	0.19
10	466.9	191.4	1916	24.4	2.44	0.24
15	370.3	118.3	1188	31.3	3.13	0.31
20	298.4	79.2	795	37.7	3.77	0.38
25	243.9	56.8	568	42.9	4.29	0.43
30	202.6	44.2	444	45.9	4.59	0.46
35	171.8	36.2	365	47.5	4.75	0.47
40	132.8	28.4	287	46.7	4.67	0.46
45	92.3	20.7	211	44.6	4.46	0.44
50	56.0	12.7	128	44.1	4.41	0.44

References

1. BMU. *Klimaschutz in Zahlen—Fakten, Trends und Impulse deutscher Klimapolitik*; BMU: Berlin, Germany, 2018.
2. Fattler, S.; Regett, A. Environmental Impact of Electric Vehicles: Influence of Intelligent Charging Strategies. In *Grid Integration of Electric Mobility 2019*; Forschungsstelle für Energiewirtschafte e.V: München, Germany, 2019.
3. European Comission. *A Clean Planet for All –A European Strategic Long-Term Vision for A Prosperous, Modern, Competitive and Climate Neutral Economy*; European Commission: Brussels, Belgium, 2018.
4. Hinterstocker, M.; Ostermann, A.; Muller, M.; Dossow, P.; von Roon, S.; Kern, T.; Pellinger, C. *Bidirectional Charging Management—Field Trial and Measurement Concept for Assessment of Novel Charging Strategies*; Forschungsstelle für Energiewirtschaft e.V.: München, Germany, 2019.
5. Pham, T. Do German renewable energy resources affect prices and mitigate market power in the French electricity market? *Appl. Econ.* **2019**, *51*, 1–14. [CrossRef]

6. EPEX SPOT SE. *Trading on EPEX SPOT*; EPEX SPOT SE: Paris, France, 2019.

7. Illing, B.; Warweg, O. Analysis of international approaches to integrate electric vehicles into energy market. In Proceedings of the 12th International Conference on the European Energy Market (EEM), Lisbon, Portugal, 19–22 May 2015.

8. Shafiullah, M.; Al-Awami, A.T. Maximizing the profit of a load aggregator by optimal scheduling of day ahead load with EVs. In Proceedings of the 2015 IEEE International Conference on Industrial Technology (ICIT), Seville, Spain, 17–19 March 2015.

9. Illing, B.; Warweg, O. Achievable revenues for electric vehicles according to current and future energy market conditions. In Proceedings of the 13th International Conference on the European Energy Market (EEM), Porto, Portugal, 6–9 June 2016.

10. Bessa, R.J.; Matos, M.A.; Soares, F.J.; Lopes, J.A.P. Optimized Bidding of a EV Aggregation Agent in the Electricity Market. *IEEE Trans. Smart Grid* **2011**, *3*, 443–452. [CrossRef]

11. Rominger, J.; Losch, M.; Steuer, S.; Koper, K.; Schmeck, H. *Analysis of the German Continuous Intraday Market and the Revenue Potential for Flexibility Options*; IEEE: Piscataway, NJ, USA, 2019.

12. Schmidt, R.; Schnittmann, E.; Meese, J.; Dahlmann, B.; Zdrallek, M.; Armoneit, T. *Revenue-Optimized Marketing of Electric Vehicles' Flexibility Options*; IEEE: Piscataway, NJ, USA, 2019.

13. Peterson, S.B.; Whitacre, J.; Apt, J. *The Economics of Using Plug-In Hybrid Electric Vehicle Battery Packs for Grid Storage*; RELX Group: London, UK, 2009.

14. Pelzer, D.; Ciechanowicz, D.; Knoll, A. Energy arbitrage through smart scheduling of battery energy storage considering battery degradation and electricity price forecasts. In Proceedings of the IEEE Innovative Smart Grid Technologies Asia (ISGT-Asia), Melbroune, Australia, 28 November–1 December 2016.

15. Hanemann, P.; Bruckner, T. Effects of electric vehicles on the spot market price. *Energy* **2018**, *162*, 255–266. [CrossRef]

16. Rodriguez, C.D.B.; Segura, J.A.J.; Ruiz, M.C.C.; Estevez, G.A.J.; Araya, P.A.M. *Evaluating the Impact of a V2G Scheme on the Demand Curve*; IEEE: Piscataway, NJ, USA, 2019.

17. Kern, T.; Hintersrocker, M.; von Roon, S. Rückwirkungen von Batterie-Vermarktungsoptionen auf den Strommarkt. In Proceedings of the 11 Internationale Energiewirtschaftstagung an der TU Wien (IEWT 2019), Vienna, Austria, 13–15 February 2019.

18. Technical Specifications BMW i3 (120 Ah). Available online: https://www.press.bmwgroup.com/global/article/attachment/T0284828EN/415571 (accessed on 7 October 2020).

19. New Renault Zoe 2017. Available online: https://www.renault.com.au/sites/default/files/pdf/brochure/ZOE-Brochure-Desktop_0.pdf (accessed on 7 October 2020).

20. infas Institut für angewandte Sozialwissenschaft GmbH. *Mobilität in Deutschland 2017–Datensatz*; infas Institut für angewandte Sozialwissenschaft GmbH: Bonn, Germany, 2019.

21. Pellinger, C.; Schmid, T. *Merit Order der Energiespeicherung im Jahr 2030—Hauptbericht*; Forschungsstelle für Energiewirtschaft e.V. (FfE): München, Germany, 2016.

22. Fattler, S.; Böing, F.; Pellinger, C. Ladesteuerung von Elektrofahrzeugen und deren Einfluss auf betriebsbedingte Emissionen. In Proceedings of the IEWT 2017, 10 Internationale Energiewirtschaftstagung Wien, Vienna, Austria, 15–17 February 2017.

23. Power Market Data. Available online: https://www.epexspot.com/en/market-data/ (accessed on 26 August 2020).

24. Day-Ahead Prices. Available online: https://transparency.entsoe.eu/transmission-domain/r2/dayAheadPrices/show (accessed on 31 August 2020).

25. BDEW. *BDEW-Strompreisbestandteile Januar 2020—Haushalte und Industrie*; BDEW Bundesverband der Energie-und Wasserwirtschaft e.V.: Berlin, Germany, 2020.

26. Platts. *WEPP Database (Europe)*; Platts: Washington, DC, USA, 2018.

27. International Energy Agency. *Energy Policies of IEA Countries France 2016 Review*; International Energy Agency: Paris, France, 2017.

28. FfE. *Hourly CO2 Emission Factors and Marginal Costs of Energy Carriers in Future Multi-Energy Systems (Germany)*; Forschungsstelle für Energiewirtschaft e. V. (FfE): München, Germany, 2019; Available online: http://opendata.ffe.de/dataset/dynamis-emission-factors/ (accessed on 28 August 2020).

29. Böing, F.; Regett, A. Hourly CO2 Emission Factors and Marginal Costs of Energy Carriers in Future Multi-Energy Systems. *Energies* **2019**, *12*, 2260. [CrossRef]

30. Beran, P.; Pape, C.; Weber, C. *Modelling German Electricity Wholesale Spot Prices with a Parsimonious Fundamental Model—Validation and Application*; House of Energy Markets and Finance: Duisburg-Essen, Germany, 2018.

31. BYD Company Limited. Battery Box 2.5. Available online: https://www.climaverd.com/sites/default/files/2019-06/ft_sb_byd_b-box_2.5-10.0_b-plus_2.5_en.pdf (accessed on 27 August 2020).

32. BMZ Energy Storage Systems. Maximum Performance for Your Independence. Available online: https://bmz-group.com/images/PDF-Downloads/Broschuere-ESS_EN_with_ESSX.pdf (accessed on 27 August 2020).

33. Ma, K.; Yang, Y.; Wang, H.; Blaabjerg, F. Design for Reliability of Power Electronics in Renewable Energy Systems. In *Use, Operation and Maintenance of Renewable Energy Systems*; Sanz-Bobi, M.A., Ed.; Springer: Basel, Switerzland, 2014.

34. EASE. *Energy Storage: A Key Enabler for the Decarbonisation of the Transport Sector*; EASE—European Association for Storage of Energy: Brussels, Belgium, 2019.

35. Conrad, J.; Pellinger, C.; Hinterstocker, M. *Gutachten zur Rentabilität von Pumpspeicherkraftwerken*; Forschungsstelle für Energiewirtschaft e.V. (FfE): München, Germany, 2014.

36. Bundesnetzagentur für Elektrizität, Gas, Telekommunikation, Post und Eisenbahnen. *Regelungen zu Stromspeichern im deutschen Strommarkt*; Bundesnetzagentur für Elektrizität, Gas, Telekommunikation, Post und Eisenbahnen: Bonn, Germany, 2020.

37. Sandelic, M.; Stroe, D.-I.; Iov, F. Battery Storage-Based Frequency Containment Reserves in Large Wind Penetrated Scenarios: A Practical Approach to Sizing. *Energies* **2018**, *11*, 3065. [CrossRef]

38. Ciechanowicz, D.; Knoll, A.; Osswald, P.; Pelzer, D. *Towards a Business Case for Vehicle-to-Grid—Maximizing Profits in Ancillary Service Markets*; Springer: Singapore, 2015.

39. Clairand, J.M. Participation of electric vehicle aggregators in ancillary services considering users' preferences. *Sustainability* **2020**, *12*, 8. [CrossRef]

40. Szinai, J.K.; Sheppard, C.J.; Abhyankar, N.; Gopal, A.R. Reduced grid operating costs and renewable energy curtailment with electric vehicle charge management. *Energy Policy* **2020**, *136*, 111051. [CrossRef]

41. Geske, J.; Schumann, D. Willing to participate in vehicle-to-grid (V2G)? Why not! *Energy Policy* **2018**, *120*, 392–401. [CrossRef]

42. BDEW. *Energy Market Germany 2020*; Bundesverband der Energie- und Wasserwirtschaft (BDEW): Berlin, Germany, 2020.

43. Besserladen. Wallbox Quasar—Bidirektionale Ladestation. Available online: https://besserladen.de/produkt/wallbox-quasar-bidirektionale-ladestation/ (accessed on 28 August 2020).

Publisher's Note: MDPI stays neutral with regard to jurisdictional claims in published maps and institutional affiliations.

Article

Electromobility and Flexibility Management on a Non-Interconnected Island

Enea Mele , Anastasios Natsis , Aphrodite Ktena * , Christos Manasis and Nicholas Assimakis

Energy Systems Laboratory, General Department, National & Kapodistrian University of Athens, 34400 Psachna, Greece; dimitrismele@uoa.gr (E.M.); anatsis@teiste.gr (A.N.); cmana@uoa.gr (C.M.); nasimakis@uoa.gr (N.A.)
* Correspondence: apktena@uoa.gr

Abstract: The increasing penetration of electrical vehicles (EVs), on the way to decarbonizing the transportation sector, presents several challenges and opportunities for the end users, the distribution grid, and the electricity markets. Uncontrollable EV charging may increase peak demand and impact the grid stability and reliability, especially in the case of non-interconnected microgrids such as the distribution grids of small islands. On the other hand, if EVs are considered as flexible loads and distributed storage, they may offer Vehicle to Grid (V2G) services and contribute to demand-side management through smart charging and discharging. In this work, we present a study on the penetration of EVs and the flexibility they may offer for services to the grid, using a genetic algorithm for optimum valley filling and peak shaving for the case of a non-interconnected island where the electricity demand is several times higher during the summer due to the influx of tourists. Test cases have been developed for various charging/discharging strategies and mobility patterns. Their results are discussed with respect to the current generating capacity of the island as well as the future case where part of the electricity demand will have to be met by renewable energy sources, such as photovoltaic plants, in order to minimize the island's carbon footprint. Higher EV penetration, in the range of 20–25%, is enabled through smart charging strategies and V2G services, especially for load profiles with a large difference between the peak and low demands. However, the EV penetration and available flexibility is subject to the mobility needs and limited by the population and the size of the road network of the island itself rather than the grid needs and constraints. Limitations and challenges concerning efficient V2G services on a non-interconnected microgrid are identified. The results will be used in the design of a smart charging controller linked to the microgrid's energy management system.

Keywords: electric vehicles; genetic algorithm; V2G services; valley filling; peak shaving; flexibility

Citation: Mele, E.; Natsis, A.; Ktena, A.; Manasis, C.; Assimakis, N. Electromobility and Flexibility Management on a Non-Interconnected Island. *Energies* **2021**, *14*, 1337. https://doi.org/10.3390/en14051337

Academic Editor: Amela Ajanovic

Received: 6 February 2021
Accepted: 22 February 2021
Published: 1 March 2021

Publisher's Note: MDPI stays neutral with regard to jurisdictional claims in published maps and institutional affiliations.

1. Introduction

Electric vehicles (EVs) have an important role in the transition towards a low-carbon economy and, more specifically, in the decarbonization of the transportation sector, which is responsible for 22% of total EU-28 greenhouse gas emissions, excluding international aviation and maritime emissions [1]. The increasing penetration of electric vehicles (EVs) reflects the technological advances in electromobility as well as the impact of the policies implemented at the EU and national levels. As a result, the market of electromobility has been growing at an accelerating pace: in 2018, the global electric vehicle fleet exceeded 5.1 million, up two million from the previous year, with the number of new electric vehicle registrations almost doubling [2].

The increasing penetration of EVs presents several challenges to and opportunities for the operation of the power grid and the electricity markets as well as to the end users. The additional electricity needed for charging the EVs could mitigate the environmental benefits if the power generation mix is fossil fuel-oriented [3]. If, however, this additional electricity demand is met by increasing the share of renewables (e.g., solar and wind

265

power), it has been shown that increased penetration levels of EVs lead to a lower carbon footprint than conventional vehicles [3,4].

Uncontrolled charging, based solely on the needs of the EV user, may lead to higher peaks in electricity demand stressing the capacity limits of the distribution grid [5–7], especially in small, isolated grids such as those of small non-interconnected islands which are also popular tourist destinations, as is the case of Greek islands.

The distribution grids of these islands rely on diesel autonomous power stations (APSs) which are progressively complemented by renewable energy sources (RESs), mainly photovoltaic (PV) plants and wind turbine (WT) parks. Due to the lack of interconnections with other electrical systems or the country's transmission system, non-interconnected islands have an increased cost of electricity generation, while they are more affected by load disturbances resulting in a greater risk for power quality problems, such as voltage and frequency stability, black-outs, and load rejection [8]. In addition, due to high fluctuations of demand on a daily and monthly basis and the seasonal peak demand due to the touristic period, there is the need for each electrical system to operate with an excessive power capacity in order to meet peak demands [9]. Note that the influx of tourists during the high season is several times that of the population. The integration of RESs in a small standalone system decreases the cost of electricity generation and the carbon footprint of the island, but the intermittent nature or solar or wind power may prove challenging for the stability of the microgrid. Electrical storage is one way to increase RES penetration, optimize the generation of the thermoelectric power plant, and address stability and power quality issues [10].

The transition towards "green" islands requires that the electricity demand is met mainly by RESs and that mobility relies mostly on EVs. This means that the power system of the island must meet the additional energy required by the island's EV fleet and any demand-side management strategies and policies must take into account the EV charging rate, stations' distribution, mobility patterns, etc. [11]. To achieve that, smart EV charging, adaptable both to the operation of the power grid and the EV users' preferences, is required. In contrast to uncontrolled charging, smart charging allows a certain level of control over the charging process [12]. A simple approach is that end users alter their charging behavior and shift the charging of their EVs from peak hours to off-peak hours (load shifting) in response to price signals or incentives [13].

A more advanced approach is a direct control mechanism, via an intermediate market entity such as an aggregator, that optimizes EV charging schedules in real time based on the needs of the power grid, the signals from the electricity market, and the preferences of the end users [12]. Scheduling EV charging so that the aggregated power demand from EVs fills the overnight valley reduces the daily cycling of the thermoelectric power plant and the operational cost of utilities [14].

On the other hand, EVs, acting as controllable and distributed storage, may be used to offer Vehicle to Grid (V2G) services and actively contribute to demand-side management. Cars, including EVs, spend the majority of their lifetime (95% on average) parked. In these periods of inactivity, EVs could charge their batteries when demand is low (valley-filling) and send power back to the grid (discharge) when demand is high (peak shaving) [15], thus becoming part of the solution and curtailing the need for costly infrastructure upgrades. V2G services may include also ancillary services (spinning reserve), active power support, and reactive power compensation [16]. In grid-connected EVs, available energy can be used as additional generation capacity in order to support the power grid in case of generation outages (spinning reserve). Furthermore, the capability of EVs to channel reactive power to the grid can be beneficial to the grid operation [17]. It has been shown that smart charging improves the saturation of grid transformers for the same number of EVs and can decrease reverse power flows from distributed generation to the transformer [12]. An overall smart and flexible management of a fleet of EVs has the potential to shave peak demand, flatten the load profile, and allow higher shares of renewable energy while accommodating more EVs to the power grid [12].

In this paper, we study the effect of EV penetration on the electricity demand of a non-interconnected island powered by a diesel-fueled autonomous power station (APS) and the flexibility offered by their storage. We use a genetic algorithm (GA) to calculate the optimum EV penetration level for the valley filling and peak shaving of electricity demand curves during both the low and high seasons. Four test cases of charging strategies are studied, taking into consideration random and specific mobility patterns, in order to evaluate the impact of the level of EV penetration on the APS generation levels and to examine the potential of EVs for flexibility services. Furthermore, according to the National Plan for Energy and Climate in accordance with the UN Agenda 2030, "for islands that are not expected to be interconnected, a significant reduction in the use of diesel for power generation is also being promoted, with the setup of state-of-the-art RES plants combined with storage technologies" [18]. To address this, we repeat the calculations for the case where the APS is complemented by a PV plant.

The structure of the paper is as follows: Section 2 gives a summary review of the relevant literature while Section 3 focuses on the formulation of the optimization problem. The GA developed to solve the problem is presented and discussed in Section 4. Section 5 is dedicated to the simulation experiments and their results, which are discussed in Section 6. Section 7 summarizes the main conclusions.

2. Literature Review

The aim of this literature review is to establish a knowledge base concerning the factors that affect EV penetration on autonomous microgrids, such as the distribution grids of small, non-interconnected islands and the charging strategies to be considered in a smart charging controller embedded or linked to the energy management system of the microgrid.

Research on the topic has focused mainly on interconnected power systems with the capacity to host large numbers of EVs. Results such as those in [19] show that a large deployment of EVs could result in violation of supply/demand matching and statutory voltage limits as well as power quality problems and voltage imbalance under certain operating conditions. It is, therefore, necessary to design charging strategies and apply charging schedules as EV penetration increases. The flexibility offered to the grid by the energy management of EV batteries is enhanced when discharging to the grid is also allowed.

Several approaches have been proposed for optimum charging and discharging strategies and schedules. The optimal charging scheduling of electric vehicles proposed in [20] employs a genetic algorithm-based optimization routine, where thermal line limits, the load on transformers, voltage limits and parking availability patterns were considered to establish an optimal load pattern for EV charging-based reliability. The results showed that a smart charging schedule for EVs led to a flattening of the load profile, to peak load shaving and to the prevention of the aging of power systems' elements. A similar approach has been adopted in [21] where an adaptive discharging and smart charging management scheme for peak shaving and load leveling in a residential distribution grid is introduced. A reference operating point is considered to flatten the load curve on a 24-hour basis by using the EV mobility characteristics and the non-EV base load. A particle swarm optimization algorithm was developed in [15] for the smart, centralized scheduling of optimum EV charging/discharging of plug-in electric vehicles in order to achieve peak shaving and valley filling of the grid load profile. Smart charging compared to uncontrolled charging is superior in terms of voltage drop and maximum loading. On the other hand, optimizing the charging cycles of an electric car in [22] using demand-side management achieves financial savings, increased demand on renewable energy, reduced demand on thermal generation plants and reduced peak load demand.

A distributed iterative algorithm for the management of the charging/discharging set-points of an EV fleet is proposed in [23], which is designed to optimize the profits of the aggregator based on the day-ahead energy forecast. The optimization problem is formulated as a mixed-integer quadratic problem. A cost-effective and an eco-friendly scheme for the centralized management of plug-in hybrid electric vehicles (PHEVs) are compared in [24]. Representative hourly driving patterns and electricity data from eight North American Electric Reliability Corporation (NERC) regions were leveraged to examine the results of the proposed schemes. The findings indicate that the management schemes proposed result in very different charging schedules: optimal cost-effective charging should take place very early in the morning and optimal eco-friendly charging later in the afternoon. As the number of PHEVs increases, charging control becomes more cost-effective and environmentally friendly. The variation in charging patterns among plug-in electric vehicle (PEV) owners as a function of charging location and charging level is studied in [25]. The results showed that the use of home, work and public charging infrastructures is interconnected, highlighting the importance of having an integrated infrastructure investment plan for different locational charging patterns among PEV owners. A study on the existing and potential charging infrastructure for PEVs in the USA [26] showed that the potential for future residential infrastructure is limited by the availability of dedicated parking locations where chargers could be installed, acting as a barrier to mainstream EV penetration. An algorithmic framework was presented in [27] for the joint optimization of the routes and charging schedules of a fleet of self-driving EVs providing on-demand mobility, taking into account the battery energy level of each vehicle and the power grid needs and constraints. The results verified the near optimality of this method, while suggesting that through vehicle to grid (V2G) operation, a 100% penetration of renewable energy could be enabled and still provide high quality mobility services.

For charging schedules and services procurement to be efficient, coordination is needed between the EV Aggregator and the System Operator [28]. Results have shown that when coordination over the charging schedule of EVs is performed, the system can accommodate a higher share of EVs without any infrastructure upgrade. The results reported concern mainly valley-filling services.

Formulations of the less-studied problem of discharging through the grid are offered, but the results refer to large, interconnected power grids with the capacity to host large EV fleets, and coordination is necessary to avoid stability and power quality problems.

Problems related to EV penetration, energy management, charging scheduling or routing also refer mainly to distribution grids connected to transmission grids. Limited results exist for EV penetration on non-interconnected grids, such as those of islands or other isolated RES microgrids. EVs are studied as a means to reduce CO_2 emissions and lower energy costs in isolated regions, as in Sao Miguel, Azores [29], where three scenarios of EV penetration have been studied. It was concluded that if at least 15% of the fleet is replaced by EVs, significant reductions in fossil fuel use and energy can be expected. However, smart charging and V2G services have not been considered. Furthermore, 15% of the fleet is very small compared to the target of the nearly 100% electrification of the transportation sector on "green islands".

According to a recent study on the potential of islands to serve as testbeds for innovative solutions [30], it is stressed that in order to enable higher penetration levels while minimizing the impact on an autonomous microgrid, an EV control system with smart charging, metering and billing functionalities must be embedded in the energy management system linked to the island's distribution grid. Such a system would aggregate the response and interface with the EVs and the grid to manage the aggregated flexibility, send messages and price signals to the EV users and offer services to the grid. It is also stressed that it is necessary to study the impact of the scaling of several existing pilot implementations which currently concern only a small number of EVs.

The main contributions of this work consist in studying the impact of EV penetration, controlled charging and V2G services on the electricity demand of a non-interconnected island, where the high-season demand is several times that of the low-season demand, and the contribution of electromobility towards the decarbonization of the island through gradually phasing out thermoelectric generators and replacing them with RESs. The results will be used in the design of a smart charging controller interconnected with or embedded in the island's energy management system. Limitations and challenges concerning efficient V2G services on a non-interconnected microgrid are identified.

3. The Optimization Problem

The optimization problem solved in order to study the effect of a fleet of N EVs on the load demand of a non-interconnected microgrid is the optimum energy available for valley filling and peak shaving for various charging schemes and EV penetration levels.

The assumptions concerning the mobility patterns and demand profiles are based on actual data from a Greek island which is normally populated by a few hundred people but in the summertime receives an influx of several thousands of tourists. As a result, the electricity demand during the summer months is several times higher than during the rest of the year (Figure 1). The peak demand in the summertime approaches 1.0 MW with the base load around 0.5 MW in the high season. On the other hand, the consumption for the rest of year is in the range of 0.2 to 0.5 MW. The load data have been provided by the Hellenic Distribution Network Operator (HEDNO) Islands Directorate.

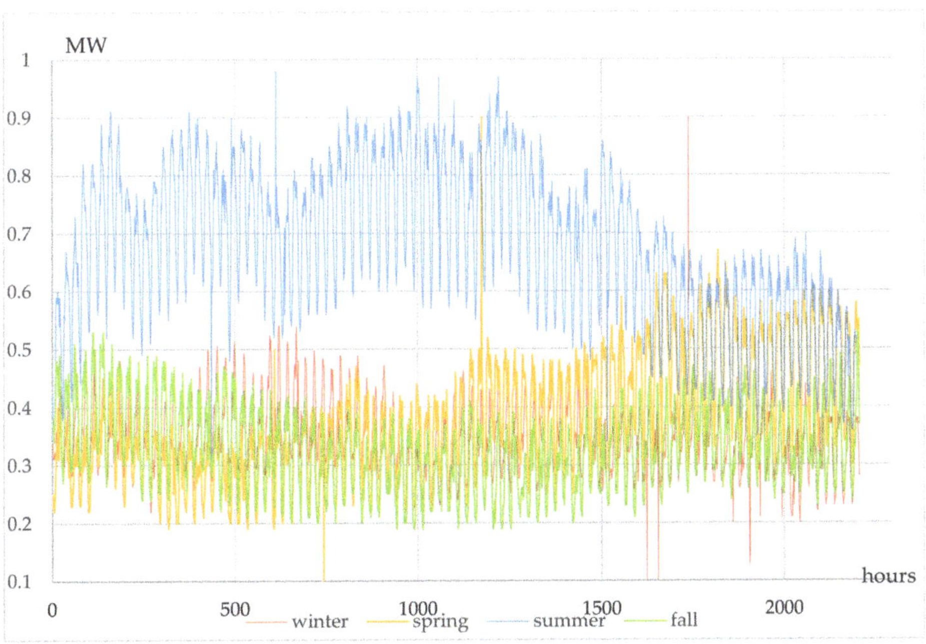

Figure 1. Hourly electricity demand for the winter, spring, summer and fall months for 2018.

The island is powered by diesel generators of approximately 1.5 MW total installed capacity [31]. The region does not have a considerable wind potential, so any RESs for electricity generation would have to rely on solar power [32]. According to a recent decision of the Greek Regulatory Authority for Energy, which, taking into account grid stability issues, redefined the renewables penetration margins in non-interconnected islands, the maximum power output from an installed PV plant on the given island has been set to 150.0 kW. Respecting this limit, a 140.0 kWp PV plant has been considered for the output curve (Figure 2) obtained for one typical winter day and one typical summer day, using PVGIS [32].

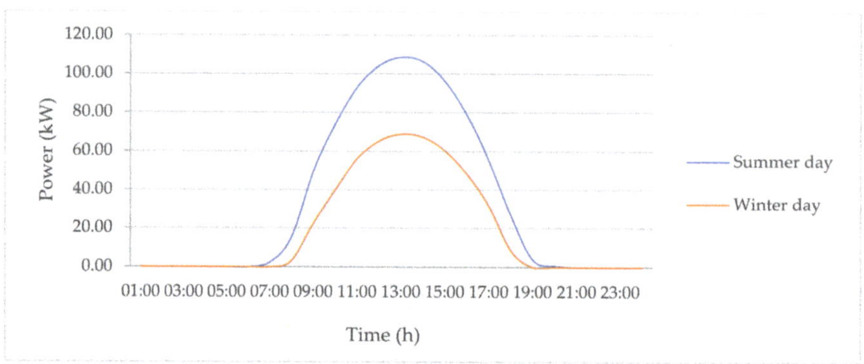

Figure 2. Output power curve of a 140.0 kWp photovoltaic (PV) plant for a typical summer and winter day.

In order to study the EV fleet impact on the electricity demand and APS output required, with and without PV generation, taking into account valley filling and peak shaving, the following assumptions are made.

The EV fleet consists of N vehicles, of which kN cars are for private use, charged at home or at public parking lots, and mN are rental cars used mostly during the tourist season, also charged in a parking lot, with $0 < m < 1$ and $k = 1 - m$. The mobility patterns considered in the test cases and simulations are dictated by the island's road network constraints. Because the island has a road network of less than 30 km, it is reasonable to assume that the kN EVs used by permanent residents will be charged, at most, once a day and mostly during the night, while the mN rental EVs may be used for valley filling or peak shaving, during idle times at the rental service parking lot, based on a smart charging schedule. According to the IEC61851-1 standard and the existing regulatory framework, the "slow" charging Mode 3 AC supported by a 55 FkVA low-voltage power supply is used in all charging facilities. This is a reasonable assumption since on small islands, due to zoning laws, even privately owned cars are parked in public parking lots. This implementation allows the aggregation of demand and flexibility offered by EVs via a functionality of the central energy management system, as proposed in [30]. Because we assume that EV charging and discharging take place at a relatively slow rate, their impact on the load flow or the stability of the grid is not taken into consideration in this study. However, the genetic algorithm developed for the optimization problem presented in the remainder of the section is designed in such a way so as to allow for load flow constraints, price signals and user preferences to be taken into account.

The optimization problem aims to flatten the demand curve.

To achieve valley filling, the batteries of N EVs are fully or partially charged. The objective function given in (1) aims to minimize the difference between average daily demand, μ, and the power that needs to be supplied by the APS to meet the demand through management of the energy flowing in or out of the EV batteries over a given time interval Δt_t [21,33]. If a PV plant supplies a portion of the required energy, the energy that needs to be generated by the APS decreases:

$$\text{minimize } f = \sum_{t=1}^{T}(\sum_{i=1}^{N} P_{i,t}^{EV} + D_t - P_t^{PV} - \mu)^2 \tag{1}$$

$$\text{s.t. } \sum_{t=1}^{T} P_{i,t}^{EV}\Delta t_t - (1 - SOC_{in,i})\frac{E_b}{\eta_c} = 0 \tag{2}$$

$$(SOC_{i,t+1} - SOC_{i,t})\frac{E_b}{\eta_c} - P_{i,t}^{EV}\Delta t_t = 0 \tag{3}$$

$$(SOC_{i,t+1} - SOC_{i,t})\frac{E_b}{\eta_c} - [D_t - \mu]\,\Delta t_t \leq 0 \tag{4}$$

$$\left|P_{i,t}^{EV}\right| - P_{ch} \leq 0 \tag{5}$$

$$0.20 \leq SOC_{i,t} \leq 0.80 \tag{6}$$

where P_t^{PV} is the power generated by the PV plant at every time step Δt_t, D_t is the demand at a given time step Δt_t, D_t is the power absorbed by the battery (positive sign) of the i-th EV at time step Δt_t, $SOC_{i,t}$ is the state of charge of the battery of the i-th EV at time step Δt_t, E_b and η_c are the storage capacity and charging efficiency of the battery, respectively, and P_{ch} is the power rating of the charging facility. The state of charge of each battery is assumed to be independent of the open-circuit voltage, the charging (or discharging) efficiency is assumed to be equal to 1 and the power rating of the charging facility depends on the charging mode following IEC61851-1.

The optimization routine is executed for a period T, i.e., for a daily demand profile with hourly data $T = 24$ and $\Delta t_t = 1$ h.

The N EVs are assumed to enter the charging facilities in a consecutive manner. The i-th EV enters the charging facility with the available remaining charge in its battery denoted by $SOC_{in,i}$. $SOC_{in,i}$ depends on the mileage d_i of the i-th EV before connecting to the charging facility and the maximum number of kilometers R_i that the EV can travel without recharging:

$$SOC_{in,i} = \left(1 - \frac{d_i}{R_i}\right)100\% \tag{7}$$

where d_i is a random variable and R_i is the range given by the manufacturer of the i-th EV; in this study, the specifications of all EVs are considered identical and match those of a commercially available medium-sized EV.

The first constraint (2) ensures that the energy absorbed by or provided to the i-th battery during the total charging interval will not exceed the capacity allowed by $SOC_{in,i}$. The second constraint (3) controls the charging of the i-th EV; the amount of energy absorbed by the battery determines the change in the SOC between two consecutive time steps. The third constraint (4) prevents the i-th EV from charging to its maximum capacity if the remaining storage capacity of the i-th EV at time t is larger than the difference between the demand D_t and the average value μ in order to ensure the flattening of the curve. The fourth constraint (5) ensures that the $P_{i,t}^{EV}$, which is the power absorbed or delivered by the battery of the i-th EV, cannot exceed the power rating of the EV charging facility, P_{ch}. The fifth constraint (5) imposes lower and upper limits to the SOC as suggested in [34–36] for longer battery lifetime.

To achieve peak shaving, the problem is similar to (1)–(6), with constraints (2) and (3) being adjusted as follows and η_d being the discharging efficiency:

$$\sum_{t=1}^{T} P_{i,t}^{EV}\Delta t - SOC_{in,i}E_b\eta_d = 0 \tag{8}$$

$$(SOC_{i,t+1} - SOC_{i,t})\,E_b\eta_d - P_{i,t}^{EV}\,\Delta t = 0 \tag{9}$$

4. The Genetic Algorithm

The optimization problem, as described by (1)–(6), is a quadratic programming problem (QP) which can be efficiently solved with a QP solver. However, a genetic algorithm (GA) [37] has been used instead to allow for future problem formulations, with non-linear constraints, user-generated data, price signals, etc.

The GA uses tournament selection and applies a penalty when a solution violates a constraint. This way, no solutions are discarded but those violating one or more constraints are less likely to be chosen. In the pseudocode of the GA given below, G is the number of generations, P is the populations, l is the chromosome length, p_x is the probability of crossover, p_m is the probability of mutation and k is the number of solutions selected as parents each time in the tournament selection:

Set random $SOC_{in,i}$ for N EVs, gc = 1 (generations counter) and pc = 1 (population counter)
For i in range (Vehicles), 1 < i < N:
For t in range (Hours), 1 < t < T:
1. Set G, P, l, p_x, p_m, k
2. If (P/2) ! = 0 P = P − 1
3. Generate P of l-sized strings
4. Select 2 parent solutions through tournament selection
5. Apply crossover p_x to obtain children solutions
6. Apply mutation p_m to children solutions
7. While pc < P:
Repeat steps 5 to 7 P/2 times, pc = pc + 1
8. Get a new generation
9. While gc < G:
Repeat steps 5 to 9 G times, gc = gc + 1
10. Choose the optimum solution in each generation, min (f)
t = t + 1
Obtain $SOC_{i,t+1}$
i = I + 1
Update D_t after i-th EV has been charged

The fitness value is given as in [38]:

$$Q\left(\vec{x}\right) = \begin{cases} F\left(\vec{x}\right) & \text{if } \vec{x} \text{ is a feasible solution} \\ F\left(\vec{x}\right) + C\left(\vec{x}\right) & \text{if } \vec{x} \text{ is unfeasible} \end{cases} \tag{10}$$

$\vec{x} = \left[P_1^{EV}, P_2^{EV}, \ldots, P_N^{EV}\right]$ is the power absorbed or delivered by each EV car. $C\left(\vec{x}\right)$ is a penalty imposed to any solution violating one or more constraints.

For the results shown below: $P = 40$, $G = 20$, $l = 5$, $k = 3$, $p_x = 1$, $p_m = 0.35$. Indicative execution times in the offline mode, using a computer equipped with an Intel Core i7-6700HQ CPU 2.60 GHz processor, 16 GB DDR4 RAM, Windows 10, range from 300 to 480 s based on the size N of the EV fleet.

5. Simulations and Results

All simulations were based on the hourly demand data of 2018. Two demand curves were used: LC1, corresponding to a typical winter day, and LC2, corresponding to a summer day, with sharp fluctuations of demand.

Four test cases have been developed reflecting various charging/discharging schemes: (a) valley filling; (b) valley filling and peak shaving with random EV mobility patterns; (c) valley filling and peak shaving with both random and predefined EV mobility patterns; (d) valley filling at predefined hours and peak shaving with random EV mobility pattern.

The size N of the EV fleet ranged from 25 to 150 vehicles for the winter curve LC1 and 50 to 250 vehicles for the summer curve LC2. The corresponding EV penetration level was calculated as the ratio of the total energy capacity of the N EVs over the total energy demand of the demand curve in each case. k, the fraction of EVs used by permanent residents, was set to 0.3, and m, the fraction of rental EVs, was set to 0.7 for all simulations.

All EVs were considered to be identical and their specifications correspond to a specific commercially available medium-sized EV: $E_b = 23.80$ kWh, $R = 145.00$ km. The charging rate was also assumed to be the same for all charging facilities and set to $P_{ch} = 7.40$ kW.

5.1. Valley Filling

The first test case allows for valley filling when the demand falls below μ.
Two cases were examined:

(a) EVs enter the charging facility randomly during the day. Charging can take place anytime during the day, as long as $D_t - \mu \leq 0$. During the winter period (LC1, Figure 3), valley filling is achieved with a fleet of 75 EVs, while during the summer period (LC2, Figure 4) the same result is obtained with a fleet of 150 EVs. Further increasing the number of vehicles would not affect the results, due to (4).

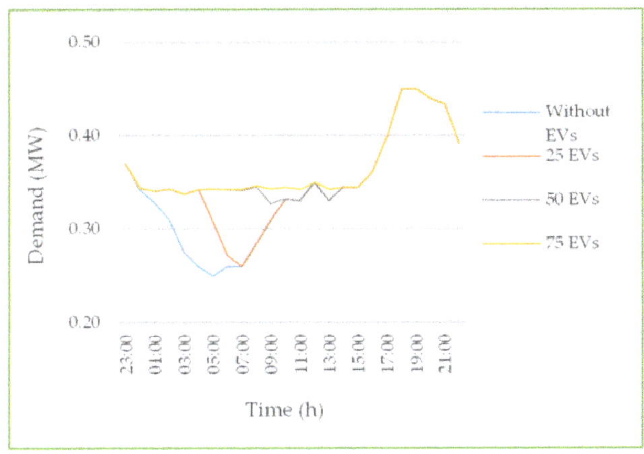

Figure 3. The effect of charging on load curve 1 (LC1) when $N = 25, 50$ and **75** cars are used for valley filling anytime during the day.

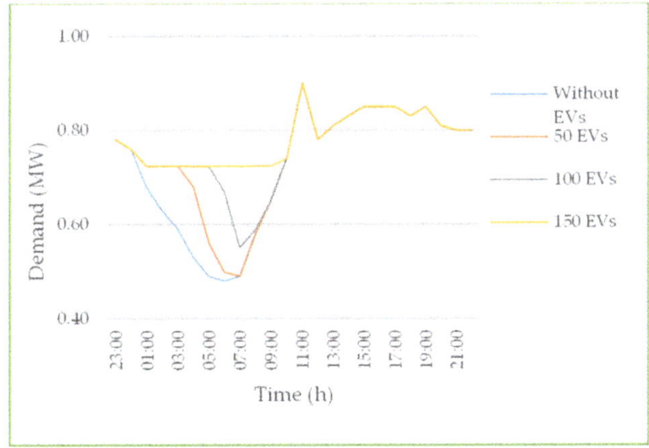

Figure 4. The effect of charging on load curve 2 (LC2) when $N = 50, 100$ and **150** cars are used for valley filling anytime during the day.

(b) All N EVs (rental and privately owned) may be charged only during a specific time interval $t_1 \leq t \leq t_2$. For the results shown here, charging is allowed only for $23:00 \leq t \leq 07:00$. During the winter period (LC1, Figure 5), valley filling is partially achieved with a fleet of 75 EVs, while during the summer period (LC2, Figure 6), the same is achieved with a fleet of 150 EVs. Further increasing the number of vehicles would not make a significant difference because of the restricted hours for charging. This charging schedule offers less flexibility.

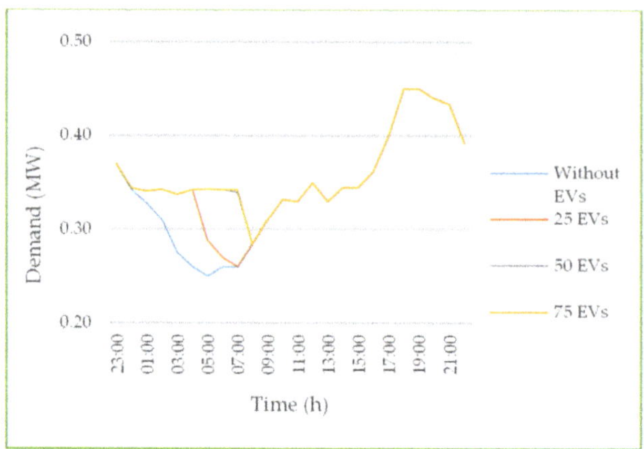

Figure 5. The effect of charging on LC1 when N = 25, 50 and **75** cars are used for valley filling between 23:00 and 07:00.

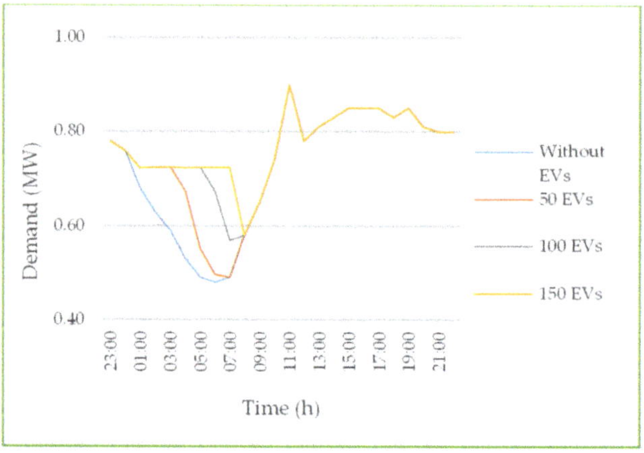

Figure 6. The effect of charging on LC2 when N = 50, 100 and **150** cars are used for valley filling between 23:00 and 07:00.

5.2. Valley Filling and Peak Shaving with Random EV Mobility Patterns

The second test case allowed for both valley filling and peak shaving to be performed at random hours of the day and in a random manner. To accomplish that, the algorithm randomly assigned a binary p value of 0 or 1 to each EV for each hour of the day: an EV with $p = 1$ at a given time interval Δt corresponds to an EV that is parked and, thus, is available for charging or discharging via the grid, while a vehicle with $p = 0$ corresponds to an EV that is moving and, therefore, unavailable for the specific time t. In this latter case, the SOC_i of a moving EV decreases according to the energy consumption rate and range R of the car. The assumption in this case is that the EV owners, following signals issued by the energy management system, will opt to charge their EVs anytime during the low-demand time and discharge to the grid anytime during high-demand time. Furthermore, they are not allowed to do the opposite.

During the winter period (LC1, Figure 7), valley filling is achieved with a fleet of merely 25 EVs. Peak shaving is observed around noon, but there is not much energy available overall for V2G services. During the summer period (LC2, Figure 8), valley filling and maximum peak shaving are achieved with a fleet of 100 EVs.

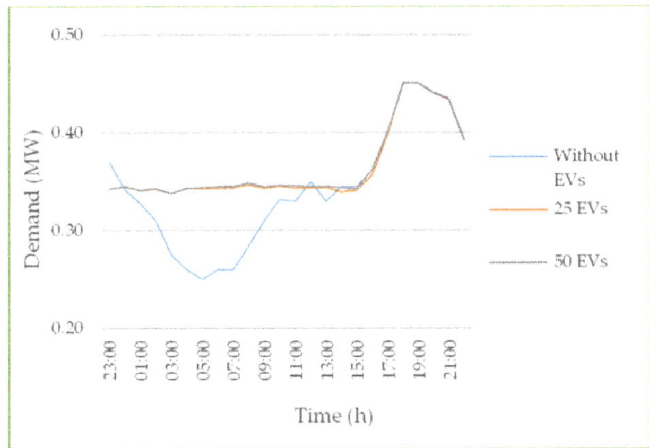

Figure 7. The effect of charging/discharging on LC1 when N = **25** and 50 cars with a random mobility pattern are used for valley filling or peak shaving.

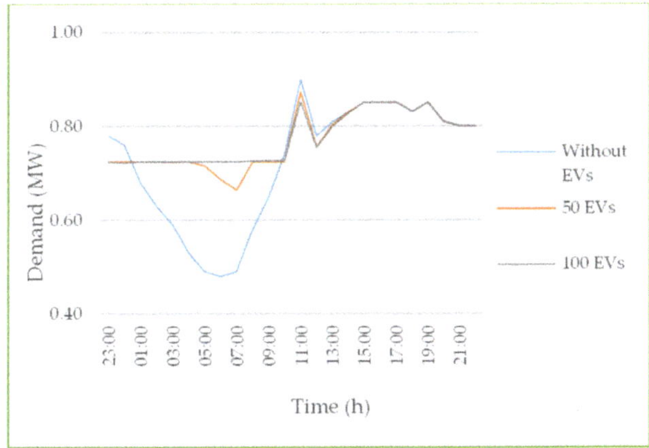

Figure 8. The effect of charging/discharging on LC2 when N = 50 and **100** cars with a random mobility pattern are used for valley filling or peak shaving.

5.3. Valley Filling and Peak Shaving with Random and Predefined EV Mobility Patterns

In the third test case, both valley filling and peak shaving are allowed, but the N EVs are now divided into two subgroups of populations, kN and mN. The first subgroup follows a predefined mobility pattern, as in test case 1: such is, for example, the case of permanent residents or workers in an office building who regularly work at predetermined hours. The predefined mobility pattern of the kN EVs, e.g., of the permanent residents, is largely shaped by the daily needs of their users: EVs are parked from 18:00 to 08:00 and are allowed to charge their battery, while in the remaining hours, they are assumed to be on the move and, therefore, unavailable. The second subgroup of mN EVs are allowed to charge or offer V2G services whenever they are parked, according to the random mobility pattern described in test case 2.

During the winter period (LC1, Figure 9), valley filling is achieved with a fleet of 50 EVs and peak shaving is only observed around noon. During the summer period (LC2, Figure 10), valley filling and maximum peak shaving are achieved with a fleet of 100 EVs.

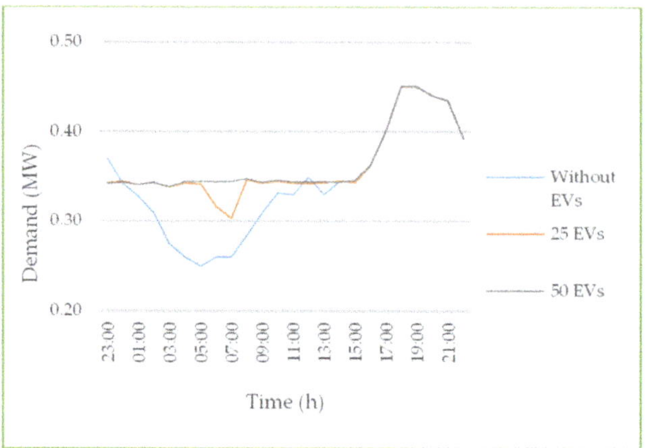

Figure 9. The effect of charging/discharging on LC1 when a fleet of N electric vehicles (EVs) is divided into subgroups of different mobility patterns and used for valley filling or peak shaving: $N = 25$ and **50**; $k = 0.3\,N$, $m = 0.7\,N$.

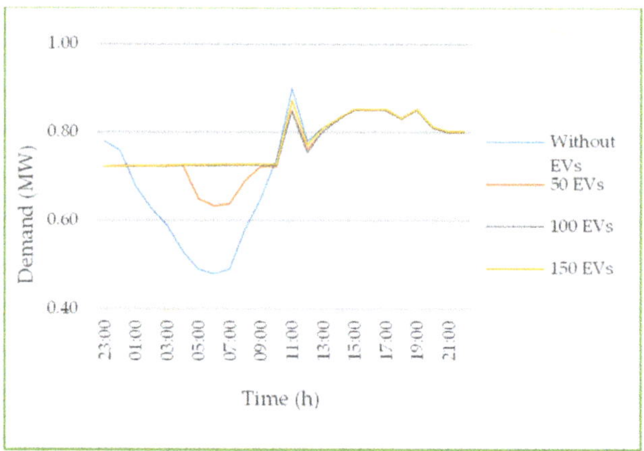

Figure 10. The effect of charging/discharging on LC2 when a fleet of N EVs is divided into subgroups of different mobility patterns and used for valley filling or peak shaving: $N = 50$ and **100**, 150; $k = 0.3\,N$, $m = 0.7\,N$.

5.4. Valley Filling at Predefined Hours and Peak Shaving with Random EV Mobility Pattern

The fourth test case dictates that all EVs must be fully charged during the hours at wich the demand is at its lowest, e.g., evening hours. This case has been designed in such a way so as to test whether the strategy of having all EVs fully charged at the beginning of each day would yield better results for peak shaving. To implement this scenario, the randomly assigned p-values, also used in the second test case, were activated: the EVs are assumed to be parked from 23:00 to 07:00, which means that $p = 1$ for these hours. During this time interval, they can either charge their battery or inject power into the grid. After 07:00, the status of the EVs, as parked or moving, is allowed to change: it is either randomly set to $p = 0$ and they are not available for charging or discharging or it remains as $p = 1$ and they can either charge their battery, inject power into the grid or move.

During the winter period (LC1, Figure 11), valley filling is achieved with a fleet of 75 EVs but there is little energy available for peak shaving. During the summer period (LC2, Figure 12), valley filling is achieved with a fleet of 150 EVs and maximum peak shaving is achieved for a fleet of 250 EVs. Overnight charging maximizes the flexibility for V2G services during the peak demand.

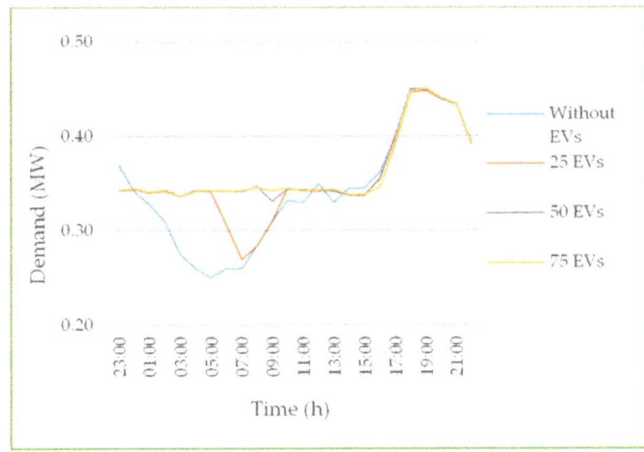

Figure 11. The effect of charging/discharging on LC1 when *N* cars have a custom mobility pattern for valley filling and random EV mobility pattern for peak shaving: *N* = 25, 50 and **75**.

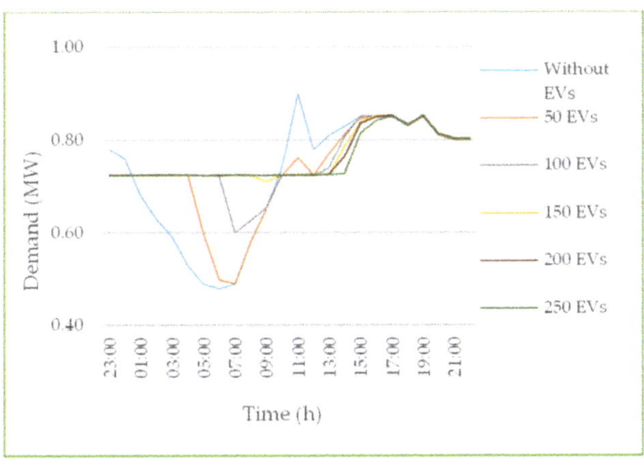

Figure 12. The effect of charging/discharging on LC2 when *N* cars have a custom mobility pattern for valley filling and random EV mobility pattern for peak shaving: *N* = 50, 100, **150**, 200, **250**.

5.5. The Effect of PV Penetration on the Power Generated by the Island's APS

The above test cases were also used to study the effect of a 140.0 kWp PV plant on the power that needs to be supplied by the island's autonomous power system (APS). Although the maximum yield of the PV plant (Figure 2) does not coincide with the peak demand of either the winter or the summer day, its operation significantly reduces the APS power output during daytime and allows for more flexibility in the shaping of the supply curve through the demand.

In test case 2, EVs have random mobility patterns and are not allowed to charge during the peak demand or discharge through the grid during the low demand. For the winter day LC1 (Figure 13), the energy generated by the PV plant enables EV charging to take place for the biggest part of the day, which also allows for more flexibility in V2G services. For the summer day LC2 (Figure 14), optimum valley filling and maximum peak shaving are achieved with a fleet of 150 EVs. More EVs are accommodated and peak shaving is enabled.

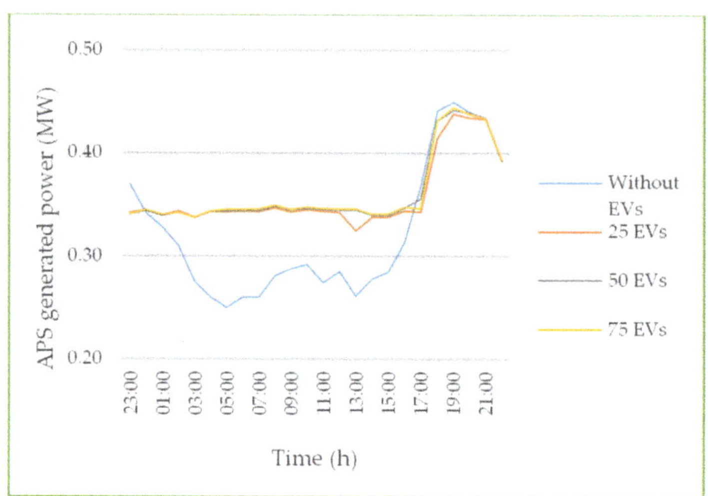

Figure 13. Autonomous power system (APS) output required in the presence of a 140.0 kWp PV plant for test case 2 and LC1.

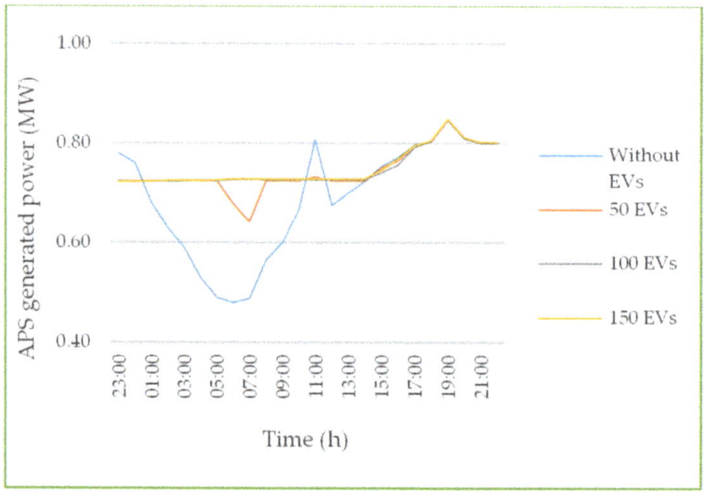

Figure 14. APS output required in the presence of a 140.0 kWp PV plant for test case 2 and LC2.

In test case 4, where all EVs are assigned random mobility patterns but must be fully charged during the night, peak shaving is achieved with a fleet of 75 EVs for the winter day LC1 (Figure 15). For the summer day LC2 (Figure 16), both valley filling and maximum peak shaving are achieved with a fleet of 200 EVs.

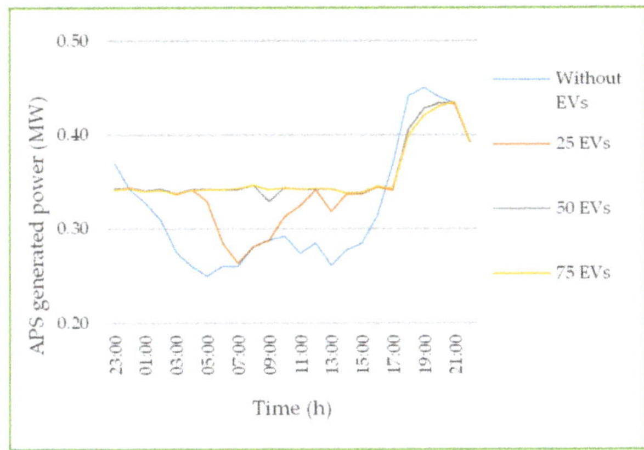

Figure 15. APS output required in the presence of a 140.0 kWp PV plant for test case 4 and LC1.

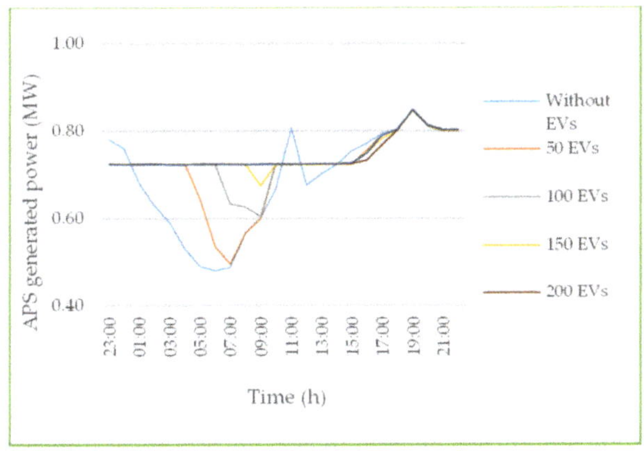

Figure 16. APS output required in the presence of a 140.0 kWp PV plant for test case 4 and LC2.

In the case of random charging during the day (test case 1a), a larger EV fleet may be charged, taking advantage of the PV energy; furthermore, a smaller number of EVs end up partially charged (Table 1). In test case 3, where the EV fleet is divided into two subgroups, one with a predefined mobility pattern and one with a random mobility pattern, the increased demand for EV charging is partly covered by PV generation and there is energy available for peak shaving services.

Table 1. Partially charged EVs for test cases 1a and 1b.

Load Curve 1 (Winter)					
Number of EVs (N)	25	50	75	100	150
Partially charged EVs (1a)	0	0	11	40	74
Partially charged EVs (1a) w/PV	0	0	0	0	43
Partially charged EVs (1b)	0	0	21	55	81
Partially charged EVs (1b) w/PV	0	0	24	45	93

Table 1. *Cont.*

Test Case 1—Load Curve 2 (Summer)					
Number of EVs (N)	**50**	**100**	**150**	**200**	**250**
Partially charged EVs (1a)	0	0	6	41	101
Partially charged EVs (1a) w/PV	0	0	0	37	85
Partially charged EVs (1b)	0	0	4	70	113
Partially charged EVs (1b) w/PV	0	0	16	55	122

6. Discussion

The genetic algorithm developed and implemented for optimum valley filling and peak shaving using the storage capacity of an EV fleet on a non-interconnected Greek island has highlighted some of the factors determining the optimum penetration level of EVs and the flexibility offered by them. The simulation experiments performed assumed various cases of mobility and charging patterns. The effect of PV generation has also been studied. In the following discussion, we first examine the case where all electricity on the island is supplied by the diesel autonomous power station (APS), and next, the effect of the PV plant.

In the first case of valley filling, which allows the EVs to charge anytime during the day, the demand curve is flattened at 22% and 28% EV penetration for the winter and summer periods, corresponding to 75 and 150 EVs, respectively. Considering that there are approximately 500 inhabitants, this level of penetration is reasonable. Increasing the size of the EV fleet will leave a number of cars partly charged (Table 1) if charging is allowed only during low-demand periods (valley filling). The presence of a PV plant allows for a larger EV fleet to be accommodated with relatively good quality of service, since the excess energy required for EV charging is supplied by the PV plant.

If a predefined time interval for charging is set, e.g., 23:00–7:00 (test case 1b), a large number of EVs will either be charged below the upper SOC limit or will not be charged at all (Table 1) if the average demand level μ, above which charging is not allowed, is not increased. Therefore, setting time restrictions for EV charging overnight is not advisable, especially when PV energy is available during the daytime.

In the second case, both peak shaving and valley filling services are offered with random EV mobility patterns. EV users, reacting to signals or pre-announced pricing options, may charge their EVs anytime during low-demand periods and discharge to the grid anytime during high-demand periods, but they are not allowed to do the opposite. Valley filling is achieved at merely 7% and 19% EV penetration for the winter and summer periods, corresponding to 25 and 100 EVs, respectively, which are reasonable numbers for the size of the island. Peak shaving, on the other hand, is not easily attainable due to the small number of EVs and the resulting storage capacity. In the wintertime, the available energy in EV batteries is not enough to both cover the EV owners' needs and offer services to the grid. In the summertime, the larger EV fleet allows some flexibility for peak shaving, with the maximum observed from 10:00 to 14:00 h, for $N = 100$ EVs; the demand peak is decreased by 6% of the original peak and the corresponding energy fed into the power grid is 0.10 MWh. The power supplied by a PV plant allows for more EVs to be accommodated and the flexibility offered by the EV storage is enhanced.

In the third test case, the EV fleet is divided into two subgroups, one with a random mobility pattern and one with a predefined mobility pattern. Valley filling is achieved at 15% and 19% EV penetration for the winter and summer periods, respectively. In the summertime, the larger EV fleet allows for some peak shaving, similar to the second test case.

To facilitate V2G services, the strategy proposed in the fourth test case was designed: EVs are fully charged overnight when demand is low in order to increase the EV capability for peak-shaving services when demand is higher. The potential for peak shaving is higher in the summer case, since a total of 0.47 MWh may be offered to the grid when $N = 250$. The maximum power shaved is 0.18 MW, which amounts to 20% of the peak demand. The flexibility for V2G services is significantly enhanced when PV generation is available.

For the case of an autonomous grid with low mobility and a restricted road network, such as the one studied here, the EV penetration level for optimum flexibility management is determined mainly by the difference between the low- and high-demand values of the demand profile.

In any case, the electricity required to charge the EVs increases the power that needs to be supplied by the island's fossil-fueled APS. However, replacing conventional cars by EVs and shaping the demand curve through valley-filling and peak-shaving operations partly counterbalances the environmental footprint of the increased APS supply.

On the road towards "green" islands, in order to phase out the local fossil-fueled power plants in favor of RESs, it is necessary to complement any RES generation with electrical storage. Except from absorbing any excess energy, electrical storage may be used to effectively shape the demand curve and minimize the energy required from the APS, leading to additional fuel and CO_2 emissions savings. However, the storage offered by EVs increases the demand and does not suffice for effective V2G services, but an appropriate charging and, when possible, discharging schedule may optimize the overall electricity demand profile.

For the case studied here, in the absence of EVs, a PV plant significantly reduces the power that needs to be supplied by the APS during daytime. However, in the evening hours, the APS must ramp up significantly to meet the high demand. In the presence of EVs, charging may be used to flatten the demand curve around a given value, thus allowing a more efficient operation of the APS even if it needs to generate more energy. The energy supplied by the PV plant contributes further towards that direction, favoring peak shaving and increasing the flexibility offered by the EV storage.

Based on the above results, a plausible strategy for EV charging and V2G services on non-interconnected islands is to either issue recommendations and price signals to EV users, using hourly forecasts and a tool such as the one depicted in Figure 17, or design daily schedules for EV charging and V2G services using day-ahead load forecasts and a tool such as the one shown in Figure 18.

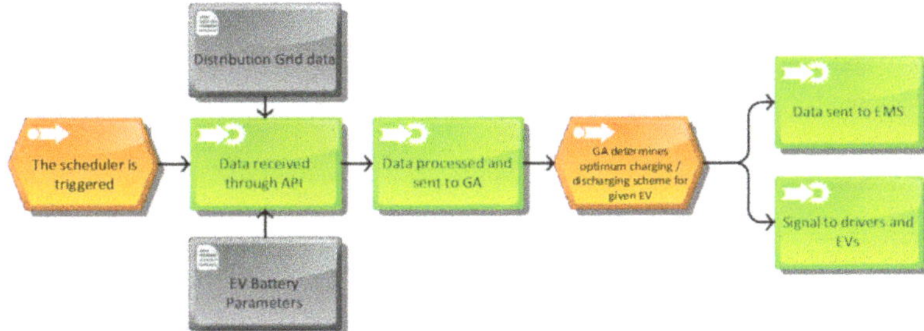

Figure 17. Genetic algorithm for online operations.

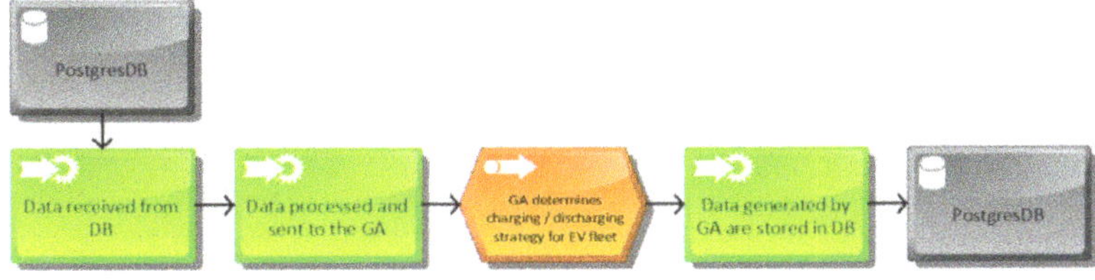

Figure 18. Genetic algorithm for offline operations.

These may be applied to all EV users regardless of their profile or, when profiling is possible, they may be adjusted to respective profiles' mobility patterns in order to enhance the impact of demand–response mechanisms.

The last part of the discussion is dedicated to the GA developed for this work. It has been designed so that it can be used in both an online and offline mode. In the online mode (Figure 17), it receives a signal from the grid every time a car is connected to it for charging or discharging (V2G) and retrieves all required data, such as SOC, energy balance and other grid-related data, from the EV battery controller and the energy management system linked to the grid. In future versions, the GA will be coupled to power flow equations and will be more detailed, AI-based and data-driven SOC models will be used. The output of the GA is signals to the driver and/or the EV battery controller, for fully automated systems, and data to the energy management system.

In the offline mode (Figure 18), the GA may be used to run simulations and design charging strategies and pricing schemes for the management of a given fleet of EV cars using next-day load forecasts or other data available in the database.

An overarching remark for the microgrid studied in this work is that increasing the EV fleet can offer more flexibility for demand-side management, but ultimately, the EV penetration levels should be subject to the mobility needs of the island and not the energy management system operations. In other words, the size of the EV fleet is limited by the population and size of the island itself. Moving towards carbon-free islands, electromobility relying on private or rental cars should not be encouraged. Instead, a strategy of small-scale mass electromobility [3], e.g., offered through frequently running minibuses, should be designed, which would serve both the locals and the visitors. In that case, the charging strategy would need to adjust accordingly, since the routes and charging schedules could be co-optimized as in [27].

7. Conclusions

The flexibility offered by an EV fleet on a Greek non-interconnected island in the Aegean Sea has been studied using a genetic algorithm to shape the demand curve and reduce the energy required of the autonomous power plant of the island. The EV flexibility is managed by an intermediate entity, e.g., an aggregator, or an energy management system linked to the distribution grid, which issues signals to EV owners or the EV battery controller.

Two operations were studied: charging for valley filling and discharging to the grid for peak shaving. The optimum size of the EV penetration which corresponds to the number of EVs participating in these two operations has been determined for four test cases with various mobility and charging patterns. Two demand curves have been examined: one for a winter day and one for a summer day, which typically has a much higher demand due to the influx of tourists.

The simulations show that time limits do not lead to better valley-filling services. Instead, more dynamic charging/discharging schemes are encouraged. The use of EVs as flexible loads and storage can flatten the demand curve, but peak shaving requires higher EV penetration levels so that the EVs have enough energy stored to also offer to the grid. The flexibility services depend on the size of the EV fleet. However, a larger EV fleet increases the total daily energy required. A PV plant allows for higher EV penetration levels and increases the peak-shaving capacity and flexibility for demand-side management operations.

Ultimately, the size of the EV fleet is subject to the mobility needs on the island and is, therefore, limited by the population and size of the road network itself. A strategy for optimum electromobility using small-scale mass transportation, such as frequently running minibuses with co-optimized routes and charging/discharging schedules, as part of a holistic approach towards "green" islands will be studied next. Certain assumptions made will be relaxed in future works to allow for the effect of time-dependent charging, pricing signals, battery degradation as well as uncertainties introduced by the stochastic nature of RESs to be studied. Load forecasting will be coupled to the genetic algorithm presented here in order to develop a tool for the design of charging schedules.

Author Contributions: Conceptualization, A.K., E.M. and A.N.; methodology, A.K., C.M. and N.A.; software, E.M.; validation, A.N. and A.K.; formal analysis C.M. and A.N.; investigation, A.N., C.M., A.K. and N.A.; resources, N.A. and C.M.; data curation, C.M. and N.A.; writing—original draft preparation, A.N., E.M., and A.K.; writing—review and editing, C.M., A.K. and N.A.; visualization, E.M., A.N., and C.M.; supervision, A.K.; project administration, A.K. All authors have read and agreed to the published version of the manuscript.

Funding: This research received no external funding.

Institutional Review Board Statement: Not applicable.

Conflicts of Interest: The authors declare no conflict of interest.

References

1. European Environment Agency. Indicator Assessment. *Greenhouse Gas Emissions from Transport in Europe*. 2019. Available online: https://www.eea.europa.eu/data-and-maps/indicators/transport-emissions-of-greenhouse-gases/transport-emissions-of-greenhouse-gases-12 (accessed on 6 December 2019).
2. IEA. *Global EV Outlook 2019: Scaling Up the Transition to Electric Mobility*; IEA: Paris, France, 2019. Available online: https://www.iea.org/reports/global-ev-outlook-2019 (accessed on 6 December 2019).
3. Ajanovic, A.; Haas, R. Electric vehicles: Solution or new problem? *Environ. Dev. Sustain.* **2018**, *20*, 7–22. [CrossRef]
4. Hacker, F.; Harthan, R.; Matthes, F.; Zimmer, W. *Environmental Impacts and Impact on the Electricity Market of a Large Scale Introduction of Electric Cars in Europe*; ETC/ACC Technical Paper 2009/4; Öko-Institut e.V.: Berlin, Germany, 2009.
5. Karakitsios, I.; Karfopoulos, E.; Hatziargyriou, N. Impact of dynamic and static fast inductive charging of electric vehicles on the distribution network. *Electr. Power Syst. Res.* **2016**, *140*, 107–115. [CrossRef]
6. Yanfei, W.; Haifeng, S.; Wenhao, W.; Yunjing, Z. The impact of electric vehicle charging on grid reliability. *IOP Conf. Ser. Earth Environ. Sci.* **2018**, *199*, 052033. [CrossRef]
7. Mousavi, S.; Flynn, D. Controlled Charging of Electric Vehicles to Minimize Energy Losses in Distribution Systems. *IFAC PapersOnLine* **2016**, *49*, 324–329. [CrossRef]
8. Díaz, A.R.; Ramos-Real, F.J.; Marrero, G.A.; Perez, Y. Impact of Electric Vehicles as Distributed Energy Storage in Isolated Systems: The Case of Tenerife. *Sustainability* **2015**, *7*, 15152–15178. [CrossRef]
9. Stavropoulou, E. The Electrical Systems of the Greek Non Interconnected Islands. In Proceedings of the 4th International Hybrid Power Systems Workshop, Heraklion, Crete, 22–23 May 2019.
10. Nieto, A.; Vita, V.; Maris, T.I. Power Quality Improvement in Power Grids with the Integration of Energy Storage Systems. *Int. J. Eng. Res. Technol.* **2016**, *5*, 438–443.
11. Kasten, P.; Bracker, J.; Haller, M.; Purwanto, J. *Electric Mobility in Europe—Future Impact on the Emissions and the Energy Systems*; Öko-Institut e.V.: Berlin, Germany, 2016.
12. *Smart Charging for Electric Vehicles*; Innovation Outlook; International Renewable Energy Agency (IRENA): Abu Dhabi, United Arab Emirates, 2019.
13. Ensslen, A.; Ringler, P.; Dörr, L.; Jochem, P.; Zimmermann, F.; Fichtner, W. Incentivizing smart charging: Modeling charging tariffs for electric vehicles in German and French electricity markets. *Energy Res. Soc. Sci.* **2018**, *42*, 112–126. [CrossRef]
14. DeForest, N.; Funk, J.; Lorimer, A.; Ur, B.; Sidhu, I.; Kaminsky, P.; Tenderich, B. *Impact of Widespread Electric Vehicle Adoption on the Electrical Utility Business-Threats and Opportunities*; Technical Report; Center for Entrepreneurship & Technology (CET), University of California: Berkeley, CA, USA, 2009.
15. Ramadan, H.; Ali, A.; Nour, M.; Farkas, C. Smart Charging and Discharging of Plug-in Electric Vehicles for Peak Shaving and Valley Filling of the Grid Power. In Proceedings of the 20th International Middle East Power Systems Conference (MEPCON), Nasr City, Egypt, 18–20 December 2018.
16. Ghofrani, M.; Detert, E.; Niromand, N.H.; Arabali, A.; Myers, N.; Ngin, P. V2G Services for Renewable Integration. In *Modeling and Simulation for Electric Vehicle Applications*; IntechOpen: Rijeka, Croatia, 2016. [CrossRef]
17. Paudyal, S.; Ceylan, O.; Bhattarai, B.; Myers, K.S. Optimal Coordinated EV Charging with Reactive Power Support in Constrained Distribution Grids. In *Proceedings of the 2017 IEEE Power & Energy Society General Meeting, Chicago, IL, USA, 16–20 July 2017*; IEEE: Piscataway, NJ, USA, 2017; pp. 1–5. [CrossRef]
18. National Energy and Climate Plan; Ministry of the Environment and Energy: Athens, Greece. 2019. Available online: https://ec.europa.eu/energy/sites/ener/files/el_final_necp_main_en.pdf (accessed on 2 February 2021).
19. Putrus, G.A.; Suwanapingkarl, P.; Johnston, D.; Bentley, E.C.; Narayana, M. Impact of Electric Vehicles on Power Distribution Networks. In Proceedings of the 2009 IEEE Vehicle Power and Propulsion Conference, Dearborn, MI, USA, 7–11 September 2009.
20. Alonso, M.; Amaris, H.; Germain, G.G.; Galan, J.M. Optimal Charging Scheduling of Electric Vehicles in Smart Grids by Heuristic Algorithms. *Energies* **2014**, *7*, 2449–2475. [CrossRef]

21. Erden, F.; Kisacikoglu, M.; Erdogan, N. Adaptive V2G Peak Shaving and Smart Charging Control for Grid Integration of PEVs. *J. Electr. Power Compon. Syst.* **2018**, *46*, 1494–1508. [CrossRef]
22. Finn, P.; Fitzpatrick, C.; Conolly, D. Demand side management of electric car charging: Benefits for consumer and grid. *Energy* **2012**, *42*, 358–363. [CrossRef]
23. Karfopoulos, E.; Hatziargyriou, N. Distributed coordination of electric vehicles for conforming to an energy schedule. *Electr. Power Syst. Res.* **2017**, *151*, 86–95. [CrossRef]
24. Kontou, E.; Yin, Y.; Ge, Y.E. Cost-effective and ecofriendly plug-in hybrid electric vehicle charging management. *Transp. Res. Rec.* **2017**, *2628*, 87–98. [CrossRef]
25. Lee, J.H.; Chakraborty, D.; Hardman, S.J.; Tal, G. Exploring electric vehicle charging patterns: Mixed usage of charging infrastructure. *Transp. Res. Part D Transp. Environ.* **2020**, *79*, 102249. [CrossRef]
26. Traut, E.J.; Cherng, T.C.; Hendrickson, C.; Michalek, J.J. US residential charging potential for electric vehicles. *Transp. Res. Part D Transp. Environ.* **2013**, *25*, 139–145. [CrossRef]
27. Boewing, F.; Schiffer, M.; Salazar, M.; Pavone, M. A Vehicle Coordination and Charge Scheduling Algorithm for Electric Autonomous Mobility-on-Demand Systems. In *Proceedings of the 2020 American Control Conference (ACC), Denver, CO, USA, 1–3 July 2020*; IEEE: Piscataway, NJ, USA, 2020; pp. 248–255. [CrossRef]
28. Ortega-Vazquez, M.; Bouffard, F.; Silva, V. Electric Vehicle Aggregator/System Operator Coordination for Charging Scheduling and Services Procurement. In Proceedings of the 2013 IEEE Power & Energy Society General Meeting, Vancouver, BC, Canada, 21–25 July 2013.
29. Camus, C.; Farias, T. The electric vehicles as a mean to reduce CO2 emissions and energy costs in isolated regions. The Sao Miguel (Azores) case study. *Energy Policy* **2012**, *43*, 153–165. [CrossRef]
30. Efthymiopoulos, I. *Islands as testbeds for Innovative Energy Solutions*; Friedrich-Ebert-Stiftung: Athens, Greece, 2015.
31. Giannakaris, P.; Ktena, A.; Manasis, C.; Assimakis, N. Hybrid power plant with storage on a non-interconnected Greek island. *B H Electr. Eng.* **2020**, *14*, 46–52.
32. European Union 2001–2012. Solar Radiation and Photovoltaic Electricity Potential Country and Regional Maps for Europe. 2012. Available online: https://ec.europa.eu/jrc/en/pvgis (accessed on 3 December 2020).
33. Karfopoulos, E. Contribution of the Management of Electric Vehicles for Their More Efficient Integration in the Electricity Grids. Ph.D. Thesis, School of Electrical and Computer Engineering (NTUA), Athens, Greece, 2017.
34. Kostopoulos, E.; Spyropoulos, G.; Kaldellis, J. Real-world study for the optimal charging of electric vehicles. *Energy Rep.* **2019**, *6*, 418–426. [CrossRef]
35. Jiang, J.; Shi, W.; Zheng, J.; Zuo, P.; Xiao, J.A.; Chen, X.; Xu, W.; Zhang, J.G. Optimized Operating Range for Large-Format LiFePO4/Graphite Batteries. *J. Electrochem. Soc.* **2013**, *161*, 336–341. [CrossRef]
36. Lu, L.; Han, X.; Li, J.; Hua, J.; Ouyang, M. A review on the key issues for lithium-ion battery management in electric vehicles. *J. Power Sources* **2013**, *226*, 272–288. [CrossRef]
37. Sheta, A.; Turabieh, H. A comparison between genetic algorithms and sequential quadratic programming in solving constrained optimization problems. *AIML J.* **2006**, *6*, 67–74.
38. Deb, K. An efficient constraint handling method for genetic algorithms. *Comput. Methods Appl. Mech. Eng.* **2000**, *186*, 311–338. [CrossRef]

Article

Modelling Stochastic Electricity Demand of Electric Vehicles Based on Traffic Surveys—The Case of Austria

Albert Hiesl *, Jasmine Ramsebner and Reinhard Haas

Energy Economics Group (EEG), Institute of Energy Systems and Electrical Drives, TU Wien, Gusshausstrasse 25-29/370-3, A-1040 Vienna, Austria; ramsebner@eeg.tuwien.ac.at (J.R.); haas@eeg.tuwien.ac.at (R.H.)
* Correspondence: hiesl@eeg.tuwien.ac.at; Tel.: +43-158801-370371

Abstract: Battery-powered electric mobility is currently the most promising technology for the decarbonisation of the transport sector, alongside hydrogen-powered vehicles, provided that the electricity used comes 100% from renewable energy sources. To estimate its electricity demand both nationwide and in individual smaller communities, a calculation based assessment on driving profiles that are as realistic as possible is required. The developed model based analysis presented in this paper for the creation of driving and thus electricity load profiles makes it possible to build different compositions of driving profiles. The focus of this paper lies in the analysis of motorised private transport, which makes it possible to assess future charging and load control potentials in a subsequent analysis. We outline the differences in demand and driving profiles for weekdays as well as for Saturdays, Sundays and holidays in general. Furthermore, the modelling considers the length distribution of the individual trips per trip purpose and different start times. The developed method allows to create individual driving and electric vehicle (EV) demand profiles as well as averaged driving profiles, which can then be scaled up and analysed for an entire country.

Keywords: motorized private transport; electric mobility; modelling electricity demand; driving patterns; battery electric vehicle; electricity demand profile

Citation: Hiesl, A.; Ramsebner, J.; Haas, R. Modelling Stochastic Electricity Demand of Electric Vehicles Based on Traffic Surveys—The Case of Austria. *Energies* **2021**, *14*, 1577. https://doi.org/10.3390/en14061577

Academic Editor: Javier Reneses

Received: 26 January 2021
Accepted: 1 March 2021
Published: 12 March 2021

Publisher's Note: MDPI stays neutral with regard to jurisdictional claims in published maps and institutional affiliations.

1. Introduction

The energy transition is widely seen as a shift in the energy sector from fossil-fuel towards renewable energy sources. Therefore, it is clear that all sectors and especially the transport sector also need to be decarbonised. Battery-driven electric mobility, along with hydrogen-powered vehicles, currently are the most promising technologies for the decarbonisation of the transport sector, as long as the electricity used is generated from 100% renewable energy sources. However, in the future, it will be necessary to rely, not only on a technological change, but also on alternative transport concepts, a different modal split, and public transport to achieve the emission targets, and avoid further congestion on the roads. Nevertheless, private transport will continue to represent a large share of the traffic volume in the future. The share of electric vehicles, which include battery electric vehicles (BEVs) and plug-in hybrid electric vehicles (PHEVs), is increasing steadily in the EU. While, in 2010 only 700 electric vehicles were newly registered, this figure increased to 550,000 vehicles in 2019. However, this still only represents a 3.5% market share of newly registered passenger cars of which about 2% were BEVs and 1% were PHEVs. The frontrunners in new registrations are Norway with 56% electric vehicles as measured by total newly registered vehicles, followed by the Netherlands with 18.53% and Iceland with 16.06%. Austria is in the middle of all EU countries with 3.43% of newly registered electric vehicles [1]. Regardless of current growth rates, a study by Eberhard and Steger-Vonmetz [2] outlines that, by 2050, the entire vehicle fleet in Austria will be electric.

To estimate the electricity demand that goes along with the substitution of fossil fuel-based private motorised transport with battery-driven electric mobility analysis on

285

both, nationwide scale and in individual smaller municipalities, appropriate modelling requires reliable data regarding the driving behaviour.

This paper shows a method for estimating the actual demand profile of individual traffic at various detail levels. To provide optimal solutions for the interaction of EVs with the electricity grid it is important to design effective charging strategies.

These need some knowledge on driving behaviour. The core objective of this paper is to model driving patterns and electricity demand profiles of (future) electric mobility based on a survey on current mobility behaviour in Austria. Typical starting times, path lengths and trip purposes are considered. It also takes into account the differences in mobility behavior between weekdays and Saturdays and Sundays. Regional distribution of electricy demand caused by electric mobility, as well as effects of e-mobility on the electricity grid are not the focus of this paper. Also, the analysis of the electric vehicle charging behavior itself is not part of this paper.

Since e-mobility has played a significant role in discussing climate-friendly transport for quite some time [3,4], there is a broad selection of literature available dealing with different approaches to model BEV load and charge profiles.

Daina et al. [5] analyse different methodologies to model load and charge profiles of electric mobility. The models are categorized according to the time scale of the electric vehicle (EV) usage patterns and according to significant methodological differences that are applied. The modelling is broken down into travel statistics models, which are based on data from conventional vehicles. Further classifications range from activity-based approaches as daily or multi-day profiles derived from differences in lifestyle and activities to Markov Chain models. A Markov model is a stochastic model used to model changing systems randomly. It assumes that future states only depend on the current state, not on the events that occurred before. Considered states of a vehicle can include driving, parking in a residential area, parking in a commercial area and parking in an industrial area. The authors conclude that there is an urgent need to develop new modelling frameworks to take into account both, long-term strategic decisions of consumers and short-term decisions on EV use, as well as the design of price and non-price incentives for behavioural change. Activity-based modelling offers an attractive starting point to achieve this goal.

The paper by Pareschi et al. [6] deals with the question whether travel surveys provide a good basis for modelling EV driving patterns. The paper shows that existing Household Travel Surveys (HTS) and other travel diaries usually provide sufficiently accurate and abundant empirical information. However, they state that there is more uncertainty regarding the future role of EVs and critical parameters in the analysis like charging losses, charging rates and powertrain design. The paper concludes that conventional HTSs are a suitable basis to generate EV insights with some critical parameters to be considered.

A further Markov chain tool to estimate EV charging behaviour is presented by Sokorai et al. [7]. This tool enables the modelling of the stochastic nature of a charging station's day-to-day usage if precise datasets of the driving behaviour are available. In addition, a case study to verify the algorithm is conducted. This study concludes that if adequate datasets on travel patterns with appropriate PEV statistics and real probability values are available as a model input, the algorithm can provide valuable stochastic information about electricity consumption at a given location.

Other papers based on the methodology of the Markov chain are presented by Schlote et al. [8] and Fischer et al. [9]. This paper presents a stochastic bottom-up model to evaluate the impact of EV's on load profiles at different parking places and their potential for load management systems. This paper also considers socio-economic, technical and spatial criteria that influence the charging of electric vehicles. The model is used to analyse the effects of uncontrolled charging of EVs on the load profile of households. They find that uncontrolled charging causes a peak load increase up to a factor of 8.5 depending on the charging infrastructure.

The work carried out by Hu et al. [10] investigates the challenges that EVs add to the electricity grid at different penetration levels, taking into account the uncertainties caused by the stochastic charging and discharging behaviour. To cope with these uncertainties, a Monte Carlo-based simulation is used to generate EV charging and discharging profiles. The results in this paper show that the specific electricity grid studied can accommodate a high penetration of EVs by limiting charging to off-peak times. Lojowska et al. [11] also propose a Monte Carlo-based methodology in their paper for estimating the demand for electric vehicles based on a stochastic approach to modelling transport patterns. The focus of this paper is on the scenario of mainly domestic charging availability. The Monte Carlo simulation was performed for 1000 EVs and then scaled to a region with one million EVs. The authors conclude that total demand can increase if there are no incentives to spread the charging demand of EVs.

The work of Paevere et al. [12] presents a methodology to obtain spatial and temporal projections of the retail electricity demand of EVs, their load shift potential and the impact on household peak loads. The paper focuses on the territory of the State of Victoria, Australia and discusses differences in the potential for EV diffusion in different regions. In addition, regional statistics are used for the length of trips and arrival times and on this basis, EV charging and discharging is calculated. They conclude that the form and extent of EV demand profiles are subject to geographical variations. Areas where commuting is dominant generally have higher peak load demand, due to relatively longer trip distances and less diversity in home arrival times.

An agent-based approach using the established modelling tool NetLogo is applied in the paper of Chaudhari et al. [13]. The aim is to closely mimic the human aggregate behaviour and its influence on the electricity demand due to EV charging. The model implemented in this paper simulates and defines each EV by its charging characteristics, mobility behaviour and vehicle type. This creates an environment in which decision-making and various circumstances are taken into account to predict the charging behaviour of individuals as well as groups. The simulations in this paper were performed over a period of 24 h and for several days. The individual and total power demand of electric vehicles were determined for different scenarios. In addition, this model should allow for both commercial and private EVs with their different driving purposes. The authors conclude that the results highlight the practical applicability of the ABM-based approach to calculate the charging demand of EVs. Another agent-based approach to estimate EV's charging demand is outlined by Lee et al. [14]. In this paper, an agent-based EV model is evaluated against real data observed during the "My Electric Avenue" project. The main finding is that, within the constraints of the available trial data, the agent model is able to replicate dominant charging pattern features.

Forecasting the electricity demand of EV's is performed in the work by Moon et al. [15], Lopez and Fernandez [16], as well as in Cama-Pinto et al. [17] and Zhang et al. [18]. The latter was conducted for autonomous vehicles. Forecasting electricity demand using big data technologies is discussed by Arias and Bae [19]. Furthermore, charging EV's in the smart city [20] and an empirically validated methodology to simulate electricity demand for electric vehicle charging are outlined in [21].

A user equilibrium model is discussed in [22]. In this paper, an approach is proposed that extends the User Equilibrium (UE) principle in order to determine, besides the flow over the network links, the service requests from the drivers to the various service stations.

In addition, there are several related publications that deal with similar topics. On-road charging of electric vehicles [23] using Contactless Power Train (CTP) where the total power demand for all the passing-by vehicles using the system is calculated and the possibility of powering the EVs directly from renewable energy sources is discussed. In [24], the optimal schedule of the charging behaviours of EVs with distinct energy consumption preferences in Smart Communities (SC) is outlined. In this paper, the authors propose a contract-based energy blockchain for secure EV charging in SC. An agent-based approach is proposed in [25] and discussed for EV deployment policies in Luxembourg. Day ahead bidding strategies for electric vehicle aggregators in uncertain electricity markets are outlined in [26], while in [27], the effects of electricity prices on the integration of high shares of photovoltaics are analysed.

The paper by Ramsebner et al. [28] directly builds on the methodology of this paper. The disaggregated demand profiles, as well as driving and parking times, were used as input for a linear optimisation model. This optimisation model aims to charge electric vehicles in a cost-optimal way, taking into account the SoC and considering different charging strategies.

As can be seen from the literature described, there are many different ways to model the demand and charging behaviour of electric mobility. Depending on the application and the systemic framework on micro or macro level, the models are able to answer a broad range of research questions. As already concluded in [5], modelling the driving behaviour of electric mobility on the basis of traffic surveys is agreed to be quite useful and sufficiently accurate. Therefore, the methodology presented in this paper is appropriate for our objectives and is also applied in the Urcharge project.

The main contribution of this paper lies in the transparent and straightforward modelling of load profiles, the easy calibration for different mobility behaviour, based on traffic surveys and in the high time resolution of the load profiles as a quarter of an hour.

The proposed methodology makes it possible to generate demand patterns for individual vehicles, different driving purposes but also for an aggregated average demand pattern including all driving purposes, which are scalable according to the share of EV's in a region or in a whole country, in this paper, demonstrated by the example of Austria. Furthermore, different regional parameters can be applied to analyse demand patterns in different seasons or to distinguish between urban and rural areas. In addition, various distributions of routes and travel times within the driving purposes, as well as the mix of driving purposes on weekdays and weekends are taken into account. By focusing on individual vehicles and driving purposes, a holistic bottom-up analysis can be carried out on an aggregated level. The methodology presented in this paper strictly distinguishes between driving demand profile and charging profiles. The latter will not be discussed further in this paper. The strict separation between the generation of demand profiles and a subsequent generation of charging profiles allows the analysis of the effects of different charging management approaches. For example, uncontrolled charging, cost-minimized charging via a linear optimization model and corresponding pricing or the evaluation of load shifting potentials on different levels of aggregation.

The remainder of this paper is organized as follows. In Section 2, we introduce the methods of modelling the stochastic distribution of starting times as well as path lengths according to the trip purpose. Furthermore, we present the assumptions on the electricity consumption in winter and summer seasons as well as the composition of trips on different types of days. Section 3 shows model results and differences between driving purposes and between individual demand profiles and average demand profiles. Section 4 discusses and concludes the paper.

2. Materials and Methods

Basically, the model is developed to analyse individual demand profiles on the one hand and aggregated or average load profiles on the other hand to estimate the electricity demand by EVs.

The fundamental procedure for creating an average load profile that can be scaled up to a municipal and countrywide level is described in Figure 1. First of all, the start times of the outward and return journey are calculated on a daily basis for different driving purposes, see Table 1, whereby, a minimum time between these two times is specified, in order to guarantee a certain parking time. These start times are then assigned to different routes associated with route times and assumed grid-to-wheel consumption based on the calculated distance per route. The daily starting times and the resulting consumption are compiled into an annual driving profile. The calculation of the annual profiles is carried out in quarter-hourly resolution. This results in a vector of the size:

35,040 (24 h/day $\times$ 4 quarter hours/hour $\times$ 365 days) $\times$ number of calculated vehicles $\times$ number driving purposes (eight for the case of Austria) (see Figure 2).

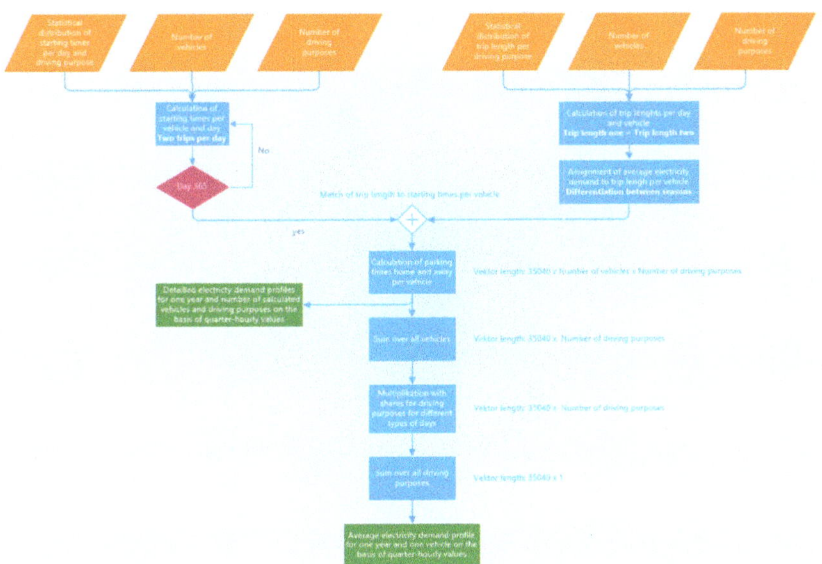

Figure 1. Fundamental methodology for creating detailed, as well as an average electricity demand profile for EV's.

Table 1. Driving purpose and type of the day used for modelling e-mobility demand pattern.

Weekday	Saturday	Sunday
	To the workplace/commuter	
	Drop-off and pick-up routes	
	Leisure	
	Business	
	Shopping	
	Visit	
	School/Education	
	Errand	

This vector could then serve as input for a subsequent linear optimization model, which aims at charging the vehicles at minimal cost. The electricity demand at each time step is combined to an average total load profile based on the shares of the individual driving purposes, resulting in a vector of 35,040 time steps and x number of EVs. Finally, an average load profile is calculated across the number of vehicles, which can then easily be scaled with the EV penetration rate assumed. If the consumption of individual driving purposes is to be analysed, this can be done before averaging and by restricting the consumption vector.

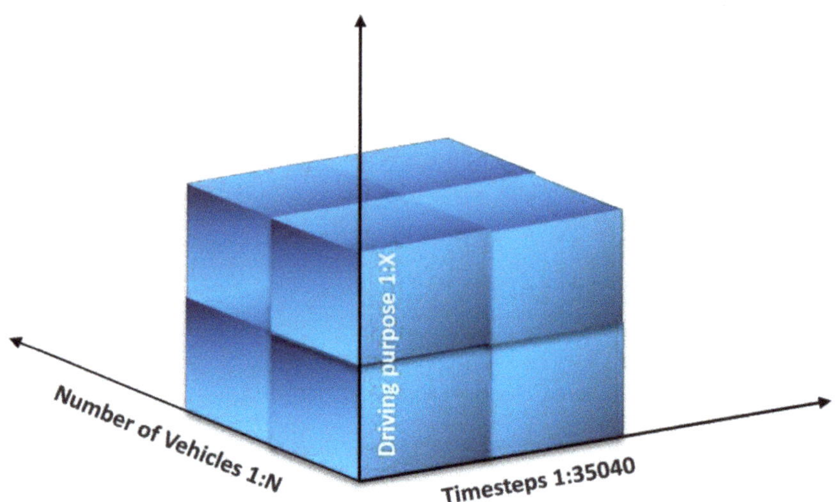

Figure 2. The resulting demand matrix considering a time resolution of a quarter of an hour, X driving purposes and N vehicles.

2.1. Differentiation between Driving Purpose and Type of the Day

In order to generate load profiles for Austria, we use the latest traffic survey from 2014 [29]. This traffic survey examined and recorded the entire mobility behaviour for Austria divided into urban and non-urban regions. In addition, a distinction was made between working days, Saturdays and Sundays and seasonal differences. The study shows that in Austria about 104 billion kilometres are driven per year, with motorised private transport accounting for about 76 billion kilometres. Public transport represents approximately 21 billion kilometres, with 11 kilometres covered by railway. Pedestrian and bicycle traffic sum up to about 4 billion kilometres.

For the calculation of demand profiles for Austria, we assume that the essential parameters such as starting times, distance travelled and the parking times for e-mobility do not change significantly and use the following data and estimations for motorised individual transport to model the demand profiles.

According to the study [29], the following driving purposes are used for modelling, see Table 1:

As outlined in Table 1, eight driving purposes have been identified for Austria in general. The composition of the total trips from these eight driving purposes is different for weekdays, Saturdays, Sundays and public holidays, as shown in Section 2.4.

2.2. Stochastic Distribution of Starting Times

The different driving purposes also show different distributions of the start times, which have to be considered in the modelling. Figure 3 shows the start time distribution of the respective trip purposes for the outward and return journey, which are modelled as Gauss curves in a next step.

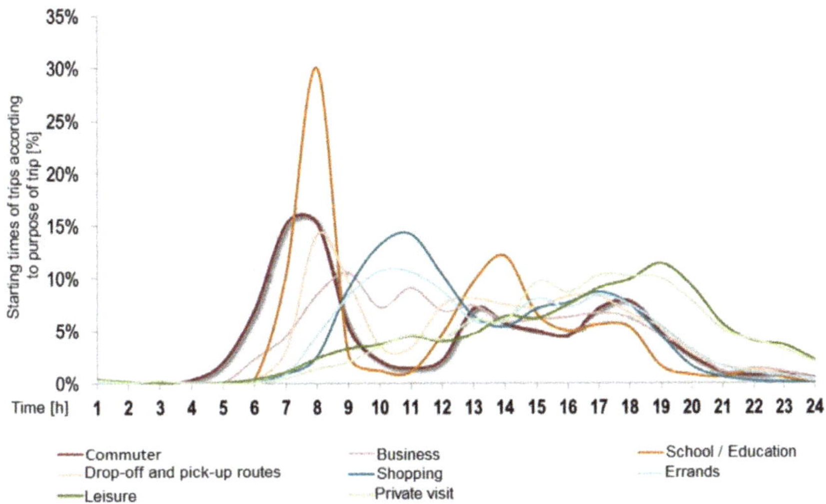

Figure 3. Distribution of starting times for outward and return journey and for different driving purposes. Reproduced from [29], Bundesministerium für Verkehr, Innovation und Technologie: 2016.

We use the Gauss distribution for a simple reproduction of the start time distribution for both the outward and return paths. For an even more precise reproduction of the curves, a superposition of several Gauss curves can be used.

The Gauss distribution is defined in Equation (1),

$$f(x|\mu,\sigma^2) = \frac{1}{\sqrt{2\pi\sigma^2}} e^{-\frac{(x-\mu)^2}{2\sigma^2}}, \tag{1}$$

using the variables:

μ Average value
σ^2 Variance
x Random value

As illustrated in Figure 4 for commuters, the actual distribution of starting times depends strongly on the number of vehicles calculated. The more vehicles are calculated, the more the starting times actually match the applied distribution. For a realistic representation of the distribution in an overall load profile, a relatively large number of vehicles must therefore be calculated per trip purpose. With a time resolution of a quarter of an hour and a calculation of 1000 vehicles (representing 1000 battery storage systems with different State of Charge (SoC)) per driving purpose, this results in—35,040 (time steps per year) × 1000 (vehicles) × 8 (driving purposes)—over 280 million calculation points.

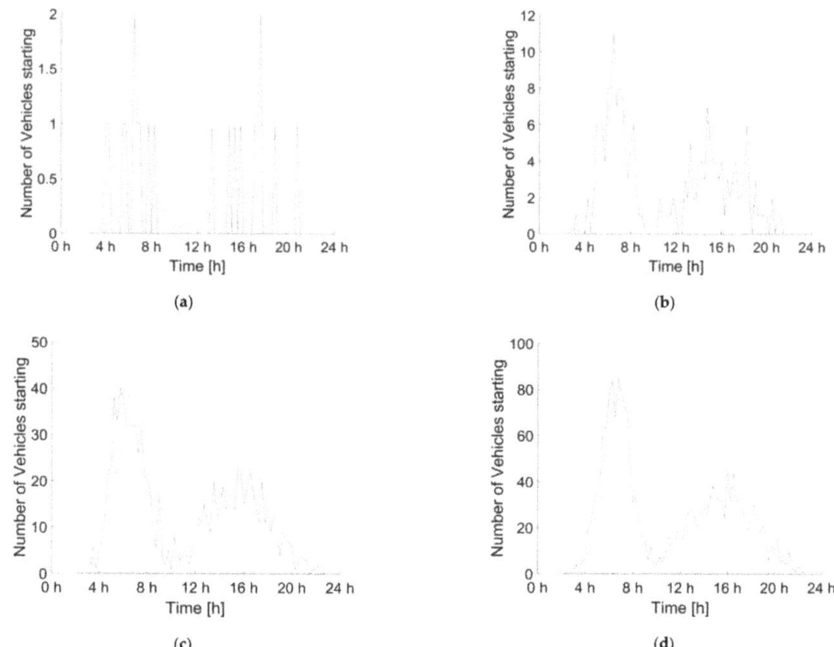

Figure 4. Modelling of starting times using Gaussian distributions for different numbers of vehicles. Comparison of perfect Gaussian curve (green line) and reproduction with different number of vehicles (blue) from top left to bottom right (**a–d**): 10 vehicles, 100 vehicles, 500 Vehicles, 1000 Vehicles.

Just as for commuters, for all other driving purposes, mean value and variance or standard deviation are defined to determine Gauss distributions for the start times of the outward and return journey. The derived parameters are listed in Table 2. The mean start time is given as a time of the day, in the model this time as well as the variance is converted into quarter-hourly values. For example, 6 a.m. is converted to the 24th quarter-hour value of the day (6 h × 4) and a standard deviation of 2.4 h is rounded to 10 quarter-hours.

Table 2. Expected value and variance of the norm distributions of the start times of the individual routes per driving purpose.

Driving Purpose	Route One	Route Two
To the workplace/Commuter	$\mu = 7:30$ am $\sigma = 1$ h	$\mu = 4:00$ pm $\sigma = 2.88$ h
Drop-off and pick-up routes	$\mu = 8:00$ am $\sigma = 0.75$ h	$\mu = 3:30$ pm $\sigma = 2.88$ h
Leisure	$\mu = 10:00$ am $\sigma = 2.5$ h	$\mu = 7:00$ pm $\sigma = 1.5$ h
Business	$\mu = 10:00$ am $\sigma = 1.5$ h	$\mu = 3:00$ pm $\sigma = 3.13$ h
Shopping	$\mu = 10:30$ am $\sigma = 1.5$ h	$\mu = 5:00$ pm $\sigma = 2.5$ h
Visit	$\mu = 11:00$ am $\sigma = 2$ h	$\mu = 5:30$ pm $\sigma = 4.25$ h
School/Education	$\mu = 8:00$ am $\sigma = 1.5$ h	$\mu = 3:30$ pm $\sigma = 4.25$ h
Errands	$\mu = 10:30$ am $\sigma = 1.5$ h	$\mu = 4:00$ pm $\sigma = 2.5$ h

2.3. Distribution of Driving Distances

In order to take into account the fact that not all trips are of the same length, we also introduce a statistical distribution of the distances travelled depending on the purpose of the trip, based on the original data.

The analysis of the empirical data results in different distributions of the distances travelled depending on the purpose of the trip and the day of the week. For the sake of simplicity, all trips per purpose of the trip were summarised and analysed according to their distribution, regardless of the type of the day. To reproduce the distribution as precisely as possible, the distances were divided into 200 classes of 1 km each, see Figure 5 as an example for the distribution of trip lengths for commuter traffic.

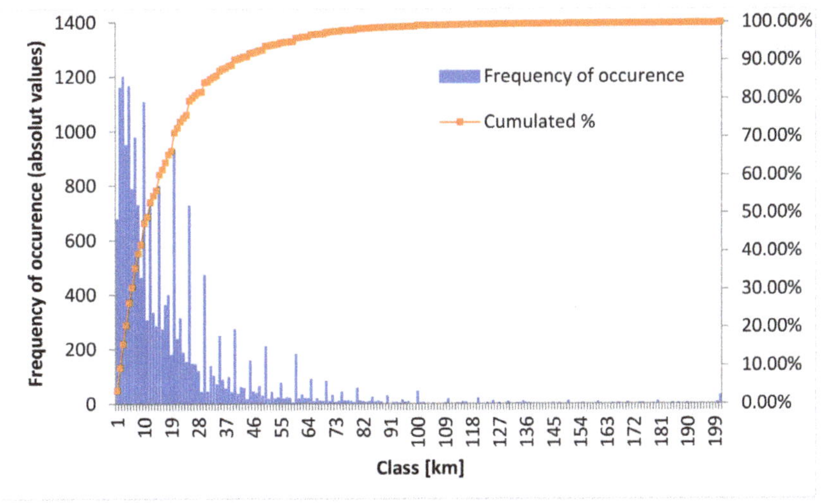

Figure 5. Probability function and distribution function for the purpose "to the workplace/commuter" Reproduced from [29], Bundesministerium für Verkehr, Innovation und Technologie: 2016.

The actual length of each route per driving purpose was then based on the class average. For the class between zero and one kilometre for example the actual length is 0.5 km. The maximum driving length was limited to 200 km because the analysis of the data showed that only few trips (e.g., 0.16% in the case of commuters) exceeded 200 km.

Since the discrete distribution of the trip lengths was determined, the multinomial distribution was chosen as a function to model the statistical allocation of the trip lengths to the individual daily trips.

The multinomial distribution is a generalisation of the binomial distribution. If mutually exclusive results are possible for a random process m and the random process is repeated n times independently, the probabilities can be calculated using the multinomial distribution. Each possible event $x1, \ldots, xk$ has a probability of occurrence of $p1, \ldots, pk$. The probability of the occurrence of a certain distance is known. The n-repeats are recalculated for each day according to the number of cars. In this way, each trip gets assigned to a random length according to the distribution.

The multinomial probability distribution is defined as follows; see Equation (2),

$$f(n_1, \ldots, n_k | N; p_1, \ldots, p_k) = \frac{n!}{n_1! \ldots n_k!} \prod_{i=1}^{k} p_i^{n_i} , \sum_{i=1}^{k} n_i = n , \sum_{i=1}^{k} p_i = 1 \qquad (2)$$

n Number of vehicles
k Number of path length

p Probability of occurrence of path length

According to the study, the average annual distance travelled by car is 13,300 km/a. With an average distance of 16 km for individual motorised traffic per trip, this means that an average of 2.28 trips are made per day.

Since we assume only 2 trips per day in the modelling, the return path is assigned the same distance as the outward path. This results in an average length of 16 km across all driving purposes. In order to be able to represent the average driving performance of 13,300 km per year, the average annual route length and thus all route lengths need to be scaled up by a factor of 1.14.

2.4. Share of Driving Purposes on Overall Trips

Although the driving purposes and start times remain the same for working days, Saturdays, Sundays and holidays, the respective shares vary considerably. For example, the share of commuter traffic on the overall trips on a working day is much higher than on a Saturday or even a Sunday or public holiday. This shift in the shares of driving purposes results in different demand profiles for weekdays, Saturdays, Sundays and holidays.

In order to calculate the respective shares of the driving purposes, one must first consider the total routes for all transport modes and then break them down to motorised individual traffic, which, however, is not outlined directly in the statistics.

Based on the number of trips per trip purpose from Table 3, the average trip length of 16 km for motorised private transport is used to calculate the kilometres travelled per day. However, since the trips are only partially covered by private motorised transport, the total distance covered per day must be aligned with the respective shares, see Table 4, in order to be able to derive the kilometres driven per day, see Equation (3). As the traffic-survey does not show whether the distribution differs between Saturdays, Sundays and holidays, the same distribution as on working days is used for the calculation of kilometres travelled.

$$D = N \times s \times l \tag{3}$$

D Distance travelled by vehicle and day [km]
N Total number of trips completed per day [1]
s Share of motorised individual traffic in total number of trips travelled [1]
l Average distance travelled per trip by motorised individual traffic [km]

Table 3. Number of total routes and share of driving purposes in total route.

Trip Purpose	Working Day	Working Day	Saturday	Saturday	Sunday	Sunday
	Share of total trips [%]	Number of trips	Share of total trips [%]	Number of trips	Share of total trips	Number of trips
Total trips/day of which	100	22,090,000	100	19,851,000	100	14,885,000
To the work-place/Commuter	26	5,743,400	7	1,389,570	5	744,250
Drop-off and pick-up routes	7	1,546,300	5	992,550	5	744,250
Leisure	15	3,313,500	29	5,756,790	47	6,995,950
Business	5	1,104,500	2	397,020	2	297,700
Shopping	16	3,534,400	29	5,756,790	3	446,550
Visit	8	1,767,200	15	2,977,650	26	3,870,100
School/Education	8	1,767,200	0.5	99,255	0.5	74,425
Errands	13	2,871,700	12	2,382,120	11	1,637,350

Table 4. Kilometres per day and respective share of motorized individual traffic on total trips.

Trip Purpose	Working Day	Working Day	Saturday	Saturday	Sunday	Sunday
	Share of motorised individual transport in total trips [%]	Distance [km]	Share of motorised individual transport in total trips [%]	Distance [km]	Share of motorised individual transport in total trips [%]	Distance [km]
Kilometres/day individual transport		162,299,648		137,003,662		95,537,884
of which						
To the work-place/Commuter	60	55,136,640	60	13,339,872	60	7,144,800
Drop-off and pick-up routes	67	16,576,336	67	10,640,136	67	7,978,360
Leisure	30	15,904,800	30	27,632,592	30	33,580,560
Business	70	12,370,400	70	4,446,624	70	3,334,240
Shopping	45	25,447,680	45	41,448,888	45	3,215,160
Visit	45	12,723,840	45	21,439,080	45	27,864,720
School/Education	9	2,544,768	9	142,927	9	107,172
Errands	47	21,595,184	47	17,913,542	47	12,312,872

From the distribution of kilometres per driving purpose for individual traffic, the shares of the respective driving purpose are deducted, see Table 5. In the end, these shares serve to generate an average demand profile for one vehicle.

Table 5. Share of driving purposes in the total routes of motorised individual traffic in %.

Trip Purpose	Working Day	Saturday	Sunday
Share of individual transport per day	100.0	100.0	100.0
of which			
To the workplace/Commuter	34.0	9.7	7.5
Drop-off and pick-up routes	10.2	7.8	8.4
Leisure	9.8	20.2	35.1
Business	7.6	3.2	3.5
Shopping	15.7	30.3	3.4
Visit	7.8	15.6	29.2
School/Education	1.6	0.1	0.1
Errands	13.3	13.1	12.9

As can be seen from Table 5, commuter traffic dominates on working days, followed by shopping and private errands. On Saturdays, shopping dominates with over a third of the trips made, followed by leisure trips and private visits. On Sundays, on the other hand, leisure trips are at the top of the list almost the same as visiting trips. See also the graphical visualization in Figure 6.

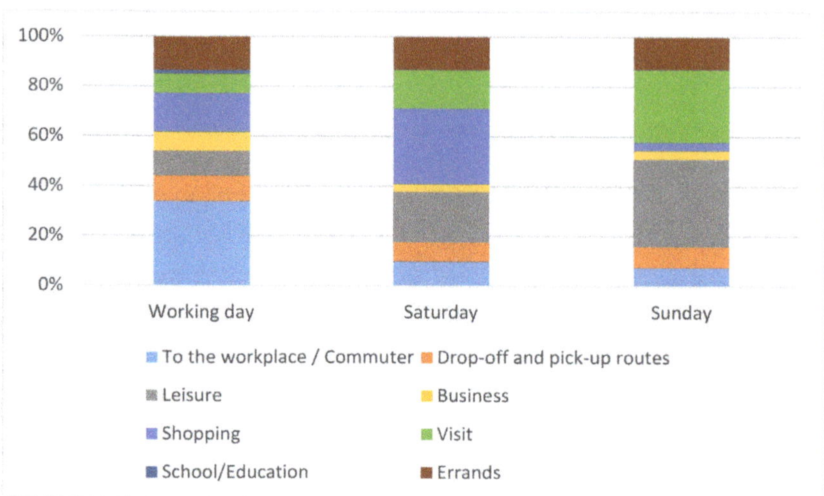

Figure 6. Share of driving purposes in the total routes of motorised individual traffic, graphical visualisation.

2.5. Electric Vehicle Consumption

The grid-to-wheel consumption of the vehicles in this paper is assumed to be the same for all driving purposes and vehicles. Even if there certainly are differences in the consumption of individual vehicles, as well as different routes, it seems justifiable for the present modelling to assume these as average consumption, as this would also be averaged in the large number of vehicles calculated. However, it is not the aim of this study to analyse the consumption of different vehicle models. Nevertheless, we consider the fact that consumption tends to be higher in winter than in summer as also outlined in the study of the American Automobile Association (2019) [30] and Iora and Tribioli (2019) [31]. These studies do not focus specifically on Austria, but on the differences in the electricity demand of electric vehicles in summer and winter. Since, to the best knowledge of the authors, there is no representative study for Austria, an average pattern was assumed, which is believed to represent the climatic conditions and circumstances in the western federal provinces with their alpine character, as well as in the eastern regions with a rather mild climate and the different types of vehicles.

For this reason, consumption throughout the year is interpolated between a maximum of 17 kWh/100 km in winter and a minimum of 15 kWh/100 km in summer, see Figure 7.

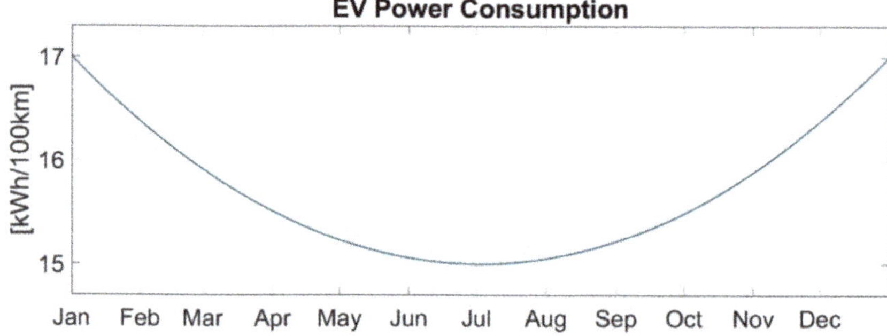

Figure 7. Development of EV grid-to-wheel consumption throughout the year. NOTE: For reasons of visualisation, the y-axis was limited between 14.5 kWh/km and 17.5 kWh/km.

3. Results

As already discussed in the previous chapters, this methodology provides an excellent starting point to answer many different research questions, especially regarding optimal load management. On the one hand, it is possible to analyse single trip purposes, single vehicles, and groups of vehicles regarding their quarter of an hour, daily or yearly electricity demand, driving distances, starting times, or even standing times at home or away from home. On the other hand, it is possible to analyse average load profiles for specific trip purposes, mixed trip purposes, and an average overall load profile that can easily be scaled up for a whole region or country, in this case Austria. In the following chapter, the modelling results are presented by means of individual and average load profiles. For this purpose, 1000 individual vehicle demand profiles were created for each trip purpose.

3.1. Individual Demand Profiles versus Average Demand Profiles

From the methodology chapters, it is quite clear that the individual driving profiles change in length per trip and the starting times, depending on the driving purpose. How this affects load profile modelling is analysed in more detail in the following sections.

Annual aggregated and averaged model results based on the input data already discussed show that the distances travelled and the annual consumption per trip purpose vary enormously (see Table 6). The shortest distance of a single-vehicle is observed for drop-off and pick-up routes with 5476 km/a, whereas the longest distance is observed for business routes (23,674 km/a). The modelled average distance per trip of 16 km is precisely in line with the study results [29]. The large number of vehicles calculated (1000) results in a relatively good statistical distribution of both the start times and the trip lengths. The annual consumption of individual vehicles varies between 690 kWh/a and 3711 kWh/a, and the variation of annual consumption depends not only on the route length but also on the different grid-to-wheel consumption in summer and winter. The average consumption per trip is 3 kWh across all driving purposes.

Table 6. Model results for eight driving purposes considering the underlying 1000 randomly generated profiles.

Trip Purpose	Min. Distance of Vehicles [km/a]	Max. Distance of Vehicles [km/a]	Average Distance of Vehicles [km/a]	Average Distance per Trip [km]	Min. Demand of Vehicles [kWh/a]	Max. Demand of Vehicles [kWh/a]	Average Demand of Vehicles [kWh/a]	Average demand per trip [kWh]
to the work-place/commuter	10,430	16,080	12,973	17.8	1627	2509	2032	2.79
drop-off and pick-up routes	5476	9426	7246	10	854	1477	1135	1.56
Leisure	8422	14,750	11,338	15.6	1320	2315	1776	2.44
Business	13,600	23,674	18,228	25	2135	3711	2855	3.92
Shopping	4442	7234	5581	7.7	690	1134	874	1.2
Visit	9526	17,332	12,806	17.6	1497	2744	2006	2.76
School/Education	13,508	21,420	16,993	23.3	2113	3362	2662	3.66
Errands	6006	11,006	8152	11.2	942	1714	1277	1.75

Depending on the analysis and the kilometres driven, these results must be adjusted or scaled to the respective regional conditions. In our case, as we will see later, we will adapt these results specifically to Austria, also considering the type of the day as an input parameter. Since we have simulated only 2 instead of 2.28 routes per day for simplicity, there is a deviation in the average annual kilometres driven (11,665 km calculated vs. 13,300 km [29]), and thus, also in the annual average consumption. If individual vehicles are to be analysed, it is possible, on the one hand, to select the vehicle (from the 1000 calculated) that comes closest in terms of distance, or on the other hand, to scale it with the number of kilometres required. If driving purposes are to be analysed more precisely, the average demand profile must be scaled with the actual distance driven or the actual consumption.

In the following, the results at vehicle level are compared to the mean value of the total calculated vehicles per trip purpose. For reasons of clarity, only the "shopping routes" and "business routes" are compared. In Figures 8 and 9, the green and blue lines represent a vehicle's consumption per day and year respectively. The orange line represents the average value per 1000 vehicles per trip purpose. The y-axes on the left side represent the demand of the individual vehicles while the y-axes on the right side of the graph represent the average vehicle's demand. The two individual vehicles were selected out of 1000 in such a way that the vehicle with the overall lowest annual consumption (green line) and the vehicle with the overall highest annual consumption (blue line) are evaluated in order to get a feeling for the differences in the respective peaks and in the frequency of the peaks that occur.

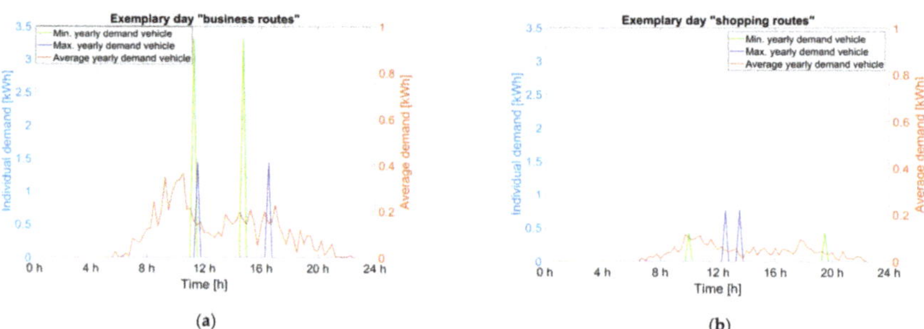

Figure 8. Electricity demand of two individual vehicles (green, blue) compared to the average demand of the respective trip purpose (orange) for an exemplary day. (**a**) business routes; (**b**) shopping routes.

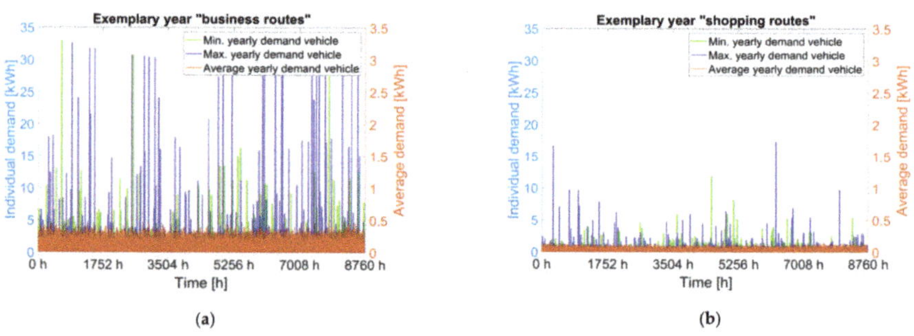

Figure 9. Electricity demand of two individual vehicles (green, blue on left axes) compared to the average demand of the respective trip purpose (orange on right axes) for an exemplary year. (**a**) business routes; (**b**) shopping routes.

The electricity demand during the trip is only assigned to one quarter of an hour in the model. However, the previously calculated distance and the average speed of 60 km/h result in corresponding trip times of one minute per km. Together with the electricity demand, the length of the journeys and the times when the vehicle is on the road, the parking times at home and away from home are calculated and assigned to the individual vehicles.

Figures 8 and 9 show how individual demand profiles and averaged demand profiles behave within a day and over a year. The daily perspective shows that due to the distribution of the start times and the distribution of the path lengths, the averaged load profile for a trip purpose has a significantly different characteristic than the individual vehicle. It becomes clear that the peak load of a vehicle can be significantly higher than the peak load of an average load profile. The average load profile tends to follow a continuous load curve, whereas an individual vehicle's profile is limited to a few points in time during the day. The peaks of individual vehicles at individual points in time are no longer as significant in an average load profile. In addition, it can be seen that a vehicle with a low annual consumption also has higher peak loads on some days than a vehicle with a generally higher annual load profile due to the distribution of the path lengths, as compared with Figure 8a. Figure 8 also shows that the heights of the peaks, as well as their occurrences, can vary greatly. In the comparison between business trips and shopping trips, it becomes clear that peaks are significantly higher in the business profile and also occur significantly more frequently. This is also reflected in the average profile, which is significantly higher. It is important to note that the current analysis does not take into account differences in weekdays and weekends. These differences will be identified in the next section.

The advantage of disaggregated modelling of demand profiles lies precisely in the fact that it is possible to extract any information about a vehicle at any time, e.g., current consumption, driving, parked at home and parked away. No information about demand peaks is lost. For subsequent precise analysis of the charging behaviour, a disaggregated demand profile is necessary to be able to calculate the individual charging power at vehicle level depending on the SoC of the battery and afterwards aggregate the total charging power of a vehicle pool. This is difficult to do with an aggregated calculation because the detail of the information is already missing and an average load profile only offers an aggregated representation. However, it is not always necessary to provide all information at a disaggregated level. It is not possible to include several 1000 individual EV demand profiles per country in large energy system models. In such cases, it definitely makes sense to use an aggregated profile.

3.2. Yearly Average Demand Profile for Austria Considering Different Types of Days

In Section 3.1, the different load profiles were presented based on the trip purpose and on individual and average load profiles.

For the analysis of the electricity demand of many vehicles in urban sections, cities or rural areas and on country level, it makes sense to create an average annual load profile, which consists of all trip purposes. Since the shares of driving purposes differ for weekdays, Saturdays and Sundays (see Section 2.4), the total demand profile also differs on these types of days.

Figure 10 shows the consumption of an average day for the different trip purposes on the left side individually and not specifically for weekday, Saturday and Sunday. In the morning, school and commuter trip drop-off and pick-up routes dominate, which of course, correlates with the selected start distribution and the path length distribution modelled according to Figure 3. We now combine the individual distributions of the individual trip purposes' load profiles, taking into account the shares from Table 5 and obtain different averaged load profiles for weekdays, Saturdays, and Sundays, see Figure 9b. Due to commuter dominance with about 34% in the morning, there is an apparent increase on weekdays compared to Saturdays or Sundays. As mentioned previously, the parking times at and away from home are calculated at vehicle level based on the start times and route lengths. This information is required to know when and for how long the vehicle is available for subsequent analyses regarding the charging management of individual vehicle groups.

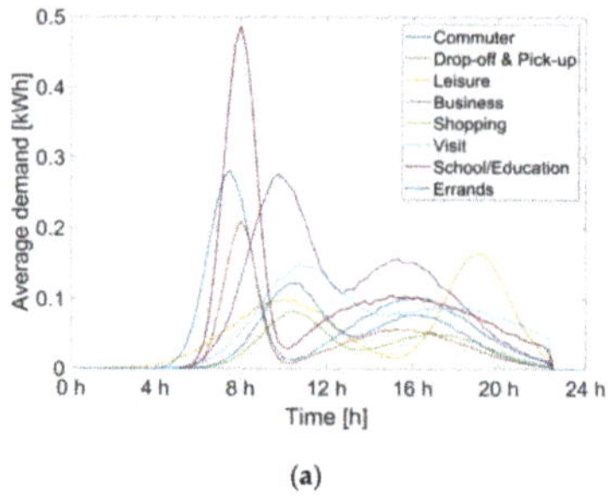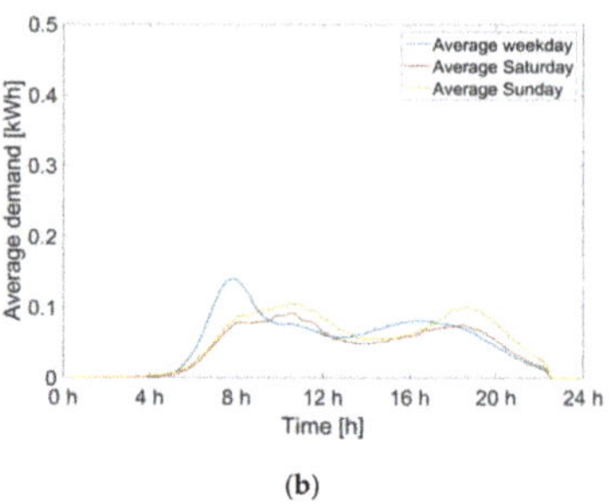

(a)　　　　　　　　　　　　　　　　　　(b)

Figure 10. Average day for an average vehicle for all driving purposes; and (**a**) average profile over all driving purposes for weekday, Saturday and Sunday considering the respective share on overall daily trips (**b**).

For the visualisation in Figure 11, however, an aggregated form of the parking times was chosen to clarify how the entirety of the vehicles behaves concerning typical parking times.

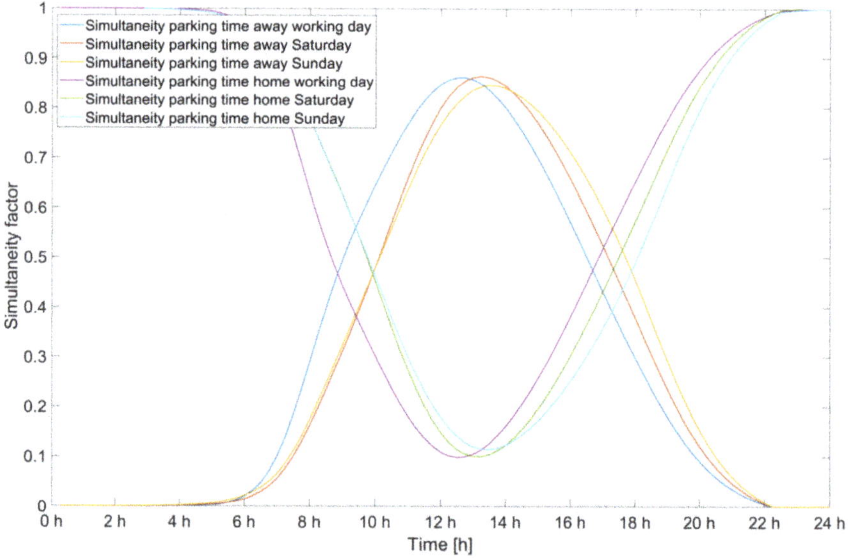

Figure 11. Simultaneity of parking times of the calculated vehicles for weekdays, Saturdays and Sundays.

As can be seen in Figure 11, there are slight differences in the availability of vehicles while parked at or away from home. On weekdays, there is a peak around lunchtime in the simultaneity of vehicles parked away from home. The gradient is also slightly higher than on Saturdays, Sundays and public holidays due to commuting. Especially in the evening and at night, almost all vehicles are parked at home. As can be seen in Figure 10, by summing up the curves at and away from home, the vehicles are parked mostly either at home or away from home, e.g., at the workplace. Only very few times a day, the vehicles are actually in motion.

The individual days' composition into a continuous annual load profile leads to an annual mileage of approx. 10,570 km/a, which is significantly lower than the mileage of an average vehicle in Austria. Furthermore, it is also less than the average distance of the combined consideration of the different path purposes in Table 6. This can be explained by the fact that the length distribution and the different shares of the path purposes in the total profile compose different shares. Therefore, the length also deviates from the previously calculated length where only an average of all path purposes was calculated. To create an Austria-wide aggregated load profile and scale the load profile for different EV penetration rates, this average load profile must first be calibrated to a travelled distance of 13,300 km. Therefore the load profile is calibrated with the factor 1.26 (13,300 km/10,570 km), resulting in the load profile in Figure 12.

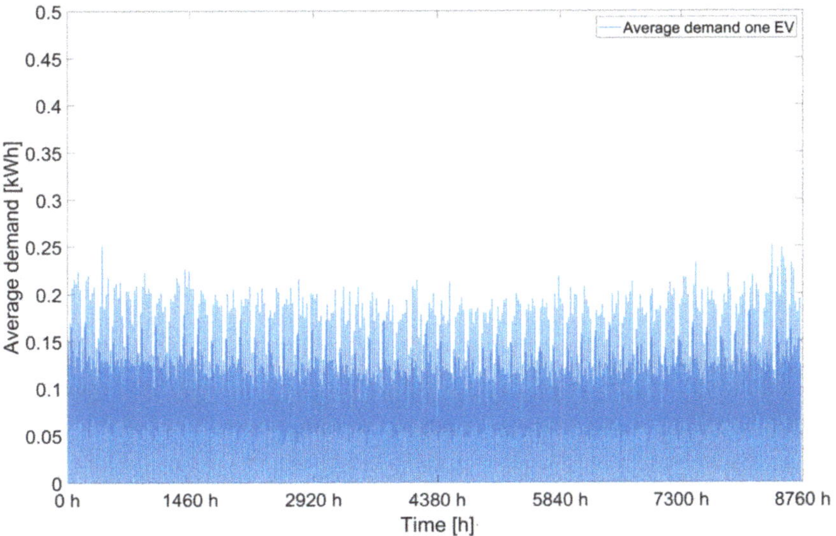

Figure 12. Average load profile of an EV considering different driving purposes, weekdays and calibrated to a distance of 13,300 km/a.

Based on this load profile, different scales can now be applied to answer different (research) questions for different regions. From the authors' point of view, the aggregated model results serve mainly as input for system modelling where no detailed individual or disaggregated analysis of demand or driving behaviour is required.

As already outlined, for detailed modelling of individual vehicles or vehicle networks' charging behaviour, the disaggregated output of the model should serve as input for further analyses. Such an analysis was implemented, for example, by Ramsebner et al. [28] where the disaggregated demand profiles, as well as driving and parking times, were used as input for a linear optimisation model. This optimisation model aims to charge electric vehicles in a cost-optimal way, taking into account the SoC and considering different charging strategies. To calculate this cost-optimal charging, a disaggregated, high-resolution demand profile is needed as input, which was created by the methodology presented in this paper.

4. Discussion and Conclusions

The methodology, presented in this paper, aimed at modelling and calibrating high-resolution EV demand profiles based on traffic surveys and at making these profiles versatile in application. These demand profiles should then be available as input parameters for energy system models.

The results presented show, that EV demand profiles can be created and evaluated in high temporal resolution per vehicle and trip purpose or on aggregated level as an average load profile. The advantage of the high resolution, disaggregated use of the load profiles is primarily that the information about individual load peaks is maintained and that start, travel and arrival, and parking times can be mapped with quarter-hour accuracy. In addition, the bottom-up approach has the advantage that a wide variety of aggregated load profiles can be created, which is de facto not possible in the other direction.

In addition, the calibration and creation of these load profiles can be conducted with other country or regional data. As we have seen, the decisive parameters are, on the one hand, the starting times, the distribution of the trip lengths, and, on the other hand, the shares of the different trip purposes among the trips per day. The more precise these data are available, the better the reality can be reproduced in the load profiles.

In order to reproduce this methodology for other countries or regions, the following data is required:

- A breakdown on the driving purpose as precise as possible
- Data on the starting times of the individual routes as accurate as possible
- Data on the distribution of trip lengths as accurate as possible
- Shares of trip purposes on different days

Alternatively, this data could be generated beforehand, or assumptions on average trip lengths must be made. If the data on starting times cannot be modelled using Gaussian distributions, for example, a different, e.g., discrete, distribution can be used to model the trip lengths.

The division of the path lengths into path classes and their distribution must also be adapted to the respective data situation. Often, only rugged ranges of the trip length distribution are given in evaluations. In this case, the path class width, as well as the path class centre must be adjusted accordingly and the trip length calculated accepting higher inaccuracy.

In any case, it is important to analyse the data in advance in order to choose the right type of distribution and to assess possible limitations of accuracy due to assumptions and data inaccuracies. The computing time and computing capacity must be taken into account. This means that it is not always possible and reasonable to calculate 1000 vehicles per trip purpose to resolve 35,040 time steps. The high time resolution as well as the amount of vehicles results in more than 280 million data points that have to be calculated. This may be necessary and feasible for the calculation of an average load profile. On a disaggregated level, for example in a subsequent linear optimisation model, the calculation of so many data points, however, reaches the limit of the computational capacity. When simultaneously calculating the cost-optimal charging of 8000 storage units with different SoC's and a resolution of 35,040 time steps, taking into account additional constraints, the calculation time can be exorbitantly high or no solution to the problem can be found. In particular, two things have to be considered before selecting the appropriate method. What purpose do the created demand profiles serve and what data is available. Furthermore, the quality of the input data plays an important role. Depending on the type and accuracy of the available data, the methodology presented here can directly be used. The probability that not all data are available on this level of detail for every analysis is high, and thus, in the case of any data gaps, corresponding assumptions or simplifications must be made, for example concerning the start times or the path length distribution.

Finally, it can be concluded that in order to develop effective charging strategies to reduce the pressure on the electricity grid and to effectively use renewable generation, electrical load profiles for EVs are needed. The methodology presented in this paper provides an excellent opportunity to establish such demand profiles on different levels of detail. Based on this paper, we find a need for future research in the consideration of plug-in hybrid vehicles, different charging strategies, as well as Vehicle to Building (V2B) and Vehicle to Grid (V2G) applications, to achieve an optimal integration of electro mobility into the energy system and use the available flexibility options efficiently.

External effects that would also change user behaviour were not taken into account in this paper. However, it is evident that the Covid pandemic and the associated lockdowns, for example, have changed mobility behaviour. This is perhaps less true for the distribution of starting times than for the average distances and lengths of trips. Due to lockdown restrictions, many people switched to home office, and commuters drive to work less frequently. Private visits and shopping trips are also limited. If one also assumes that the modal split will change in the future and public transport will be increasingly expanded and incentivised, the average distances travelled by car will tend to shorten. This will lead to a general reduction in peak loads for EV electricity demand, fewer kilometres driven, and a subsequent reduction in annual electricity demand.

Author Contributions: Conceptualization, A.H.; methodology, A.H.; software, A.H.; validation, A.H., J.R. and R.H.; formal analysis, A.H.; investigation, A.H.; writing—original draft preparation, A.H.; writing—review and editing, J.R., R.H. and A.H.; visualization, A.H.; supervision, R.H.; project administration, J.R.; funding acquisition, J.R. and R.H. All authors have read and agreed to the published version of the manuscript.

Funding: This research was carried out within the cooperative R&D project URCHARGE powered by the Austrian Climate and Energy Fund within the program "Zero Emission Mobility" with a focus on IS.

Institutional Review Board Statement: Not applicable.

Informed Consent Statement: Not applicable.

Conflicts of Interest: The authors declare no conflict of interest. The funders had no role in the design of the study; in the collection, analyses, or interpretation of data; in the writing of the manuscript, or in the decision to publish the results.

References

1. Newly Registered Electric Cars by Country—European Environment Agency. Available online: https://www.eea.europa.eu/data-and-maps/daviz/new-electric-vehicles-by-country-3#tab-dashboard-01 (accessed on 16 February 2021).
2. Eberhard, T.; Steger-Vonmetz, C. *Laden im Wohnbau-Was ist bis 2030 zu tun*; AustriaTech: Vienna, Austria, 2019.
3. Chan, C.C. An overview of electric vehicle technology. *Proc. IEEE* **1993**, *81*, 1202–1213. [CrossRef]
4. Eaves, S.; Eaves, J. A cost comparison of fuel-cell and battery electric vehicles. *J. Power Sources* **2004**, *130*, 208–212. [CrossRef]
5. Daina, N.; Sivakumar, A.; Polak, J.W. Modelling electric vehicles use: A survey on the methods. *Renew. Sustain. Energy Rev.* **2017**, *68*, 447–460. [CrossRef]
6. Pareschi, G.; Küng, L.; Georges, G.; Boulouchos, K. Are travel surveys a good basis for EV models? Validation of simulated charging profiles against empirical data. *Appl. Energy* **2020**, *275*, 115318. [CrossRef]
7. Sokorai, P.; Fleischhacker, A.; Lettner, G.; Auer, H. Stochastic Modeling of the Charging Behavior of Electromobility. *World Electr. Veh. J.* **2018**, *9*, 44. [CrossRef]
8. Schlote, A.; Crisostomi, E.; Kirkland, S.; Shorten, R. Traffic modelling framework for electric vehicles. *Int. J. Control.* **2012**, *85*, 880–897. [CrossRef]
9. Fischer, D.; Harbrecht, A.; Surmann, A.; McKenna, R. Electric vehicles' impacts on residential electric local profiles–A stochastic modelling approach considering socio-economic, behavioural and spatial factors. *Appl. Energy* **2019**, *233–234*, 644–658. [CrossRef]
10. Hu, Q.; Li, H.; Bu, S. The Prediction of Electric Vehicles Load Profiles Considering Stochastic Charging and Discharging Behavior and Their Impact Assessment on a Real UK Distribution Network. *Energy Procedia* **2019**, *158*, 6458–6465. [CrossRef]
11. Lojowska, A.; Kurowicka, D.; Papaefthymiou, G.; Van Der Sluis, L. Stochastic Modeling of Power Demand Due to EVs Using Copula. *IEEE Trans. Power Syst.* **2012**, *27*, 1960–1968. [CrossRef]
12. Paevere, P.; Higgins, A.; Ren, Z.; Horn, M.; Grozev, G.; McNamara, C. Spatio-temporal modelling of electric vehicle charging demand and impacts on peak household electrical load. *Sustain. Sci.* **2014**, *9*, 61–76. [CrossRef]
13. Chaudhari, K.S.; Kandasamy, N.K.; Krishnan, A.; Ukil, A.; Gooi, H.B. Agent-Based Aggregated Behavior Modeling for Electric Vehicle Charging Load. *IEEE Trans. Ind. Inform.* **2019**, *15*, 856–868. [CrossRef]
14. Lee, R.; Yazbeck, S.; Brown, S. Validation and application of agent-based electric vehicle charging model. *Energy Rep.* **2020**, *6*, 53–62. [CrossRef]
15. Moon, H.; Park, S.Y.; Jeong, C.; Lee, J. Forecasting electricity demand of electric vehicles by analyzing consumers' charging patterns. *Transp. Res. Part D Transp. Environ.* **2018**, *62*, 64–79. [CrossRef]
16. López, F.C.; Fernández, R. Álvarez Predictive model for energy consumption of battery electric vehicle with consideration of self-uncertainty route factors. *J. Clean. Prod.* **2020**, *276*, 124188. [CrossRef]
17. Cama-Pinto, D.; Martínez-Lao, J.A.; Solano-Escorcia, A.F.; Cama-Pinto, A. Forecasted datasets of electric vehicle consumption on the electricity grid of Spain. *Data Brief* **2020**, *31*, 105823. [CrossRef]
18. Zhang, C.; Yang, F.; Ke, X.; Liu, Z.; Yuan, C. Predictive modeling of energy consumption and greenhouse gas emissions from autonomous electric vehicle operations. *Appl. Energy* **2019**, *254*, 113597. [CrossRef]

19. Arias, M.B.; Bae, S. Electric vehicle charging demand forecasting model based on big data technologies. *Appl. Energy* **2016**, *183*, 327–339. [CrossRef]
20. Shuai, W.; Maille, P.; Pelov, A. Charging Electric Vehicles in the Smart City: A Survey of Economy-Driven Approaches. *IEEE Trans. Intell. Transp. Syst.* **2016**, *17*, 2089–2106. [CrossRef]
21. Harris, C.B.; Webber, M.E. An empirically-validated methodology to simulate electricity demand for electric vehicle charging. *Appl. Energy* **2014**, *126*, 172–181. [CrossRef]
22. Ferro, G.; Minciardi, R.; Robba, M. A user equilibrium model for electric vehicles: Joint traffic and energy demand assignment. *Energy* **2020**, *198*, 117299. [CrossRef]
23. Stamati, T.-E.; Bauer, P. On-road charging of electric vehicles. In Proceedings of the 2013 IEEE Transportation Electrification Conference and Expo (ITEC), Detroit, MI, USA, 16–19 June 2013; pp. 1–8.
24. Su, Z.; Wang, Y.; Xu, Q.; Fei, M.; Tian, Y.-C.; Zhang, N. A Secure Charging Scheme for Electric Vehicles with Smart Communities in Energy Blockchain. *IEEE Internet Things J.* **2019**, *6*, 4601–4613. [CrossRef]
25. Querini, F.; Benetto, E. Agent-based modelling for assessing hybrid and electric cars deployment policies in Luxembourg and Lorraine. *Transp. Res. Part A Policy Pract.* **2014**, *70*, 149–161. [CrossRef]
26. Zheng, Y.; Yu, H.; Shao, Z.; Jian, L. Day-ahead bidding strategy for electric vehicle aggregator enabling multiple agent modes in uncertain electricity markets. *Appl. Energy* **2020**, *280*, 115977. [CrossRef]
27. Schwarz, M.; Auzépy, Q.; Knoeri, C. Can electricity pricing leverage electric vehicles and battery storage to integrate high shares of solar photovoltaics? *Appl. Energy* **2020**, *277*, 115548. [CrossRef]
28. Ramsebner, J.; Hiesl, A.; Haas, R. Efficient Load Management for BEV Charging Infrastructure in Multi-Apartment Buildings. *Energies* **2020**, *13*, 5927. [CrossRef]
29. Tomschy, R.; Herry, M.; Sammer, G.; Klementschitz, R.; Riegler, S.; Follmer, R.; Gruschwitz, D.; Josef, F.; Gensasz, S.; Kirnbauer, R.; et al. *Österreich Unterwegs 2013/2014-Ergebnisbericht zur Österreichweiten Mobilitätserhebung, Österreich Unterwegs 2013/2014*; Im Auftrag von: Bundesministerium für Verkehr, Innovation und Technologie, Autobahnen- und Schnellstraßen-Finanzierungs-Aktiengesellschaft, Österreichische Bundesbahnen Infrastruktur AG, Amt der Burgenländischen Landesregierung, Amt der Niederösterreichischen Landesregierung, Amt der Steiermärkischen Landesregierung und Amt der Tiroler Landesregierung; Bundesministerium für Verkehr, Innovation und Technologie: Vienna, Austria, 2016.
30. American Automobile Association. AAA Electric Vehicle Range Testing. February 2019. Available online: https://www.aaa.com/AAA/common/AAR/files/AAA-Electric-Vehicle-Range-Testing-Report.pdf (accessed on 15 January 2021).
31. Iora, P.; Tribioli, L. Effect of Ambient Temperature on Electric Vehicles' Energy Consumption and Range: Model Definition and Sensitivity Analysis Based on Nissan Leaf Data. *World Electr. Veh. J.* **2019**, *10*, 2. [CrossRef]

Article

Electric Shared Mobility Services during the Pandemic: Modeling Aspects of Transportation

Katarzyna Turoń [1],* , Andrzej Kubik [1],* and Feng Chen [2]

1   Department of Road Transport, Faculty of Transport and Aviation Engineering,
    Silesian University of Technology, 8 Krasińskiego Street, 40-019 Katowice, Poland
2   Sino-US Global Logistics Institute, Shanghai Jiao Tong University, Shanghai 200240, China; fchen@sjtu.edu.cn
*   Correspondence: katarzyna.turon@polsl.pl (K.T.); andrzej.kubik@polsl.pl (A.K.)

Abstract: The global spread of the COVID-19 virus has led to difficulties in many branches of the economy, including significant effects on the urban transport industry. Thus, countries around the world have introduced different mobility policies during the pandemic. Due to government restrictions and the changed behaviors of transport users, companies providing modern urban mobility solutions were forced to introduce new business practices to their services. These practices are also apparent in the context of the electric shared mobility industry. Although many aspects and problems of electric shared mobility have been addressed in scientific research, pandemic scenarios have not been taken into account. Noticing this research gap, we aimed to update a previously developed model of factors that influence the operation of electric shared mobility by incorporating aspects related to the COVID-19 pandemic and its impact on this industry. This article aims to identify the main factors influencing the electric shared mobility industry during the COVID-19 and post-lockdown periods, together with their operation areas and the involved stakeholders. The research was carried out on the basis of expert interviews, social network analysis (SNA), and the use of the R environment. The article also presents sustainable transport management recommendations for cities and transport service operators, which can be implemented after a lockdown caused by an epidemic. The results in this paper can be used to support transport modeling and the creation of new policies, business models, and sustainable development recommendations. The contents will also be helpful to researchers worldwide in preparing literature reviews for articles related to sustainable management in the COVID-19 pandemic reality.

Keywords: COVID-19 pandemic; electric mobility; transportation modeling

Citation: Turoń, K.; Kubik, A.; Chen, F. Electric Shared Mobility Services during the Pandemic: Modeling Aspects of Transportation. *Energies* **2021**, *14*, 2622. https://doi.org/10.3390/en14092622

Academic Editor: Robert H. Beach

Received: 2 February 2021
Accepted: 30 April 2021
Published: 3 May 2021

Publisher's Note: MDPI stays neutral with regard to jurisdictional claims in published maps and institutional affiliations.

1. Introduction

The spread of the SARS-CoV-2 virus has significantly influenced users of transport services. With the threat posed by the pandemic, the number of trips has decreased significantly, which has reduced the demand for transport. Data published by Apple showed that fewer individuals requested travel options and directions to a given place between the end of March and May 2020 compared with January, with the following average decreases [1]:

- 54% in Italy,
- 51% in the United Kingdom,
- 8% in Germany,
- 14% in the United States of America.

The implementation of restrictions on public transport has significantly reduced its use around the world. For example, in Poland, the number of available buses, trams, and trains decreased by 50% [2]. In the United States of America and Germany, it decreased by 34% and 42%, respectively [2]. These restrictions and fear of contracting the virus from public transport have meant that society is much more inclined to choose a car for everyday travel. Simultaneously, however, problems arose for service providers of vehicles for shared transport, i.e., all forms of so-called shared mobility solutions (rental offices that provide short-term rentals of different kinds of vehicles) and ride-sharing solutions (such as Uber or Grab).

On the one hand, many residents who did not have private vehicles wanted to use shared transport services; on the other hand, they were afraid to do so because of sanitary and safety issues. There were also issues related to the approaches taken by local and government authorities with respect to shared mobility vehicles. New mobility service providers were forced to adapt their business practices to the new market situation. Furthermore, these circumstances also had a strong impact on the shared electro-mobility market. Some forecasts even indicated that the pandemic would end the development of electric mobility and destroy practices in the field of sustainable urban mobility created over the years [3,4]. As a result, there was considerable chaos worldwide in the electric and traditional shared mobility industries.

The changing needs of society have led to the emergence of new or improved services offered in electric shared mobility systems. All of these changes were the result of improved business models. We identified a gap in the current research related to changes in services offered during a pandemic. From a scientific point of view, we identified a need to update the model describing factors that influence the development of shared mobility services because many aspects, problems, and challenges defined before the pandemic have now changed. This article aims to identify the main factors influencing the electric shared mobility industry during the COVID-19 and post-lockdown periods, as well as their operation areas and involved stakeholders. The research was carried out on the basis of expert interviews, social network analysis (SNA), and the use of the R environment. The article also presents sustainable transport management recommendations for cities and transport service operators, which can be implemented after a lockdown caused by an epidemic. The results in this paper can be used to support transport modeling and the creation of new policies, business models, and sustainable development recommendations. The contents will also be helpful to researchers worldwide in preparing their literature reviews for articles about sustainable management in the COVID-19 pandemic reality.

2. Theoretical Background

Electric shared mobility services can be analyzed from many different perspectives. This type of analysis is often used to model new services or optimize services currently offered in urban transport systems [5–11]. These factors describe elements that are taken into account at different stages of the electric shared mobility service life cycle [12–14]. These stages include the planning, implementation, operation, and maintenance of the service. This type of approach is used for a variety of electric shared mobility services available on the market, such as e-bike-sharing, e-scooter-sharing, and e-car-sharing services. A literature analysis shows that the main factors related to electric shared mobility services are fees, parking, fleets, costs, types of systems, and electric vehicle power supply. However, there are more detailed subfactors depending on the kind of system [9–30]. Following a literature review, a summary of previous research is presented in Table 1.

Table 1. Factors for modeling and optimizing electric shared mobility systems.

System Type	Aspects	Literature
Electric bike-sharing system	Basic costs of electric bike-sharing systems (operational, infrastructure)	[9,10]
	Electric bike parameters (cost of purchase, safety, range)	[9,11]
	Location of the station (bike, docking)	[11,13]
	Weather conditions	[5]

Table 1. Cont.

System Type	Aspects	Literature
Electric scooter and moped-sharing system	Basic costs of electric scooter- and moped-sharing systems (operational, infrastructure)	[15,16]
	Weather conditions	[5]
	Electric bike parameters (cost of purchase, safety, range, battery capacity)	[18,19]
	Charging and docking station (location, policy, battery charging speed)	[19]
	Weather conditions	[5]
Electric car-sharing systems	Basic costs of electric car-sharing systems (operational, infrastructure)	[6,12]
	Electric bike parameters (cost of purchase, fleet size, safety, range, battery capacity)	[14,20,21]
	Charging and parking locations (location, policy, battery charging speed)	[22–30]
	Models of systems (rental rules, vehicle relocation time)	[29,30]

From the perspective of sustainable transport development, the best services rely on electric vehicles or encourage drivers to transition from owning a vehicle to using a shared car. Until the coronavirus pandemic outbreak, services relied heavily on the demand for mobility and the provision of transport alternatives within urban areas. Most improvements to sustainable transport services focused on aspects such as replenishing fleets and introducing an appropriate environmental management policy (recycling and storage of batteries from electric vehicles, recycling of unused vehicles) or on social issues in terms of eliminating transport barriers and social discrimination resulting from a lack of available vehicles or lack of funds for their purchase or maintenance [31,32]. Such changes are usually derived from appropriate transport modeling for locating or relocating vehicles, optimizing their routes, or optimizing fleets [33–36], as these factors have emerged as the main aspects of planning transport systems based on shared mobility services. However, the outbreak of the epidemic led to significant and unexpected changes in the city logistics market. For example, in [37], the authors reviewed articles on the COVID-19 epidemic and found that issues related to rethinking and redesigning business models for sustainability are still under-researched. Other authors [38–41] emphasized that an appropriate multicriteria analysis of implemented policies (including those related to corporate social responsibility [32]) for sustainable development is key when designing new services for the public and ensuring appropriate resilience. In turn, transport analyses in the context of COVID-19 have mainly focused on issues related to the social approach and the changing transport behaviors of passengers [40,41], unprecedented changes in travel habits [42], and technical issues related to passenger flows [42]. However, there has been a lack of focus on holistic approaches to studying aspects associated with modeling electric shared mobility after COVID-19; for this reason, comprehensive analyses were performed in this study.

3. Materials and Methods

The steps of this study included conducting qualitative research (expert interviews), performing social network analysis (to produce a network of factors analyzed in the context of the COVID-19 pandemic), analyzing results, and providing proposals and recommendations.

The first step was to refer to a list of aspects that affect the development of electric shared mobility services, as previously proposed by the authors in [43]. The list of factors influencing shared mobility services resulted from an international workshop on electric shared mobility, and it was used to develop a mathematical model for e-shared mobility services. The list of factors that influence electric shared mobility systems is presented in Table 2.

Table 2. Aspects (problems) that affect electric shared mobility.

Aspects Abb.	Aspect Characteristics
P1	Vehicles in the system operation zone (designation of the system operation area, vehicle fleet size)
P2	Application for system users: availability and possible functions offered to the user
P3	New user data verification process (through traditional customer service and through the application)
P4	Rules for renting vehicles: types of payments for trips made (i.e., subscription, payment after each trip, non-cash payments via the application)
P5	Laws and regulations on the movement of individual types of vehicles available under the shared mobility system
P6	Parking spaces: location, number of dedicated spaces for the electric shared mobility system, possible fees
P7	Technical condition of the fleet (indicators describing the vehicle's reliability, the process of its wear)
P8	Security and monitoring of vehicles (the type of antitheft security measures used, vehicle monitoring via GPS, vehicle equipment with recorders that enable video and sound recording)
P9	User safety: the safety level of vehicles offered in the system (i.e., according to European New Car Assessment Program (NCAP)), maintenance of the vehicle's appropriate technical condition by the service operator, eco-driving bonuses
P10	Infrastructure for electric vehicles: number of chargers, number of possible simultaneous vehicles charging via the charger, number of parking spaces under the charger
P11	User data security: user data security, payment data security, mobile application use security
P12	Weather conditions: variability in weather conditions, the possible occurrence of weather anomalies
P13	Cost of use: available subscriptions, account activation, prepayment, available discounts, possible fines (e.g., for traffic offenses), payment method
P14	Maintaining the system's profitability: relocation of vehicles, number of employees, required education, employee qualifications
P15	Operators: cooperation with a local government entity
P16	Privileges for operators: possible reductions in local taxes and the elimination of some or all local government fees
P17	Integration of electric shared mobility systems with currently available city systems, creation of a unified system (a mobile application for all available services)
P18	Technical safety of vehicles: monitoring drivers' driving and the control of trips made
P19	Basic data on system operation: the area of system operation, number of available vehicle types, number of possible operation zones for the system

Source: [43].

The list presented in Table 2 was developed as a holistic set of factors for modeling transport systems. These factors were taken into account prior to the outbreak of the COVID-19 pandemic. The list of factors was used as an element of the questionnaire used during expert interviews. Forty-six representatives of co-mobility service providers were invited to participate in the research project "New mobility and the COVID-19 pandemic". The project included 20 meetings with representative stakeholders of the shared mobility industry. The meetings were organized and moderated by scientists from the Faculty of Transport and Aviation Engineering of the Silesian University of Technology. Meetings were held in December 2020 via the Internet. Of the 46 invited guests, 20 respondents accepted the invitation. The 20 shared mobility operators represented five e-bike suppliers, three e-scooter suppliers, four e-moped suppliers, and eight e-car-sharing providers. The operators who participated in the survey offer their services in the largest Polish urban centers, i.e., in Warsaw, Katowice, Kraków, and Gdańsk.

This research was carried out using deliberate sampling. The use of a nonrandom selection of respondents for the study was based on the desire to learn the opinions of enterprises representing the shared mobility industry.

The qualitative component of the study was conducted through individual in-depth interviews (IDIs). Interviews were conducted in face-to-face meetings with scientists acting as moderators and interviewers, and meetings were taped. Following the research technique, an interview scenario was developed to gain in-depth knowledge of the impact of COVID-19 on various aspects of shared mobility services. The purpose of conducting in-depth interviews was to eliminate the influence of group members and obtain information that was unaffected by interactions with other people, especially because they performed the same industry activity.

Moreover, the use of individual interviews was motivated by the significant possibility that respondents had formed personal views on the COVID-19 pandemic. Furthermore, individual interviews allowed for the relatively free expression of controversial issues, especially for such a personal issue as the impact of the pandemic on the company's suitability and profitability. The study aimed to elicit answers regarding the specific operation areas of shared mobility systems during the pandemic. The goal was not to extract creative solutions or concepts from participants but to obtain answers to update the mathematical model.

The research sample size was confirmed to be statistically representative. The minimum number of respondents R_{min} necessary to perform expert-based research was determined using Equation (1) [44].

$$R_{min} = 0.5 * \left(\frac{3}{s} + 5 \right) = min.\ 18\ respondents \tag{1}$$

where s denotes statistical compliance and has a value of 10%.

The interviews were structured according to a specific scenario with the goal of answering questions about the impact of the COVID-19 pandemic on the electric shared mobility industry. The qualitative research in this study used the method of field reconnaissance, which allowed the main aspects of the process to be selected, and the quantitative part was based on the assignment of relevant aspects to the areas of system and stakeholder operations. During the research, the LearningApps computer application was used, which is a program for designing games. The application of supporting and projection techniques using visual materials through games allowed for a thorough interpretation of the respondent's statements. Furthermore, the techniques enabled respondents to answer questions in a more accessible and pleasant way and prevented their fatigue.

The steps of the in-depth interview process and the interview scenario are presented below. Steps of in-depth interviews:

1. Definition of the study group (operators of electric shared mobility services);
2. Planning of the research setting (online meetings using the widely available Zoom platform);
3. Development of an in-depth interview scenario;
4. Recruitment of participants by sending e-mails, including reminders, about the meeting;
5. Performance of interviews;
6. Transcription, transfer of notes, and information digitization;
7. Analysis of the obtained data.

The in-depth interview scenario was divided into three parts: introduction, main questions, and summary. A detailed interview scenario is included in Appendix A.

On the basis of the respondents' answers, a list of factors influenced by the COVID-19 pandemic was compiled. Interestingly, the indicated factors were the same for all types of electric shared mobility services. A summary of responses is presented in Table 3.

Table 3. Electric shared mobility aspects during the COVID-19 pandemic.

Aspects	COVID-19 Influence
P1. Location of vehicles	✓
P2. Mobile apps	-
P3. Registration process	✓
P4. Rental process	✓
P5. Legal requirements	✓
P6. Location of parking spaces	✓
P7. Fleet condition	✓
P8. Security of vehicles	-
P9. Safety of users	✓
P10. Infrastructure	-
P11. User security	-
P12. Weather	-
P13. Prices	✓
P14. Vehicle replacements	✓
P15. Privileges for operators	✓
P16. Surcharges for operators	✓
P17. Urban traffic accessibility	-
P18. Serviceability safety	✓
P19. Rental area	✓

The next step was to identify and describe new services that have been implemented in the represented companies in response to the COVID-19 pandemic. The identified practices were assigned to relevant factors (designated P1–P19). The responses are summarized in Table 4.

Table 4. The list of business practices introduced during the COVID-19 pandemic by electric shared mobility operators.

System Type	Practice Description	Aspects
Electric bike-sharing systems	Offered free memberships to essential workers	13, 15
	Replaced docking stations near hospitals	14, 15, 19
	Implemented antibacterial handlebars and provided additional disinfection of bicycles	7, 9
Electric scooter and moped sharing	Offered long-term rentals, for example, daily	13
	Offered new locations where the vehicle can be returned	14, 19
	Implemented antibacterial grips and handlebars in vehicles, provided additional disinfection of vehicles	7, 9
	Introduced additional disinfection in the form of vehicle cleaning, ozonation, and foil coating on the steering wheel or seats	7, 9
Electric car-sharing systems	Increased the number of cars	1
	Launched services in new cities	19
	Offered long-term rentals, for example, weekly or monthly, for up to 3 months	13
	Offered the choice of kilometer or minute-kilometer systems	13
	Strengthened cooperation with business-to-business partners	13

After collecting all of the respondents' answers, the social network analysis (SNA) method, which describes an individual's behavior at the micro and macro levels and the interactions between them, was used [45–47]. This method enables the comprehensive analysis of interactions by assessing the relationship between different entities. This method was invented and used by Moreno as early as the 1930s [45–47]. SNA is used in many research disciplines (e.g., social sciences, computer science, technology forecasting, production) [48–52]. SNA characterizes network structures in terms of nodes with connections and indicates the direction of each relationship by arrows between nodes [47]. Relationships between nodes in SNA may show linkages and indicate the influence of individual elements or a combination thereof [48–52]. The features of network structure elements are described by the power of the interaction between elements and the frequency and nature of the relationship [47]. The main determinant is the place occupied by an element in the network structure, as it reflects the status of individual units [52–55]. The strength of the SNA method is its ability to visualize complex dependencies in the phenomenon under consideration. SNA can be used to determine the influence of particular groups of elements on processes taking place in the analyzed network [56]. Detailed stages of the analysis are presented in the next chapter of the article.

According to the principles of SNA, we identified the main groups of stakeholders who represent the electric shared mobility market. Three groups of stakeholders were taken into account [57–59]:

- Electric shared mobility users, "U": a group that includes all individual users of shared mobility services, both active participants of the journey and passive participants registered in the service provider's data systems;
- Decision-makers, "G": a group that includes local, regional, central, and international authorities who have an impact on decisions made regarding the development or recession of shared mobility services;
- Electric shared mobility service providers, "O": a group that includes all providers of electric shared mobility services, including operators of e-bike-sharing, e-scooter-sharing, e-moped-sharing, and e-car-sharing services.

The three defined stakeholder groups have the greatest impact on electric shared mobility [57–59]. Identifying the interactions between the selected stakeholder groups can improve the current situation of the electric shared mobility market. Furthermore, they may also have a substantial impact on the development of electro-mobility and sustainable urban mobility plans or the development of regulations to create new mobility services [57–59].

Following the SNA method, the next step was to define key areas that are directly related to the electric shared mobility market, from the service development stage to their operations on the market. According to the interview responses of experts, the following four areas were defined [43]:

- Area 1: Development of electric shared mobility services;
- Area 2: Market introduction of electric shared mobility services;
- Area 3: Operation of the electric shared mobility services market;
- Area 4: Administration and management of electric shared mobility services.

The indicated areas are related to the life cycle of electric shared mobility services on the market, from the concept of its planning to its operation with appropriate management and administration or marketing of the services offered. These areas are always considered when developing business plans or performing risk analyses for the operation of shared mobility services.

On the basis of the list of problems that occur in electric shared mobility presented in [43], the respondent's task in the expert interview was to define the relationship between problems and the stakeholders and areas under consideration. After determining the factors influenced by the COVID-19 pandemic (Tables 2 and 3), they were summarized, as presented in Table 5.

Table 5. Electric shared mobility influencing factors (problems) with corresponding stakeholders and areas.

Stakeholder Node-Aspect Number	Area Number
U-P1	A3
U-P3	A3
U-P4	A3
O-P5	A4
U-P5	A4
G-P5	A4
O-P6	A4
U-P6	A4
G-P6	A4
O-P7	A1
U-P7	A1
O-P7	A2
U-P7	A2
O-P7	A3
U-P7	A3
U-P9	A4
O-P13	A1
U-P13	A1
G-P13	A1
O-P13	A3
U-P13	A3
G-P13	A3
O-P14	A3
U-P14	A3
G-P14	A3
O-P15	A4
O-P16	A4
G-P16	A4
O-P18	A3
U-P16	A3
U-P19	A1
U-P19	A2

4. Results

The results were analyzed using the R software and the igraph library. The generated network of connections was made up of 22 nodes with 56 connections. The topography of the created network is shown in Figure 1. The objects labeled "U", "O", and "G" represent categories of stakeholders, and those marked "P" represent problems. The arrows that connect the individual nodes indicate a relationship between a pair of stakeholders and aspects. The direction of the arrow characterizes the relationship between a pair of nodes. The aspects that affect a greater number of links are situated in the center of the network under consideration. Conversely, nodes with fewer links are near the edge of the network.

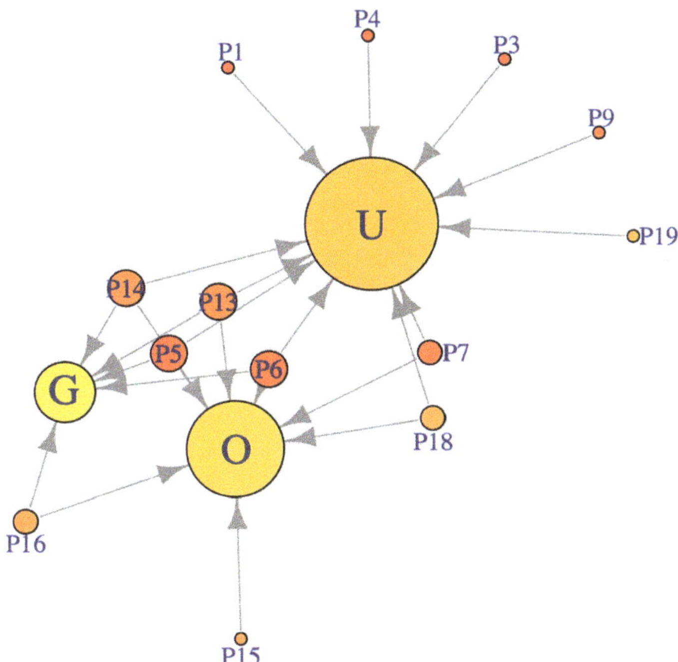

Figure 1. The structure of stakeholder-associated aspects, where P1–P19 are analyzed aspects (see Table 3); "O" denotes electric shared mobility operators, "G" denotes decision-makers, and "U" denotes e-shared mobility users.

The network overview allows for identifying relationships between stakeholders and problems in the network under study and selecting the main problems in the network topography. Problems that have many connections (Figure 1) reflect their multifactorial nature and the multiple stakeholders that are affected by these relationships. Changing one aspect changes the rest of the network structure, as well as stakeholder relationships. Table 6 summarizes the number of output connections for each aspect.

Table 6. The number of output connections with related aspects.

Aspects	The Number of Output Connections
"P1"	1
"P3"	1
"P4"	1
"P5"	3
"P6"	3
"P7"	6
"P9"	1
"P13"	6
"P14"	3
"P15"	1
"P16"	2
"P18"	2
"P19"	4

The proximity centrality values of the individual network elements were successively calculated. Centrality is a measure of the importance of nodes in the network and identifies the most closely related nodes.

The centrality value can be used for the following purposes:

- Assessment of a node's impact on other nodes in the network;
- Identification of the nodes that are most influenced by other nodes;
- Observation of the flow or spread of something in the network (information, objects, phenomena);
- Analysis of the effectiveness with which phenomena spread in the network via individual nodes;
- Identification of nodes that block or prevent the spread of phenomena.

$$C(x) = \frac{1}{\Sigma_y d(x,y)} = 0.339 \tag{2}$$

The values for each aspect are shown in Table 7. Values greater than the calculated value of $C(x)$ highlight the aspects that have the greatest impact on the network under investigation. An aspect whose value is lower than $C(x)$ is of less significance and will not affect the operation of the entire network. The calculated values highlight the network's complexity and the aspects that need to be considered first to ensure the network's effective operation during the COVID-19 pandemic. Aspects whose nodes are located in the center of the network need to be considered simultaneously.

Table 7. The average node distances between aspects.

Aspects	Average Node Distances
1	0.172
3	0.172
4	0.172
5	0.376
6	0.376
7	0.686
9	0.172
13	0.970
14	0.515
15	0.125
16	0.250
18	0.343
19	0.436

The values of individual aspects (see Table 7) indicate that the most important problem is aspect 13, followed by aspects 7, 14, and 19, which obtained the next highest values. These results identify aspects that are of particular importance when modeling an electric shared mobility system. Problems P18, P5, and P6 obtained similar values, and they should be considered a common group of factors that influence the network. All other issues, the values of which are less than $C(x) = 0.339$, are of very little statistical importance; thus, they can be omitted during modeling. On the basis of these research results, a new network structure was developed, which indicates links between areas and stakeholder problems during the SARS-CoV-2 pandemic. The network structure is shown in Figure 2.

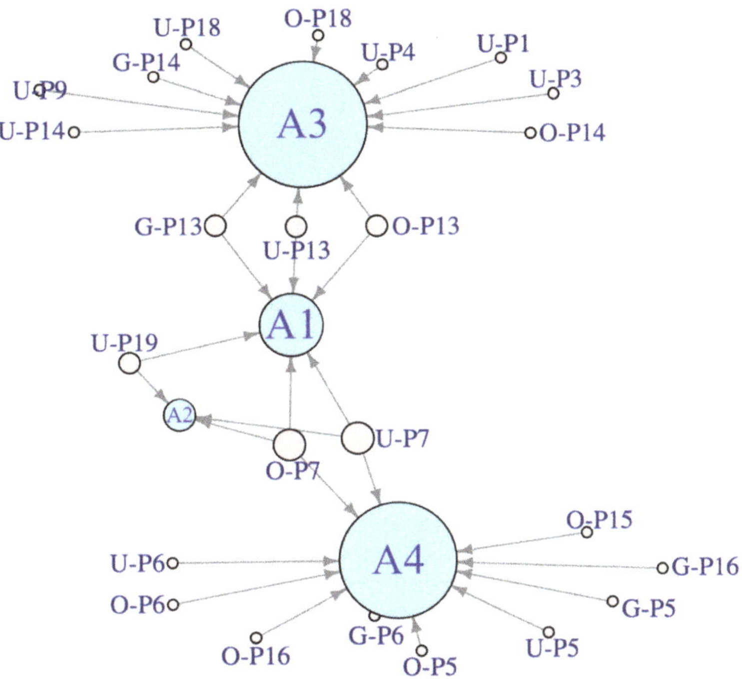

Figure 2. The new network for the four main areas, where O/U/G-P1–P19 are analyzed aspects (see Table 5); "O" denotes electric shared mobility operators, "G" denotes decision-makers, and "U" denotes e-shared mobility users.

Knowledge of the network topography allows for much more effective management of individual elements in the network during the COVID-19 pandemic. The size of the circle of an individual aspect in SNA indicates the scope of the problem associated with it (Figure 2). Areas 3 (operation) and 4 (governance) have the greatest impact on the whole network. Nevertheless, without area A1 (planning), areas 3 and 4 cannot appropriately manage their activities, which is shown directly in the structure nodes. Area 2 (implementation) has a small impact on the network's structure. An essential observation is that only aspects 19 and 7 directly affect area 2. Figure 3 shows the number of connections of each area.

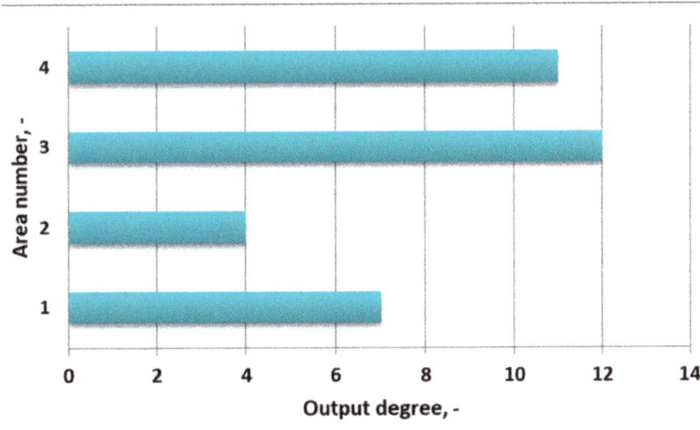

Figure 3. The number of output degrees for the area.

Knowing how specific aspects are linked to areas will allow for appropriate and effective management of the electric shared mobility system during the COVID-19 pandemic.

5. Discussion

The analyses performed using the SNA method made it possible to identify the most important aspects affecting the operation of the electric shared mobility network during the COVID-19 pandemic. Additionally, the areas that need to be given the most attention for the system to function optimally were determined. Table 6 details the most important aspects, such as prices, vehicle replacement, operation area, legal requirements, location of parking spaces, and serviceability safety.

The results obtained on the basis of the SNA method are applicable to the entire electric shared mobility market. According to the performed analysis, it can be concluded that, despite the use of various types of vehicles in electric shared mobility systems, aspects related to management and functionality obtain similar values. It should be emphasized that the research carried out was representative of the business practices used by the entire electric shared mobility industry. The use of the R language enabled a graphical representation of the network structure (SNA) and the visualization of the final results.

As can be seen from the results, the aspect that has the greatest impact is aspect 13, i.e., price. The importance of this factor is widely understood in the market reality. The business practices of service providers of electric shared mobility indicate that operators focused on developing new pricing schemes during the pandemic [60]. First, they started to offer more packaged services, for example, long-term packages (weekly or monthly). Importantly, the ability to rent a vehicle for a longer period of time has also become available for scooter rental services. The added scooters were only available at "per minute" rates. During the pandemic, a 24 h or, for example, 3–5-day rental system was introduced [61]. In addition, the price of using the "stop" function was reduced. Special discounts were also introduced for people delivering food with shared mobility vehicles, as well as special rates for doctors and social assistance for people infected with COVID-19 [60,61]. Interestingly, in the service model developed before the COVID-19 pandemic [43], pricing issues were not the decisive factor that needed to be taken into account. It is also worth noting that the price factor in the P13 problem (price) has a much higher $C(x)$ value relative to the others. Therefore, there is a need to conduct research on the economic aspects of transport changes in electromobility during the COVID-19 pandemic.

Aspect 7, which describes the technical condition of the fleet, is also important. This aspect is particularly important given that, in the transport models presented in the literature thus far, only the number of vehicles, not their technical condition, was taken into account. The fleet's technical condition is also of particular significance for protecting against the pandemic and other viruses. Of particular importance are all actions taken to disinfect strategic places in vehicles, such as the protection of the steering wheel, gearshift lever, and door handles by wrapping them [60,61].

The other two aspects are P14 (vehicle replacement) and P19 (rental area). These aspects are closely related not only thematically but also in terms of the value of the analysis. From a practical point of view, it is worth mentioning that, during the COVID-19 pandemic, the business models of electric shared mobility systems have changed. Many service providers have incorporated new business concepts into their services, e.g., the ability to prebook a vehicle, additional relocation, replacement of the vehicle in the door-to-door system, and changes in system operations [60,61]. Interestingly, many service providers have also expanded their zones of operation to include additional cities or districts that were previously excluded from these systems. Such solutions have enabled it to become an alternative to public transport, as well as allowed increasing social distance in the vehicle. Notably, in the previous transport model, which does not take the pandemic into account [43], these were not the primary determinants of the efficient operation of the system.

In further analysis of the results, it is worth focusing on aspects that should be regularly updated, such as local policy (aspect 5), parking spaces (aspect 6), and the technical safety of vehicles (aspect 18). Due to their similar values, these aspects should be considered together. During the pandemic, many local legal restrictions had a strong impact on the development of electromobility. After analyzing the recommendations of city authorities regarding mobility during and after the COVID-19 pandemic, it can be concluded that all suggestions are focused on cycling and walking [62,63]. Recommendations for bicycles include ensuring the ability of individual users and users of shared bicycles to navigate bicycle paths. Local authorities do not place great emphasis on car-sharing services, and very little attention is paid to shared mobility and electromobility services in their decisions [64,65].

Other transport services with negligible effects are taxis; however, they are strictly a substitute for traveling in individual vehicles. In our opinion, the practices introduced by electric shared mobility providers constitute the most comprehensive solutions implemented during the pandemic in urban transport. By excluding new mobility services from health and safety recommendations, society is losing a network of improvements that has been built over the years in furtherance of urban resilience and sustainable transport mobility. Moreover, strategies for the development of electromobility in different types of urban areas are not reflected in the current mobility recommendations; such actions may lead to a recession in the market position established by electromobility in recent years.

Aspects 1 (location of vehicles), 3 (registration process), 4 (rental process), 9 (safety of users), 15 (privileges for operators), and 16 (surcharges for operators) identify problems that slightly disrupt the system network, and they can be taken into account at a later stage of further analysis. It should be noted that no studies were found that specifically dealt with the subject of modeling transport services during a pandemic, which would be based on a similar type of research and network analysis. The analyzed articles on transport during the pandemic focused on various topics related to shared mobility services; for example, in [58], the authors analyzed the movement patterns of a population in Sicily. The results showed that the surveyed part of the population traveled via shared mobility much more often than walking or cycling. Respondents positively assessed the operation of electric shared mobility during the COVID-19 pandemic [66].

Hungary is another example where shared mobility was further developed during the pandemic. Research carried out in Budapest clearly showed that bike sharing increased significantly [67]. The author stated that this development was influenced by the fact that users were looking for a safe means of transport to avoid the COVID-19 virus.

The results of another study indicated that bike sharing is now more likely to become the preferred mobility option for people who have previously commuted in private cars as passengers and those who were already registered users of bike sharing [68]. The research results highlighted how important it is for the system to function during the COVID pandemic to raise public awareness and promote alternative transport, such as e-bikes and e-scooters [68].

From the transport administration perspective, during the pandemic, approaches to restrictions varied greatly from country to country. For example, [69] showed that there is no clear-cut approach to leadership. Many transport issues have been treated very generally by placing limitations on all individuals and their daily lives. There is also an ill-considered approach to energy issues, including electromobility, and disregard for specific stakeholder groups [69]. As emphasized in [70], the shared economy sector has been left most exposed to the COVID-19 pandemic. The author further emphasized that the constraints on the sharing economy may be deliberate, and cities affected by the pandemic may impose new regulations and taxes on shared mobility services to protect public transport [70]. In turn, in [71], the authors suggested that governments gain a better understanding of transport behavior and effective transport planning, as this would allow for better organization of both public transport and shared mobility services. Moreover, as discussed in [72], changes in business models after the COVID-19 crisis will be directed toward digital, automatic, and contactless activities. All activities in the field of shared mobility fit perfectly into this context.

Our research had some limitations, and the results presented in this article suggest directions for further analyses. The materials that we collected were sourced from different electric car-sharing service providers. Their number and opinions are statistically representative; however, it should be emphasized that the obtained results came from companies operating in Poland. Although these companies maintain standards in line with foreign trends, there may be slight differences related to the operation of the system and, for example, local legal requirements in the field of electromobility.

6. Conclusions

In summary, this article identified aspects and areas that should be considered in the context of a pandemic when modeling and optimizing energy services for shared mobility. The authors aimed to show differences between the classical transport models developed before the pandemic and the situation after its occurrence. The approach proposed in the article fills the research gap resulting from missing guidelines for electromobility transport models during a pandemic. The article highlights the possibilities of using the SNA method as an essential tool in the process of designing and operating an electric shared mobility system during the COVID-19 pandemic. The SNA method is used to determine the importance of individual nodes making up a network. As shown in Table 6, mean node distances greater than the value of $C(x)$ indicate problems outside the area of proximity centrality. Aspects with values greater than the calculated range disturb the effective operation of the network. These values indicate that the network is extensive and that the factors are closely related. This method is a simple approach to creating and presenting the topography of an analyzed network. The results presented in this article show that this technique can perfectly complement the conducted research and can significantly facilitate a multi-aspect analysis of the considered phenomenon.

The analyses showed that the most important factors in the operation of the electric shared mobility market are prices, the condition of the fleet, the replacement of vehicles, rental area, legal requirements, the location of parking spaces, and operational safety. The results presented in this article can support both transport modeling and the creation of new policies or restrictions related to the pandemic. The developed approach can also be used by electric shared mobility service providers to update their business models and develop new practices.

The added value of using the SNA method to model transport processes is that the results can be taken into account during the transformation of electric shared mobility systems.

These results can be used to support scientists who need clearly defined factors as input values when building models and conducting analyses from a scientific point of view. The results of this article can, therefore, contribute to any sustainable transport plan.

In future research, we plan to conduct detailed analyses of electrical shared mobility service business models following the second wave of COVID-19. The obtained data can be compared with the implications indicated in this study. The next phase of research and further SNA analyses will update the proposed model for the transport of electric shared mobility services in the post-pandemic period.

Author Contributions: Conceptualization, K.T.; methodology, K.T. and A.K.; validation, F.C.; resources, K.T., A.K., and F.C. data curation, K.T. and A.K.; writing—original draft preparation, K.T., and A.K.; writing—review and editing, K.T. and A.K.; supervision, F.C.; project administration, K.T.; funding acquisition, K.T. All authors have read and agreed to the published version of the manuscript.

Funding: The research and analyzes performed for this article were funded by the Silesian University of Technology grant for support to start activities in a new research topic, as part of the Initiative-Excellence program Research University 12/010/SDU/10-22-01. The APC was funded by Silesian University of Technology.

Institutional Review Board Statement: Not applicable.

Informed Consent Statement: Not applicable.

Data Availability Statement: The data presented in this study are available on request from the corresponding author.

Conflicts of Interest: The authors declare no conflict of interest.

Appendix A

Appendix A.1. Introduction

- Self-introduction of the researcher.
- Open conversation (goal: relaxation of the respondent).
- Thank you for finding time (goal: positive emotions).
- Informing what we want to find out.
- Information that the answers are anonymous.
- The question of whether we can record an interview for our scientific needs.
- Asking if the participant has additional questions about the study.

Appendix A.2. Main Part

- Warming up: What electric shared mobility services does your company provide? How long has your company been operating on the Polish market?
- Opening up: How has the COVID-19 pandemic overall affected your company's market situation?
- Development (from general to detail): Has the pandemic affected the following 19 aspects (presented in the Table 2)? In order to get to know the aspects better, each of them was discussed with the respondent. Then, the element of the wheel of fortune was used. All 19 aspects were randomly selected, and the respondent's task was to answer whether a given aspect of his company's operation was affected by the COVID-19 pandemic. After determining which aspects were affected by the pandemic, the following questions were asked: How? What has been limited? What was minimized, and what was maximized? A fragment of the game during the interviews is presented in Figure A1.

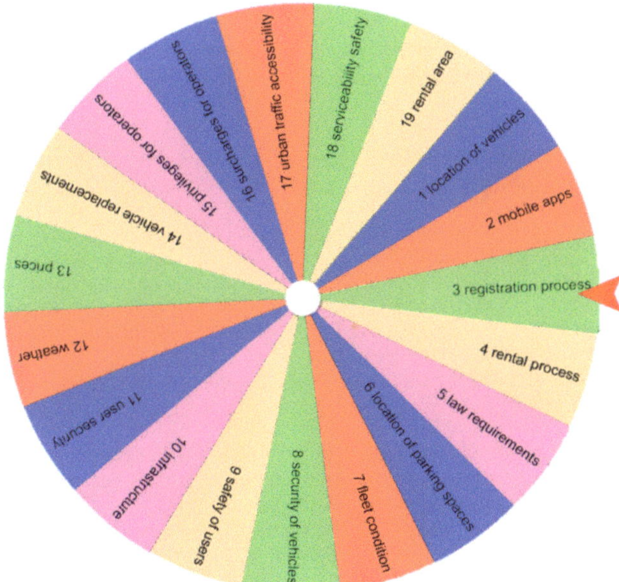

Figure A1. Fragment of the wheel of fortune game using during the interviews.

- Next, the respondents were asked to indicate which of the 19 aspects presented in Table 2 are considered the most important for the functioning of services in the business arena. For this purpose, the LearningApps application was used, where the respondents sorted the problems from the most important to the least important in their company. An exemplary view of the LearningApps application is shown in Figure A2.

The researchers asked for a list of the most important factors so that, in the next step, it would be easier for respondents to consider what types of electric shared mobility services were implemented due to the COVID-19 pandemic.

- The last step was to assign the practices provided by the respondent to the list of 19 aspects of electric shared mobility systems' functioning.

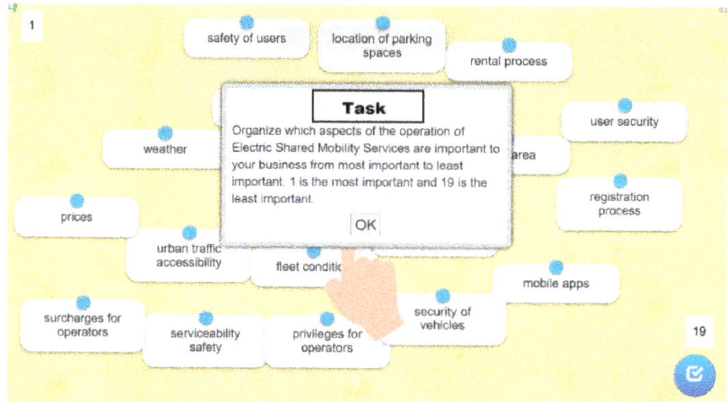

Figure A2. An exemplary view of the LearningApps application.

Appendix A.3. Summary

Emphasize that the researcher has no more questions. Is there anything else you would like to say? Would you like to highlight something before we finish?

Thank you for your time. Your answers will greatly help us when modeling new mobility services during the COVID-19 pandemic.

References

1. Apple Incorporation Mobility Report. Available online: https://www.apple.com/covid19/mobility (accessed on 18 November 2020).
2. Mobility during COVID-19. Available online: https://www.gstatic.com/covid19/mobility/2020-05-02_PL_Mobility_Report_en.pdf (accessed on 10 May 2020).
3. Amblard, M. How will the Covid-19 Crisis Impact the Electrification of Mobility? *The Urban Mobility Daily*. Available online: https://urbanmobilitydaily.com/how-will-the-covid-19-crisis-impact-the-electrification-of-mobility/ (accessed on 19 January 2021).
4. Neil, K. The Effects of COVID-19 on Electro-Mobility. *Autvista Group*. Available online: https://autovistagroup.com/news-and-insights/effects-covid-19-electromobility (accessed on 19 January 2021).
5. Hu, Y.; Zhang, Y.; Lamb, D.; Zhang, M.; Jia, P. Examining and optimizing the BCycle bike-sharing system—A pilot study in Colorado, US. *Appl. Energy* **2019**, *247*, 1–12. [CrossRef]
6. Shaheen, S.; Martin, E.; Totte, H. Zero-emission vehicle exposure within U.S. car-sharing fleets and impacts on sentiment toward electric-drive vehicles. *Transp. Policy* **2020**, *85*, 23–32. [CrossRef]
7. Wang, J.; Lindsey, G. Do new bike share stations increase member use: A quasi-experimental study. *Transp. Res. Part A Policy Pract.* **2019**, *121*, 1–11. [CrossRef]
8. Campbell, A.; Cherry, C.; Ryerson, M.; Yang, X. Factors influencing the choice of shared bicycles and shared electric bikes in Beijing. *Transp. Res. Part C Emerg. Technol.* **2016**, *86*, 399–414. [CrossRef]
9. Pérez, V.C. *Simulation of a Public E-Bike Sharing System*; Universitat Politechnica Di Catalugna: Barcelona, Spain, 2016.
10. Wu, L.; Gu, W.; Fan, W.; Cassidy, M. Optimal design of transit networks fed by shared bikes. *Transp. Res. Part B Methodol.* **2020**, *13*, 63–83. [CrossRef]

11. Awad-Núñez, S.; Julio, R.; Gomez, J.; Moya-Gómez, B.; Sastre González, J. Post-COVID-19 travel behaviour patterns: Impact on the willingness to pay of users of public transport and shared mobility services in Spain. *Eur. Transp. Res. Rev.* **2021**, *13*, 20. [CrossRef]

12. Feng, X.; Sun, H.; Wu, J.Z.; Lv, Y. Trip chain based usage patterns analysis of the round-trip car-sharing system: A case study in Beijing. *Transp. Res. Part A Policy Pract.* **2020**, *140*, 190–203. [CrossRef]

13. Weinert, J.; Ma, C.; Yang, X.; Cherry, C. The transition to electric bikes in china: Effect on travel behavior, mode shift, and user safety perceptions in a medium-sized city. *Transport. Res. Rec. J. Transport. Res. Board* **2007**, *1938*, 62–68. [CrossRef]

14. Huang, K.; An, K.; de Almeida Correia, G.H. Planning station capacity and fleet size of one-way electric car-sharing systems with continuous state of charge functions. *Eur. J. Oper. Res.* **2020**, *287*, 1075–1091. [CrossRef]

15. Lee, M.; Chow, J.; Yoon, G.; He, B.Y. Forecasting e-Scooter Competition with Direct and Access Trips by Mode and Distance in New York City. Available online: https://arxiv.org/ftp/arxiv/papers/1908/1908.08127.pdf?fbclid=IwAR0qZkxTDj-8qYeerGpHtqnacR6ylFfgHsmYisQ1hrXxpp1qkTT0XylO5Mw (accessed on 10 September 2020).

16. Aguilera-García, A.; Gomez, J.; Sobrino, N. Exploring the adoption of moped scooter-sharing systems in Spanish urban areas. *Cities* **2020**, *96*, 102424. [CrossRef]

17. Moran, M.E.; Laa, B.; Emberger, G. Six scooter operators, six maps: Spatial coverage and regulation of micromobility in Vienna, Austria. *Case Stud. Transp. Policy* **2020**, *8*, 658–671. [CrossRef]

18. Chen, Y.-W.; Cheng, C.-Y.; Li, S.-F.; Yu, C.-H. Location optimization for multiple types of charging stations for electric scooters. *Appl. Soft Comput.* **2018**, *67*, 519–528. [CrossRef]

19. Younes, H.; Zou, Z.; Wu, J.; Baiocchi, G. Comparing the Temporal Determinants of Dockless Scooter-share and Station-based Bike-share in Washington, D.C. *Transp. Res. Part A Policy Pract.* **2020**, *134*, 308–320. [CrossRef]

20. Grüger, F.; Dylewski, L.; Robinius, M.; Stolten, D. Carsharing with fuel cell vehicles: Sizing hydrogen refueling stations based on refueling behavior. *Appl. Energy* **2018**, *228*, 1540–1549. [CrossRef]

21. Bieliński, T.; Dopierała, Ł.; Tarkowski, M.; Ważna, A. Lessons from Implementing a Metropolitan Electric Bike Sharing System. *Energies* **2020**, *13*, 6240. [CrossRef]

22. Carrese, S.; D'Andreagiovanni, F.; Giacchetti, T.; Nardin, A.; Zamberlan, L. An optimization model for renting public parking slots to carsharing services. *Transp. Res. Procedia* **2020**, *45*, 499–506. [CrossRef]

23. Abbasi, S.; Ko, J.; Kim, J. Carsharing station location and demand: Identification of associated factors through Heckman selection models. *J. Clean. Prod.* **2021**, *279*, 123846. [CrossRef]

24. Repoux, M.; Kaspi, M.; Boyacı, B.; Geroliminis, N. Dynamic prediction-based relocation policies in one-way station-based car-sharing systems with complete journey reservations. *Transp. Res. Part B Methodol.* **2019**, *130*, 82–104. [CrossRef]

25. Clemente, M.; Fanti, M.P.; Iacobellis, G.; Ukovich, W. Modeling and simulation of an electric car sharing system. In Proceedings of the 25th European Modeling and Simulation Symposium, Athens, Greece, 23–25 September 2013; pp. 587–592.

26. Benarbia, T.; Omari, A.; Aour, B.; Labadi, K.; Hamaci, S. Electric cars-sharing systems modeling and analysis. In Proceedings of the International Conference on Control, Automation and Diagnosis (ICCAD), Hammamet, Tunisia, 19–21 January 2017; pp. 30–35.

27. Boyacı, B.; Zografos, K.G. Modelling user preferences in one-way electric car-sharing systems. In Proceedings of the 60th Annual Conference of the Operational Research Society, Lancaster, UK, 11–13 September 2018.

28. Hamroun, A.; Labadi, K.; Lazri, M. Modelling and Performance Analysis of Electric Car-Sharing Systems Using Petri Nets. In Proceedings of the E3S Web of Conferences, Constanta, Romania, 26–27 June 2020; p. 03001.

29. Axsen, J.; Sovacool, B.K. The roles of users in electric, shared and automated mobility transitions. *Transp. Res. Part D Transp. Environ.* **2019**, *71*, 1–21. [CrossRef]

30. Gambella, C.; Malaguti, E.; Masini, F.; Vigo, D. Optimizing relocation operations in electric car-sharing. *Omega* **2018**, *81*, 234–245. [CrossRef]

31. Bieliński, T.; Kwapisz, A.; Ważna, A. Bike-sharing systems in Poland. *Sustainability* **2019**, *11*, 2458. [CrossRef]

32. Wiederkehr, P.; Gilbert, R.; Crist, P.; Caïd, N. Environmentally Sustainable Transport (EST): Concept, Goal, and Strategy—The OECD's EST Project. *Eur. J. Transp. Infrastruct. Res.* **2004**, *4*, 11–25.

33. Elkington, J. Enter the triple bottom line. In *The Triple Bottom Line*; Routledge: London, UK, 2013; p. 23.

34. Bansal, P. The corporate challenges of sustainable development. *Acad. Manag. Perspect.* **2002**, *16*, 122–131. [CrossRef]

35. Nilsson, M.; Griggs, D.; Visbeck, M. Policy: Map the interactions between Sustainable Development Goals. *Nat. News* **2016**, *534*, 320–322. [CrossRef]

36. Okraszewska, R.; Romanowska, A.; Wołek, M.; Oskarbski, J.; Birr, K. Integration of a multilevel transport system model into sustainable urban mobility planning. *Sustainability* **2018**, *10*, 479. [CrossRef]

37. Di Vaio, A.; Boccia, F.; Landriani, L.; Palladino, R. Artificial Intelligence in the Agri-Food System: Rethinking Sustainable Business Models in the COVID-19 Scenario. *Sustainability* **2020**, *12*, 4851. [CrossRef]

38. Molina, J.A.; Giménez-Nadal, J.I.; Velilla, J. Sustainable Commuting: Results from a Social Approach and International Evidence on Carpooling. *Sustainability* **2020**, *12*, 9587. [CrossRef]

39. Song, K.-H.; Choi, S. A Study on the Behavioral Change of Passengers on Sustainable Air Transport after COVID-19. *Sustainability* **2020**, *12*, 9207.

40. Tian, H.; Liu, Y.; Li, Y.; Wu, C.H.; Chen, B.; Kraemer, M.U.; Wang, B. An investigation of transmission control measures during the first 50 days of the COVID-19 epidemic in China. *Science* **2020**, *368*, 638–642. [CrossRef]

41. Sidorchuk, R.; Lukina, A.; Markin, I.; Korobkov, S.; Ivashkova, N.; Mkhitaryan, S.; Skorobogatykh, I. Influence of Passenger Flow at the Station Entrances on Passenger Satisfaction Amid COVID-19. *J. Open Innov. Technol. Mark. Complex.* **2020**, *6*, 150. [CrossRef]

42. Yuan, Z.; Xiao, Z.; Dai, Z.; Huang, J.; Zhang, Z.; Chen, Y. Modelling the effects of Wuhan's lockdown during COVID-19, China. *Bull. World Health Organ.* **2020**, *98*, 484–494. [CrossRef] [PubMed]

43. Turoń, K.; Kubik, A.; Chen, F.; Wang, H.; Łazarz, B. A Holistic Approach to Electric Shared Mobility Systems Development—Modelling and Optimization Aspects. *Energies* **2020**, *13*, 5810. [CrossRef]

44. Mishin, V.M. *Research of Control Systems*; Textbook for Universities; Unity-Dana: Moscow, Russia, 2003.

45. Yu, T.; Shen, G.Q.; Shi, Q.; Lai, X.; Li, C.Z.; Xu, K. Managing social risks at the housing demolition stage of urban redevelopment projects: A stakeholder-oriented study using social network analysis. *Int. J. Proj. Manag.* **2017**, *35*, 925–941. [CrossRef]

46. Hatala, J.P. Social network analysis in human resource development: A new methodology. *Hum. Resour. Dev. Rev.* **2006**, *5*, 45–71. [CrossRef]

47. Caniato, M.; Vaccari, M.; Visvanathan, C.; Zurbrügg, C. Using social network and stakeholder analysis to help evaluate infectious waste management: A step towards a holistic assessment. *Waste Manag.* **2014**, *34*, 938–951. [CrossRef]

48. El-adaway, I. *Analyzing Trac Layout Using Dynamic Social Network Analysis*; Technical Report; Mississippi State University—National Center for Intermodal Transportation for Economic Competitiveness: Starkville, MS, USA, 2014; pp. 1–39.

49. Rudikowa, L.; Myslivec, O.; Sobolevsky, S.; Nenko, A.; Savenkov, A. The development of a data collection and analysis system based on social network users' data. *Procedia Comput. Sci.* **2019**, *156*, 194–203. [CrossRef]

50. Lee, K.; Jung, H. Dynamic semantic network analysis for identifying the concept and scope of social sustainability. *J. Clean. Prod.* **2019**, *233*, 1510–1524. [CrossRef]

51. Yuan, J.; Chen, K.; Li, W.; Ji, C.; Wang, Z.; Skibniewski, M. Social network analysis for social risks of construction projects in high-density urban areas in China. *J. Clean. Prod.* **2018**, *108*, 940–961. [CrossRef]

52. Li, Z.; Sun, Z.; Geng, Y.; Dong, H.; Ren, J.; Liu, Z.; Tian, X.; Yabar, H.; Higano, Y. Examining industrial structure changes and corresponding carbon emission reduction effect by combining input-output analysis and social network analysis: A comparison study of China and Japan. *J. Clean. Prod.* **2017**, *162*, 61–70. [CrossRef]

53. Freeman, L.C. Centrality in social networks conceptual clarification. *Soc. Netw.* **1979**, *1*, 215–239. [CrossRef]

54. Hagen, L.; Neely, S.; Robert-Cooperman, C.; Keller, T.; DePaula, N. Crisis communications in the age of social media: A network analysis of zika-related tweets. *Soc. Sci. Comput. Rev.* **2018**, *36*, 523–541. [CrossRef]

55. Kim, J.; Rasouli, S.; Timmermans, H.J.P. Social networks, social influence and activity-travel behaviour: A review of models and empirical evidence. *Transp. Rev.* **2017**, *38*, 499–523. [CrossRef]

56. Kawa, A. Analiza sieci przedsjebiorstw z wykorzystaniem metody SNA. *Przeds. Zarz.* **2013**, *14*, 39–49.

57. Le Pira, M.; Ignaccolo, M.; Inturri, G.; Pluchino, A.; Rapisarda, A. Modelling stakeholder participation in transport planning. *Case Stud. Transp. Policy* **2016**, *4*, 230–238. [CrossRef]

58. Dörry, S.; Decoville, A. Governance and transportation policy networks in the cross-border metropolitan region of Luxembourg: A social network analysis. *Eur. Urban Reg. Stud.* **2016**, *23*, 69–85. [CrossRef]

59. Le Pira, M.; Marcucci, E.; Gatta, V.; Ignaccolo, M.; Inturri, G.; Pluchino, A. Towards a decision-support procedure to foster stakeholder involvement and acceptability of urban freight transport policies. *Eur. Transp. Res. Rev.* **2017**, *9*, 54. [CrossRef]

60. How Have Shared Mobility Operators Responded to COVID? Available online: https://www.automotiveworld.com/articles/how-have-shared-mobility-operators-responded-to-covid/ (accessed on 19 November 2020).

61. Transport Arrangements for the Singapore COVID-19 Circuit Breaker Measures. Available online: https://landtransportguru.net/transport-arrangements-for-the-singapore-covid-19-circuit-breaker-measures/ (accessed on 13 May 2020).

62. Paris To Create 650 Kilometers of Post-Lockdown Cycleways, Forbs. Available online: https://www.forbes.com/sites/carltonreid/2020/04/22/paris-to-create-650-kilometers-of-pop-up-corona-cycleways-for-post-lockdown-travel/#1796664854d4 (accessed on 7 May 2020).

63. NYC to Close 40 Miles of Streets to Give Walkers More Space. Bloomberg. Available online: https://www.bloomberg.com/news/articles/2020-04-27/nyc-to-hire-1-000-people-to-trace-contacts-of-covid-19-positive (accessed on 7 May 2020).

64. COVID-19 Transportation Experiences. Available online: https://www.transformative-mobility.org/news/combating-covid-19-the-shenzhen-bus-groups-experience (accessed on 10 May 2020).

65. Chinese Transport after COVID-19. The Paper China. Available online: https://www.thepaper.cn/newsDetail_forward_5914134 (accessed on 9 May 2020).

66. Campisi, T.; Basbas, S.; Skoufas, A.; Akgün, N.; Ticali, D.; Tesoriere, G. The Impact of COVID-19 Pandemic on the Resilience of Sustainable Mobility in Sicily. *Sustainability* **2020**, *12*, 8829. [CrossRef]

67. Bucsky, P. Modal share changes due to COVID-19: The case of Budapest. *Transp. Res. Interdiscip. Perspect.* **2020**, *8*, 100141. [CrossRef]

68. Nikiforiadis, A.; Ayfantopoulou, G.; Stamelou, A. Assessing the Impact of COVID-19 on Bike-Sharing Usage: The Case of Thessaloniki, Greece. *Sustainability* **2020**, *12*, 8215. [CrossRef]

69. Chapman, A.; Tsuji, T. Impacts of COVID-19 on a Transitioning Energy System, Society, and International Cooperation. *Sustainability* **2020**, *12*, 8232. [CrossRef]

70. Mokter, H. The effect of the Covid-19 on sharing economy activities. *J. Clean. Prod.* **2020**, *280*, 124782.
71. Love, P.; Ika, L.; Matthews, J.; Fang, W. Shared leadership, value and risks in large scale transport projects: Re-calibrating procurement policy for post COVID-19. *Res. Transp. Econ.* **2020**, 100999. [CrossRef]
72. Seetharaman, P. Business models shifts: Impact of Covid-19. *Int. J. Inf. Manag.* **2020**, *54*, 102173. [CrossRef] [PubMed]

Article

Electric Vehicles Ready for Breakthrough in MaaS? Consumer Adoption of E-Car Sharing and E-Scooter Sharing as a Part of Mobility-As-A-Service (MaaS)

Paula Brezovec * and Nina Hampl

Sustainable Energy Management Unit, Department of Operations, Energy and Environmental Management, Alpen-Adria-Universität Klagenfurt, Universitätsstraße 65-67, 9020 Klagenfurt am Wörthersee, Austria; nina.hampl@aau.at

* Correspondence: paula.brezovec@aau.at; Tel.: +43-463-2700-4085

Abstract: Current mobility trends indicate that the popularity of privately-owned cars will decrease in the near future. One reason for this development is the diffusion of mobility services such as car or bike sharing, or Mobility-as-a-Service (MaaS) bundles. Especially, MaaS bundles have the potential to respond to environmental issues and provide reliable mobility to users, thus illustrating the possibilities of being mobile without owning a car. Most of the past research on MaaS bundles, however, has focused on bigger cities that already have good infrastructural bases. Building on previous work in the MaaS field, we conducted a choice-based conjoint survey ($n = 247$) in Austria to investigate consumer preferences for MaaS packages in a suburban area. Further, we gathered data on the consumers' willingness to pay, especially for including electric vehicles in the form of e-car sharing and e-scooter sharing in MaaS packages. The results highlight the importance of package price as the attribute with the highest impact on purchase intention. Further, participants in our study most preferred MaaS packages that included e-car sharing to ones with e-scooter sharing. Using latent class analysis, we classified the respondents into three different segments with varying preferences for MaaS bundle features, and conducted market simulations.

Keywords: mobility-as-a-Service; electric vehicles; e-car sharing; e-scooter sharing; sustainable mobility; suburban area; choice-based conjoint; latent class analysis; willingness to pay; market simulations

Citation: Brezovec, P.; Hampl, N. Electric Vehicles Ready for Breakthrough in MaaS? Consumer Adoption of E-Car Sharing and E-Scooter Sharing as a Part of Mobility-As-A-Service (MaaS). *Energies* **2021**, *14*, 1088. https://doi.org/10.3390/en14041088

Academic Editor: Amela Ajanovic
Received: 2 December 2020
Accepted: 4 February 2021
Published: 19 February 2021

Publisher's Note: MDPI stays neutral with regard to jurisdictional claims in published maps and institutional affiliations.

1. Introduction

Currently, the majority of cars in use are household owned. However, on average, these cars are parked 92% of the time, therefore their capacity is not nearly optimally used [1]. Although the mobility sector is rapidly changing and the trend is to move away from car ownership, it is still unclear how this market will develop [2]. Car sharing, as an alternative to car ownership, is gaining momentum [3,4]. Generally, according to a recent market forecast, the car sharing business will grow at an annual rate of more than 24% between 2020 and 2026 to more than USD 9 billion [5]. Besides the urban context, more marginal suburban locations are showing the same trend [6]. Another trend likely to have a high impact on future transport systems, is electrification [7]. Largely, electric cars appear to be taking the lead as a green alternative to private cars [8,9]. Overall, several studies indicate that electric vehicles (EVs) can positively address environmental concerns e.g., [10–12]. Thus, new mobility services that build on sharing and electric mobility such as Mobility-as-a-Service (MaaS) offerings could potentially provide green and non-ownership alternatives to meet future customers' estimated mobility needs [13]. One way of fostering MaaS adoption is to combine different mobility services (e.g., PT, car sharing, bike sharing) in so-called "multimodal mobility or MaaS packages" to increase, e.g., convenience, flexibility and cost reduction for customers e.g., [13–15]. Combining such sustainable mobility solutions, Gould et al. [16] see MaaS as an opportunity not only to

325

decarbonize the transport sector, but also to foster the diffusion of EVs by including such alternative means of transport in multimodal mobility bundles/MaaS packages [17]. In the context of MaaS, Karlsson et al. [18], and other researchers [2,19] pointed out that to meet users' needs, more studies on large-scale implementation and detailed analysis of potential users' preferences regarding MaaS are essential. All pilot projects and studies regarding MaaS packages to date have been conducted in big cities like Amsterdam, Helsinki, London and Sydney [14,20–25]. Additionally, extant research has dedicated more attention to studying best practices than to following a "bottom-up approach" that could identify potential users' preferences e.g., [26–28]. Our research aims to contribute to the discourse on MaaS and attempts to close the research gap by identifying preferences related to multimodal mobility packages' features and respondents' willingness to pay (WTP) in less densely populated suburban districts. To reach this goal, we conducted a choice-based conjoint (CBC) study ($n = 247$) including latent class analysis (LCA) and market simulations (sensitivity analysis) to investigate potential consumers' preferences for MaaS packages in a suburban district of Klagenfurt, the capital of the Austrian federal state Carinthia. In Klagenfurt's Harbach district a real estate project for residential living, "hi Harbach", is currently being developed with a new multimodal mobility point ("hi MOBIL"; the project "hi MOBIL—Multi-modal mobility node Klagenfurt-Harbach" is funded by the Austrian Ministry for Transport, Innovation and Technology. Project leader: City of Klagenfurt; other project partners: Diakonie de la Tour, Family of Power, Klagenfurt Mobil GmbH, Landeswohnbau Kärnten, Vorstädtische Kleinsiedlung; Duration: 10/2018–09/2021.), the first of its kind in Klagenfurt and in Carinthia. Once implemented, it will provide residents with MaaS offerings. To our knowledge, only a few studies so far have analyzed consumer preferences regarding multimodal mobility packages in similar contexts, i.e., in a suburban area and/or residential settlement [29–31]. Besides this general aspect, our study is the first to analyze EV inclusion, i.e., e-car sharing and e-scooter sharing, in MaaS packages.

The remainder of this paper is organized as follows: after briefly introducing the topic in Section 1, Section 2 gives an overview of the scientific literature focusing on service bundling in the mobility sector and related to MaaS. Section 3 introduces the sample and the research method, and Section 4 provides the results. Finally, Section 5 discusses the results, draws conclusions and suggests potential future research topics.

2. Literature Review

2.1. Service Bundling in the Mobility Sector

As new transportation modes (e.g., electric cars, e-scooters) and concepts (shared mobility, MaaS) enter the market, their diffusion can be stimulated by bundling. Originally developed in Helsinki [32], such MaaS packages or bundles are seen as a promising new mobility concept that, in recent years, especially practitioners have given much attention. The main idea of MaaS is to fulfil public mobility needs without people needing to own a private car. An associated key issue to be addressed in MaaS relates to various kinds of travel cards issued by different public transport (PT) companies or mobility service providers [13,33]. Combining different transport modes and services within one journey, i.e., using more than one of the PT, car sharing, ride sharing, taxi and bicycle options, is not a new idea. However, with the most advanced and integrated form of MaaS the whole package can be booked and (pre-)paid with, e.g., only one mobile application or one card [2]. In the context of the above-mentioned trends, digitalization plays a central role as it provides enhanced interconnectivity and gives users additional benefits, such as saving time by giving alternative transport possibilities in organizing a trip. Further, the new concept should not only facilitate the use of existing transport modes; it should offer additional value, e.g., by simultaneously providing a higher service level or a lower price [33].

2.2. Overview of MaaS Initiatives, Concepts and Projects

Sweden and Finland with UbiGo and Whim, have acted as pioneers in introducing and developing MaaS. UbiGo, a project piloted in Gothenburg, Sweden, in 2013 is often referred to as the first real-life MaaS demonstration [34]. At the time, they offered 151 pilot users a mobility package with several transport modes integrated in a single subscription [34–36]. By the end of the pilot project almost all participants expressed an interest in using UbiGo. Reportedly, the trial users also developed more negative attitudes towards private car use, and became more positive towards alternative modes of transport [37]. Further, Whim, developed and launched in Finland in 2016, is operated as a MaaS initiative [38]. At Whim, customers can choose between three types of bundles: the classic "pay-as-you-go" option, the "Whim Urban" (€49 per month) that includes unlimited urban PT and discounted taxi prices, and the "Whim Unlimited" (€499 per month) that offers unlimited access to PT, taxis or a (shared) car. Besides these two promising MaaS concept implementations (UbiGo and Whim), there are multiple other (pilot) MaaS projects that have been launched, especially in Europe, but also further afield. [23] provide a good overview of different studies summarizing the features of and lessons learnt from integrated mobility solutions focusing particularly on practice examples [39–42]. For instance, they mention Schad et al. [41] whose work shows that multimodal mobility packages help people to change their habits and rethink their mobility behavior. Nearly 90% of all users in their study's sample started using their own car less.

We took Kamargianni et al. [23] as a basis for this study, specifically the index (the MaaS integration index) they developed, which enables a comparison between different MaaS initiatives, concepts and projects. The index has three levels of integration related to the different transport modes that are part of a MaaS package. The first level is ticket integration which refers to the number of modes that can be accessed via a single ticket or smart card. The second level is the integration of information and communication technologies (ICT) via an application or online interface that can relate to two different functions: journey planning and booking. The third level includes full integration of the mobility modes in one single package where customers do pre-payment, thereby purchasing a combination of mobility services for a specific amount (time or distance). According to this MaaS integration index the two above-mentioned initiatives, UbiGo and Whim, achieved the highest scores. Another initiative that scored highly on this index is SHIFT. Initiated in 2013 in Las Vegas, SHIFT includes shuttle buses, bike sharing, car rental, car sharing, and a valet service. In contrast to UbiGo and Whim, SHIFT is completely institutionally integrated, offering unique mobility packages to customers. Further, SHIFT owns all its vehicles, many of which are EVs, thereby following not only the sharing concept, but also the electrification trend, on which our study also focusses.

Further, Kamargianni et al. [23] provide a good overview of different integration forms in the personal mobility sector. They distinguish two levels of integration in their review, namely partial integration and advanced integration (with or without mobility packages). The partial integration category is illustrated by a system introduced in Belgium, built on a concept developed in Brussels. The car-sharing company, Cambio, cooperates with STIB (Société des Transports Intercommunaux de Bruxelles), a multimodal mobility operator for PT, bike sharing and taxi services. The two service providers (Cambio and STIB) designed a common smart card for users of both systems. Additionally, Cambio members enjoy discounts if they subscribe to the STIB service. However, payments are not integrated, nor is the ICT. To illustrate advanced integration, in Germany a few different concepts were similarly brought together as mobility services: Qixxit, Moovel and Switchh. Qixxit has included rail, urban PT, car sharing, car rental, bike sharing and taxi services, as well as flights and coach trips on a national level. Through a smart app the user has access to different services such as planning, booking, real-time information, and personalized trip advice. Moovel, another national mobility integration service, in contrast to Qixxit, gives ICT integration. Moovel's single smartphone platform provides booking and payment for nearly all services, which include PT, car sharing, bike sharing, national rail, taxi services, etc. [23]. Although, Switchh, developed for Hamburg (Germany), also represents advanced integration in relation to the other two concepts, it still has potential for further integration because currently it has no single invoice for payment of the services.

The advanced integration level includes the same basic components as those in the mobility solutions identified as having partial integration; additionally, it has three particular identifying elements: ticket sales, payment options, and ICT integration. To illustrate, Hannovermobil (2.0) provides a service in which users receive one bill that integrates all the mobility services they used at the end of each month. A similar solution introduced in Montpellier (France), is known as EMMA (EMMA is an integrated mobility platform in the city of Montpellier, France. Transports de l'Agglomération de Montpellier (TAM) is the key operator of EMMA services.). However, in contrast to Hannovermobil (2.0), EMMA users can also purchase monthly or annual mobility contracts (including all the TAM services). Depending on their needs, or their affiliation to different user groups, contracts have different designs and payment structures (e.g., EMMA Young, EMMA Senior) incorporating all three above-mentioned elements of advanced integration [23]. In addition to these examples, the Netherlands has developed different concepts (e.g., Mobility Mixx, NS-Business Card), as have different German speaking countries; however, they are still in a research phase (Austria/Vienna: Smile; Germany/Berlin: BeMobility). The MaaS concepts Kamargianni et al. (2016) have classified as offerings with the highest level of integration are the ones already mentioned: UbiGo (Gothenburg), Whim (Helsinki) and SHIFT (Las Vegas).

2.3. Consumer Preferences toward MaaS Packages

Huwer's study [43] represents one of the first studies related to the subject of MaaS. Very recently, scholars e.g., [2,19,21,44,45] advanced this field in studies on potential consumers' preferences towards services included in MaaS packages and their characteristics. Methodologically, most of these studies used stated preference experiments, as does this study [21,22,25,45–47].

Matyas and Kamargianni [24,47] conducted the first study in this field that used a stated preference experiment to investigate consumer preferences for features of MaaS bundles. They collected preference data in the Greater London area for different types of MaaS subscription plans; fixed and flexible ones, the latter with a menu option to configure mixed bundles. The bundles included PT, bike sharing, car sharing and taxi services, as well as additional features such as bike sharing rental time of up to 60 min at a time or a pooled taxi as an option. Their findings reveal that the type and number of transport modes in the plans are important determinants of users' willingness to subscribe to such plans. Specifically, the respondents preferred PT to the other transport modes (e.g., bike sharing or car sharing) in their plans. However, their preferences for sharing options were moderated by their previous use of such services. Kamargianni et al. [23] and Guidon et al. [45] also showed a high preference for PT in MaaS packages. To elaborate, [23] revealed the highest preference for PT in combination with e-car sharing, on demand busses, and e-bike sharing, in that order. In contrast, they found that potential customers showed less preference for PT bundled with taxi services, ride sharing, and car rental services. Guidon et al. [45] found the highest WTP for bundles that include complementary mobility services related to PT such as car sharing and park-and-ride services. Ho et al. [22,46] report on two stated preference studies relating to MaaS subscription plans conducted in Sydney, Australia, and Tyneside, UK. These studies' key findings include that individuals are not willing to pay for bike sharing as part of MaaS packages, and that overall, discounts should be offered on the bundled mobility services to increase MaaS package adoption. Further, the high impact of subscription price on users' purchase intention has also been shown by [21,25]. Mulley et al. [25] reveal that what users are willing to pay for MaaS bundles is even lower than the providers' unit costs of providing the service included in the bundle. More generally, [21,22] report a rather low overall intention of their respondents to subscribe to MaaS offerings. In this regard, scholars show that habit (e.g., daily car or public transport usage) has a strong impact on users' willingness to subscribe to MaaS offerings e.g., [21,22,48], which is also reflected in general literature on transport mode choice and modal diversion e.g., [49,50].

Our literature review on consumer preferences and WTP for MaaS bundles using stated preference experiments highlights that nearly all earlier studies are focused on urban areas. To the best of our knowledge only a few studies so far have analyzed consumer preferences regarding MaaS packages in suburban areas and/or residential settlements [29–31]. Wright et al. [31], for instance, investigated MaaS bundles including PT and carpooling services in four European suburban regions or areas (Canton Ticino, Brussels, Zagreb and Ljubljana). The study was based on a pilot project called 'RideMyRoute', which included an application to help users better plan their journey involving PT as well as carpooling. In this study with relatively limited focus, surveys were administered before and after the pilot project. Further, none of the studies we found in the literature investigated the inclusion of EVs as part of MaaS bundles, such as e-car sharing or e-scooter sharing. Thus, considering these research gaps, we seek to answer the following research questions:

- Research Question 1: What are the most attractive attributes of MaaS bundles in suburban areas/residential settlements?
- Research Question 2: How high is the potential users' purchase intention of MaaS bundles that include EVs (i.e., e-car sharing and/or e-scooter sharing)?
- Research Question 3: What differences regarding respondents' socio-demographic (e.g., gender, age, income), psychological (e.g., climate change concerns) and behavioral (e.g., car usage) characteristics can be identified between (non-)adopter groups in suburban areas/residential settlements?

3. Methodology and Data

3.1. Data Collection and Sample

This paper builds on a web-based survey ($n = 247$) conducted in Austria in 2019 in course of the research project "hi MOBIL—Multi-modal mobility node Klagenfurt-Harbach". The hi MOBIL project's goal is to install a multimodal mobility point as part of the real estate project "hi Harbach" in Klagenfurt, Carinthia, that will provide residents with MaaS offerings. Our sample consists of individuals who, at the point of testing, were registered at the two involved real estate companies as prospective hi Harbach residents and current residents in the Harbach district, which would also have the possibility of using the multimodal mobility point (At the point of testing, 600 prospective residents of hi Harbach were registered with the real estate companies. Due to data privacy issues the real estate companies sent out the survey invitation (as well as two reminders) per email. Of the prospective residents, 149 participated in our survey, which was a response rate of 24.8% for this subgroup of the sample. The residents in the Harbach district within a radius of 1 km of the hi Harbach project (approx. 5943 individuals) were invited to participate in the survey via mass mailing (including the survey link, which was directly indicated in the text and alternatively accessible via QR code). Of these residents. 98 participated in the survey, which was a response rate of 1.6% for this subgroup of our sample. We checked for statistically significant differences in preferences between these two subgroups but did not find any.). The questionnaire included a section on travel behavior and a conjoint experiment to investigate preferences and WTP for different features of the multimodal mobility offering, which included transport modes (e.g., PT, e-car sharing), contract termination modes, modes of access to the MaaS offering, and price per month. To incentivize survey participation, the respondents could be included in a raffle offering prizes such as a hotel voucher.

3.2. Choice-Based Conjoint

To identify the relative importance of the factors included in the study testing the willingness to purchase MaaS bundles, we used conjoint analysis, or specifically, a CBC experiment. This method is widely used in marketing research and practice (e.g., studying product design or pricing; conducting market segmentation based on stated preference data, etc.) [51]. It counts as mainstream in research on decision-making [52,53] especially due to its indirect approach to determining preferences for specific decision attributes and attribute levels. The indirect questioning approach consists of several choice tasks where respondents have to choose one of a number (typically 3 to 4) of alternative options (e.g., existing or hypothetical products or services). The respondents' choices are taken as the dependent variable in the estimation model. The options in the choice tasks are described along a set of pre-defined attributes with levels varying on a random basis between the options within and between the choice tasks. These attributes and attribute levels are treated as independent variables for which part-worth utilities are estimated based on the assumption that these part-worth utilities add up to the overall utility of the product or service [52–54]. This indirect, decompositional approach to measuring preferences has several advantages over other methods (e.g., interview, survey). For instance, CBC allows for a real-time collection of data and, thus, for a more direct and accurate investigation of consumer decision models, which deals with issues typically associated with methods based on self-reports such as social desirability and hindsight biases e.g., [55]. Further, this research method is highly suitable for products or services that are not in the market yet or of which the market diffusion is still limited [56,57]. This is the case in our research context where we study preferences for MaaS bundle features in Austria.

3.3. Survey Design

The survey consisted of two parts. First, we administered a questionnaire assessing the mobility behavior of respondents (asking, e.g., "How often do you use the following transport modes?") and collecting data on their socio-demographic (e.g., gender, age, income) and psychological (e.g., climate change concerns) characteristics (see Tables 1 and 2). Second, we included the CBC experiment in the survey. Before the CBC experiment started, the respondents received an introduction which gave them details on the attributes and attribute levels that described the MaaS bundle options in the choice tasks (e.g., different transport modes, modes of access) and provided an illustrative example (see Figure 1). The study was pre-tested with the hi MOBIL project partners, experts from the mobility sector, and potential users. Table 1 below summarizes the selected attributes and attribute levels we used in this study.

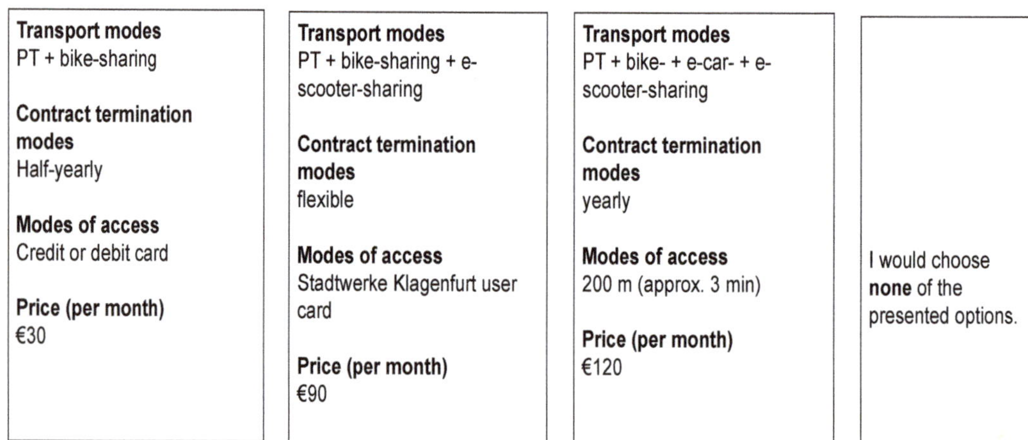

Figure 1. Example of choice tasks.

Table 1. Attributes and attribute levels included in the choice-based conjoint (CBC) experiment.

Attributes	Description	Levels
Transport modes	Transport modes included in the package (incl. a specific number of free hours or mileage [1])	PT + bike sharing PT + bike sharing + e-scooter sharing PT + bike sharing + e-car sharing PT + bike sharing + e-car sharing + e-scooter sharing
Contract termination modes	Possibility of changing the package or cancel the contract	Annually Bi-annually Monthly Flexible
Modes of access	Possibility of activating the individual mobility offers for use	Mobility card [2] Stadtwerke Klagenfurt user card Credit or debit card Credit or debit card Smartphone (via app)
Price (per month)	Total monthly cost for all transport modes included in the package	€30 €60 €90 €120

[1] PT: Includes all busses of the Klagenfurt Mobil GmbH within the Klagenfurt central traffic area. Bike sharing: Only valid for the services offered by nextbike within Klagenfurt. The bundle includes the basic tariff (€4/month). The first 30 min are free of charge for each rental; each additional 30 min are charged with a €1 tariff. The 24 h-tariff starts with the 5th rental hour and will be charged at €9. E-car sharing: Only valid for the services offered by FAMILY OF POWER. The bundle includes 10 h/month. Each additional hour will be charged at €4.80. E-scooter sharing: Valid for all services offered by a specific provider (not defined yet at the time of data collection). The bundle includes up to 10 rides (incl. unlocking) and max. 200 min/month. For additional services, unlocking will be charged at €1, and at €0.15 per additional minute. [2] An own mobility card for using the MaaS bundle services independent of any specific service provider.

To investigate the impact of including EV sharing modes (i.e., e-car sharing and/or e-scooter sharing) in MaaS bundles on the potential users' purchase intention we used the following four levels for the attribute transport modes: (1) PT + bike sharing (baseline offering), (2) PT + bike sharing + e-scooter sharing (package that includes e-scooter sharing only to determine part-worth utility for this specific EV sharing mode), (3) PT + bike sharing + e-car sharing (package that includes e-car sharing only to determine part-worth utility for this specific EV sharing mode), and (4) PT + bike sharing + e-car sharing + e-scooter sharing (package that includes both e-scooter sharing and e-car sharing, to determine part-worth utility for combining these two EV sharing modes).

The conjoint experiment comprised 12 choice tasks, which is in line with other studies in this field e.g., [58–60]. In general, the recommended number of choice tasks depends on different parameters such as the number of attributes and attribute levels or the targeted sample size [61]. Each of the choice tasks in this study comprised three options of hypothetical MaaS bundles described by our four pre-defined attributes. In order to mirror real purchase situations, we added a fourth alternative to each choice task, i.e., a so-called "none option", that respondents could choose if they did not prefer any of the other three presented options (see Figure 1). In each choice task, the respondents had to select which option of the four they preferred most. We designed our CBC experiment by following a full-profile approach (all attributes were shown in each of the choice tasks) and a balanced-overlap design strategy (balanced level of overlap of attribute levels between alternatives within choice tasks) [61,62].

3.4. Data Analysis

We used Sawtooth Software's Lighthouse Studio to design the CBC experiment and to analyze the stated preference data. We applied a hierarchical Bayes (HB) model implemented in Lighthouse Studio to estimate the part-worth utilities for the overall sample. The HB procedure estimates robust part-worth utilities on an individual level, which makes it superior to alternatives such as multinomial logit models that only estimate utilities on an aggregated level [63,64]. For more details on the HB estimation procedure see [61]. We used the average root likelihood (RLH) value (geometric mean of all predicted probabilities) as a measure of the goodness of fit of our model [61]. The average RLH value for our model was 0.67, which means that our model was 2.7 times better than the chance model (0.25, i.e., the predicted probability of randomly choosing one of the four options (1/4) within each choice task).

Further investigating the results from the HB estimation, we used another well-established method, LCA, that clusters the overall sample (n = 247) into largely homogeneous groups to gain more insight on the preference heterogeneity among respondents and to identify different customer segments [65,66]. As LCA uses different starting points at each computation, [66] recommends re-running the model several times and estimating a number of different group solutions. Following this advice, we estimated two-group to five-group-solutions and retained for each group solution the estimate with the highest chi-square score. Besides the chi-square score, we also used other quality criteria—percent certainty and the Consistent Akaike Information Criterion (CAIC)—to identify the best segmentation result (see Table 2).

Table 2. Summary from latent class analysis highlighting best replications.

Groups	Percent Certainty	CAIC	Chi-Square
2	36.43	5467.10	2993.70
3	40.45	5262.61	3324.11
4	42.79	5196.31	3516.33
5	43.89	5232.10	3606.46

The scores for percent certainty and chi-square can be interpreted as "the higher, the better" as they indicate how much better a solution is in comparison to a "no segments" solution. CAIC, on the contrary, needs to be minimized [67,68]. Table 2 shows that percent certainty and chi-square advocated for a five-group-solution, but CAIC was at its minimum with a four-group-solution. However, in order to define the optimal number of segments, we additionally checked the group sizes (very large and very small groups should be avoided), as well as the interpretability of the part-worth utilities and relative importance scores of the groups, checking whether the groups and differences between them can be described meaningfully, as the literature recommends [66]. Thus, the model we finally chose was the three-group-solution (for further details on the customer segments, see Section 4.3).

Further, we conducted a sensitivity analysis based on market simulations using part-worth data by assessing different variations of MaaS bundles. We applied a randomized first choice model to estimate the share of preference per customer segment for each of the MaaS bundle scenarios [69].

4. Results

4.1. Part-Worth Utilities and Relative Importance Scores

In this section, we present the results regarding the relative importance of attributes included in the CBC experiment (e.g., transport modes, price (per month)) and the part-worth utilities per attribute level. The relative importance scores were determined by the difference between the highest and lowest part-worth utility per level for each attribute and represent the attribute's impact on the dependent variable, i.e., purchase intention. Due to standardization over all attributes the scores sum up to 100% making it possible to compare the scores between attributes. Our results showed that the attribute price (per month) was the most important consideration for the respondents in our sample (49.10%). After price of the package, in terms of importance ranking, respondents rated transport mode (27.28%), contract termination mode (13.19%), and modes of access (10.42%). These findings indicated that monthly price had the largest effect on the overall utility of the MaaS bundles and, thus, on respondents' purchase intention. The attributes termination mode and modes of access each had a relatively marginal effect on purchase intention.

Further, Table 3 gives an overview of the part-worth estimation results per attribute level with corresponding standard deviations and 95% confidence intervals. The part-worth utility of an attribute level can only be compared to those of other levels of the same attribute. It represents the desirability of a specific attribute level for the respondents (e.g., the higher the utility score, the more positive its effect on the intention to purchase the MaaS bundle). The attribute levels with the highest utility scores per attribute were as follows: €30/month (lowest bundle price), bundles including all offered transport modes (PT + bike sharing + e-car sharing + e-scooter sharing), flexible cancellation options, and the Stadtwerke Klagenfurt user card as mode of access to the MaaS services. That the respondents in our sample preferred the highest service for the lowest bundle price with flexible cancellation options seems obvious. However, that the respondents preferred the PT user card instead of an app was an interesting finding. In the questionnaire used in the survey, we also directly asked the respondents to indicate their most preferred level for each of the attributes (except for monthly price, of which the direct WTP was assessed in a separate question) included in our conjoint design (build-your-own (BYO) exercise). The results of the BYO exercise (see Table 7) show that direct preference for an app is twice as high as for the Stadtwerke Klagenfurt user card (These different results of the direct and indirect questioning approaches might be explained by this attribute, modes of access, having the lowest relative importance (10.42%). This typically happens when respondents do not pay close attention to variations related to an attribute in the choice tasks.). For the other attributes, the BYO showed results similar to the CBC analysis.

Table 3. The hierarchical Bayes (HB) model estimation of mean utility values and mean relative importance scores (*n* = 247).

Attributes and Attribute Levels	Mean	Standard Deviation	Lower 95% CI [1]	Upper 95% CI [1]
Transport Modes (m = 27.28; SD = 12.69)				
PT + bike sharing	−45.12	39.10	−49.99	−40.24
PT + bike sharing + e-scooter sharing	−18.92	32.14	−22.93	−14.91
PT + bike sharing + e-car sharing	26.50	42.70	21.18	31.83
PT + bike sharing + e-car sharing + e-scooter sharing	37.54	28.92	33.93	41.14
Contract Termination Modes (m = 13.19; SD = 5.69)				
Annually	−22.51	20.62	−25.08	−19.94
Bi-annually	−4.57	16.40	−6.62	−2.52
Monthly	11.33	14.47	9.52	13.13
Flexible	15.75	14.86	13.90	17.60
Modes of Access (m = 10.42; SD = 6.14)				
Mobility card	−0.16	21.52	−2.84	2.52
Smartphone (via app)	−1.37	19.65	−3.82	1.09
Credit or debit card	−3.83	15.29	−5.73	−1.92
Stadtwerke Klagenfurt user card	5.35	16.18	3.33	7.37
Price (per Month) (m = 49.10; SD = 13.62)				
€30	101.62	44.88	96.02	107.21
€60	25.23	20.11	22.72	27.74
€90	−37.97	26.62	−41.29	−34.65
€120	−88.88	28.87	−92.48	−85.28
None	111.13	141.74	93.46	128.81

[1] Confidence interval.

4.2. Willingness-to-Pay

To define the respondents' WTP for the different attributes of MaaS packages included in our CBC experiment, we transformed the part-worth utilities per attribute level into monetary values. As we treated all independent variables (attribute levels, including the levels of the price attribute) as categorial variables in our estimation [70], we determined the WTP for this part-worth model using the following formula [71,72]:

$$\text{WTP}\,(u_{ij}) \;=\; (u_{ij} - u_{ij\,\text{Default}}) * \frac{p_{max} - p_{min}}{u_{pj\,max} - u_{pj\,min}} \tag{1}$$

This formula defines WTP as the difference between the part-worth utility (u_{ij}) of an attribute's (i) level (j) and a default part-worth utility ($u_{ij\,\text{Default}}$) (i.e., the least preferred attribute level in the same attribute) multiplied by the price of one utility unit (i.e., difference between the highest (p_{max}) and lowest (p_{min}) level of the price attribute divided by the utility difference between the highest and lowest price level ($u_{pj\,max}-u_{pj\,min}$)). Figure 2 shows the WTP values indirectly calculated based on the results of the CBC experiment.

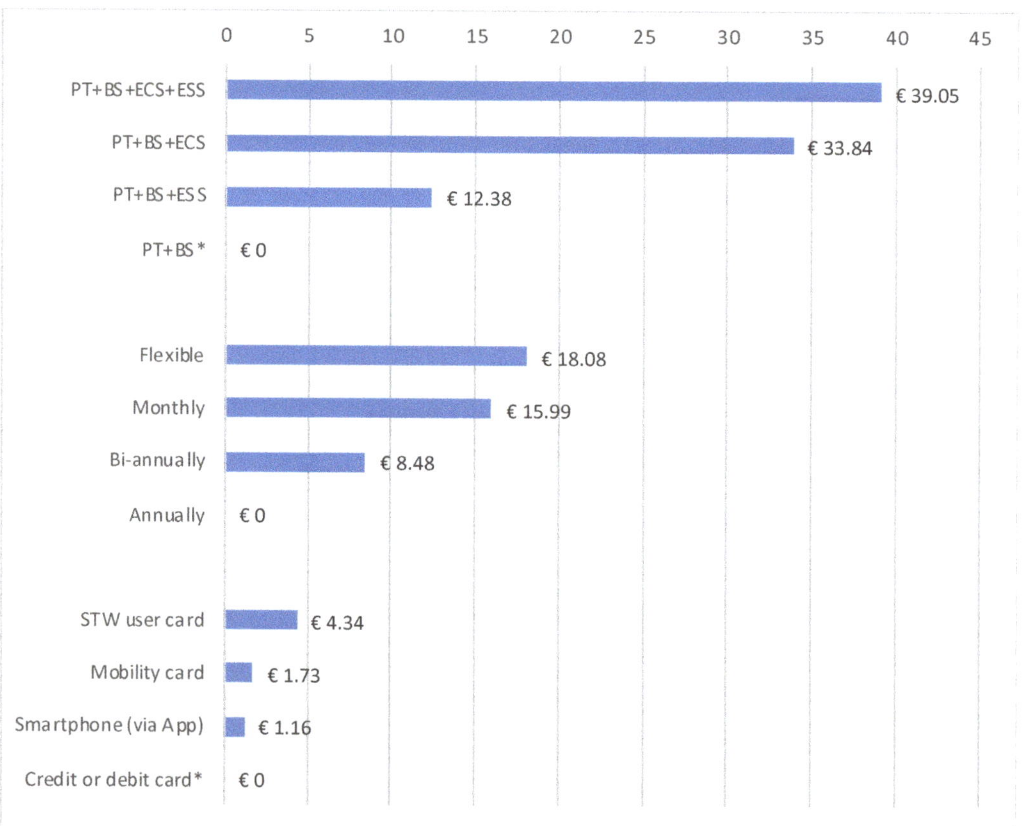

Figure 2. Indirect willingness to pay (WTP) for attribute levels of Mobility-as-a-Service (MaaS) packages (relative to default). Note: Bike sharing (BS), e-car sharing (ECS), e-scooter sharing (ESS), Stadtwerke Klagenfurt (STW).

In order to provide additional insight on the respondents' WTP, we also included a question in the questionnaire part of the survey asking directly how much the respondents would be willing to pay for a MaaS package that meets their preferences. Overall, the results indicated that study participants, if directly asked, were willing to pay approx. €52 for such a MaaS package. Figure 3 shows, based on the answers to this open question, how many respondents in our sample (in percentage) were willing to buy a MaaS package at a given price level. For instance, the market potential of a product with a price of €25 would be 79%.

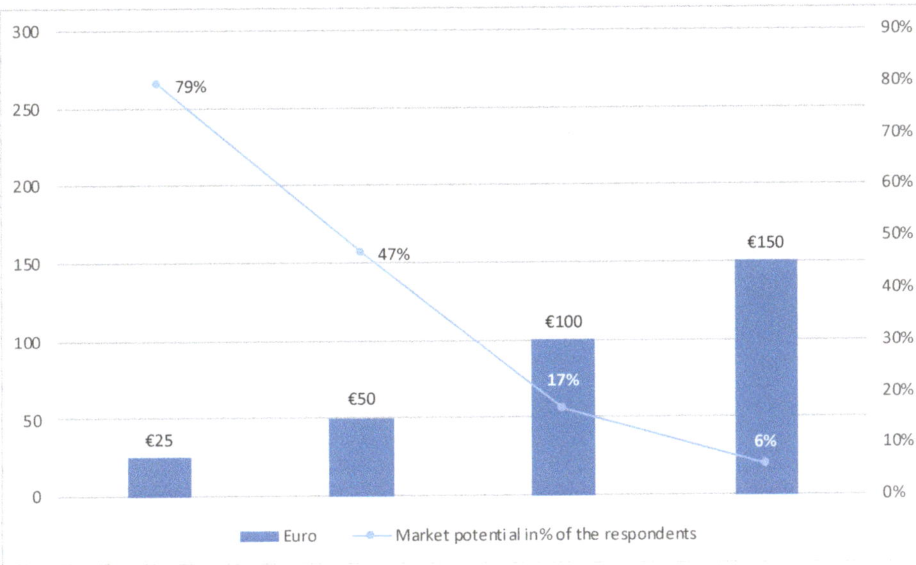

Figure 3. Direct WTP and market potential for MaaS packages.

4.3. Customer Segments

In a further step, we applied LCA to further explore the heterogeneity of preferences in the study sample. As described in Section 3.4, we chose a three-group-solution as our final segmentation result applying multiple criteria. Tables 4 and 5 give detailed overviews of the differences in the part-worth utilities as well as relative importance scores of the attributes and attribute levels between the three customer segments. Building on insights presented in the tables, we labelled these segments as convenience seekers, price-sensitive worriers, and non-adopters.

Considering the respondents' stated preferences on segment level, we labeled the first potential adopter group convenience seekers ($n = 77$, 31.2% of the respondents). In general, they rated price as the most important attribute (57.1%), as did respondents in the other two segments. The attribute they rated as second most important, is transport modes (31.0%). Considering the part-worth utilities for that attribute we found that the respondents in this segment most preferred offerings including PT + bike sharing + e-car sharing + e-scooter sharing. Contrary to the other two segments, this group preferred an own mobility card to other modes of accessing the mobility services included in the bundle. Investigating socio-demographic differences between the segments showed that convenience seekers were significantly younger than non-adopters ($p = 0.004$) (see Tables 1 and 3). Regarding household income (net/month) this group indicated the highest value on average (€2664.8); however, their differences to the other two groups were statistically not significant. Further, almost all respondents in this group (93.5%) owned at least one car which, compared to the other segments, costed the most per month (approx. €260 for fuel, insurance, etc.). In comparison to the other groups, a higher share of respondents in this segment used their car on a daily basis (13% versus 9.7% of the price-sensitive worriers and 2.6% of the non-adopters; see Tables 1 and 4). This group generally was the most interested in and willing to adopt multimodal services, i.e., more than 71.4% of the respondents in this segment indicated that they (most) likely would purchase a MaaS bundle that meets their preferences. This was also reflected in the negative value of the none option in our part-worth model (-150.90). Their directly indicated WTP for a MaaS package that met their preferences was the highest among the three groups, i.e., €62.3, on average. In this regard, the convenience seekers significantly differed from the price-sensitive worriers (€54.8, $p < 0.001$) and the non-adopters (€38.0, $p < 0.001$). Finally, the results showed a significant difference regarding the subjective norm (peer group effect) between the segments ($p < 0.001$). Regarding this variable, the convenience seekers (3.13) significantly differed in relation to the other two segments, i.e., the non-adopters (2.61, $p < 0.001$) and price-sensitive worriers (2.73, $p = 0.004$).

Table 4. HB model estimation of mean utility values per segment.

	Segment 1: Convenience Seekers	Segment 2: Price-Sensitive Worriers	Segment 3: Non-Adopters
Segment size	$n = 77$ (31.2%)	$n = 93$ (37.6%)	$n = 77$ (31.2%)
Transport Modes			
PT + bike sharing	−63.41	−52.83	5.44
PT + bike sharing + e-scooter sharing	−33.47	−25.11	−40.48
PT + bike sharing + e-car sharing	36.18	40.29	1.10
PT + bike sharing + e-car sharing + e-scooter sharing	60.70	37.65	33.95
Contract Termination Modes			
Annually	−7.23	−22.18	−75.28
Bi-annually	−11.31	−0.17	−6.97
Monthly	8.02	12.30	25.09
Flexible	10.53	10.05	57.16
Modes of Access			
Mobility card	18.02	−0.72	−6.04
Credit or debit card	−7.69	−2.41	−5.58
Smartphone (via app)	−6.61	−7.88	−0.83
Stadtwerke Klagenfurt user card	−3.72	11.01	12.45
Price (per Month)			
€30	106.25	131.43	116.19
€60	50.60	35.75	−58.44
€90	−34.78	−45.10	−17.18
€120	−122.08	−122.08	−40.57
None	−150.90	114.21	353.73

Table 5. Attribute importance scores per segment.

Segment	Segment 1: Convenience Seekers	Segment 2: Price-Sensitive Worriers	Segment 3: Non-Adopters
Segment size	$n = 77$ (31.2%)	$n = 93$ (37.6%)	$n = 77$ (31.2%)
Transport modes	31.0	23.3	18.6
Contract termination modes	5.5	8.6	33.1
Modes of access	6.4	4.7	4.6
Price (per month)	57.1	63.4	43.7
Total	100%	100%	100%

The second potential adopter segment was labeled price-sensitive worriers ($n = 93$, 37.6%). The respondents in this group paid the most attention of all groups to the monthly price of the bundles in the choice tasks (63.4%). Further, they most preferred a monthly contract termination mode in comparison to the other two groups that preferred a flexible one. Regarding the mobility services included in the bundle, they most preferred a combination of PT + bike sharing + e-car sharing; rather than a package with the largest range of services, i.e., one also including e-scooter sharing, they preferred the slightly reduced choice, although the difference between these two part-worth values was small (40.29 and 37.65, respectively). These findings are in line with the results of our BYO exercise, which shows that this group found no added value if e-scooter sharing was included in the package. Regarding age, the price-sensitive worriers were significantly younger than the non-adopters ($p = 0.001$). Further, they expressed a significantly higher willingness to adopt MaaS packages (61.3%) than the non-adopters group (36.4%, $p < 0.001$).

We labeled the third adopter segment non-adopters ($n = 77$, 31.2%). Our part-worth estimates revealed the highest value of the none option in this group (353.73), indicating that these respondents most often chose the none option in the choice tasks, i.e., they did not prefer any of the three presented MaaS bundle variations. Further, in our direct question on the general willingness to adopt MaaS bundles, this group was the least likely of all three segments (approx. 36%) to adopt a MaaS package. They also showed a significantly lower directly indicated WTP in relation to the other two segments (convenience seekers $p = 0.001$; price-sensitive worriers $p = 0.023$). Although, this segment scored the lowest on the willingness to adopt MaaS bundles, some other relevant criteria were insightful and distinctive. This group stood out regarding the relatively high importance it assigned to the attribute contract termination modes (33.1%) and its desire for flexibility in contract termination. In contrast to the other segments, this group's respondents were significantly older ($p < 0.001$), with an average age of 52.2. Overall, compared to the other two segments this group had the lowest share of car owners (83.1%), those who did own cars use them the least often (only 2.6% indicated using their cars on a daily basis), and, apparently due to low mobility needs, had the lowest monthly costs for a car (approx. €226), which might partly explain their low interest in MaaS bundles.

4.4. Sensitivity Analysis on Segment Level

Building on the LCA results, we conducted a sensitivity analysis to test for differences in the segments' preferences (randomized first choice) regarding variations in the MaaS bundle features. Following here, we present the results of four market simulations and graphically summarize these results in Figure 4, below.

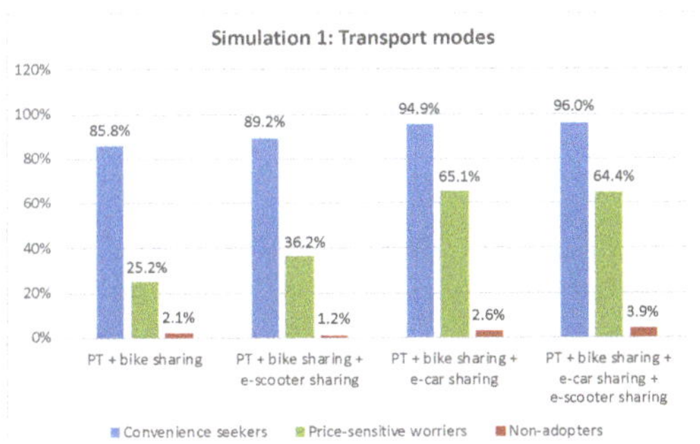

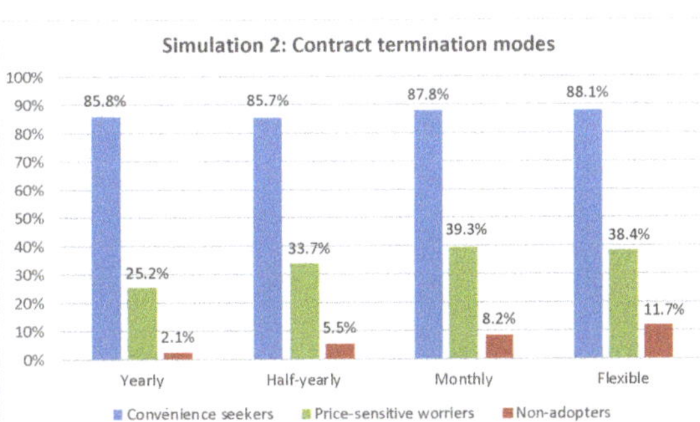

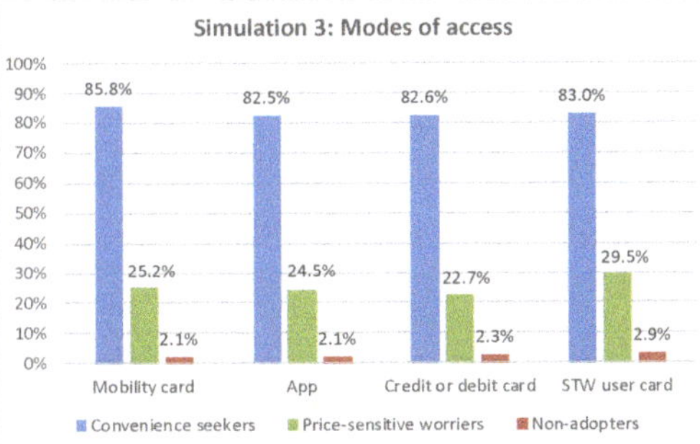

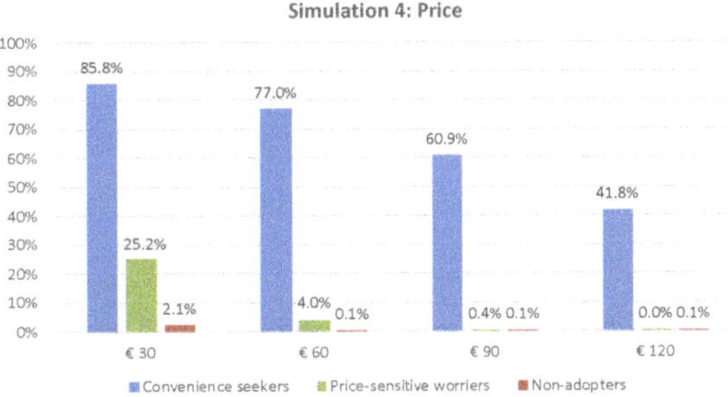

Figure 4. Summary of market simulations on segment level (% of respondents).

4.4.1. Simulation 1: Transport Modes

For the first simulation we designed MaaS packages to evaluate the effect of different transport mode combinations on the segments' choice. All created packages differed only regarding the transport mode attribute: (1) PT + bike sharing, (2) PT+ bike sharing + e-scooter sharing, (3) PT+ bike sharing + e-car sharing, and (4) PT+ bike sharing + e-car sharing + e-scooter sharing. The other attributes were held constant at an annual subscription, an own mobility card as access mode, and a price of €30 per month (This price level seems to be realistic for a baseline offering of mobility services (i.e., PT + bike sharing) as a monthly ticket at the Stadtwerke Klagenfurt would cost a city of Klagenfurt resident €25 (see https://www.stw.at/privat/mobilitaet/tickets-tarife-und-abos/#30-tag e-karte-umweltschutz), while they charge €4 per month for using nextbike's bike sharing services. An annual subscription mode would allow for lower prices per month if more transport modes were added to the bundle, and an own mobility card for accessing the services would allow a higher level of independence from only one provider.). The results showed that the largest share of respondents in the convenience seekers' segment (96%) would prefer a bundle that includes PT+ bike sharing + e-car sharing + e-scooter sharing, all else being equal. However, the difference between the most and second most preferred levels (PT+ bike sharing + e-car sharing) was quite small (1.1 percentage points), which suggested that e-scooter sharing provided these respondents only marginal value. Overall this segment showed a very high preference (> 85% of the respondents in this segment would choose this bundle in the market) for each of the bundle combinations, which underlined their high willingness to adopt MaaS offerings. The group of price-sensitive worriers most preferred a package including a combination of PT+ bike sharing + e-car sharing as mobility services (65.1%). In contrast, the non-adopters showed consistently low preference for all the package combinations in this first simulation ($< = 3.9\%$).

4.4.2. Simulation 2: Contract Termination Modes

In the second simulation, we tested for preference share sensitivity regarding the contract termination mode. We varied the attribute according to its four levels (annually, bi-annually, monthly, flexible), while other dimensions were held stable as in the first simulation (the attribute transport modes remained constant at PT + bike sharing). The second simulation results revealed similar tendencies to those of the first simulation. The convenience seekers most preferred a package with a flexible contract termination mode and the price-sensitive worriers one with a monthly subscription. However, the difference between the most and the second most preferred levels of the convenience seekers (88.1% and 87.8% respectively) and the price-sensitive worriers (39.3% and 38.4%, respectively) groups was only marginal. Again, this second market simulation highlighted the significant willingness of respondents in the convenience seekers segment (>85%) to adopt a MaaS package.

4.4.3. Simulation 3: Modes of Access

For our third simulation we varied the modes of access (mobility card, app, credit or debit card, and Stadtwerke Klagenfurt user card) to our baseline MaaS offering, keeping all other attributes stable. The results indicated that the mobility card was the most preferred access mode among the respondents of the convenience seekers segment (85.8%), which was the access mode we chose for our baseline MaaS bundle. The price-sensitive worriers most preferred the Stadtwerke Klagenfurt user card, where 29.5% would be willing to buy a MaaS package with such an access mode. The BYO exercise disclosed that the most attractive access mode for all groups, would be the app. If the mode of access were switched to this level (app), then the two likely adopter groups' share of preference would drop to 82.5% and 24.5%, respectively.

4.4.4. Simulation 4: Price (Per Month)

The last sensitivity analysis varies the price level (€30, €60, €90 and €120), keeping all other variables equal. This simulation reflected the tremendous importance of price for the respondents in our sample. Specifically, the convenience seekers' share of preference for our baseline product (85.8%) would significantly drop to nearly half of its value, i.e., 41.8%, if the price were to increase from €30 to €120. However, compared to the other two groups, this segment remained the one with the highest willingness to pay in monetary terms, as even at a price level of €120, a share of more than 40% of the respondents in this sample would opt for such a MaaS bundle. The share of price-sensitive worriers would drop from 25.2% for the reference product to 4.0% if the price of the MaaS package were doubled to €60.

4.5. Insights on Electric Vehicles as a Part of the MaaS Packages

Besides considering the respondents' preferences for various features of MaaS packages, this study took a closer look at the role of EVs, i.e., e-car sharing and e-scooter sharing, in MaaS bundle adoption. Our CBC experiment's results suggest that the respondents in our sample strongly prefer a MaaS package including e-car sharing to one including e-scooter sharing. The differences between the part-worth utilities of the transport mode levels clearly suggested this in that, compared to PT + bike sharing (reference level) utility increased by 26.20 points if e-scooter sharing was added to the bundle (see Table 3), whereas it increased by 71.62 points if e-car sharing was added. Additionally, the difference between the level PT + bike sharing + e-car sharing and the level which included all transport modes, was only 11.03 utility points. This finding was also reflected in our CBC experiment's segment level results for our two potential adopter groups. For the price-sensitive worriers, part-worth utility even decreased from 40.29 to 37.65 (see Table 4) if e-scooter sharing was added to a bundle that has PT + bike sharing + e-car sharing. The results of the BYO exercise provided an even clearer picture (Table 7): 14.6% of the respondents would opt for a bundle including PT + bike sharing + e-scooter sharing and about one third (34.8%) would opt for the same bundle that had e-car sharing, instead.

Further, we asked the respondents in the questionnaire part of our survey to indicate what transport modes they would include in a MaaS bundle. The survey participants rated several transport modes on a scale ranging from 1 = "should not be included at all" to 4 = "should definitely be included". Table 6 below shows the share of respondents that selected each of these scale values, and Table 6 gives an overview of the rating means of the segments compared to the overall sample. In the last column of this table, we summed up the shares for the two highest preference levels (should definitely be included and should be included). The results show that e-car sharing was selected by 68.8% of the respondents to be (definitely) included in a MaaS bundle, and thus ranked third after two modes representing PT. Respondents preferred e-car sharing even more than conventional car sharing (63.2%). Interestingly, bike sharing and e-bike sharing were equally preferred (58.3%). Of the respondents, 43.3% selected e-scooter sharing to be included in a MaaS bundle, whereas only 27.5% of the respondents selected e-load-cycle sharing that we had also included in the list.

Table 6. Results for direct question on preferred transport modes to be included in MaaS bundle.

	Should Definitely Be Included	Should Be Included	Should Not Be Included	Should Not Be Included at All	Sum of Columns "Should (Definitely) Be Included"
PT (e.g., bus, tram, metro)	78.14%	14.57%	2.43%	4.86%	92.71%
Train (e.g., regional train)	42.11%	30.77%	14.57%	12.55%	72.87%
E-car sharing	38.06%	30.77%	11.34%	19.84%	68.83%
Car sharing	33.20%	29.96%	15.38%	21.46%	63.16%
Bike sharing	29.55%	28.74%	15.38%	26.32%	58.30%
E-bike sharing	29.55%	28.74%	14.98%	26.72%	58.30%
Micro-PT	21.86%	22.27%	19.84%	36.03%	44.13%
E-scooter sharing	23.08%	20.24%	15.38%	41.30%	43.32%
Taxi	19.84%	18.62%	15.38%	46.15%	38.46%
E-load-cycle sharing	12.15%	15.38%	22.27%	50.20%	27.53%
Other	4.17%	5.56%	6.94%	83.33%	9.72%

5. Discussion and Conclusions

Based on the fact that the transport sector is the highest contributor to greenhouse gas emissions [73], Collins et al. (p. 640, [74]) noted that selecting travel modes is "among the most environmentally-significant decisions faced by individuals". Thus, in the context of the present study and following recent calls in the literature e.g., [21,45], we investigated individuals' preferences for MaaS bundles that include more sustainable transport modes (i.e., PT, bike sharing, e-car sharing, e-scooter sharing) with the ultimate goal of decreasing individual car ownership and usage. In contrast to existing studies that have mainly focused on urban areas, we aimed to gain insight on potential adopters' preferences for MaaS packages in sub-urban areas. Further, the study intended to shed light on the impact EVs, i.e., e-cars and e-scooters, included in MaaS bundles would have on the adoption of such multimodal mobility services.

5.1. Discussion of Results and Implications

The study's findings highlight the importance of a MaaS bundle's price regarding its acceptance, as other researchers have shown in the past, studying different geographical contexts e.g., [21,22,25,46]. The CBC results reveal that price contributes nearly 50% (49.10%) to the overall utility of the average respondent in our sample. Directly asked, the average study participant would be willing to pay approx. €52 for a MaaS package that meets their preferences. Although the LCA results show that there are differences between the segments in the relative importance values they assign to the price attribute, they are still at a high level (e.g., the lowest relative importance value is 43.7% for the non-adopters group). The convenience seekers segment shows the highest WTP in both the CBC analysis and the direct question implemented in the questionnaire part of the survey. The differences in the part-worth values for the levels €30 (106.25) and €60 (50.60) for the convenience seekers is much smaller (55.65 utility points) than for the other potential adopter group, i.e., the price-sensitive worriers, (95.67 utility points). The differences between the two groups are quite similar in the part-worth utilities for the higher price levels. This group difference in price sensitivity is also reflected in the simulation results. Moreover, average direct WTP indications significantly differ ($p < 0.001$) between the convenience seekers (€62.3) and the price-sensitive worriers (€54.8). The implications of these findings are that MaaS package providers need to consider varying WTP thresholds between different potential user segments, such as offering price-differentiated bundles (e.g., services or features that are included or excluded in the different bundles). Further, MaaS bundles' price could be decreased by government subsidies (e.g., for using PT, as is already the case in Vienna where an annual PT pass at €365 costs comparatively much less than a 24 h-ticket at €8, and as is currently under discussion for roll-out in a similar way nationally, as the 1-2-3-ticket). However, recent scholarly work suggests that price sensitivity related to MaaS adoption is more pronounced in the early stages of market diffusion and might decrease with the maturity of the market and MaaS technology (e.g., an app) due to commuters' increased experience with the new offerings [75].

The findings regarding the overall price the respondents in our study are willing to pay for MaaS services are related to the question of what the respondents are willing to pay for specific features of a MaaS offering. Our findings show that the second most important attribute in our stated preference analysis is transport mode (27.28%), i.e., what (type and quantity) of mobility services are included in the MaaS bundle. Overall, our analysis shows that the respondents prefer larger bundles (i.e., highest part-worth utility for PT + bike sharing + e-car sharing + e-scooter sharing; 37.54 part-worth utility) to smaller bundles (i.e., PT + bike sharing; −45.12 part-worth utility). On a more detailed level, our results suggest that the respondents do not perceive as much value in e-scooter sharing as in e-car sharing. If asked directly, almost 70% of the respondents in our sample would choose e-car sharing while only about 43% would choose e-scooter sharing as a basic component of MaaS packages they would purchase. Besides the fact that e-car sharing is a better substitute to private car ownership than e-scooter sharing, we found three reasons to be quite plausible explanations of these findings. First, e-scooter sharing is still quite novel in Austria, and more so in Klagenfurt. Although literature on product or service bundling suggests that product spill-over effects from more mature to less mature products or services increase the acceptance of the overall bundle [76–78], we cannot find such an effect in our survey results. Generally, the experience with sharing services is rather low in our sample, although the experience with (e-)scooter sharing is comparatively higher than for the other sharing modes (see Table 5). Second, an explanation for the lower preference for e-scooter sharing might be that e-scooters suffer from the same disadvantages as bikes such as in being dependent on weather conditions. Thus, as bike sharing, together with PT, was already included as baseline elements of the MaaS bundle, the respondents in our sample might not have perceived additional value in the mobility services provided by e-scooter sharing. Third, research on product/service bundling emphasizes the importance of value-adding complementarity of the products or services in the bundles, because otherwise consumers are more likely to purchase the products or services separately e.g., [77,79]. This agrees with Guidon et al.'s [45] findings, which showed that complementary mobility services such as PT and car sharing achieved the highest WTP among their respondents. The particular reasons at play regarding the lower preference for e-scooter sharing and similar transport modes (e.g., e-bike sharing) does, however, need further research. Nevertheless, these findings have important implications for the configuration of MaaS packages.

We found the other attributes we included in the CBC experiment (contract termination modes, modes of access) of marginal relative importance regarding respondents' willingness to adopt MaaS bundles (13.19% and 10.42%, respectively); however, we noticed some interesting differences between the three segments. For instance, one third (33.1%) of the overall utility of the average respondent in the non-adopters group can be attributed to the contract termination mode. This segment specifically prefers flexible cancellation options, which could be a potential lever to convince current non-adopters to try out MaaS offerings, as they would be able to terminate the subscription easily and on short notice if they do not find it useful or practical. Further, the mode of access seems to be of marginally higher importance to the convenience seekers who would prefer an own mobility card rather than the Stadtwerke Klagenfurt user card, which was the most preferred level among the respondents of the other two groups. However, as we found contradictory results in our BYO exercise revealing the highest preference for accessing the services via a smartphone app (43.7% of the respondents), we encourage scholars to conduct further research in this field.

Overall our results show that the willingness to adopt MaaS bundles is rather high in our sample (56.7% indicate that they would (very) likely adopt a MaaS bundle that meets their preferences). However, whether the respondents in our sample would actually decrease individual car usage by adopting MaaS packages, remains an open question, which can only be answered in a real-life setting and by applying a longitudinal study design. Studies that investigated the success of MaaS pilot projects report that, at least, the immediate effects on the reduction of car usage are positive e.g., [41], but the sustainability of such behavioral changes is still unclear. Further, our study confirms the high preference for PT to be included in MaaS packages, as other scholars have also found in the past e.g., [21,24,45,47]. For instance, 92.7% of the respondents indicated that PT (e.g., bus, tram, metro) and 72.9% indicated that trains (e.g., regional trains) should (definitely) be included in a MaaS bundle.

Finally, we also tested for differences between the identified segments in socio-demographic, behavioral (mobility behavior), and psychological characteristics. Surprisingly, we did not find statistically significant differences between the two potential adopter groups and the non-adopters regarding sustainability-related psychological or behavioral variables investigated in our study, such as climate change concern, pro-environmental attitude or energy-saving behavior. This finding seems to refer to the often cited "attitude-behavior gap" or "attitude-intention-behavior gap" (for a recent literature review see e.g., [80]), explaining why pro-environmental attitudes often do not transform into pro-environmental behavior, or in our case, into purchase intention of environmentally friendly products. Further, in line with our results, Schikofsky et al.'s [81] qualitative study reports that pro-environmental attitudes are of "subordinated relevance" for the intention to adopt MaaS packages. However, they found that psychological needs such as autonomy, competence and relatedness to a social peer group are good predictors of the intention to use MaaS offerings. Thus, future research might further investigate the socio-psychological determinants of the intention to adopt MaaS offerings, as such insights could have important implications for how service providers should design offerings and marketing campaigns and what motives and attitudes of potential adopters they seek to address.

5.2. Limitations and Further Research

As with any research, the results of our study are subject to some limitations. First, it is important to mention that we drew the study sample from a sub-urban district in Klagenfurt, Carinthia. Thus, our analysis is limited to a specific regional section of the Austrian population. In future, research should test our findings in other (sub-urban) regions in Austria and in different demographic and cultural contexts. Second, looking more closely at methodological limitations, although the MaaS packages we tested were developed to be as realistic as possible, respondents' decisions would still not be the same as real-life decisions in purchasing such packages and actually paying for these services. Future studies could compare stated preference data, like ours, with revealed preferences for MaaS offerings actually implemented in the market. Third, in the conjoint part of our study only a limited number of attributes and attribute levels were included. Building on a literature review, as well as on interviews with experts, stakeholders, potential adopters and the likes, there is a wide pallet of factors potentially influencing travel mode choice generally, and affecting purchase intention of MaaS packages, specifically (e.g., availability of transport modes within a certain distance to home/work, transfer time between modes, employers' financial support of MaaS bundles, etc.). Further, we would encourage future research that combines preference data with data from secondary sources, such as transport sector-related characteristics of the cities or rural areas the respondents live in. For instance, [30] used local characteristics of 15 European cities, such as weather, PT satisfaction, density, city structure and the likes, to design multimodal mobility packages customized to the particularities of the cities. Additional analyses could further investigate potential supply-side effects of different local or regional MaaS offerings (e.g., implications for raw material and energy demand, settlement and housing patterns, etc.) [82]. Finally, although, we included a few psychological variables in our study, this still represents only a fraction of the individual characteristics that could be important for understanding potential MaaS package adopters' preferences and for distinguishing between potential adopters and non-adopters. Future research should take this into account, and could use this study as a point of departure, considering a range of other factors not included here but found to be relevant in general literature on travel mode choice (e.g., affective factors such as comfort, symbolic factors such as status and prestige, etc.). For instance, we would anticipate that respondents who pay high attention to status symbols are less likely to adopt MaaS packages e.g., [83].

Author Contributions: Conceptualization, P.B. and N.H.; methodology, P.B. and N.H.; formal analysis, P.B.; investigation, N.H.; data curation, P.B.; writing—original draft preparation, P.B.; writing—review and editing, N.H.; supervision, N.H.; project administration, P.B.; funding acquisition, N.H. All authors have read and agreed to the published version of the manuscript.

Funding: This study was conducted in the course of the hi MOBIL project, which received funding from the Austrian Ministry for Climate Action, Environment, Energy, Mobility, Innovation and Technology.

Data Availability Statement: The data presented in this study are available on request from the corresponding author. The data are not publicly available due to privacy reasons.

Acknowledgments: We would like to thank Isabell Lehnertz for her valuable support in the course of this study and two anonymous reviewers for their detailed comments and suggestions.

Conflicts of Interest: The authors declare no conflict of interest. The funding agency had no role in the design of the study; in the collection, analyses, or interpretation of data; in the writing of the manuscript, or in the decision to publish the results.

Appendix A

Table 1. Description of the sample and the segments.

Variable	Sample	Convenience Seekers	Price-Sensitive Worriers	Non-Adopters
Demographic Variables				
Age	46.6	44.47	43.68	52.21
Gender (man)	34.0%	32.5%	35.5%	33.0%
Education				
Compulsory school	6.1%	5.2%	2.2%	11.7%
Vocational training	39.3%	44.2%	39.8%	33.8%
High school	25.5%	24.7%	26.9%	24.7%
Collage/university	29.1%	26.0%	31.2%	29.9%
Income	€2513.1	€2664.75	€2601.24	€2255.12
Mobility Behavior Related Variables				
WTP for MaaS	€51.88	€62.32	€54.77	€37.95
WTA [1] MaaS ((most) likely)	56.7%	71.4%	61.3%	36.4%
Car ownership (at least one car)	89.5%	93.5%	91.4%	83.1%
Daily car usage (as driver)	8.5%	13%	9.7%	2.6%
Cost for car (per month)	€238.52	€257.38	€230.42	€225.94
Socio-Psychological Variables				
Climate change concerns	3.86	3.87	3.88	3.82
Pro-environmental attitude	3.36	3.40	3.30	3.41
Values (Schwartz's value scale)	3.80	3.95	3.72	3.74
Energy-saving behavior	3.02	3.07	2.97	3.03
Subjective norm	2.82	3.13	2.73	2.61
Intention towards using PT	2.63	2.78	2.60	2.52

[1] Willingness to adopt.

Table 2. Psychographic measurement scales and items with origin.

Scheme	Psychological Items	Origin
Climate change concerns (1 = not worried at all; 5 = very worried)	How worried are you, if at all, about climate change?	Leiserowitz et al. (2019) (N/A)
Pro-environmental attitude (α = 0.89) (1 = not agree at all; 4 = fully agree)	I would say of myself that I am environmentally conscious. Being environmentally friendly is an important part of my personality. I would describe myself as someone who cares about the environment.	Whitmarsh and O'Neill (2010) (α = 0.70)
Values ((short) Schwartz's value scale) (α = 0.70) (1 = not important at all; 6 = very important)	POWER (social power, authority, wealth) ACHIEVEMENT (success, capability, ambition, influence on people and events) HEDONISM (gratification of desires, enjoyment in life, self-indulgence) STIMULATION (daring, a varied and challenging life, an exciting life) SELF-DIRECTION (creativity, freedom, curiosity, independence, choosing one's own goals) UNIVERSALISM (broad-mindedness, beauty of nature and arts, social justice, a world at peace, equality, wisdom, unity with nature, environmental protection) BENEVOLENCE (helpfulness, honesty, forgiveness, loyalty, responsibility) TRADITION (respect for tradition, humbleness, accepting one's portion in life, devotion, modesty) CONFORMITY (obedience, honoring parents and elders, self- discipline, politeness) SECURITY (national security, family security, social order, cleanliness, reciprocation of favors)	Schwartz (1992, 1996) (α = 0.89)
Energy-saving behavior (α = 0.68) (1 = never; 4 = always)	Turn off lights, the computer and other electronic devices when they are not needed. Using warmer clothes at home instead of heating more. Disconnect the phone and other devices to be charged as soon as they are fully charged. Walk short distances or use the bike. Choose holiday destinations that do not require a flight. Consume seasonal, local and organic foods. If possible, avoid the consumption of meat products. Engage for energy saving measures writing letters politicians and/or the employer, as well as similar actions.	Spence, Leygue, Bedwell and O'Malley (2014) (α = 0.60)
Subjective norm (α = 0.87) (1 = not agree at all; 4 = fully agree)	Most people who are important to me would support me using public transport for daily travel. Most people who are important to me think that I should use public transport for daily travel. Most people who are important to me think it is good if I would give up on car.	Bamberg, Rölle and Weber (2003) (N/A)
Intention towards using PT (1 = not agree at all; 4 = fully agree)	My intention to use public transportation on my regular trips (shopping, leisure, university, work) is strong.	Friedrichsmeier, Matthies and Klöckner (2013)

Table 3. ANOVA summary table of differences between segments [1,2].

Source	Sum of Squares	df	Mean Square	F	*p*
Age [a,b]	3565.88	2	1782.94	8.12	0.000
WTA [c,d]	32.86	2	16.43	12.04	0.000
WTP [e,f]	24,124.66	2	12,062.33	7.09	0.001
Values [g,h]	2.73	2	1.36	5.56	0.004
Subjective norm [i,j]	11.25	2	5.62	8.94	0.000

[a] Convenience seeker vs. non-adopters—*p* = 0.004, [b] Non-adopters vs. price-sensitive worriers—*p* = 0.001, [c] Convenience seeker vs. non-adopters—*p* = 0.000, [d] Non-adopters vs. price-sensitive worriers—*p* = 0.001, [e] Convenience seeker vs. non-adopters—*p* = 0.001, [f] Non-adopters vs. price-sensitive worriers—*p* = 0.023, [g] Convenience seeker vs. non-adopters—*p* = 0.023, [h] Convenience seekers vs. price-sensitive worriers—*p* = 0.006, [i] Convenience seeker vs. non-adopters—*p* = 0.000, [j] Convenience seekers vs. price-sensitive worriers—*p* = 0.004. [1] Post hoc test after Turkey HDS. [2] Variables not listed in this table did not reveal statistically significant differences between the segments.

Table 4. Summary table for usage frequency of different transport modes (% of the respondents).

Usage Frequency of Different Transport Modes	Daily	Min. Once a Week	Min. Once a Month	Few Times per Year	Rarely or Never
Car (as driver)	8.5	20.6	6.5	12.6	51.8
Car (as co-driver)	4.9	12.6	11.3	17.8	53.4
PT (e.g., bus, tram, metro)	12.6	8.9	17.8	27.9	32.8
Train (e.g., regional train)	1.2	2.0	12.1	33.6	51.0
Motorbike	1.2	4.0	3.2	2.0	89.5
Bike	25.9	21.9	12.1	16.2	23.9
E-bike	2.0	2.8	1.2	2.8	91.1
Walking	59.5	23.5	10.1	2.0	4.9
Micro-PT	—	—	—	5.3	92.7
Taxi	—	0.8	6.1	38.5	54.7
(E-)load-cycle	—	-	-	1.2	98.8
(E-)scooter	—	1.6	3.6	6.1	88.7

Table 5. Summary table for usage frequency of different sharing modes (% of the respondents).

Usage Frequency of Different Sharing Modes	Daily	Min. Once a Week	Min. Once a Month	Few Times per Year	Rarely or Never
Car sharing	0.8	0.8	2.0	3.6	92.7
E-car sharing	—	—	—	2.0	98.0
Bike sharing	—	—	1.6	2.8	95.5
E-bike sharing	—	—	—	1.6	98.4
(E-)load-cycle sharing	—	—	0.4	—	99.6
(E-)scooter sharing	—	0.8	3.2	4.5	91.5

Table 6. Results for direct question on preferred transport modes to be included in MaaS bundle (mean).

	Sample	Convenience Seeker	Price-Sensitive Worrier	Non-Adopter
PT (e.g., bus, tram, metro)	3.68	3.57	3.78	3.65
Train (e.g., regional train)	3.03	2.88	2.91	3.42
E-car sharing	2.88	2.90	2.83	2.94
Car sharing	3.08	3.18	3.06	2.96
Bike sharing	2.79	2.76	2.99	2.52
E-bike sharing	2.81	2.83	2.87	2.69
Micro-PT	2.37	2.32	2.27	2.60
E-scooter sharing	2.07	1.88	2.09	2.31
Taxi	1.99	1.96	1.92	2.15
E-load-cycle sharing	2.39	2.49	2.53	2.02
Other	1.30	1.41	1.29	1.20

Table 7. Results of BYO exercise (% of the respondents).

Attributes	Sample	Connivence Seekers	Non-Adopters	Price-Sensitive Worriers
Transport Modes				
PT + bike sharing	23.9	23.4	37.7	12.9
PT + bike sharing + e-scooter sharing	14.6	18.2	10.4	15.1
PT + bike sharing + e-car sharing	34.8	28.6	28.6	45.2
PT + bike sharing + e-car sharing + e-scooter sharing	26.7	29.9	23.4	26.9
Contract Termination Modes				
Annually	5.7	6.5	6.5	4.3
Bi-annually	8.9	10.4	3.9	11.8
Monthly	35.6	35.1	31.2	39.8
Flexible	49.8	48.1	58.4	44.1
Modes of Access				
Mobility card	20.6	22.1	22.1	18.3
Smartphone (via app)	43.7	46.8	33.8	49.5
Credit or debit card	13.8	14.3	13.0	14.0
Stadtwerke Klagenfurt user card	21.9	16.9	31.2	18.3

References

1. Ellen MacArthur Foundation. Towards a Circular Economy-Economic and Business Rationale for an Accelerated Transition. Greener Management International, 97. Available online: https://doi.org/2012-04-03 (accessed on 25 November 2019).
2. Hensher, D.A. Future bus transport contracts under a mobility as a service (MaaS) regime in the digital age: Are they likely to change? *Transp. Res. Part. A Policy Practice* **2017**, *98*, 86–96. [CrossRef]
3. Bert, J.; Collie, B.; Gerrits, M.; Xu, G. What's Ahead for Car Sharing? The New Mobility and Its Impact on Vehicle Sales. Available online: https://trid.trb.org/view/1399142 (accessed on 29 November 2020).
4. Grosse-Ophoff, A.; Hausler, S.; Heineke, K.; Möller, T. *How Shared Mobility Will Change the Automotive Industry*; McKinsey & Company: Washington DC, USA, 2017.
5. Global Market Insights. Car Sharing Market. Statistics-Global Growth Trends 2026. Available online: https://www.gminsights.com/industry-analysis/carsharing-market (accessed on 29 November 2020).
6. Shaheen, S.A.; Cohen, A.P. Carsharing and personal vehicle services: Worldwide market developments and emerging trends. *Int. J. Sustain. Transp.* **2013**, *7*, 5–34. [CrossRef]
7. Cooper, P.; Tryfonas, T.; Crick, T.; Marsh, A. Electric vehicle mobility-as-a-service: Exploring the "Tri-Opt" of novel private transport business models. *J. Urban. Technol.* **2019**, *26*, 35–56. [CrossRef]
8. Gnann, T.; Plötz, P.; Funke, S.; Wietschel, M. What is the market potential of plug-in electric vehicles as commercial passenger cars? A case study from Germany. *Transp. Res. Part. D Transp. Environ.* **2015**, *37*, 171–187. [CrossRef]
9. Paffumi, E.; de Gennaro, M.; Martini, G.; Scholz, H. Assessment of the potential of electric vehicles and charging strategies to meet urban mobility requirements. *Transportmetrica A Transport. Sci.* **2015**, *1*, 22–60. [CrossRef]
10. Creutzig, F.; Ravindranath, N.H.; Berndes, G.; Bolwig, S.; Bright, R.; Cherubini, F.; Fargione, J. Bioenergy and climate change mitigation: An assessment. *Gcb Bioenerg.* **2015**, *7*, 916–944. [CrossRef]
11. Lutsey, N. Global climate change mitigation potential from a transition to electric vehicles. *Int. Counc. Clean Transp.* **2015**, *2015*, 5.
12. Turcksin, L.; Mairesse, O.; Macharis, C. Private household demand for vehicles on alternative fuels and drive trains: A review. *Eur. Transport. Res. Rev.* **2013**, *5*, 149–164. [CrossRef]
13. Ambrosino, G.; Nelson, J.D.; Boero, M.; Pettinelli, I. Enabling intermodal urban transport through complementary services: From flexible mobility services to the shared use mobility agency: Workshop 4. Developing inter-modal transport systems. *Res. Transp. Econ.* **2016**, *59*, 179–184. [CrossRef]
14. Ho, C.; Hensher, D.A.; Mulley, C.; Wong, Y. Prospects for switching out of conventional transport services to mobility as a service subscription plans–A stated choice study. In Proceedings of the International Conference Series on Competition and Ownership in Land Passenger Transport, (Thredbo 15), Stockholm, Sweden, 13–17 August 2017.
15. Li, Y.; Voege, T. Mobility as a service (MaaS): Challenges of implementation and policy required. *J. Transp. Technol.* **2017**, *7*, 95–106. [CrossRef]
16. Gould, E.; Wehrmeyer, W.; Leach, M. Transition pathways of e-mobility services. *WIT Trans. Ecol. Environ.* **2015**, *194*, 349–359.
17. Smith, G.; Sarasini, S.; Karlsson, I.M.; Mukhtar-Landgren, D.; Sochor, J. Governing Mobility-as-a-Service: Insights from Sweden and Finland. In *The Governance of Smart Transportation Systems*; Springer: Cham, Germany, 2019; pp. 169–188.

18. Karlsson, I.C.M.; Mukhtar-Landgren, D.; Smith, G.; Koglin, T.; Kronsell, A.; Lund, E.; Sochor, J. Development and implementation of Mobility-as-a-Service–A qualitative study of barriers and enabling factors. *Transp. Res. Part. A Policy Practice* **2020**, *131*, 283–295. [CrossRef]

19. Mulley, C.; Nelson, J.D.; Wright, S. Community transport meets mobility as a service: On the road to a new a flexible future. *Res. Transp. Econ.* **2018**, *69*, 583–591. [CrossRef]

20. Audouin, M.; Finger, M. The development of mobility-as-a-service in the Helsinki metropolitan area: A multi-level governance analysis. *Res. Transp. Bus. Manag.* **2018**, *27*, 24–35. [CrossRef]

21. Caiati, V.; Rasouli, S.; Timmermans, H. Bundling, pricing schemes and extra features preferences for mobility as a service: Sequential portfolio choice experiment. *Transp. Res.Part. A Policy Practice* **2020**, *131*, 123–148. [CrossRef]

22. Ho, C.Q.; Hensher, D.A.; Mulley, C.; Wong, Y.Z. Potential uptake and willingness-to-pay for Mobility as a Service (MaaS): A stated choice study. *Transp. Res. Part. A Policy Practice* **2018**, *117*, 302–318. [CrossRef]

23. Kamargianni, M.; Li, W.; Matyas, M.; Schäfer, A. A critical review of new mobility services for urban transport. *Transp. Res. Proced.* **2016**, *14*, 3294–3303. [CrossRef]

24. Matyas, M.; Kamargianni, M. The potential of mobility as a service bundles as a mobility management tool. *Transportation* **2018**, *46*, 1–18. [CrossRef]

25. Mulley, C.; Ho, C.; Balbontin, C.; Hensher, D.; Stevens, L.; Nelson, J.D.; Wright, S. Mobility as a service in community transport in Australia: Can it provide a sustainable future? *Transp. Res. Part A Policy Practice* **2020**, *131*, 107–122. [CrossRef]

26. Goodall, W.; Dovey, T.; Bornstein, J.; Bonthron, B. The rise of mobility as a service. *Deloitte Rev.* **2017**, *20*, 112–129.

27. Kostiainen, J.; Tuominen, A. Mobility as a service—Stakeholders' challenges and potential implications. In *Towards User-Centric Transport in Europe*; Springer: Cham, Germany, 2019; pp. 239–254.

28. Utriainen, R.; Pöllänen, M. Review on mobility as a service in scientific publications. *Res. Transp. Bus. Manag.* **2018**, *27*, 15–23. [CrossRef]

29. Aapaoja, A.; Eckhardt, J.; Nykänen, L.; Sochor, J. MaaS service combinations for different geographical areas. In Proceedings of the 24th world congress on intelligent transportation systems, Montreal, QC, Canada, 29 October–2 November 2017.

30. Esztergár-Kiss, D.; Kerényi, T. Creation of mobility packages based on the MaaS concept. *Travel Behav. Soc.* **2019**, *21*, 307–317. [CrossRef]

31. Wright, S.; Nelson, J.D.; Cottrill, C.D. MaaS for the suburban market: Incorporating carpooling in the mix. *Transp. Res. Part. A Policy Practice* **2020**, *131*, 206–218. [CrossRef]

32. Heikkilä, S. Mobility as a Service–A Proposal for Action for the Public Administration, Case Helsinki. Master's Thesis, Aalto University, Espoo, Finland, 2014.

33. Giesecke, R.; Surakka, T.; Hakonen, M. *Conceptualising Mobility as a Service*; IEEE: Piscataway, NJ, USA, 2016; pp. 1–11.

34. Sochor, J.L.; Strömberg, H.; Karlsson, M. An innovative mobility service to facilitate changes in travel behavior and mode choice. In Proceedings of the 22nd World Congress on Intelligent Transportation Systems, Bordeaux, France, 5–9 October 2015.

35. Sochor, J.; Strömberg, H.; Karlsson, I.C.M. Implementing Mobility as a Service challenges in integrating user, commercial, and societal perspectives. *Transp. Res. Record* **2015**, *4*, 1–9. [CrossRef]

36. Sochor, J.; Strömberg, H.; Karlsson, I.C.M. Trying out mobility as a service experiences from a field trial and implications for understanding demand. In Proceedings of the 95th Transportation Research Board Annual Meeting, Washington, DC, USA, 1 January 2016.

37. Karlsson, I.M.; Sochor, J.; Strömberg, H. Developing the 'Service'in Mobility as a Service: Experiences from a field trial of an innovative travel brokerage. *Transp. Res. Proced.* **2016**, *14*, 3265–3273. [CrossRef]

38. Whim. All Your Journeys. Available online: https://whimapp.com (accessed on 2 April 2019).

39. Axhausen, K.W.; Simma, A.; Golob, T.F. Pre-commitment and usage: Season tickets, cars and travel. *Arbeitsb. Verk. Raumplan.* **2000**, *495*, 24.

40. Lathia, N.; Capra, L. *How Smart is your Smartcard? Measuring Travel Behaviours, Perceptions, and Incentives*; ACM: Beijing, China, 2011; pp. 291–300.

41. Schad, H.; Flamm, M.; Wagner, C.; Frey, T. *New, Integrated Mobility Services, NIM*; Project A3 of the National Research Programme (NRP): Bern, Switzerland, 2005.

42. Thøgersen, J. Promoting public transport as a subscription service: Effects of a free month travel card. *Trans. Policy* **2009**, *16*, 335–343. [CrossRef]

43. Huwer, U. Public transport and car-sharing—benefits and effects of combined services. *Transport. Policy* **2004**, *11*, 77–87. [CrossRef]

44. Durand, A.; Harms, L.; Hoogendoorn-Lanser, S.; Zijlstra, T. Mobility-as-a-Service and Changes in Travel Preferences and Travel Behaviour: A Literature Review. KiM | Netherlands Institute for Transport Policy Analysis: Hague, The Netherlands, 2018.

45. Guidon, S.; Wicki, M.; Bernauer, T.; Axhausen, K. Transportation service bundling–for whose benefit? Consumer valuation of pure bundling in the passenger transportation market. *Transp. Res. Part. A Policy Pract.* **2020**, *131*, 91–106. [CrossRef]

46. Ho, C.Q.; Mulley, C.; Hensher, D.A. Public preferences for mobility as a service: Insights from stated preference surveys. *Transp. Res. Part. A Policy Practice* **2020**, *131*, 70–90. [CrossRef]

47. Matyas, M.; Kamargianni, M. Survey design for exploring demand for Mobility as a Service plans. *Transportation* **2019**, *46*, 1525–1558. [CrossRef]

48. Alonso-González, M.J.; Hoogendoorn-Lanser, S.; van Oort, N.; Cats, O.; Hoogendoorn, S. Drivers and barriers in adopting Mobility as a Service (MaaS)–A latent class cluster analysis of attitudes. *Transp.Res. Part. A Policy Pract.* **2020**, *132*, 378–401. [CrossRef]

49. Diana, M. From mode choice to modal diversion: A new behavioural paradigm and an application to the study of the demand for innovative transport services. *Technol. Forecast. Soc. Change* **2010**, *77*, 429–441. [CrossRef]

50. Gardner, B. Modelling motivation and habit in stable travel mode contexts. *Transp. Res. Part. F Traffic Psychol. Behav.* **2009**, *12*, 68–76. [CrossRef]

51. Wittink, D.R.; Cattin, P. Commercial Use of Conjoint Analysis: An Update. *J. Market.* **1989**, *53*, 91–96. [CrossRef]

52. Green, P.E.; Srinivasan, V. Conjoint analysis in consumer research: Issues and outlook. *J. Consum. Res.* **1978**, *5*, 103–123. [CrossRef]

53. Green, P.E.; Srinivasan, V. Conjoint analysis in marketing: New developments with implications for research and practice. *J. Market.* **1990**, *54*, 3–19. [CrossRef]

54. Louviere, J.J.; Street, D.; Burgess, L.; Wasi, N.; Islam, T.; Marley, A.A. Modeling the choices of individual decision-makers by combining efficient choice experiment designs with extra preference information. *J. Choice Modell.* **2008**, *1*, 128–164. [CrossRef]

55. Golden, B.R. The past is the past—or is it? The use of retrospective accounts as indicators of past strategy. *Acad. Manag. J.* **1992**, *35*, 848–860.

56. Adamowicz, W.; Louviere, J.; Williams, M. Combining revealed and stated preference methods for valuing environmental amenities. *J. Environ. Econ. Manag.* **1994**, *26*, 271–292. [CrossRef]

57. Louviere, J.J.; Hensher, D.A.; Swait, J.D. *Stated Choice Methods: Analysis and Applications*; Cambridge University Press: Cambridge, UK, 2000.

58. Hampl, N.; Loock, M. Sustainable development in retailing: What is the impact on store choice? *Bus. Strategy Environ.* **2013**, *22*, 202–216. [CrossRef]

59. Kubli, M.; Loock, M.; Wüstenhagen, R. The flexible prosumer: Measuring the willingness to co-create distributed flexibility. *Energy Policy* **2018**, *114*, 540–548. [CrossRef]

60. Priessner, A.; Hampl, N. Can product bundling increase the joint adoption of electric vehicles, solar panels and battery storage? Explorative evidence from a choice-based conjoint study in Austria. *Ecol. Econ.* **2020**, *167*, 106381. [CrossRef]

61. Sawtooth Software. *The CBC/HB System for Hierarchical Bayes Estimation Version 5.0 Technical Paper*; Sawtooth Software: Sequim, WA, USA, 2009.

62. Chrzan, K.; Orme, B. An overview and comparison of design strategies for choice-based conjoint analysis. *Sawtooth Softw. res. Pap. Ser.* **2000**, *98382*, 360.

63. Rossi, P.E.; Allenby, G.M. Bayesian statistics and marketing. *Market. Sci.* **2003**, *22*, 304–328. [CrossRef]

64. Huber, J.; Train, K. On the similarity of classical and Bayesian estimates of individual mean partworths. *Market. Lett.* **2001**, *12*, 259–269. [CrossRef]

65. Magidson, J.; Vermunt, J.K. *Latent class models. The Sage Handbook of Quantitative Methodology for the Social Sciences*; SAGE Publications: Thousand Oaks, CA, USA, 2004; pp. 175–198.

66. Sawtooth Software. *The CBC Latent Class Technical Paper (Version 3)*; Sawtooth Software: Sequim, WA, USA, 2004.

67. Bozdogan, H. Model selection and Akaike's information criterion (AIC): The general theory and its analytical extensions. *Psychometrika* **1987**, *52*, 345–370. [CrossRef]

68. Ramaswamy, V.; de Sarbo, W.S.; Reibstein, D.J.; Robinson, W.T. An empirical pooling approach for estimating marketing mix elasticities with PIMS data. *Market. Sci.* **1993**, *12*, 103–124. [CrossRef]

69. Orme, B.K.; Chrzan, K. *Becoming an Expert in Conjoint Analysis: Choice Modelling for Pros*; Sawtooth Software: Sun Valley, ID, USA, 2017.

70. Eggers, F.; Sattler, H.; Teichert, T.; Völckner, F. *Choice-based conjoint analysis. Handbook of Market Research*; Springer: Cham, Germany, 2018.

71. Orme, B. *Getting Started with Conjoint Analysis: Strategies for Product Design and Pricing Research*, 2nd ed.; Research Publishers LLC: Madison, WI, USA, 2010.

72. Salm, S.; Hille, S.L.; Wüstenhagen, R. What are retail investors' risk-return preferences towards renewable energy projects? A choice experiment in Germany. *Energy Policy* **2016**, *97*, 310–320. [CrossRef]

73. European Comission. Road Transport: Reducing CO_2 Emissions from Vehicles. 2019. Available online: https://ec.europa.eu/clima/policies/transport/vehicles_en (accessed on 20 January 2019).

74. Collins, C.M.; Chambers, S.M. Psychological and situational influences on commuter-transport-mode choice. *Environ. Behave.* **2005**, *37*, 640–661. [CrossRef]

75. Mola, L.; Berger, Q.; Haavisto, K.; Soscia, I. Mobility as a service: An exploratory study of consumer mobility behaviour. *Sustainability* **2020**, *12*, 8210. [CrossRef]

76. Choi, J.P. Bundling new products with old to signal quality, with application to the sequencing of new products. *Int. J. Ind. Organ.* **2003**, *21*, 1179–1200. [CrossRef]

77. Reinders, M.J.; Frambach, R.T.; Schoormans, J.P. Using product bundling to facilitate the adoption process of radical innovations. *J. Product Innov. Manag.* **2010**, *27*, 1127–1140. [CrossRef]

78. Simonin, B.L.; Ruth, J.A. Bundling as a strategy for new product introduction: Effects on consumers' reservation prices for the bundle, the new product, and its tie-in. *J. Bus. Res.* **1995**, *33*, 219–230. [CrossRef]

79. Harris, J.; Blair, E.A. Consumer preference for product bundles: The role of reduced search costs. *J. Acad. Market. Sci.* **2006**, *34*, 506–513. [CrossRef]
80. ElHaffar, G.; Durif, F.; Dubé, L. Towards closing the attitude-intention-behavior gap in green consumption: A narrative review of the literature and an overview of future research directions. *J. Clean. Prod.* **2020**, *275*, 122556. [CrossRef]
81. Schikofsky, J.; Dannewald, T.; Kowald, M. Exploring motivational mechanisms behind the intention to adopt mobility as a service (MaaS): Insights from Germany. *Transp. Res. Part. A Policy Practice* **2020**, *131*, 296–312. [CrossRef]
82. Pinto, J.; Morales, M.E.; Fedoruk, M.; Kovaleva, M.; Diemer, A. Servitization in support of sustainable cities: What are steel's contributions and challenges? *Sustainability* **2019**, *11*, 855. [CrossRef]
83. Hiscock, R.; Macintyre, S.; Kearns, A.; Ellaway, A. Means of transport and ontological security: Do cars provide psycho-social benefits to their users? *Transp. Res. Part. D Transp. Environ.* **2002**, *7*, 119–135. [CrossRef]

 *energies*

Article

Selecting E-Mobility Transport Solutions for Mountain Rescue Operations

Christian Wankmüller [1,*], Maximilian Kunovjanek [1], Robert Gennaro Sposato [1] and Gerald Reiner [2]

[1] Department of Operations, Energy, and Environmental Management, Universitaet Klagenfurt, 9020 Klagenfurt, Austria; maximilian.kunovjanek@aau.at (M.K.); robert.sposato@aau.at (R.G.S.)
[2] Institute for Production Management, Vienna University of Economics and Business, 1020 Vienna, Austria; gerald.reiner@wu.ac.at
*   Correspondence: christian.wankmueller@aau.at

Received: 11 November 2020; Accepted: 10 December 2020; Published: 15 December 2020

Abstract: This study introduces e-mobility for humanitarian purposes and presents the first investigation of innovative e-mobility transport solutions (e.g., e-bike, e-stretcher, and drone) for mountain rescue. In practice, it is largely unclear which e-mobility transport solutions might be suitable and what selection attributes are to be considered. The subsequent study supports the technology selection process by identifying and measuring relevant selection attributes to facilitate the adoption of e-mobility in this domain. For the purpose of this study, a multi-method research approach that combines qualitative and quantitative elements was applied. In the first step, results of a systematic search for attributes in literature were combined with inputs gained from unstructured expert interviews and discussions. The perceived importance of the identified selection attributes was then measured by analyzing survey data of 341 rescue workers using the best-worst scaling methodology. Finally, the results were reiterated in another expert discussion to assess their overall validity. Study results indicate that e-mobility transport solutions need to primarily enhance operational performance and support the safety of mountain rescue personnel. Surprisingly, economic and sustainability aspects are less of an issue in the process of technology selection.

Keywords: e-mobility; mountain rescue operations; emergency response; multi-method-research; best–worst scaling

---

1. Introduction

E-mobility, which includes every means of transport powered by an electric powertrain [1], has recently been gaining momentum due to its promising potential for tackling the ecologic problems of today's society. It has the capability to reduce greenhouse gas emissions, increase energy efficiency, and foster renewable power production, which are well-recognized properties by governments and policymakers [2]. European countries regularly play a leading role in the wide-ranging implementation of e-mobility [3]. This trend is slowly taking hold on a global scale. Various countries and organizations pursue an intensified adoption of e-mobility, which is not only limited to automobiles. In many other application areas, performance gains can also be achieved in terms of logistics and transportation due to novel e-mobility transport solutions [4]. In this regard, the fields of disaster relief and humanitarian logistics are focused on not only harnessing the ecologic benefits of e-mobility but also creating entirely new technological solutions to improve their performance when supporting people in need [5–7], as do mountain rescue (MR) services.

MR services are the primary responders in accident cases in alpine areas as well as in humanitarian disasters, where they become an integral part in the alleviation of human suffering. Prominent examples of such cases are earthquakes, such as the Amatrice earthquake in 2016 [8], large-scale avalanches, landslides and floods. During their operations, the MR services are often exposed to a multitude of challenges when taking care of and evacuating patients from remote locations. Relevant operations include patient localization and transportation, which regularly take place in isolated and potentially dangerous terrains and under harsh weather conditions [9]. These challenges negatively affect the overall logistics and transportation performance of MR services, often resulting in relatively time-consuming ascents and descents to the prior location of the patient [10]. Furthermore, no two emergencies are the same; rather, they can be described as heterogeneous and dynamic in nature, involving numerous and varied combinations of actors, skills, equipment, and environmental conditions. With the equipment currently available, rescue services often reach their operational limits when it comes to efficiently handling these drivers of complexity [11]. Additionally, alpine leisure activities are rising in popularity due to increased accessibility to remote mountainous areas [12]; thus, they now attract millions of people annually [13]. This growing enthusiasm for active pursuits in mountainous areas, in turn, leads to an even steeper increase in related accidents [14]. The corresponding logistical challenges are highly diverse and tackling them demands scientific assessments and the implementation of novel approaches. Therefore, modern technology is investigated and applied in most professional MR services [15].

The implementation of e-mobility transport solutions may be a viable option to achieve required performance gains. In the case of MR services, electric drones are a sound alternative for the emergency transport of medical equipment [16,17], and several other approaches, such as e-bikes and e-stretchers, also exist. E-bikes (i.e., bicycles with an additional electric propulsion) can be used for performing reconnaissance, locating patients, or getting faster access to emergency sites. Similarly, e-stretchers also make use of electric propulsion technology and are a variant of the common stretchers used to transport patients or operational equipment. Here, in particular, differences in the altitude and distance between an accident site and emergency vehicle can be more easily overcome. Furthermore, this technology positively affects the stabilization of the stretcher, ensuring safer transport conditions for patients while sparing the physical strength of the mountain rescuer [18].

The starting point of this scholarly effort is ascertaining that practitioners and various other stakeholders involved in MR cannot fully rely on the decision-making experience with respect to novel transport modes and technologies. Thus, it is largely unclear which e-mobility transport solutions might be suitable and what attributes are to be considered when selecting an adequate e-mobility transport solution, in accordance with the requirements of the MR services. As public views on e-mobility and new modes of electric transportation for civilian and commercial purposes have been evaluated in recent years, it follows naturally that their application to other purposes should come under consideration too. According to [19], technology transfer in humanitarian emergencies is a critical issue, but, so far, no conclusive investigation has been undertaken regarding the applicability of e-mobility transport solutions in humanitarian logistics. The study at hand aims at extending scientific knowledge to the relatively new sphere of e-mobility for humanitarian purposes and offers a first exploration in that direction based on the case of MR services. With a special focus on the technology selection process, we formulated the following research questions (*RQs*):

- RQ1: What are the decision-relevant attributes for selecting e-mobility transport solutions for mountain rescue personnel?
- RQ2: What is the perceived importance of the identified attributes for selecting e-mobility transport solutions for mountain rescue personnel?

The research questions were formulated during the course of the joint Interreg project Smart Test for Alpine Rescue Technology (SMART) involving Italian and Austrian MR services. For the corresponding analysis, a multi-method research approach consisting of qualitative and quantitative elements was chosen. Firstly, results of a systematic search for attributes in literature were combined with inputs gained from unstructured expert interviews and discussions with representatives of the participating MR services. Then, the perceived importance of the identified selection attributes was measured by analyzing survey data of 341 MR personnel using the best–worst scaling (BWS) methodology. Finally, the results were reiterated in another expert discussion to assess their overall validity. The applied research approach as well as the achieved results can provide a good orientation for both academics and practitioners in the context of transport technology assessment in (mountain) rescue operations and related fields. The article continues with the setting of the scientific background by analyzing the existing body of knowledge in the field. Then, the applied research methods are described, and the results of the analysis are presented and discussed. The article concludes with an overview about the limitations and implications of this study and potential future lines of research.

2. Related Work

Due to the novelty of the topic under consideration, specific research in the field of e-mobility for MR is sparse. Consequently, this section instead offers a general outline of the present study's scholarly context where we first delve into the current body of knowledge regarding the general adoption of e-mobility transport solutions. Then, closing in on the core topic of the present study, we present literature on the selection of e-mobility for various purposes and then connect this literature to the MR context, illustrating scholarly work that discusses the selection of equipment in this domain. This, consequently, guides us to the apparent gap in the existing body of knowledge that lays the foundation for subsequent analysis.

In recent years, various studies on the adoption of e-mobility for different applications have been published. A comprehensive overview about recent literature in that regard is presented by [20]. Here, special attention is dedicated to papers that assess factors that are linked to the successful adoption of electric vehicles. Papers in this research domain, for instance, have investigated the impact of innovation policies on the future development of international electric vehicle industries and markets [21,22]. Research results reveal that purchase subsidies, purchase restrictions and driving restrictions are the most effective policies to push e-mobility adoption. Ref [23] provided insights into incentives that promote the purchase and use of electric vehicles in the Norwegian market. The authors pointed out that the diffusion of electric vehicles is largely driven by economic incentives (e.g., exemption from toll charges) set by the government. Filtering adequate policies and incentives is only one of the many possible ways of guaranteeing an accelerated transition towards e-mobility. Similarly, quantitative modelling studies by [24,25] have underlined the importance of policies that address the visibility and familiarity of e-mobility in society and thus lead to increased acceptance and adoption of e-mobility transport solutions. As market penetration of e-mobility increases, a range of studies clearly shows that selection processes will extend beyond the above-mentioned mostly economical aspects to include considerations of environmental aspects and sustainability [26–34]. It is not surprising to find that the list of relevant aspects to be considered in the adoption of electric vehicles will obviously include the characteristics of the vehicles themselves. In this respect, ref [35] were able to show that the price and range of electric vehicles have an impact on the adoption of this technology, whereby price is a more significant barrier than vehicle range. This then inevitably links to perceptions of the individual, as not all features of an electric vehicle might be of the same importance to prospective consumers.

Several studies have investigated individual level predictors of the selection process with respect to electric vehicles. For example, ref [36] investigated customers' evaluations of electric vehicles for daily use. They found that human–machine interaction (i.e., design of displays and charging systems), traffic and safety implications and ecological aspects play an important role in the evaluation of electric vehicles. Ref [37] analyzed the factors that influence the selection process of an individual with certain technical background or knowledge when adopting electric vehicles. They found that general interest in technology, the distance driven, appreciating both the looks and environmental qualities of an electric vehicle positively affected the intention to adopt, while perceptions of electric vehicles being slow were negatively associated with the intention to adopt. Another study focusing on the perceptions of passengers towards electrified buses is presented by [38]. Through an online survey and subsequent BWS analysis, the authors found that safety, eco-friendliness, and ride comfort are important when using electrified buses. Extending this research beyond certain features of an electric vehicle, scholars have included personal variables such as attitudes and social determinants relying on social-psychological theories. Ref [39], for example, have analyzed personal factors that influence the selection of electric vehicles using the theory of planned behavior, finding that subjective norms and attitudes towards technology are significant drivers. Other studies by [40–44] provided a deeper analysis of the critical personal dimensions that impact the selection of e-mobility transport solutions from the customers' perspective, underlining the importance of individual attitudes, experience, information level, acquisition costs and social environment (e.g., family and friends) in the process of e-mobility selection.

Only a handful of scientific articles have provided insights into the selection of equipment in the MR domain. Through surveys, ref [45] analyzed the equipment of medical backpacks in MR operations and pointed out several important selection criteria, with a special focus on equipment properties and quality standards. Others have further recognized equipment weight as important in this context [46–48]. In detail, ref [46] assessed MR casualty bags using an experimental evaluation process with a special focus on applicability in cold and windy environments. According to the authors, the equipment selection process is not only restricted by weight and bulk of the used equipment but also by the necessity to withstand extreme climatic conditions. Ref [47] conducted a usability trial (including focus groups and surveys) of MR stretchers. Aside from the light weight, the equipment should furthermore be easy to use and be able to transport and feature a high payload. In the usability assessment of specialized medical equipment, ref [48] identified, through questionnaire analysis, the weight of the equipment and the time taken to apply it as important factors for an eventual equipment selection.

Combining the literature on the adoption and selection of e-mobility concepts and that of equipment for MR organizations, this study, to the best of our knowledge, is the first to address the selection of e-mobility transport solutions for MR. Our study is unique in the sense that it is the first that introduces e-mobility to the MR domain. With this, we aim to add to the establishment of a new strain of research by studying the applicability of e-mobility for humanitarian purposes.

3. Study Context and Methodology

The study at hand was conducted as part of the joint Interreg START project with Austrian and Italian MR services. The main objective of the project is to strengthen tactical cooperation and improve coordination between MR services in cross-border emergencies. Aside from this, the project pursues the objective of identifying, evaluating and implementing innovative e-mobility transport solutions that offer potential approaches to enhance the operational performance of MR teams in challenging response missions [49]. Driven by the novelty of electrification in the MR domain, key decision makers require support in the selection of suitable e-mobility transport solutions. Here, special interest lies in e-mobility transport solutions that facilitate easier access of MR teams to isolated patient locations while sparing the physical strength of the rescue personnel. Furthermore, e-mobility transport solutions should speed up the transportation of injured people, keeping in mind the environmental aspects.

3.1. Research Design

A multi-method research approach was employed for this study, combining qualitative and quantitative empirical research methods under the methodological guidance of Louviere and Islam [50]. This method allows for a comprehensive, bottom-up approach involving practitioners' perspectives, with the aim of providing applicable insights guided by the exigencies of the targeted stakeholder group. To answer the first research question (*RQ1*), attributes for selecting e-mobility transport solutions for MR services had to be identified. Therefore, as is visible in the process map depicted in Figure 1, we started our research with a systematic search for attributes in literature to elicit already documented attributes that are relevant in the context of e-mobility and mountain rescue. The identified attributes were then used in unstructured interviews with two MR experts to gather feedback from their applied perspectives and further elaborate on the evaluation attributes. A subsequent expert discussion reversed the initial broadening scope in an effort to arrive at a shortlist of attributes, which were then quantitatively tested using a BWS survey in the fourth and penultimate empirical phase, aiming to answer *RQ2*. This collection of primary data becomes, according to [51], increasingly important in the context of humanitarian research. Returning to a qualitative level of analysis, the last methodological step reiterated the results obtained from the previous steps in another expert discussion. The following paragraphs provide an in-depth look at the methodological aspects of these concatenated methodological components.

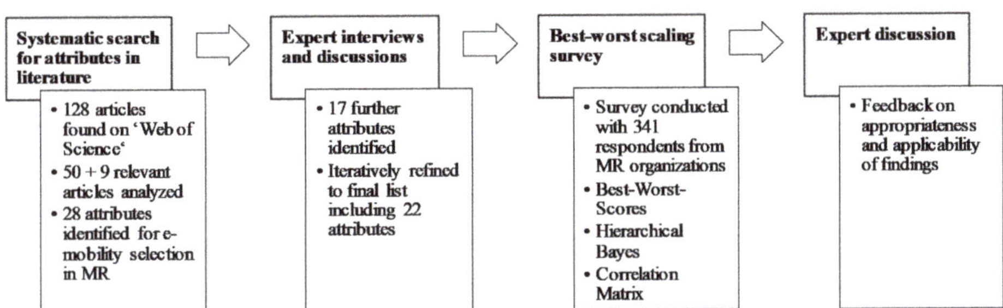

Figure 1. Process map of the applied multi-method research approach.

3.2. Systematic Search for Attributes in Literature

While Section 2, "Related Work", served to position our paper in the research landscape and to justify our research aim, we now start out with the systematic search for attributes in literature, as the second stage of the applied methodology. Ref [52] recommends starting a systematic search by designing a search string to develop a reproducible and transparent mode of finding relevant literature sources. In the case of this systematic search, literature concerning the assessment of e-mobility transport solutions as well as that focusing on technology assessment in the MR domain was scanned to find characteristic attributes. As scientific literature specifically focusing on MR is sparse, the search was expanded to include literature regarding other rescue services as well. Therefore, we identified adequate search terms and combined them using Boolean operators (e.g., "AND"; "OR") as suggested by [53] to form the final search string as shown in Appendix A. A systematic article title search (denoted by "TI" in the search string) of the "Web of Science" database, concluded in September 2019, yielded 128 articles.

After the initial search, all article titles and abstracts were scanned for their suitability to the subject under review and to eliminate irrelevant articles, as recommended by [54]. Ref [55] explained that clearly defining inclusion criteria helps to transparently identify relevant literature during a systematic search process. The applied inclusion criteria for this study are likewise listed in Appendix A. Articles that did not fit these inclusion criteria were subsequently excluded. Examples of excluded articles encompass but are not limited to those focusing on route selection problems for e-vehicles, charging station location optimization problems or biological articles that were inadvertently found by the search string (i.e., six articles including the biologic terms "entomobryidae" or "elasmobranch" that will also be returned by the search term "e*mob*"). This step was independently performed by two researchers to reduce bias, as recommended by [56]. After the independent evaluation of the articles, inconsistencies were jointly assessed, and the corresponding articles were reexamined to arrive at a pre-ultimate list of the 50 articles that remained.

Then, we expanded this limited body of literature with further peer-reviewed articles published in contextually relevant journals, a strategy that was previously applied by [57] as well as [58] who have also conducted systematic searches in the realm of humanitarian logistics and supply chain management. Therefore, journals such as *"High Altitude Medicine & Biology"* and *"Wilderness & Environmental Medicine"* were explicitly scanned for technology selection articles from the mountain rescue domain. This resulted in nine additional articles being considered for the identification of the evaluation attributes. The final 59 articles were subjected to a full-text analysis to extract decision-relevant attributes concerning the technology selection process. Based on this body of sources, a list of 28 attributes was generated, 14 of which were related to e-mobility and 14 to (mountain) rescue operations.

3.3. Unstructured Expert Interviews and Discussion

Similarly to [59], the next step served to collect feedback on and expand the list of attributes. Therefore, unstructured expert interviews and open discussions were held with two representatives from the MR services. The interviews and subsequent discussions lasted for approximately three hours and were conducted in the campus of the university at the end of September 2019. One of the involved experts was the head of operations of the local MR service for 40 years and the other one held a leading position within the service and was responsible for new technology implementation at the national level since 2005. Both have long-term experience in the MR domain and are still involved in response missions on the ground. Furthermore, both experts have been actively involved in the work package "E-mobility", led by the research team as part of the project described above. It is, therefore, safe to assume that both experts were highly knowledgeable in the study context and, as involved partners, were motivated to contribute to the study to the best of their abilities. The interviews were conducted by two researchers and held separately with each expert. Initially, the interviewees were introduced to the study's background, aims and research design. After this first input from the research team, the list of 28 attributes obtained from the systematic search for attributes in literature was handed out in printed form. The experts were invited to review the attributes in sequence and share their individual judgments and opinions with respect to their relevance for the selection of e-mobility concepts for MR services. In addition to commenting on the list of attributes, they were further invited to manually extend or reduce the list if necessary, according to their evaluation. In case the expert pointed to a yet undocumented attribute, this new attribute was noted, and the expert was asked to describe it in more detail. Both experts independently mentioned further MR-specific relevant attributes that were not yet included in the list. In total, 17 additional attributes (5 attributes for e-mobility and 12 for MR) were added, resulting in a total of 45 attributes.

Following these independent assessments, the two experts then joined a discussion with the two researchers to critically re-evaluate the 45 attributes from scientific as well as applied perspectives. This helped us arrive at the condensed final list of attributes to be presented to the survey participants involved in the next methodological phase. Specifically, the aim was to arrive at a list composed of decision-relevant attributes that are fully understandable, relevant, clearly delimited and completely evaluable. A reflection on the individually contributed attributes from the experts should therefore guarantee that every respondent had the same understanding of the presented attributes [60]. Every attribute was scrutinized taking these requirements into consideration; upon that, decisions were made concerning which attributes should be included, merged, or excluded. Attributes that were merged are for instance, "Environmental footprint" and "CO_2 emission" to "Enhances sustainability"; "Topography", "Disposition" and "Infrastructure" to "Applicable in every terrain"; "Lighting conditions" and "Usable at day and night" to "Applicable under all light conditions"; or "Low basic training effort" and "Low post-implementation training effort" to "Low training effort". Some attributes, such as "Recharging infrastructure", were deemed irrelevant, as the MR personnel transports all the necessary equipment to the operation site and back. Therefore, recharging can easily be performed at the base-station of the MR unit. Furthermore, the establishment of recharging stations in high alpine environments seems impractical. Through the joint re-evaluation, the initial 45 attributes were reduced to a total and final list of 22 attributes (see Table 1) that formed the basis for the BWS in the next research stage. The involved researchers acted as moderators and documented relevant statements accordingly.

Table 1. Relevant attributes as identified in scientific literature and expert discussion sessions.

Category	Attribute	Description	Source
E-mobility	High range	A high distance range that can be covered with the e-vehicle; subject to technologic, communication or legal limitations.	[20,61–64]
	High payload	A high amount of payload that can be transported or towed as well as seating capacity and trunk space.	[65,66]
	Long battery life	High battery capacity and long runtime with one charging.	[63]
	Low purchase costs	Costs associated with purchasing e-mobility transport solution are low.	[20,65–68]
	Enhances sustainability	The applied e-mobility transport solution enhances sustainability related aspects. Especially concerning the CO_2 emissions and raw material sourcing.	[62,64,66], [67,69–73]
	Low noise generation	Noise emission generated during the usage of the e-mobility transport solution is low.	[63,64]
	Conforms to legal requirements	Legal frameworks and (developing) requirements for the application of e-mobility transport technology are conformed.	Expert discussion; [45]
Mountain rescue service	Light weight	Weight of e-mobility transport solution.	[45–48]
	Low training effort	Amount of training effort associated with acquiring skills to handle new e-mobility transport solutions.	Expert discussion; [20,63]
	Ready-to-use	E-mobility transport solution can be used spontaneously. There is no need for expansive planning before usage during actual response missions (e.g., due to charging or assembly).	Expert discussion; [20,47,62]
	Meets quality certification	E-mobility transport solution meets quality certification requirements (e.g., CE or ISO).	Expert discussion
	High application variety	E-mobility transport solution can be used for multiple application purposes.	Expert discussion
	Easy to transport	Transportability of e-mobility transport solution.	[45,47]
	Long usability	The duration the e-mobility transport solution can remain in use in the mountain rescue service (i.e., product-life-cycle).	Expert discussion
	Easier access to remote locations	E-mobility transport solution facilitates the access to remote locations.	[74]
	Applicable in every terrain	E-mobility transport solution can be used in challenging terrain.	Expert discussion; [66]
	Applicable under all weather conditions	Technological reliability of the e-mobility transport solution in challenging weather conditions.	[46,66]
	Applicable under all light conditions	E-mobility transport solution is applicable under all light conditions.	Expert discussion; [75]
	Supports safety of MR personnel	Safety impacts for mountain rescue personnel concerning operational activities as well as technological aspects.	Expert discussion; [63,64]
	Provides speed advantage	Acceleration and speed of e-mobility transport solution.	[61]
	Supports mission documentation	The e-mobility transport solution enables enhanced documentation efforts during mountain rescue missions.	Expert discussion; [76]
	Compatible with other equipment	E-mobility transport solution is compatible with already existing equipment.	Expert discussion

3.4. Best–Worst Scaling Survey

To facilitate a discussion of the above-listed evaluation attributes and answer *RQ2*, a survey following the BWS approach was conducted. BWS is applied in a wide range of different research areas, including marketing [77], health care [78] and international business research [79]. However, in the context of transportation and logistics, BWS can be regarded as a fairly new research approach [80]. Ref [60], for example, used it to analyze attributes that customers prefer from a third-party logistics provider, while [81] applied BWS to measure the relative importance of the six initially equally weighted logistics performance indicators introduced by the World Bank in 2007. Based on the BWS results, significant differences in attribute importance were found.

BWS is underpinned by the random utility theory, which assumes that an individual's relative preference for object A over object B is a function of the relative frequency with which A is chosen as better than or preferred to B [82]. This methodology involves a cognitive process by which respondents repeatedly choose, from varying sets of attribute combinations, the attributes that they believe exhibit the largest perceptual difference in an underlying continuum of interest [83]. This is performed by observing the best and worst choices in a set of multiple options by repetitively combining the two choices. In a simple example of three choices "A", "B" and "C", "A" can be considered the best and "C" the worst. This ranking implies that "A" should be chosen for the pairs "AB" and "AC", and "B" should be chosen for the third pair "BC". Thus, the best and worst choices provide information that can be expanded to several pairs of choices [77]. This approach is called the case 1 BWS [84] and has previously been applied in the e-mobility domain in a study conducted by [85], wherein BWS was used to assess the importance of complementary mobility services to consumer behavior. Although contextually different, the study served as methodological guidance for the BWS analysis as presented in this article. Additionally, we followed the BWS steps as proposed by [86]. Accordingly, the first step comprised the setting of the study context followed by the identification of attributes. Then, the experimental design was formulated, and a survey was created and conducted. Ultimately, the acquired data were analyzed.

To answer *RQ2*, an online questionnaire comprising the 22 identified attributes, which is a reasonable number of attributes for a BWS according to [77], was developed using the software Sawtooth. The design of the German language questionnaire included 14 sets, each displaying six attributes in varying combinations. We showed each attribute 4 times per respondent ($n = 4$) in order to ensure that every attribute is visible to the respondent with the same frequency. The number of sets was calculated following Equation (1), where K is the total number of attributes ($K = 22$) in the survey and k is the number of items ($k = 6$) per set [87].

$$Number\ of\ sets = \ n \times \frac{K}{k} \tag{1}$$

The selection of attributes for each set is not conducted manually, instead the design algorithm of the software follows predefined guidelines, which assure that all of the attributes appear in combinations that serve as reliable representations of all possible combinations [87]. We distributed the survey to the personnel in the MR service of Tyrol (Austria), Carinthia (Austria) and South Tyrol (the northernmost Italian–German speaking province of Italy). The Tyrolian MR service is divided into 91 subdivisions with independent administration, the Carinthian MR service into 19 subdivisions and the MR service of South Tyrol into 35 subdivisions. The BWS survey (for an example, see Figure 2) was sent to each head of operations. In addition to answering the survey themselves, the heads of operations were asked to function as contact persons and pass on the survey to their respective group of mountain rescuers. After an interval of 14 days, the data acquisition was concluded following a predefined closing date; in summary, 341 completed questionnaires were collected. The group of respondents was composed of 319 (93.5%) male and 22 (6.5%) female MR personnel. They were, on average, 42.6 years old (Standard deviation (StdDev) = 13.4; min = 18, max = 79) and have been working in MR for 16.8 years on average (StdDev = 14.1; min ≤ 1; max = 60). Regarding their position within the MR service, 45 (13.2%) were in training and education, 181 (53.1%) were general operational staff without leading positions and 115 (33.7%) were in leading operational positions. Here, a person in a leading position takes over coordinative activities and the communication with representatives of other rescue organizations in large-scale operations (i.e., head of operations, officer-in-charge). Furthermore, they perform administrative tasks, which involve organizing training, negotiating annual budgets, processing internal settlements, and communicating with externals (such as local authorities). Non-leading personnel are not directly involved in such processes but have a consulting role in the decision making of leading personnel. Furthermore, non-leading personnel constitute the main body of active members that execute the operations on the ground. To ensure representativeness of the sample, we gathered data on the average age of the rescuers' population in the three regions, which closely matched the age structure represented in the sample.

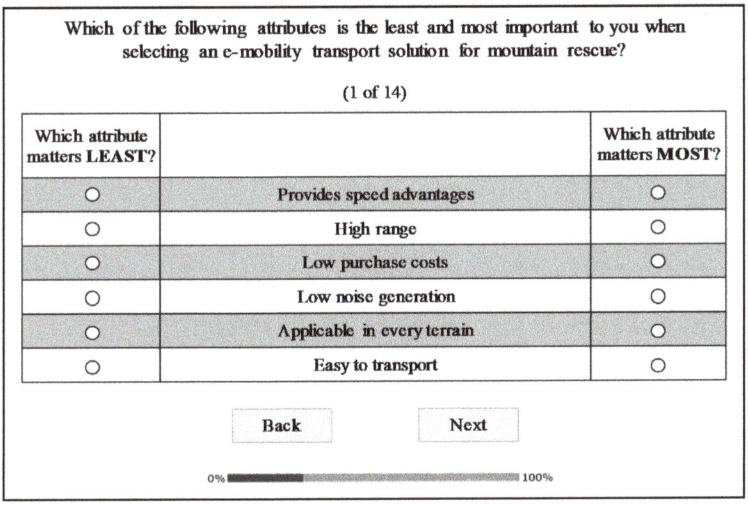

<table>
<tr><td colspan="3" align="center">Which of the following attributes is the least and most important to you when selecting an e-mobility transport solution for mountain rescue?

(1 of 14)</td></tr>
<tr><td>Which attribute matters LEAST?</td><td></td><td>Which attribute matters MOST?</td></tr>
<tr><td align="center">O</td><td align="center">Provides speed advantages</td><td align="center">O</td></tr>
<tr><td align="center">O</td><td align="center">High range</td><td align="center">O</td></tr>
<tr><td align="center">O</td><td align="center">Low purchase costs</td><td align="center">O</td></tr>
<tr><td align="center">O</td><td align="center">Low noise generation</td><td align="center">O</td></tr>
<tr><td align="center">O</td><td align="center">Applicable in every terrain</td><td align="center">O</td></tr>
<tr><td align="center">O</td><td align="center">Easy to transport</td><td align="center">O</td></tr>
<tr><td colspan="3" align="center">Back   Next</td></tr>
<tr><td colspan="3" align="center">0% ▇▇▇▇▇▇▇▇▇▇▇▇▇▇▇▇▇ 100%</td></tr>
</table>

Figure 2. Example of the best–worst combination.

4. Results and Discussion

For the purpose of preliminary analysis, ref [84] recommend calculating the so-called best–worst scores to gain an initial understanding of the survey results. Therefore, for each option, the number of times it was chosen as best in the study minus the number of times it was chosen as worst are divided by the total number of options that appeared (Equation (2)).

$$Best - Worst\ Scores = \frac{Times(Best) - Times(Worst)}{Number\ of\ option\ appears} \tag{2}$$

According to [83], one of the main advantages of BWS, as compared to other common rating-based evaluation methods, is that it also accounts for information about negative evaluations. Hence, a positive best–worst score means that an attribute was evaluated as the most important attribute more often than it was as the least important, while the inverse leads to a negative best–worst score. The best–worst scores were calculated for the leading and non-leading rescue workers (rescue workers in training and general operational staff) separately in order to gain an initial idea about potential differences in attribute selections. A graphical representation of the survey results featuring the best–worst scores can be found in Figure 3.

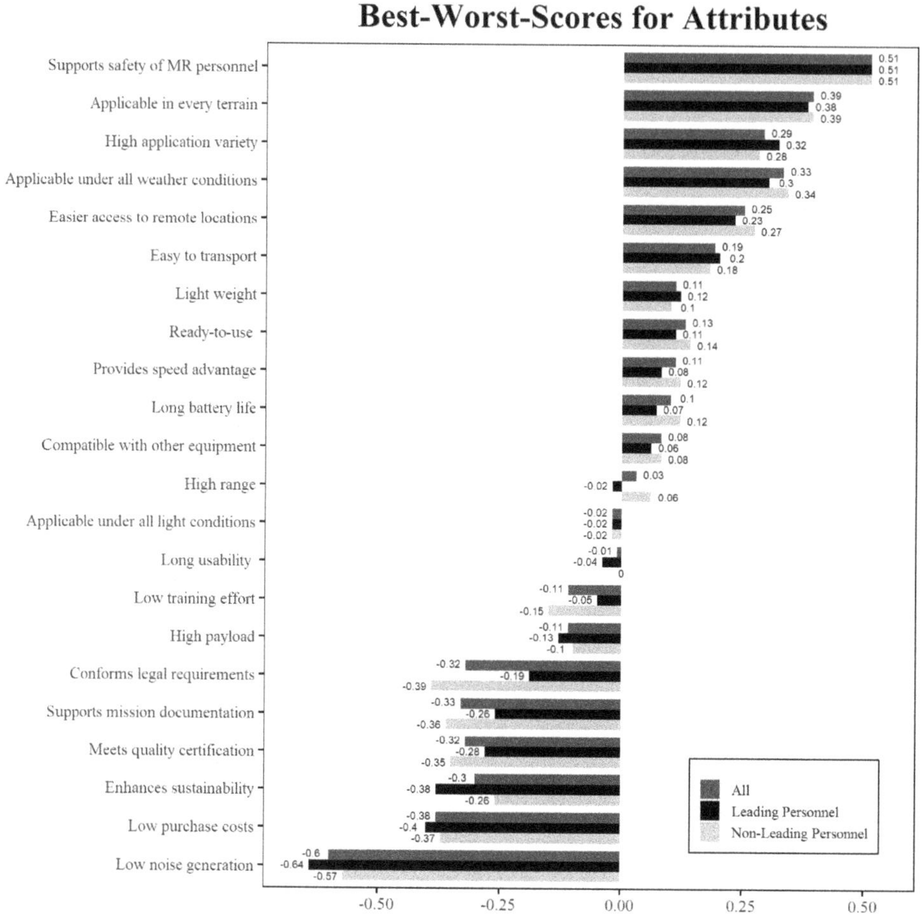

Best-Worst-Scores for Attributes

Figure 3. Graphical representation of the best–worst scores.

Several interesting implications can be derived from the calculation of the best–worst scores, we start by analyzing the results of the entire sample ("All"). First, the safety of the MR personnel is by far the most important attribute when assessing e-mobility transport solutions (best–worst score = 0.512). This means that the attribute "Supports safety of MR personnel" was chosen as most important in three out of four relevant attribute combinations. This seems intuitive, but previous studies have not paid too much attention to this aspect to date. Hence, the identification of the requirement for an e-mobility transport solution to actually support the safety of the MR personnel can be regarded as a major finding of this study. The next main implication is that the MR personnel regard it as highly important that the e-mobility transport solution is applicable under different conditions, because "Applicable in every terrain" (best–worst score = 0.387), "Applicable under all weather conditions" (best–worst score = 0.327) and "High application variety" (best–worst score = 0.294) were the next most important attributes. This desire might be grounded in the circumstances that MR operations often take place in harsh conditions, and that predictions about the terrain, the weather conditions or the actual intended use of the technology are hard to make in advance. The two attributes "Easy to transport" (best–worst score = 0.187) and "Light weight" (best–worst score = 0.108) are highly correlated (Correlation (corr) = 0.87 at $p < 0.001$), and they are both well-established factors in the realm of equipment assessment for (mountain) rescue services. Attributes related to technical specifications of e-mobility transport solutions are, according to the results, less important when compared to the operational attributes described above. Here, the attributes "Long battery life" (best–worst score = 0.102), "Compatible with other equipment" (best–worst score = 0.077) and "High range" (best–worst score = 0.033) are of lower relative importance when it comes to the selection process of adequate e-mobility solutions. On the other end of the table of best–worst scores, it is interesting to find that "Low purchase costs" (best–worst score = −0.380) is the second least important attribute. This contradicts the common picture of MR services (and other voluntary organizations) facing limited financial resources that might impede costly investments. Considering this, the results point towards the conclusion that attributes relevant to rescue operations are far more important than possible economic considerations. The same holds true for sustainability aspects that are somehow related to the core benefits of e-mobility. Both "Low noise generation" (best–worst score = −0.596) and "Enhances sustainability" (best–worst score = −0.302) also rank at the bottom of the list of attribute importance. This finding indicates that the ecological advancements of e-mobility (e.g., lower CO_2 emissions), which, according to [80], actually represent one of the main sales arguments for commercial applications, are not critical drivers for selection in the context of MR. Surprisingly, this points out that the actual project intention to also meet ecological dimensions when implementing e-mobility transport solutions is less important than initially expected. Concerning the differences between leading and non-leading MR personnel, it is surprising to observe in Figure 3 that results are rather homogeneous. To better assess potential variations among the groups concerning significance, the subsequent analyses provide more detailed insights.

In consequence, survey results were used to identify attribute weights [50,85]. For this task, the hierarchical Bayes (HB) approach was applied as supported by the Sawtooth software package, which was used to create and conduct the survey at hand. During the application of Bayes' rule, posterior probabilities are produced by updating prior probabilities with likelihoods obtained from the data [88]. This means that, instead of estimating each respondent's utilities individually, the algorithm estimates how different each respondent's utilities are from those of the other respondents in the study. It estimates the average utilities for the entire sample and then uses the respondent's individual data to determine how each respondent differs from the sample average. The algorithm then adjusts each respondent's utilities so that they reflect the optimal mix of the individual respondent choices and sample averages [89]. This procedure is conducted in a hierarchical manner with two levels: on the higher level, the individual's parameters are described by a multivariate normal distribution, and on the lower level, the individual's parameters are governed by a particular model, such as multinomial logit or linear regression [90]. The HB approach has to be run over multiple iterations, as every time the individual utilities are updated, the sample average needs to be updated as well until the model converges at the final values [89]. Table 2 shows the results of the HB analysis for the entire sample as well as separately for leading and non-leading personnel, comprising 30,000 iterations applied to the collected survey data. Table 2 further indicates whether differences between the leading and non-leading MR members regarding attribute weights are statistically significant, reporting t- and p-values for mean comparisons. t-tests were conducted by applying boot strapping with bias-corrected accelerated confidence intervals.

Results indicate that most attribute weights did not differ significantly between leading and non-leading personnel. The only two significant differences were found for "High range" ($t(256.64) = 2.74$; $p = 0.01$) and "Conforms to legal requirements" ($t(187.38) = −2.48$; $p = 0.02$). For the attribute "High range", non-leading personnel yielded a higher average attribute weight ($M = 3.91$, $SE = 0.16$) than leading personnel ($M = 3.20$, $SE = 0.20$). The opposite was true for "Conforms to legal requirements", where, on average, attribute weights were higher for leading personnel ($M = 1.17$, $SE = 0.17$) than for non-leading personnel ($M = 0.69$, $SE = 0.09$). Furthermore, in the HB analysis, the attribute importance rankings did not change considerably compared to the best–worst score ranking. The computed attribute weights, however, were found to yield great benefits when intending to work with these results in potential future practical or scientific undertakings. They can be used for the actual assessment of e-mobility transport solutions for MR practitioners, serve as input variables for related simulation studies or even offer guidance for similar studies in related fields.

Table 2. Comparison of average attribute weights as derived from hierarchical Bayes (HB) analysis of the online best–worst scaling (BWS) survey between leading and non-leading mountain rescue (MR)-members including *t*-statistic and *p*-value.

Attributes	All	Columns Graph	Leading	Non–Leading	*t*	*p*
		Attribute Weights			**Mean Difference**	
Supports safety of MR personnel	11.51		11.54	11.49	−0.12	0.91
Applicable in every terrain	10.6		10.6	10.6	0	1
Applicable under all weather conditions	9.33		8.91	9.54	1.78	0.08
High application variety	8.62		8.91	8.47	−1.1	0.29
Easier access to remote locations	8.08		7.94	8.16	0.53	0.58
Easy to transport	6.67		6.92	6.54	−0.97	0.34
Ready-to-use	5.67		5.53	5.75	0.68	0.48
Provides speed advantage	5.59		5.28	5.75	0.95	0.36
Light weight	5.17		5.34	5.08	−0.62	0.52
Long battery life	4.97		4.52	5.19	2	0.05
Compatible with other equipment	4.8		4.9	4.75	−0.36	0.72
High range *	3.67		3.2	3.91	**2.74**	**0.01**
Applicable under all light conditions	3.08		2.98	3.13	0.48	0.66
Low training effort	3.01		3.6	2.71	−1.83	0.06
Long usability	2.97		2.85	3.04	0.72	0.47
High payload	1.41		1.33	1.44	1.15	0.28
Meets quality certification	1.29		1.62	1.12	−1.54	0.11
Enhances sustainability	1.13		1.03	1.17	0.64	0.55
Supports mission documentation	0.95		1.16	0.84	−1.56	0.12
Conforms to legal requirements *	0.85		1.17	0.69	**−2.48**	**0.02**
Low purchase costs	0.51		0.55	0.49	−0.66	0.54
Low noise generation	0.14		0.13	0.14	0.51	0.62
Total	**100**		**100**	**100**		

* Attributes differ significantly at *p* < 0.05.

To further deepen our understanding of the achieved results, correlations between the attributes were analyzed using the programming language "R". In Figure 4, a correlation matrix of the six most important attributes according to the BWS analysis and three attributes concerning participant characteristics can be found. The circle size stands for the magnitude of the correlation, and the hue indicates if it is positive or negative.

In Figure 4, we included the attribute "Gender", and a negative correlation (corr = −0.25 at *p* < 0.001) between this attribute and the position within the MR service exists. This means that women are more likely to be in a lower ranking position within the MR service than men. Taking a closer look at this, however, reveals that women only have a "Service Time" of 5.2 years on average, while their male counterparts have been in MR for 17.3 years on average. In this context, it has to be pointed out that, in the sample, there were only 22 female respondents, and 13 of them were relatively new to MR (service time ≤ 3 years). Additionally, of interest is the position of the "Supports safety of MR personnel" attribute. The applicability of the e-mobility transport solution in every terrain correlates (corr = 0.39 at *p* < 0.001) with the perceived safety support. Additionally, "Easier access to remote locations" (corr = 0.24 at *p* < 0.001) features a similar relationship. Another interesting finding from analyzing attribute correlations is that "Position" and "Conforms to legal requirements" (corr = 0.28 at *p* < 0.001) are positively correlated. This means that personnel in leading positions assign more value to this attribute than the operational staff or the personnel in training.

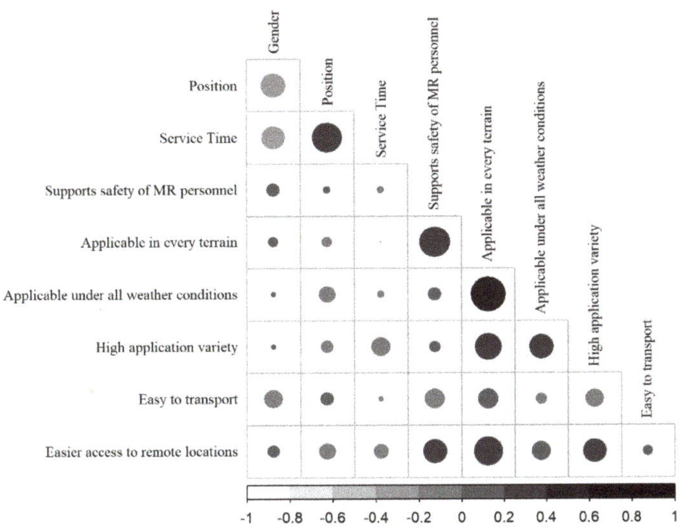

Figure 4. Correlation matrix of most important attributes and participant characteristics.

Expert Discussion

The final methodological step comprised a second expert discussion with experts familiar with the topic under study to collect feedback on the results of the BWS and to assess the validity of the overall findings [91]. Special attention was dedicated to gathering insights on the appropriateness and applicability of the findings from the practitioners' points of view. During practical tests for drone technology at the end of July 2020 at Brenner Pass (Austria), the research team arranged for a group of five high-ranking functionaries from the MR services of South Tyrol and Tyrol. It comprised the head and deputy head of the MR service of South Tyrol, two heads of MR bases and the chief financial officer of the MR service of Tyrol. All of them have long-term experience in the MR domain and have been substantially involved in the development of the respective MR services including technological advancements and purchasing decisions. They are still involved in response missions on the ground. The involved researchers acted as moderators and documented the relevant statements. The expert discussion opened with the researchers giving a comprehensive overview of the BWS and the HB results. Then, the experts were asked for their opinion on the fact that the attribute "Low purchase costs" was ranked as the second least important. According to the experts' statements, costs are less of an issue in the equipment selection. This was underlined by one expert mentioning, *". . . if the technology supports safety, is applicable in every terrain and under all weather conditions, offers a high application range and enables easier access to remote locations, then costs are completely irrelevant"*. This indicates that the findings of the BWS exactly reflect the real situation and can therefore adequately explain this observation. Furthermore, the experts are highly aware that equipment in the context of MR is not mass produced, preventing potential lower purchase costs. Additionally, MR services receive ample annual budgets, which allow them to make equipment decisions relatively independently from monetary dimensions. As this is, to a certain degree, in contrast to other humanitarian organizations that often face resource constraints, a reflection of potential contextual factors revealed two possible reasons. First, the study regions in Italy and Austria are economically dependent on mountaineers and a safe alpine environment. This is of high relevance for the responsible authorities, which therefore commonly provide sufficient financial support. Second, due to the perceived importance of MR services in the focal regions, voluntary financial support is common. In the region of Tyrol alone, more than 25,000 private individuals donate to the local MR service on an annual basis, which might stem from a living culture surrounding alpine life and mountaineering. As a reference, the regarded e-mobility solutions range in price from EUR 400–3000 for standard portable drones and EUR 15,000–60,000 for high-quality and high-performance drones with specialized equipment such as night vision or thermal cameras. The considered e-bikes, which were specifically designed for MR operations, have a starting price of EUR 6000. The e-stretcher is more expensive than a standard wheel-bearing stretcher but when considering rescue costs, it yields the potential to even decrease costs overall.

The discussion was then continued with a reflection on why "Provides speed advantage" is not among the most important attributes. Two experts pointed out that, aside from speed gains provided by a new technology, the safety and applicability of the technology still have top priority in response missions. The general perception is that safety and high applicability to different circumstances will always outweigh speed advantages. This shared culture may also explain the relative unimportance of the attribute "Enhances sustainability" compared with other attributes. In this regard, one of the experts stated that *"sustainability can only be considered, when there is no negative impact on the performance of the entire rescue mission"*. Another expert concurred and added that *"emergency organizations' main objective should not be put on sustainability, but on efficiently designing response operations"*. Next, the experts' attention was steered to the differences between leading and non-leading personnel in the evaluation of the attributes "Conforms legal requirements" and "High range". Concerning the legal requirements, the experts were in complete agreement that the difference basically stems from the lower level of involvement and responsibility of the operational staff in associated administrative and organizational activities, which are generally the subjects for leading personnel. The difference in the attribute weight of "High range" can, according to the expert judgment, be explained by the more intense involvement of non-leading personnel in the actual handling of technologies during response missions.

Afterwards, the experts were asked about the appropriateness of the results. Overall, they confirmed the appropriateness of the results; however, one expert noted that costs do still play a role, especially for investments at the national level. Neglecting costs is, according to his opinion, only reasonable when comparing and selecting among specific technological solutions. Finally, remarks concerning the applicability of the findings in practice and other decision-making problems were collected. While for one expert the results were rather specific in the context of e-mobility transport solutions, others pointed towards their generalizability and adoption to alternative decision-making contexts. As a whole, they saw the list of attributes and corresponding weights as transparent and objective support in their decision making, which was up to now mainly driven by intuition and personal experience.

5. Conclusions and Future Research

In this study, decision making in the selection process of e-mobility transport solutions in MR was analyzed. From the practitioners' points of view, several interesting insights were gained. First, the most decision-relevant attributes for selecting e-mobility transport solutions in MR services were identified. These attributes offer guidance on what MR services should consider when selecting new e-mobility technology. Here, we can say that e-mobility transport solutions need to support the safety of MR personnel and have to be applicable in many different environments. Economic and sustainability concerns are less of an issue in this context. Second, the perceived importance of the identified attributes was identified. This can help in creating evaluation tools or other decision-support tools when facing a selection problem for e-mobility transport solutions. Providers of e-mobility transport solutions may also benefit from the results as we support them in addressing the specific needs of the MR services. We make them aware of the fact that MR services devote a lot of attention to the flexible use of technologies and to a lesser extent to ecological attributes.

From an academic point of view, a major contribution of this study is that it provides the first analysis of e-mobility for humanitarian and emergency purposes. Our findings enrich the scientific literature concerning e-mobility selection and lay out an insightful starting point for intensified research in the field of humanitarian research. In this regard, we provide a comprehensive list of decision-relevant attributes that equally incorporates aspects of e-mobility and MR services. The BWS analysis enabled us to identify attribute weights, which can be used by humanitarian researchers to assess other technological or equipment-related decision problems in the context of the (mountain) rescue domain. Furthermore, the separate analyses of the leading and non-leading MR personnel provided insights into the differences in decision making within voluntary organizations, which is another clear academic contribution of this study.

Limitations of this study include the fact that the analyzed sample is culturally rather homogenous and limited to one geographic region. Repeating such technological assessments with other (humanitarian) organizations in a different cultural or geographical setting might provide further valuable insights. Furthermore, the circumstance that the focal MR organizations do not face stringent budgetary constraints might be a limiting factor concerning the generalizability of the achieved results. Additionally, the vast majority of the participants of this study were male, and potential variations in the gender distribution in MR services may lead to different results when it comes to the valuation of assessment attributes. Future work should address this matter by collecting more gender-balanced samples. Furthermore, the shared values among members in MR services might be slightly different compared to other emergency organizations, potentially impacting the set of attributes for selecting e-mobility transport solutions. This could be the subject of future research where results from similar application studies stemming from other fields, such as first aid, firefighting, naval rescue, etc., would increase the reach of the derived implications. Additional research topics may also comprise the analysis of the actual effects of e-mobility transport solutions on the performance of MR services during field applications and rescue missions once the new technologies are in use. Further developments in battery capacity and battery weight are needed in order to increase the attractiveness of e-mobility for MR services. Battery weight plays a crucial role because mountain rescuers have to carry all necessary equipment on their own. Using light-weight batteries for e-mobility transport solutions can minimize total equipment weight and extend the operational performance of MR personnel.

Author Contributions: C.W.; M.K.; R.G.S.; G.R.; conceptualization: C.W., M.K. and G.R.; methodology: C.W., M.K. and R.G.S.; software: C.W. and M.K.; validation: C.W. and M.K.; formal analysis: C.W. and M.K.; investigation: all; resources: C.W. and M.K.; data curation: C.W. and M.K.; writing—original draft preparation: all; writing—review and editing: all; visualization: M.K.; project administration: C.W. and G.R.; funding acquisition: C.W. and G.R. All authors have read and agreed to the published version of the manuscript.

Funding: This work was supported by the European Union Fund for regional development and Interreg V-A Italy Austria 2014–2020. The second author was employed within the Interreg project: Smart Test for Alpine Rescue Technology (SMART).

Acknowledgments: We thank Nina Hampl for assisting with the methodological aspects of the research presented here and her precise and insightful comments that greatly improved the manuscript.

Conflicts of Interest: The authors declare no conflict of interest.

Appendix A

Table A1. Search string used for initial attribute selection.

Search Criteria	Inclusion Criteria
TI = ("E*mob*" OR "electric vehicle*" OR "electric mob*")	Articles clearly focusing on e-mobility.
AND TI = ("selection" OR "adoption" OR "best*worst" OR "maximum difference" OR "rescue" OR "humanitarian" OR "first response" OR "first aid" OR "emergency" OR "mountain" OR "alpine")	Articles that focus on either technology selection, the BWS research methodology, rescue operations, or the technology application in alpine areas.
AND LANGUAGE = English AND DOCUMENT TYPES = Article	Only scientific articles published in English.
INDEXES = ("SCI-EXPANDED" OR "SSCI" OR "A&HCI" OR "ESCI")	Articles must be published in journals listed in an established index.
TIMESPAN = All years	No restriction on the year of publication.

Supporting data will be made available upon request by the authors.

References

1. Harrison, G.; Gómez Vilchez, J.J.; Thiel, C. Industry strategies for the promotion of E-mobility under alternative policy and economic scenarios. *Eur. Transp. Res. Rev.* **2018**, *10*, 438. [CrossRef]
2. Lo Schiavo, L.; Delfanti, M.; Fumagalli, E.; Olivieri, V. Changing the regulation for regulating the change: Innovation-driven regulatory developments for smart grids, smart metering and e-mobility in Italy. *Energy Policy* **2013**, *57*, 506–517. [CrossRef]
3. Egnér, F.; Trosvik, L. Electric vehicle adoption in Sweden and the impact of local policy instruments. *Energy Policy* **2018**, *121*, 584–596. [CrossRef]
4. Klumpp, M. Electric mobility in last mile distribution. In *Efficiency and Innovation in Logistics*; Springer: Berlin/Heidelberg, Germany, 2014; pp. 3–13.
5. Adderly, S.A.; Manukian, D.; Sullivan, T.D.; Son, M. Electric vehicles and natural disaster policy implications. *Energy Policy* **2018**, *112*, 437–448. [CrossRef]
6. Austrian Red Cross. Wiener Rotes Kreuz: Auf die Segways–fertig–los! Start für Rasche Hilfe auf Zwei Rädern. Available online: https://www.roteskreuz.at/news/datum/2011/08/17/wiener-rotes-kreuz-auf-die-segways-fertig/ (accessed on 14 May 2020).
7. Ustun, T.S.; Cali, U.; Kisacikoglu, M.C. Energizing microgrids with electric vehicles during emergencies—Natural disasters, sabotage and warfare. In Proceedings of the IEEE International Telecommunications Conference (INTELEC), Osaka, Japan, 18–22 October 2015; pp. 1–6. [CrossRef]
8. Foti, F.; Milani, M. Earthquake in Amatrice (Italy), 24 August 2016: The Role of the Medical Teams of the National Alpine Rescue Corp (CNSAS). *Prehosp. Dis. Med.* **2017**, *32*, S112–S113. [CrossRef]
9. Podsiadło, P.; Darocha, T.; Kosiński, S.; Sałapa, K.; Ziętkiewicz, M.; Sanak, T.; Turner, R.; Brugger, H. Severe Hypothermia Management in Mountain Rescue: A Survey Study. *High Alt. Med. Biol.* **2017**, *18*, 411–416. [CrossRef]
10. Li, Y.-y.; Dong, X. Mountain Disaster Incidents and Corresponding Emergency Rescue Measures. *Proc. Eng.* **2014**, *71*, 207–213. [CrossRef]
11. Shimansky, C. *Accidents in Mountain Rescue Operations*; Mountain Rescue Association: Evergreen, CO, USA, 2008.
12. Apollo, M. The true accessibility of mountaineering: The case of the High Himalaya. *J. Outdoor Recreat. Tour.* **2017**, *17*, 29–43. [CrossRef]
13. UNWTO. *Sustainable Mountain Tourism—Opportunities for Local Communities*; World Tourism Organization (UNWTO): Madrid, Spain, 2018; ISBN 9789284420261.
14. Soulé, B.; Lefèvre, B.; Boutroy, E. The dangerousness of mountain recreation: A quantitative overview of fatal and non-fatal accidents in France. *Eur. J. Sport Sci.* **2017**, *17*, 931–939. [CrossRef]
15. Yarwood, R. Risk, rescue and emergency services: The changing spatialities of Mountain Rescue Teams in England and Wales. *Geoforum* **2010**, *41*, 257–270. [CrossRef]
16. Wankmüller, C.; Truden, C.; Korzen, C.; Hungerländer, P.; Kolesnik, E.; Reiner, G. *Optimal Allocation of Defibrillator Drones in Mountainous Regions*; OR Spectrum: Basel, Switzerland, 2020; pp. 1–30.
17. Tatham, P.; Ball, C.; Wu, Y.; Diplas, P. Long-endurance remotely piloted aircraft systems (LE-RPAS) support for humanitarian logistic operations. *J. Hum. Logist. Sup Chain Manag.* **2017**, *7*, 2–25. [CrossRef]
18. Greischberger Rescue System. Retten mit System. Available online: http://www.grs-rescue.at/wp/wp-content/uploads/2019/05/GRS_Prospekt.pdf (accessed on 10 February 2020).
19. Santos, A.L.R.; Wauben, L.S.G.L.; Goossens, R.; Brezet, H. Systemic barriers and enablers in humanitarian technology transfer. *J. Hum. Logist. Supply Chain Manag.* **2016**, *6*, 46–71. [CrossRef]
20. Coffman, M.; Bernstein, P.; Wee, S. Electric vehicles revisited: A review of factors that affect adoption. *Transp. Rev.* **2017**, *37*, 79–93. [CrossRef]
21. Su, Y.-S.; Lin, C.-J.; Li, C.-Y. An assessment of innovation policy in Taiwan's electric vehicle industry. *Int. J. Technol. Manag.* **2016**, *72*, 210–229. [CrossRef]
22. Yu, J.; Yang, P.; Zhang, K.; Wang, F.; Miao, L. Evaluating the effect of policies and the development of charging infrastructure on electric vehicle diffusion in China. *Sustainability* **2018**, *10*, 3394. [CrossRef]
23. Aasness, M.A.; Odeck, J. The increase of electric vehicle usage in Norway—Incentives and adverse effects. *Eur. Transp. Res. Rev.* **2015**, *7*, 2018. [CrossRef]

24. Silvia, C.; Krause, R.M. Assessing the impact of policy interventions on the adoption of plug-in electric vehicles: An agent-based model. *Energy Policy* **2016**, *96*, 105–118. [CrossRef]

25. Mirhedayatian, S.M.; Yan, S. A framework to evaluate policy options for supporting electric vehicles in urban freight transport. *Transp. Res. Part D Transp. Environ.* **2018**, *58*, 22–38. [CrossRef]

26. Faria, R.; Moura, P.; Delgado, J.; de Almeida, A.T. A sustainability assessment of electric vehicles as a personal mobility system. *Energy Convers. Manag.* **2012**, *61*, 19–30. [CrossRef]

27. Lave, L.B.; MacLean, H.L. An environmental-economic evaluation of hybrid electric vehicles: Toyota's Prius vs. its conventional internal combustion engine Corolla. *Transp. Res. Part D Transp. Environ.* **2002**, *7*, 155–162. [CrossRef]

28. Jochem, P.; Babrowski, S.; Fichtner, W. Assessing CO_2 emissions of electric vehicles in Germany in 2030. *Transp. Res. Part A Policy Pract.* **2015**, *78*, 68–83. [CrossRef]

29. Prud'homme, R.; Koning, M. Electric vehicles: A tentative economic and environmental evaluation. *Transp. Policy* **2012**, *23*, 60–69. [CrossRef]

30. Lemme, R.F.F.; Arruda, E.F.; Bahiense, L. Optimization model to assess electric vehicles as an alternative for fleet composition in station-based car sharing systems. *Transp. Res. Part D Transp. Environ.* **2019**, *67*, 173–196. [CrossRef]

31. Hawkins, T.R.; Singh, B.; Majeau-Bettez, G.; Strømman, A.H. Comparative environmental life cycle assessment of conventional and electric vehicles. *J. Ind. Ecol.* **2013**, *17*, 53–64. [CrossRef]

32. Noshadravan, A.; Cheah, L.; Roth, R.; Freire, F.; Dias, L.; Gregory, J. Stochastic comparative assessment of life-cycle greenhouse gas emissions from conventional and electric vehicles. *Int. J. Life Cycle Assess.* **2015**, *20*, 854–864. [CrossRef]

33. Ahmadi, P.; Cai, X.M.; Khanna, M. Multicriterion optimal electric drive vehicle selection based on lifecycle emission and lifecycle cost. *Int. J. Energy Res.* **2018**, *42*, 1496–1510. [CrossRef]

34. Das, M.C.; Pandey, A.; Mahato, A.K.; Singh, R.K. Comparative performance of electric vehicles using evaluation of mixed data. *OPSEARCH* **2019**, *56*, 1067–1090. [CrossRef]

35. Adepetu, A.; Keshav, S. The relative importance of price and driving range on electric vehicle adoption: Los Angeles case study. *Transportation* **2017**, *44*, 353–373. [CrossRef]

36. Cocron, P.; Bühler, F.; Neumann, I.; Franke, T.; Krems, J.F.; Schwalm, M.; Keinath, A. Methods of evaluating electric vehicles from a user's perspective-the MINI E field trial in Berlin. *IET Intell. Transp. Syst.* **2011**, *5*, 127–133. [CrossRef]

37. Egbue, O.; Long, S.; Samaranayake, V.A. Mass deployment of sustainable transportation: Evaluation of factors that influence electric vehicle adoption. *Clean Technol. Environ. Policy* **2017**, *19*, 1927–1939. [CrossRef]

38. Kwon, Y.; Kim, S.; Kim, H.; Byun, J. What Attributes Do Passengers Value in Electrified Buses? *Energies* **2020**, *13*, 2646. [CrossRef]

39. Alzahrani, K.; Hall-Phillips, A.; Zeng, A.Z. Applying the theory of reasoned action to understanding consumers' intention to adopt hybrid electric vehicles in Saudi Arabia. *Transportation* **2019**, *46*, 199–215. [CrossRef]

40. Schmalfuß, F.; Mühl, K.; Krems, J.F. Direct experience with battery electric vehicles (BEVs) matters when evaluating vehicle attributes, attitude and purchase intention. *Transp. Res. Part F Traffic Psychol. Behav.* **2017**, *46*, 47–69. [CrossRef]

41. Rahmani, D.; Loureiro, M.L. Assessing drivers' preferences for hybrid electric vehicles (HEV) in Spain. *Res. Transp. Econ.* **2019**, *73*, 89–97. [CrossRef]

42. Morton, C.; Anable, J.; Nelson, J.D. Assessing the importance of car meanings and attitudes in consumer evaluations of electric vehicles. *Energy Effic.* **2016**, *9*, 495–509. [CrossRef]

43. Carley, S.; Siddiki, S.; Nicholson-Crotty, S. Evolution of plug-in electric vehicle demand: Assessing consumer perceptions and intent to purchase over time. *Transp. Res. Part D Transp. Environ.* **2019**, *70*, 94–111. [CrossRef]

44. Bühne, J.-A.; Gruschwitz, D.; Hölscher, J.; Klötzke, M.; Kugler, U.; Schimeczek, C. How to promote electromobility for European car drivers? Obstacles to overcome for a broad market penetration. *Eur. Transp. Res. Rev.* **2015**, *7*. [CrossRef]

45. Elsensohn, F.; Soteras, I.; Resiten, O.; Ellerton, J.; Brugger, H.; Paal, P. Equipment of medical backpacks in mountain rescue. *High Alt. Med. Biol.* **2011**, *12*, 343–347. [CrossRef]

46. Grant, S.J.; Dowsett, D.; Hutchison, C.; Newell, J.; Connor, T.; Grant, P.; Watt, M. A Comparison of Mountain Rescue Casualty Bags in a Cold, Windy Environment. *Wilderness Environ. Med.* **2002**, *13*, 36–44. [CrossRef]

47. Hignett, S.; Willmott, J.W.; Clemes, S. Mountain rescue stretchers: Usability trial. *Work* **2009**, *34*, 215–222. [CrossRef]

48. Runcie, H.; Greene, M. Femoral Traction Splints in Mountain Rescue Prehospital Care: To Use or Not to Use? That Is the Question. *Wilderness Environ. Med.* **2015**, *26*, 305–311. [CrossRef] [PubMed]

49. Mountain Rescue Tyrol. START-Smart Test of Alpine Rescue Technology. Available online: https://bergrettung.tirol/php/interreg,1411.html (accessed on 6 March 2019).

50. Louviere, J.J.; Islam, T. A comparison of importance weights and willingness-to-pay measures derived from choice-based conjoint, constant sum scales and best–worst scaling. *J. Bus. Res.* **2008**, *61*, 903–911. [CrossRef]

51. Kovács, G.; Moshtari, M.; Kachali, H.; Polsa, P. Research methods in humanitarian logistics. *JHLSCM* **2019**, *9*, 325–331. [CrossRef]

52. *Research Methods for Operations Management*, 2nd ed.; Karlsson, C., Ed.; Routledge: New York, NY, USA, 2016; ISBN 9781138945425.

53. Sauer, P.C.; Seuring, S. Sustainable supply chain management for minerals. *J. Clean. Prod.* **2017**, *151*, 235–249. [CrossRef]

54. Jahangirian, M.; Eldabi, T.; Naseer, A.; Stergioulas, L.K.; Young, T. Simulation in manufacturing and business: A review. *Eur. J. Oper. Res.* **2010**, *203*, 1–13. [CrossRef]

55. Petticrew, M.; Roberts, H. *Systematic Reviews in the Social Sciences*; A Practical Guide, 12th Print; Blackwell Publishing: Malden, MA, USA, 2012; ISBN 978-1-4051-2110-1.

56. Seuring, S.; Müller, M.; Westhaus, M.; Morana, R. Conducting a Literature Review—The Example of Sustainability in Supply Chains. In *Research Methodologies in Supply Chain Management: With 67 Tables*; Kotzab, H., Seuring, S., Müller, M., Reiner, G., Eds.; Physica-Verl.: Heidelberg, Germany, 2005; pp. 91–106, ISBN 978-3-7908-1636-5.

57. Kunz, N.; Reiner, G. A meta-analysis of humanitarian logistics research. *J. Hum. Logist. Sup. Chain Manag.* **2012**, *2*, 116–147. [CrossRef]

58. Wankmüller, C.; Reiner, G. Coordination, cooperation and collaboration in relief supply chain management. *J. Bus. Econ.* **2020**, *90*, 239–276. [CrossRef]

59. Cheung, K.L.; Wijnen, B.F.M.; Hollin, I.L.; Janssen, E.M.; Bridges, J.F.; Evers, S.M.A.A.; Hiligsmann, M. Using Best-Worst Scaling to Investigate Preferences in Health Care. *Pharmacoeconomics* **2016**, *34*, 1195–1209. [CrossRef]

60. Coltman, T.R.; Devinney, T.M.; Keating, B.W. Best-worst scaling approach to predict customer choice for 3PL services. *J. Bus. Logist.* **2011**, *32*, 139–152. [CrossRef]

61. Ewing, G.O.; Sarigöllü, E. Car fuel-type choice under travel demand management and economic incentives. *Transp. Res. Part D Transp. Environ.* **1998**, *3*, 429–444. [CrossRef]

62. Haustein, S.; Jensen, A.F. Factors of electric vehicle adoption: A comparison of conventional and electric car users based on an extended theory of planned behavior. *Int. J. Sustain. Transp.* **2018**, *12*, 484–496. [CrossRef]

63. Langbroek, J.H.M.; Cebecauer, M.; Malmsten, J.; Franklin, J.P.; Susilo, Y.O.; Georén, P. Electric vehicle rental and electric vehicle adoption. *Res. Transp. Econ.* **2019**, *73*, 72–82. [CrossRef]

64. Priessner, A.; Sposato, R.; Hampl, N. Predictors of electric vehicle adoption: An analysis of potential electric vehicle drivers in Austria. *Energy Policy* **2018**, *122*, 701–714. [CrossRef]

65. DellaValle, N.; Zubaryeva, A. Can we hope for a collective shift in electric vehicle adoption? Testing salience and norm-based interventions in South Tyrol, Italy. *Energy Res. Soc. Sci.* **2019**, *55*, 46–61. [CrossRef]

66. Dua, R.; White, K.; Lindland, R. Understanding potential for battery electric vehicle adoption using large-scale consumer profile data. *Energy Rep.* **2019**, *5*, 515–524. [CrossRef]

67. Chakraborty, A.; Kumar, R.R.; Bhaskar, K. A game-theoretic approach for electric vehicle adoption and policy decisions under different market structures. *J. Oper. Res. Soc.* **2020**, 1–18. [CrossRef]

68. Kim, S.; Lee, J.; Lee, C. Does Driving Range of Electric Vehicles Influence Electric Vehicle Adoption? *Sustainability* **2017**, *9*, 1783. [CrossRef]

69. Clinton, B.C.; Steinberg, D.C. Providing the Spark: Impact of financial incentives on battery electric vehicle adoption. *J. Environ. Econ. Manag.* **2019**, *98*, 102255. [CrossRef]

70. Jenn, A.; Laberteaux, K.; Clewlow, R. New mobility service users' perceptions on electric vehicle adoption. *Int. J. Sustain. Transp.* **2018**, *12*, 526–540. [CrossRef]

71. Sierzchula, W.; Bakker, S.; Maat, K.; van Wee, B. The influence of financial incentives and other socio-economic factors on electric vehicle adoption. *Energy Policy* **2014**, *68*, 183–194. [CrossRef]

72. Yang, S.; Cheng, P.; Li, J.; Wang, S. Which group should policies target? Effects of incentive policies and product cognitions for electric vehicle adoption among Chinese consumers. *Energy Policy* **2019**, *135*, 111009. [CrossRef]

73. Chatzikomis, C.I.; Spentzas, K.N.; Mamalis, A.G. Environmental and economic effects of widespread introduction of electric vehicles in Greece. *Eur. Transp. Res. Rev.* **2014**, *6*, 365–376. [CrossRef]

74. Deeb, J.G.; Walter, N.; Carrico, C.; Gašperin, M.; Deeb, G.R. Helicopter Mountain Rescue in Slovenia from 2011 to 2015. *Wilderness Environ. Med.* **2018**, *29*, 5–10. [CrossRef] [PubMed]

75. Perusich, K.; Tepper, H.; Marca, J.R.B.D.; Lefevre, R.; Baseil, R. Linking technologists and humanitarians. *IEEE Technol. Soc. Mag.* **2009**, *28*, 25–31. [CrossRef]

76. Pfau, L.; Blanford, J.I. Use of Geospatial Data and Technology for Wilderness Search and Rescue by Nonprofit Organizations. *Prof. Geogr.* **2018**, *70*, 434–442. [CrossRef]

77. Louviere, J.; Lings, I.; Islam, T.; Gudergan, S.; Flynn, T. An introduction to the application of (case 1) best-worst scaling in marketing research. *Int. J. Res. Mark.* **2013**, *30*, 292–303. [CrossRef]

78. Flynn, T.N.; Louviere, J.J.; Peters, T.J.; Coast, J. Best-worst scaling: What it can do for health care research and how to do it. *J. Health Econ.* **2007**, *26*, 171–189. [CrossRef]

79. Buckley, P.J.; Devinney, T.M.; Louviere, J.J. Do managers behave the way theory suggests? A choice-theoretic examination of foreign direct investment location decision-making. *J. Int. Bus. Stud.* **2007**, *38*, 1069–1094. [CrossRef]

80. Beck, M.J.; Rose, J.M.; Greaves, S.P. I can't believe your attitude: A joint estimation of best worst attitudes and electric vehicle choice. *Transportation* **2017**, *44*, 753–772. [CrossRef]

81. Rezaei, J.; van Roekel, W.S.; Tavasszy, L. Measuring the relative importance of the logistics performance index indicators using Best Worst Method. *Transp. Policy* **2018**, *68*, 158–169. [CrossRef]

82. Thurstone, L.L. A law of comparative judgment. *Psychol. Rev.* **1927**, *34*, 273. [CrossRef]

83. Finn, A.; Louviere, J. Determining the Appropriate Response to Evidence of Public Concern: The Case of Food Safety. *J. Public Policy Mark.* **1992**, *11*, 12–25. [CrossRef]

84. Marley, A.A.J.; Flynn, T. Best Worst Scaling: Theory and Practice. *Int. Encycl. Soc. Behav. Sci.* **2015**, *2*, 548–552. [CrossRef]

85. Hinz, O.; Schlereth, C.; Zhou, W. Fostering the adoption of electric vehicles by providing complementary mobility services: A two-step approach using Best–Worst Scaling and Dual Response. *J. Bus. Econ.* **2015**, *85*, 921–951. [CrossRef]

86. Garver, M.S.; Williams, Z.; LeMay, S.A. Measuring the importance of attributes in logistics research. *Int. J. Logist. Manag.* **2010**, *21*, 22–44. [CrossRef]

87. Sawtooth Software, Inc. Sawtooth Software Technical Paper Series: The MaxDiff System Technical Paper. 2020. Available online: https://sawtoothsoftware.com/resources/technical-papers/maxdiff-technical-paper (accessed on 28 July 2020).

88. Johnson, R. *Understanding HB: An Intuitive Approach*; Sawtooth Software, Inc.: Sequim, WA, USA, 2000.

89. Howell, J. *CBC/HB For Beginners*; Sawtooth Software, Inc.: Sequim, WA, USA, 2009.

90. Orme, B. Hierarchical Bayes: Why All the Attention? Sawtooth Software, Inc.: Sequim, WA, USA, 2000.

91. Brink, H.I.L. Validity and reliability in qualitative research. *Curationis* **1993**, *16*, 35–38. [CrossRef]

Publisher's Note: MDPI stays neutral with regard to jurisdictional claims in published maps and institutional affiliations.

Article

Second Life Batteries Used in Energy Storage for Frequency Containment Reserve Service

Lukáš Janota * , **Tomáš Králík** and **Jaroslav Knápek**

Faculty of Electrical Engineering, Czech Technical University in Prague, Technická 1902/2, 166 27 Praha 6, Czech Republic; tomas.kralik@fel.cvut.cz (T.K.); knapek@fel.cvut.cz (J.K.)
* Correspondence: janota.lukas@fel.cvut.cz

Received: 4 November 2020; Accepted: 24 November 2020; Published: 3 December 2020

Abstract: The new Li-ion battery systems used in electric vehicles have an average capacity of 50 kWh and are expected to be discarded when they reach approximately 80% of their initial capacity, because they are considered to no longer be sufficient for traction purposes. Based on the official national future development scenarios and subsequent mathematical modeling of the number of electric vehicles (EVs), up to 400 GWh of storage capacity in discharged batteries will be available on the EU market by 2035. Therefore, since the batteries still have a considerable capacity after the end of their first life, they could be used in many stationary applications during their second life, such as support for renewables, flexibility, energy arbitrage, peak shaving, etc. Due to the high output power achieved in a short time, one of the most promising applications of these batteries are ancillary services. The study assesses the economic efficiency of the used batteries and presents several main scenarios depending on the likely future development of the interconnected EU regulatory energy market. The final results indicate that the best results of second-life batteries utilization lie in the provision of Frequency Containment Reserve Service, both from a technical and economic point of view. The internal rate of return fluctuates from 8% to 21% in the realistic scenario, and it supports the idea that such systems might be able to be in operation without any direct financial subsidies.

Keywords: second life of batteries; electromobility; battery energy system storage; ancillary service; frequency regulation; economic evaluation

---

1. Introduction

The global warming crisis is becoming one of the most dangerous threats that we will have to face worldwide in this century. The ever-accelerating growth rate of the world economy and industrial production has led to steadily increased greenhouse gas (GHG) production. Since the year 1997, when the Kyoto Protocol was signed, which is the first international treaty for limiting and reducing GHG emissions, the signed state parties have started their journey for sustainable development and a cleaner environment. The latest legal extension of the Kyoto Protocol framework is the Paris Agreement from 2015. Within this agreement, the long-term temperature goal to keep the average world temperature increase at least below 2 °C in comparison with preindustrial levels (approx. 1880) was negotiated and set [1].

The European Union (EU), since the Kyoto Protocol was signed, has established itself as an international leader in the energy transition from fossil fuels towards clean energy, and this approach is reflected in its ambitious energy and climate policy [2]. The current EU general energy policy framework from 2019 sets out to achieve the following goals by 2030: (1) reduction of at least 40% in GHG compared to 1990 levels (currently, there is an ongoing EU discussion about the increase up to 55–60%), (2) increase to 32% of the share of renewable energies in the final energy consumption, (3) improvement of 32.5% in energy efficiency, and (4) the interconnection of at least 15% of the EU's electricity transmission systems [3]. The strict climate–energy policy set in this way will continue in the future and should lead to a carbon-free economy in 2050. To successfully achieve the main goal state of charge (SoC) in 2050, the GHG emissions and energy consumption must also be significantly reduced from the transport sector, which accounts for almost a quarter of the total environmentally hazardous emissions in Europe and is responsible for 33% of the final energy consumption (353 Mtoe) [4]. Within the sector, road transport is responsible for more than 70% of all GHG emissions from transport in the EU and is individually responsible for pollution in densely populated areas [5].

Figure 1 shows that the production of GHG emissions from the transport sector does not show the same decline trend over the years as the emissions produced from other sectors in the EU28. For the first time, the production of emissions from transport began to decline in 2007. At the end of 2018, however, GHG emissions produced by transportation again reached the level of 2007, almost 1100 Mt CO_2e.q. In 1990, the production of emissions was 862 Mt CO_2e.q. [6]. The EU answers to this unsatisfactory development and the challenge of reducing transport emissions with urgent and irreversible transitions to zero-emission mobility. By the year 2050, GHG emissions from transport (including aviation but excluding international maritime) will need to be reduced by at least 60% compared to 1990 [5]. The European Commission has set an average car fleet CO_2 target 95g CO_2/km for all new cars to accelerate the reduction of emissions from the road transport sector. This limit is expected to be further rapidly tightened; by the year 2030, a stringent standard at 70 g CO_2/km is planned [7]. Emission standards like that will no longer be able to meet by conventional combustion cars, so car manufacturers will have to raise their low/zero emissions vehicle shares significantly [8].

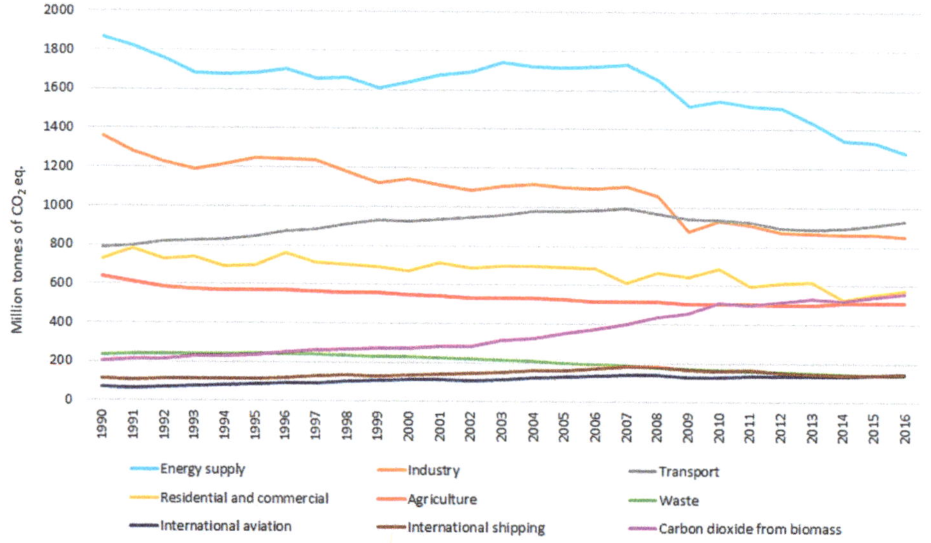

Figure 1. Evolution of greenhouse gas emissions by sector in the EU28 [5].

According to the International Energy Agency, at the end of 2019, over 1.5 million electric vehicles (EVs) (including battery electric vehicles (BEVs) and plug-in hybrid electric vehicles (PHEVs)) were registered in the EU [9]. The total number of registered EVs in the EU is expected to accelerate its growth due to the falling prices of battery technology (battery cost breakdown) and government financial subsidies and support [10]. According to estimates of the cumulative number of EVs based on optimal development, EU countries will have more than 10 million electric vehicles by 2030 [11].

In the next few years, the rapidly expanding EV market will start producing a large amount of waste, especially battery waste, which is not easily recyclable [12]. This fact is also confirmed by the analysis that the Global Composite Annual Growth Rate of Lithium-Ion Batteries (batteries most commonly used as energy sources in EV) will grow by 36% between 2015 and 2020 [13].

Batteries are usually discarded from electric vehicles when they reach a level of about 80% of their original capacity after their first life cycle in EVs, which is usually about 8 to 10 years [14,15]. However, these discarded batteries are, due to their technical and safety parameters, still suitable for use in stationary applications with less demanding load profiles [16]. Hence, instead of collecting and recycling used batteries, batteries are collected and sent through the repurposing process. Within this process, the state of health (SoH) of discarded battery cells is checked, and batteries are being prepared for their second lives [17]. According to a study "New life for used EV batteries as stationary storage" by Bloomberg New Energy Finance, the global cumulative discarded capacity could reach 26 GWh by the year 2025.

Through this repurposing process, the total battery lifespan is prolonged, the current amount of waste is minimized, and the idea of the circular EU economy is fulfilled. Further usage of discarded battery cells will also bring significant financial savings, savings in GHG emissions, and a reduction of primary energy consumption that would otherwise arise from the production process of new battery cells [18]. The idea of a secondary use of discarded batteries could also help accelerate the development of the already accessible high-capacity battery storage, the development of which is still hampered by the high cost of battery technology [14]. The current price of new battery Li-ion technologies ranges from 200 to 300 USD/kWh and, according to initial studies and analyses, the price of discarded battery cells should not exceed 35% of the price of new battery cells of the same technology [15].

Second-life batteries (SLBs) could be used in a wide variety range of stationary applications for the transmission and distribution grid but, also, for commercial purposes such as load shifting, peak shaving, black start, backup, and grid deferral [19]. Due to the specific technical parameters of batteries, such as providing high performance (high output power) with a short response and activation time, these discarded batteries could find applications—primarily, in stationary applications supporting grid stability control and the integration of increasing numbers of grid-scale decentralized energy sources (DERs) [20]—as their output power is variable and uncontrollable over time and whose installation is necessary for the fulfillment of defined climate and energy targets EU [15].

The fast development of distributed generation brings new challenges, especially in the field of voltage and frequency regulation of the power grid [21]. Battery systems are becoming crucial providers of fast frequency control services, also known as fast frequency containment process (FCP) [22]. FCP is an ancillary service provided to the transmission system operator (TSO) with the shortest reaction time, in the order of tens of seconds, and the time until full activation of the provided power backup does not exceed 30 s. That is the reason why this service is dedicated to cover fast and small changes in transmitted power in the power grid. In Germany, there is currently installed more than 400 MW of output power and a capacity of about 550 MWh of large-scale storage systems (LSS), which operate mainly in the frequency containment reserve (FCR) market [23].

The technical condition after the decommissioning of traction batteries from EVs and their possibility for usage in LSS for network control and support was already analyzed in several research papers and studies [24]. Jeremy S. et al. from the National Renewable Energy Laboratory, within their study, focused on SLB degradation, and their SoH status concluded that, with a proper energy management system (EMS) and optimal operating strategy, repurposed automotive batteries can last 10 years or more in stationary applications [25]. The method for optimal EMS design and the optimal sizing of the battery energy system storage (BESS) energy capacity using SLBs for enhancing renewable energy grid integration was studied and proposed in [26]. However, in reference [27], four different application scenarios on a real stationary SLB storage were simulated and tested (support EV charges, area frequency regulation, self-consumption, and grid investment deferral), and each of these applications differed dramatically in the length of its secondary life. In the frequency area regulation scenario, the secondary lifespan was only 4.7 years until the BESS was no longer able to meet the technical requirements for FCP providers, and these investments will be economically unprofitable. According to this study and others, the lifespan of BESS is most affected by the depth of discharge (DoD) and the average state of charge (SoC) [28–30]. Additionally, in [21], the operation of SLBs in BESS are simulated in three network support model cases—namely, peak shaving, voltage control in power systems, and ancillary services. During the study, the authors found a positive impact on the operation of the energy system at all voltage levels in the use of BESS (no economic evaluation included in this study).

With this identified knowledge in mind, the focus of this paper was to propose a complex mathematical model based on real historical electricity frequency data obtained from the TSO to investigate the technical suitability and economic efficiency of SLB usage in BESS. The simulation model was developed and tested in MATLAB (R2020b, MathWorks, Natick, MA, USA). In order to achieve optimal and relevant results, according to the analyzed critical technical lifespan parameters, the emphasis was placed on the optimal BESS capacity sizing and the design of an optimal strategy for the SoC and DoD battery management within the operation. A modeled SLB BESS was used for the provision of FCP services using the harmonized European Balancing Guidelines, and it was verified in the environment of the Czech regulating energy (RE) and ancillary services market. For the correct economic evaluation of the BESS operation during the whole secondary battery lifetime, several different scenarios were created following the impact of the newly emerging interconnected European market with RE [31].

2. Complex BESS Model Structure and Principle

The proposed methodology of modeling the usage of SLBs in a BESS operation, which is shown in Figure 2, can be divided into two essential parts for simplification:

1. Mathematical-technical model of BESS operation (Blocks 1–4)
2. Economic model of the BESS operation (Blocks 5–8)

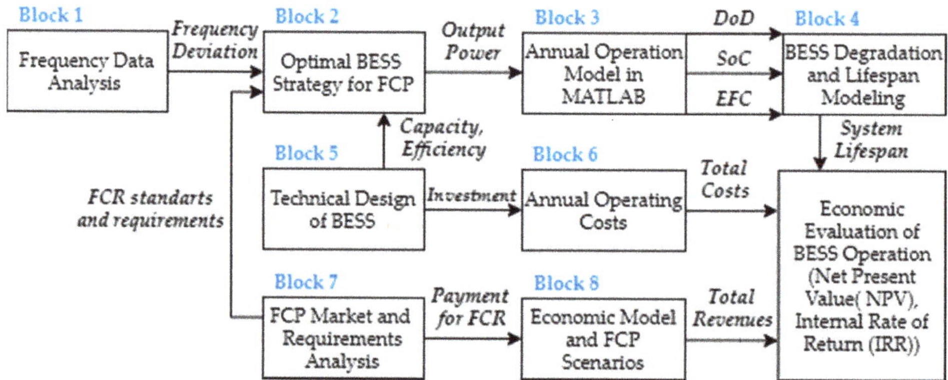

Figure 2. Block diagram representing the proposed methodology and individual parts of the model for evaluating the use of second-life batteries (SLBs) for the battery energy system storage (BESS). FCP: frequency containment process, FCR: frequency containment reserve, DoD: depth of discharge, SoC: state of charge, and EFC: equivalent full cycles.

Through the obtained analyses and simulated outputs (system lifespan, total cost, and total revenues) from both main parts of the methodology, a final critical evaluation of the suitability and economy of the secondary traction batteries utilization in the stationary storage was performed.

2.1. Mathematical-Technical Model of BESS Operation

The first step in the mathematical-technical part of the methodology is a detailed analysis of the historical data of the frequency in the selected regulating region (RR). Through this initial analysis in Block 1, we calculated samples of frequency deviation during the years and statistical frequency deviation distribution, and we obtained crucial information for BESS management strategy design about whether there is more often a lack or excess of electricity in the power grid. Within Block 2, an optimal operation strategy to maintain an ideal SoC level of BESS, based on the input parameters from Blocks 1, 5, and 7, was proposed. This strategy is essential to meet the requirements and conditions for FCR providers given by TSOs. The operation strategy, with respect to all operating conditions, gives the required output power for each frequency deviation sample. Based on this output power performance, a complex annual operation was simulated in Block 3. In this methodology part, the total amount of electricity injected and extracted from the power grid is counted, and based on this supplied/consumed amount of energy, we calculated the average operating parameters of BESS during the years (SoC and DoD). To simplify the estimation of the system lifespan within Block 4, we converted the sum value of all battery cycles performed during the provision of the FCP to equivalent full cycles (EFC). The use of the EFC methodology is generally recommended for determining the battery cell degradation process in many papers [29,30].

2.2. Economic Model of BESS Operation

In the second part of the proposed methodology, we focused on a comprehensive evaluation of the economics of BESS. Since the main intention of this study was to verify the possibility of providing the ancillary service by BESS with SLBs, it was necessary to identify within Block 7 the requirements of the FCP service providers. Additionally, at this stage of the procedure, we analyzed the impact of the emerging internal European balancing energy market (EBEM) on the price of reserved balancing reserves, which is pay-off for the settlement between the TSO and balancing service provider (BSP). Based on the purchase prices of used storage components (transformer and converter), acquisition costs, and an estimation of the discarded batteries market price, we determined in Block 5 the total cost of purchasing the second life LSS. In the following Blocks 6 and 8, formed on the initial investment outlay, the results of annual MATLAB simulations, and established operating assumptions respecting the essence of the EBEM, we calculated all other economic necessary parameters (total operation costs and total operation revenues). To verify the validity of the obtained results complex, various scenarios of possible future developments in the balance energy internal market were created and predicted within Block 6.

3. Frequency Regulation in European Standards

The frequency regulation process—often called simply balancing—includes all actions and processes through which TSOs, on an ongoing basis, ensure the maintenance of system frequency within a predefined stability range around the nominal value of the system frequency [32]. The nominal frequency value (f_N) in the synchronously working area within the Union for the Coordination of Transmission of Electricity (UCTE) is 50 Hz. The difference between the actual system frequency $f(t)$ and nominal frequency value is called the frequency deviation, Δf, as shown in Equation (1). Frequency deviation occurs in the transmission grid when the total electricity generation does not equal to total electricity consumption within the controlled region of the TSO. By the frequency deviation definition, it is apparent that the deviation can have both negative and positive values. Negative values correspond to the demand being greater than the electricity generation, and positive values are linked with the situation when the generation exceeds demand.

$$\Delta f = f(t) - f_N, \tag{1}$$

Currently, TSOs within Europe use slightly different processes and products to ensure the balanced state of the grid due to disparate historical developments and balancing philosophies. In order to achieve the highest future security and reliability of interconnected transmission network operations, an internal market for sharing the balancing energy and power reserves within the synchronously operating area was created on the basis of a cooperation agreement signed by participants of the European Network of Transmission System Operators for Electricity (ENTSO-E) [29]. Figure 3 illustrates the proposed harmonized sequence of regulation processes to achieve cooperation and a successful integration of the common European balance energy market within the load frequency control (LFC) area or block.

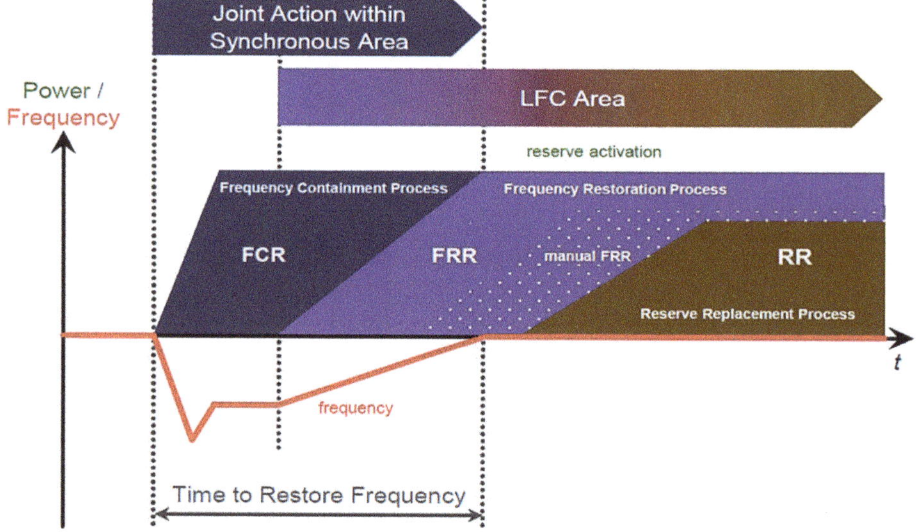

Figure 3. The harmonized sequence of balancing processes for full frequency restoration [33]. LFC: load frequency control, FRR: frequency restoration reserve, and RR: regulating region.

The harmonized frequency recovery process consists of four consecutive provided power reserves. For each type of balancing reserve in the process, different technical requirements are set in the Network Code of ENTSO-E. Balancing service providers (BSPs) offer bids on the balancing market to TSOs of their available power reserve or energy capacity within the services listed below. The price of accepted bids is determined on the basis of the marginal price of all BSP bids and total procured volume for each trading interval of the market.

1. frequency containment reserve (FCR)
2. automatic frequency restoration reserve (aFRR)
3. manual frequency restoration reserve (mFRR)
4. replacement reserve (RR)

After the occurrence of an imbalance, the FCP service power reserves are activated first to contain the system frequency. The FCP service is the one with the shortest activation time, and it also provides the lowest amount of active power reserves per unit. These are the main reasons why this service is suitable for providers offering a performance within the limited storage capacities of BESS. In this study, we designed and verified the operation of a stand-alone BESS, which does not have a backup resource unit. Therefore, we focused only on providing reserves within the FCR.

3.1. Frequency Containment Process

Within the frequency regulation process, the total provided unit power reserve (P_N) is activated automatically by a primary frequency control device. The controller secures the real-time system deviation measurement and a constant ratio between the system frequency disturbance and activated output power. The expression for the regulation control mechanism of the required activated reserve (ΔP_R) is given as follows:

$$\Delta P_R = -\frac{100}{S_G}\frac{P_N}{f_N}\Delta f, \tag{2}$$

The generator drop S_G is a ratio (without dimension), and it represents the ability to regulate the required activated reserve continuously. The ratio is generally expressed as a percentage:

$$S_G = \frac{-\Delta f / f_N}{\Delta P_R / P_N} \text{ in } \%, \tag{3}$$

The amount of automatically activated power reserve is linearly proportional to a system frequency deviation. As mentioned earlier, the frequency system variety could be both positive and negative, so FCP is a symmetric service, and both positive and negative FCR are required within it. The minimum volume of provided FCR by one unit is 1 MW, and the maximum is 25 MW, with the step of 1 MW for market bidding. The maximum FCR capacity is activated in the case of a system frequency deviation ±200 mHz. According to the ENTSO-E Guideline on Electricity Transmission System Operation (GETSO) [34], each BSP connected to the TSO shall ensure that the FCR offered fulfills the operation requirements and meets the technical properties. The reaction time of full FCR activation has to comply with the following requirements:

- At least 50 % of the FCR power reserve must be delivered no later than 15 s after the request.
- The full value of the FCR power reserve must be delivered no later than 30 s after the request.
- If the frequency deviation in the power grid is higher than 200 mHz, the increase in activated FCR power must be at least linear in the range of 15 to 30 s.

By the FCR report part of the TenneT Network Code [35], the state when the frequency deviation in the power network is in the range [−50 mHz, +50 mHz] is considered as a normal operating state of the system. During this time, the availability of each unit providing the FCR to the grid must be 100%. If the grid frequency exceeds the value of ±50 mHz from f_N, a warning state of the frequency recovery process is recognized whenever at least one of the following situations occurs:

- Δf exceeds ±50 mHz for a longer time than 15 min
- Δf exceeds ±100 mHz for a longer time than 5 min
- whenever Δf exceeds the maximum allowed system frequency deviation of ±200 mHz

If the frequency recovery process enters the warning regime, all FCR units must be able to deliver their maximum contracted power (P_N) for 30 min. In the remaining parts of this study, we call this condition the 30-min criterion.

3.2. SoC Management Strategy

Due to the physical nature of secondary battery cells and their limited storage capacity, it is crucial to provide optimal an BESS SoC management strategy and to keep the charge level within the optimal operating range to meet the 30-min criterion at any time without interruption of the FCR supply. In Czech TSO's Network Code, there is no section devoted to providing FCR through stand-alone energy storage. For the purposes of this study and realistic assumptions for the harmonization of the requirements and rules for the provision of the FCP service, the rules set by the transmission system operators in Germany will be used. Allowed operation strategies in Germany to maintain the SoC within the optimal charged range have been already analyzed by several studies aimed at providing balancing reserves and ancillary services through LSS [36,37]. These strategies aimed to ensure that the SoC level will not exceed the operating range are called degrees of freedom (DoFs). Electricity obtained through all DoF strategies is free of charge.

3.2.1. Over-Fulfillment (OF)

The current unit output power set by the FCP regulation controller could be optionally increased by up to 20% at any time. Over-fulfillment is the main DoF to maintain the SoC, as it can provide effective additional BESS charging and discharging during providing balancing service.

3.2.2. Dead Band (DB)

The range of ±10 mHz around the nominal frequency is considered as the dead band (DB), within which the provision of FCR is not mandatory. DB management of the SoC is allowed only when the correction power is not in the direction of the current frequency deviation and does not cause an additional imbalance in the system.

3.2.3. Gradient Controller Increase (GC)

The controller gradient represents the minimum required activated power ramp slope, which is given by the requirement of a fully activated power reserve within 30 s. As BESS are able to change their output power very quickly, in the order of hundreds of milliseconds, the power change gradient over time can be changed and used to control the SoC.

3.2.4. Market-Based Energy Trade

The three DoF strategies mentioned in the previous subsections are excellent support tools for maintaining the SoC level in the permitted operation area, but in cases of high network frequency deviations, their potential is insufficient. To ensure the SoC limit remains sufficient for FCR provision, the energy needed to charge and discharge the battery is traded on the intraday energy market (IEM). There are two possible ways to use this strategy:

- energy trading realized using the output power offset (FCP provision is not interrupted) and
- charging or discharging using energy trading after the 30-min criterion (FCP is not provided), and BESS has 2 h to restore its SoC into the permitted range. The SoC parameter is very closely connected with the actual size of the chosen battery system. It is a multicriteria task, which is introduced within Section 4.4. Technical BESS Design.

3.3. FCR Power to Frequency Characteristic

In the following Figure 4, we depict a graphic representation of all possible operating characteristics during providing a reserve for the FCP service via the BESS. The basic ΔP-Δf (commonly known as power to frequency ratio) output characteristic (black dashed curve) is given by Equation (2), and its curve shape confirms that the activated power reserve is linearly proportional to a system frequency deviation.

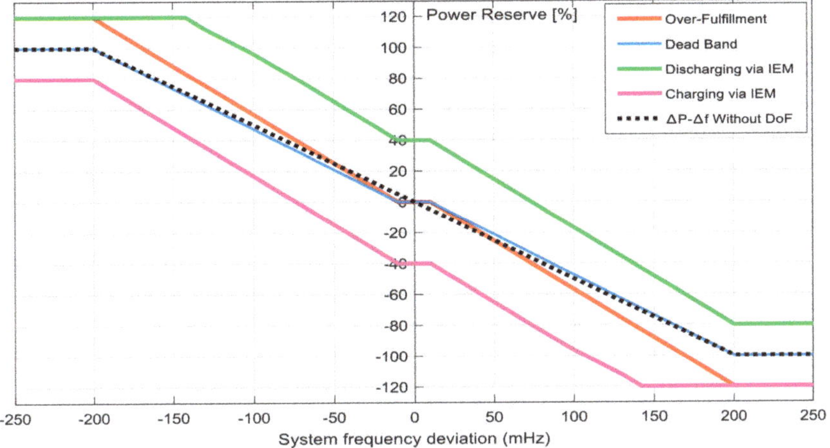

Figure 4. Power to frequency ration (ΔP-Δf) characteristics for the provisioned FCP service with demonstrating possible individual degree of freedom (DoF) strategies for the BESS SoC maintenance. IEM: intraday energy market.

3.4. Frequency Containment Reserves Market

Within this study to evaluate the technical possibility and economic efficiency of SLB usage in BESS for participating in the FCR Market, we made the assumption that the Czech TSO is already an active participant of the ENTSO-E Project FCR Cooperation. Currently, Czech Republic TSO is only in the role of an observer, but we are expected to become a full member in the upcoming years. This precondition allows us in the economic model to calculate with the historical prices of settlement FCR bids. FCR Cooperation aims to integrate the balancing market to foster effective competition, increase liquidity, and support the speed of deployment of renewable energy sources. The Austrian, Belgian, Dutch, French, German, and Swiss TSOs currently procure their FCR in a common market, which means that they have the same conditions, auction period, and same symmetric product [38].

The FCR cooperation secures power reserves through the common FCR market with daily D-1 (day-ahead) tender type and standard 4-h FCR products. A D-1 tender type means that every FCR provider posts their bids to the market platform the day before the delivery at 8:00 h. The FCR settlement price is determined for every tender period (4 h) and is calculated by merit order pricing methodology, which is represented in Figure 5.

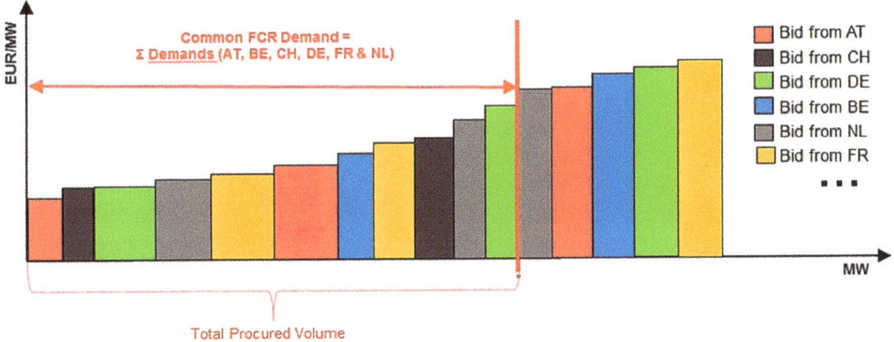

Figure 5. FCR cooperation market merit order pricing methodology, [38]. FCR: Frequency Containment Reserve AT: Austria, CH: Switzerland, DE: Deutschland, BE: Belgium, NL: Netherlands, FR: France.

3.5. FCR Provision Requirements Summarized

In Section 3 of this study, we analyzed the ancillary services, its market, and defined all necessary conditions and requirements that must be met when offering the FCP regulation power reserves capacity. In Table 1 below, we summarize the most important technical or operational parameters that will be further used in BESS modeling and the optimal SoC control strategy design.

Table 1. Summary of the central requirements and conditions for the provision of the frequency containment reserve (FCR) service within the synchronously operating EU system. FCP: frequency containment process and SoC: state of charge.

Required FCP Parameter	Value
Minimum Offered Power Reserve	1 MW
Maximum Offered Power Reserve	25 MW
Minimum Reserve Bid Size	±1 MW
Full Reserve Activation Frequency	±200 mHz
Maximum Allowed Dead Band	±10 mHz
Full Power Reserve Activation Time	30 s
Minimum Full Activation Period	30 min
SoC Restoration Time (after 30 min crit.)	2 h
Normal State Deviation Range	±50 mHz
Availability Within Normal Grid State	100%
FCR Product Tender Period	4 h

4. Mathematical Operation BESS Model

In the following section, the proposed methodology of BESS modeling in MATLAB is presented. The simulated BESS model is assumed to be based on second-life traction Li-ion batteries, so we adapted the technical operation parameters of the model to this fact. With this in mind, we wanted to achieve safety FCP provision and the longest possible lifespan.

4.1. Frequency Data Analysis

The grid frequency data from 2015 to 2018 were analyzed for the purpose of the main inputs of the proposed model. The historical frequency datasets were taken from the annual evaluation of the transmission system operation on the Czech TSO ČEPS website (only the Czech version, ČEPS, a.s., Prague, Czech Republic). During the analysis process, some inconsistent frequency samples $f_{i,Error}$ were identified. Values such as 0 Hz or no data measured were replaced by the calculation method shown in (4).

$$f_{i,Error} = \frac{\sum_{i-1}^{i-3} f_i + \sum_{i+1}^{i+3} f_i}{6} \tag{4}$$

Within this statistical analysis, we examined whether the BESS during the FCP provision will more often inject or extract the electrical energy from the power grid. In the following Figure 6, we present a graphical representation of the mathematical and statistical functions of the analyzed set of frequency deviation samples in the Czech transmission network in 2018.

The probability density function is skewed right, which points to the fact that the frequency deviation is, more often, positive. This state is the same for all datasets from the analyzed years. According to the power to frequency ratio (ΔP-Δf) characteristics (Figure 4), we made a presumption that the BESS will be extracting energy more often from the grid, which, in battery logic, means more frequent charging. In Table 2, we present numerical values from the performed frequency data analysis, which will be crucial in the design of the optimal BESS operating strategy.

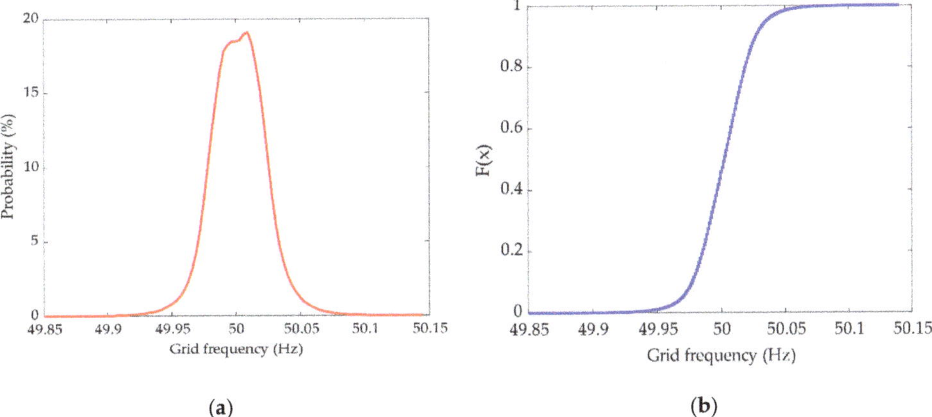

(a)   (b)

Figure 6. (**a**) Probability density function of the grid frequency data in 2018. (**b**) Cumulative distribution function of the grid frequency data in 2018.

Table 2. Numerical outputs of the historical frequency data analysis. Δf: frequency deviation and f_N: nominal frequency value.

Analyzed Parameter	2015	2016	2017	2018
Maximum Negative Δf (mHz)	147	127	133	155
Maximum Positive Δf (mHz)	126	125	133	140
Time Δf higher that f_N (min)	276,615	277,660	277,857	277,314
Time Δf lower that f_N (min)	241,091	241,525	240,694	240,945
Time Δf higher than f_N, DB activate (min)	171,986	172,711	179,413	177,494
Time Δf lower than f_N, DB activate (min)	137,403	137,750	144,342	143,014
Time Δf within DB $\pm$ 10 mHz (min)	216,211	215,139	201,845	205,092

4.2. Optimal Operation Strategy

The design of an optimal battery system management strategy is key to obtaining the system's lifespan under real operating conditions and to perform the final technical-economic evaluation of BESS operation. We based the design of the optimal strategy on the technical parameters of Li-ion battery cells, the requirements for FCR service providers (see Table 1. Summary of the central requirements and conditions for the provision of the frequency containment reserve (FCR) service within the synchronously operating EU system), and the results of the statistical analysis of historical frequency data.

As we examine the possibility of using discarded battery cells in this paper, lithium nickel manganese cobalt (NMC) battery cell technology will be used in the proposed and simulated LSS. NMC battery technology was chosen as it had a more than 65% electric vehicle (EV) market share in 2019 [39].

4.2.1. NMC Battery Cell Parameters

NMC battery cells have a longer cycle life, more stable operation, and higher energy density than other battery technologies (nickel cobalt aluminum (NCA), lithium iron phosphate (LFP), and lithium cobalt oxide (LCO)) used on the EV market nowadays. NMC batteries also do not suffer from such high self-discharge during the nonoperation or low output power time, which is advantageous for small-frequency deviations in the grid. In Table 3 below, we present the basic technical parameters of the chosen battery cell chemistry. The table data were analyzed on the website of Battery University and from the article [40].

Table 3. Lithium nickel manganese cobalt (NMC) battery cell chemistry technical parameters.

Parameter	Value
Voltage operation range	3.0–4.2 V
Nominal Voltage	3.7 V
Specific energy (Capacity)	150–230 Wh/kg
Charge C-rate	0.7–1 C
Discharge C-rate	1 C, 2 C
Cycle life	~2500
Coulombic efficiency	>99%, due to C-rate

From the table above, the critical parameters for designing the operational strategy are the allowed voltage operation range and both C-rate values. The C rate is a dimensionless quantity that determines the permitted battery power and can be expressed as follows:

$$C_{rate} = \frac{Power_{max}}{Capacity_{initial}} \; (-, hour) \tag{5}$$

The C rate gives us the ratio between the maximum continuous output power and the battery's total installed capacity. In the discharge C-rate row in Table 3, we state two values; the higher one (2C) can be used only for a short period of the time. Otherwise, the battery degradation process will be accelerated. We decided not to use the possible 2C discharge rate during our modeling during the grid's normal operation state. The only case when we decided to allow the BESS to give 2C discharge output power was when the TSO declared a state of emergency.

Therefore, in our annual simulated model, we further operated with three possible cases of setting the maximum limit of the BESS output power to reach the maximum lifespan during the operation.

$$P_{charging, \; normal \; grid \; state} = \frac{Capacity_{installed}}{C_{rate, \; charge}} = \frac{Capacity_{installed}}{1} \; (MW) \tag{6}$$

$$P_{dicharging, \; normal \; grid \; state} = \frac{Capacity_{installed}}{C_{rate, \; discharge}} = \frac{Capacity_{installed}}{1} \; (MW) \tag{7}$$

$$P_{dicharging, emergency \; grid \; state} = \frac{Capacity_{installed}}{C_{rate, \; discharge,peak}} = \frac{Capacity_{installed}}{2} \; (MW) \tag{8}$$

The FCR power-to-frequency basic principle (Figure 5), Equation (6) will limit the BESS model's output power in the case of positive frequency deviations in the grid. On the other hand, we have to consider Equations (7) and (8) in times of negative frequency deviations.

The voltage operating range is shown in Figure 7 and is given by charging and discharging the NMC battery cell curve. From the courses of these characteristics, we determined the assumption that we will consider the course of the charging and discharging curves in the range from 10% SoC to 90% SoC as linear. Regarding the study of the effects of cycling on lithium-ion battery hysteresis and overvoltage [41], we decided not to operate the BESS outside the linear range. Operating the battery in these areas means overvoltage or undervoltage of the battery cell. These operating conditions place additive stress on the battery cell chemical structure and a consequent rapid decline in life. The SoC and stored energy limitation in the proposed BESS FCR model are determined as follows:

$$SoC_{min} = 0.1; \; E_{min} = SoC_{min} * Capacity_{installed} \; (MWh) \tag{9}$$

$$SoC_{max} = 0.9; \; E_{max} = SoC_{max} * Capacity_{installed} \; (MWh). \tag{10}$$

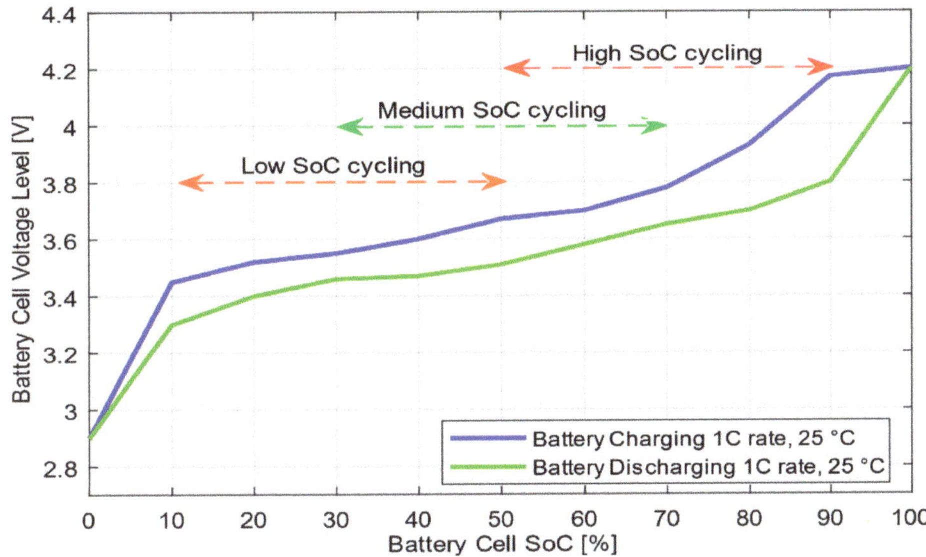

Figure 7. Charging and discharging curve of the lithium nickel manganese cobalt (NMC) battery cell.

The average operating SoC and SoC of the battery inactivity affect the resulting battery life. According to Battery University and the data in Table 3, the average lifespan of NMC batteries is 2500 equivalent full cycles with an estimated 80% DoD cycles calculated for the optimal operating conditions, which are the following: operation temperature around 25 °C and medium SoC cycling [42]. In response to this knowledge, and the fact that we are integrating discarded batteries that are more prone to degradation, the optimal strategy design will focus on maintaining the battery charge level between 40% and 60% SoC to prolong the battery lifetime.

$$SoC_{optimal,high} = 0.6; \; E_{optimal,high} = SoC_{optimal,high} * Capacity_{installed} \; (MWh) \tag{11}$$

$$SoC_{optimal,low} = 0.4; \; E_{optimal,low} = SoC_{optimal,low} * Capacity_{installed} \; (MWh). \tag{12}$$

Outside of the set optimal SoC BESS level, the correction DoF strategies will be used, either individually or mixed according to the current BESS SoC level, the present value of the system's frequency deviation, and the grid's operation state.

4.2.2. Permitted SoC Range

In the previous section, we analyzed and set the optimal operating parameters of the performance and charge level BESS to reach the optimal battery cell's technical lifespan. This section will design the battery system's optimum size based on the desired parameters and FCP provisions of the services.

When designing the optimal size of the BESS, we considered meeting the 30-min criteria and the availability of 100% of the time during the normal state of the network operation. We also considered that the power reserve in the FCP service is provided in both directions. The BESS must be ready to supply energy within the 30-min criterion in the up and down directions. We also had to consider the limitation of 10% and 90% SoC given by the discharge and charging curve of the battery NMC cell.

The results of these considerations are represented in the graph in Figure 8. The permitted continuous working area is determined by the ratio between the installed BESS capacity and the maximum possible reserve within the FCP service. The ratio is calculated for the NMC battery cell technology, and we consider here the discharging and charging C rate equal to 1.

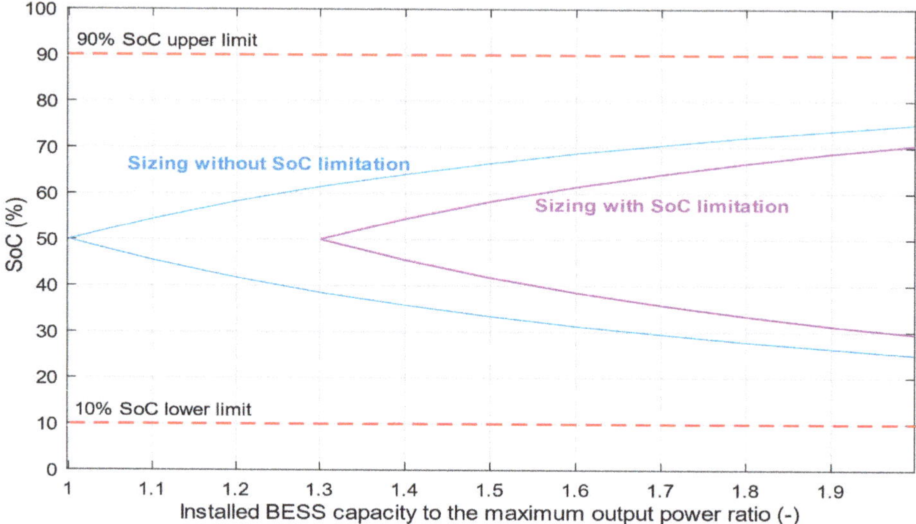

Figure 8. Permitted SoC BESS working area.

According to Equations (11) and (12), the optimal interval of the SoC (40% and 60%) of the working area was determined based on the analyzed optimal technical operating parameters. Following these values, we further simulate the BESS operations with a capacity-to-power ratio at least higher than 1.6 to meet the defined optimal work area.

4.3. Annual Operation MATLAB Model Design

In this section, the simulated annual BESS operation during the FCP provision is designed and tested. During the simulation, the necessary correction strategies were activated based on the current frequency deviation in the grid and the SoC level of the storage system. The size of the battery storage's output power and the composition changes over time may consist of the following possible composition power performance. All degrees of freedom analyzed in earlier sections of this paper are summarized together in the BESS SoC correction degrees of freedom output power P_{DOFs}.

$$P_{DOFs} = \pm\, P_{DB} \pm P_{OF} \pm P_{GC}\ (MW) \tag{13}$$

$$P_{DOFs} = \pm\frac{0.1\cdot P_{BESS,max}}{\Delta f_{Deadband}} \pm 1.2\cdot P_{FCP} \pm \frac{P_{FCP}\cdot 30}{\Delta T_{ramp,\,GC}}\ (MW) \tag{14}$$

$$P_{BESS\_charging} = -\eta_{BESS_{charging}}\left(-P_{FCP} \pm P_{DOFs} - P_{ET}\right)(MW) \tag{15}$$

$$P_{BESS\_discharging} = -\frac{1}{\eta_{BESS_{discharging}}}\left(+P_{FCP} \pm P_{DOFs} + P_{ET}\right)(MW) \tag{16}$$

The BESS output power consists of the main power within the delivery of the FCR power reserve (P_{FCP}), power of the correction strategy within the dead band (P_{DB}), and power within over-fulfillment (P_{OF}). Power obtained with the gradient controller increase strategy (P_{GC}) is given by the time of the preset ramp ($\Delta T_{ramp,GC}$), which is smaller than 30 s. The amount of output power given by trading on the intraday energy market (P_{ET}) was determined by whether the 30-min criterion was held. In the case that 30 min criterion is held, the BESS has 2 h to restore the optimal SoC level, so the minimum P_{ET} is given as half of the power of FCR provided (in our case 5 MW). The resulting output power was determined by the overall energy supply efficiency of the battery system. We determined to use for academic purposes a 93% total discharge efficiency and 95% total charge efficiency regarding a discussion with colleagues from the Battery Technology Department and the results in a study by S. Madani [40]. In these percentages, we included a coulombic efficiency of 99% of the NMC battery cells as well.

The size of each power outputs within the individual SoC level correction strategies and their activations were decided on the basis of the actual frequency of the grid and BESS parameters. The following Table 4 represents the combinations of allowed power operating states that may occur during the simulation and the possibilities of what SoC correction strategies are available at these times. The information of the current frequency deviation and the prediction of its near-development is essential, as none of the permitted correction strategies must go against the purpose of maintaining the system stability.

Table 4. Operation conditions for correction power degree of freedom DoF strategies.

SoC Level	P_{DB}		P_{OF}		P_{GC}		P_{ET}	
	$\Delta f < 0$	$\Delta f > 0$	$\Delta f < 0$	$\Delta f > 0$	$\Delta f < 0$	$\Delta f > 0$	$\Delta f < 0$	$\Delta f > 0$
SoC ≤ 10%	FCP service is not provided, and P_{DOFs} are not available.						- -	- -
SoC ≤ 30%	✕	-	✕	-	✕	-	-	-
SoC ≤ 45%	✕	-	✕	-	✕	-	✕	✕
SoC ϵ (45%, 55%)	±	±	✕	✕	✕	✕	✕	✕
SoC ≥ 55%	+	✕	+	✕	+	✕	✕	✕
SoC ≥ 70%	+	✕	+	✕	+	✕	+	+
SoC ≥ 90%	FCP service is not provided, and P_{DOFs} are not available.						+ +	+ +

Within the proposed annual simulation, we used and calculated the following variables and BESS operation technical parameters. The value of each of them was determined for every timestamp of the analyzed historical frequency data.

The state of charge of the BESS (SoC_t) is the function of actual output BESS power, previous SoC level, time of the power activation, and installed BESS capacity. In the annual model, we use a timestamp equal to 1 min, as given by the historical measured frequency datasets. The current SoC level is equivalent to the following relation:

$$SoC_t = (SoC_{t-1} \cdot \frac{1 - SD_{NMC,day}}{TS_{day}}) + \left(\frac{P_{BESS,t}}{60}\right) \cdot \frac{1}{Capacity_{installed}} \ (\%) \tag{17}$$

The BESS stored energy level ($E_{BESS,t}$) is calculated simultaneously through slightly different relations:

$$E_{BESS,t} = (E_{BESS,t-1} \cdot \frac{1 - SD_{NMC,day}}{TS_{day}}) + \left(\frac{P_{BESS,t}}{60}\right) \ (MWh) \tag{18}$$

$SD_{NMC,day}$ in the equation above represents the self-discharge mechanism of the NMC battery cells, and the variable TS_{day} is determined as the number of samples during the day. According to the time samples of the network frequency, we consider its value of 1440 per day. In our simulation of NMC cell degradation, we determined to calculate with 0.15% self-discharging per 24 h due to the acknowledgements and methods from the study of the self-discharge principles of NMC Li-ion cells by Thomas Deutschen et al. [43].

The curve represents the amount of energy stored in the BESS during the annual modeled operation. We calculated the whole year's energy flow through the battery storage system. This information is crucial in calculating the battery's estimated degradation time using the equivalent full cycle (EFC) methodology. In our model, we worked according to the SoC set working limits, with a maximum allowed DoD of 80%.

$$E_{BESS,annual} = \int_{t=1}^{T} |P_{BESS,t}| dt \ (MWh) \tag{19}$$

We determined the number of performed EFCs during the annual simulation through the volume of energy corresponding to the full discharge cycle with an 80% depth.

$$EFC_{annual} = \frac{E_{BESS,\ annual}}{DoD \cdot Capacity_{installed}} \ (-) \tag{20}$$

To describe and illustrate the proposed model and the annual simulation principle, a flow chart was created, which is shown in the Figure 9.

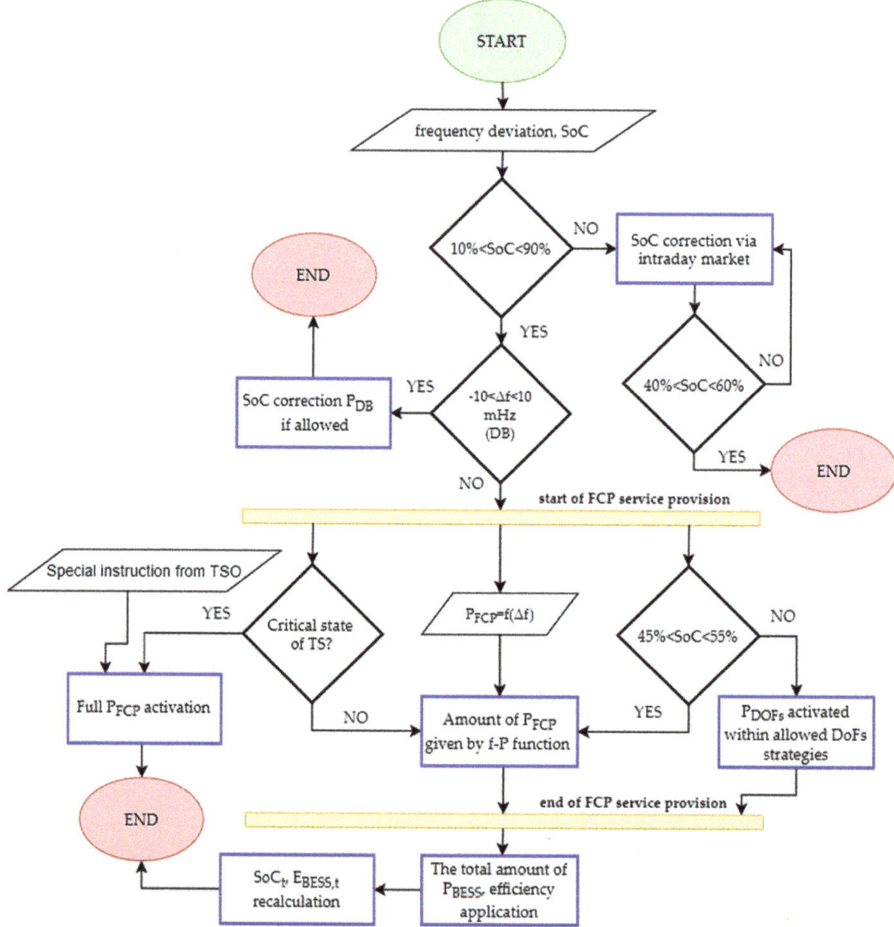

Figure 9. Flow chart representing the proposed MATLAB code of the annual BESS operation model. P_{FCP}: Power provided within the FCP service given by Equation (2), $E_{BESS,t}$: Total energy flow through the BESS within year t calculated by Equation (19), TSO: Transmission System Operator.

4.4. Technical BESS Design

The proper technical design of the main battery storage components and the final selected BESS capacity is an integral part of our complex model, since it directly affects the possible range of power reserve provided within the FCP service.

For this simulation, and the optimal choice between the revenues from the provision of the network support service and the amount of investment in the BESS, we decided to provide 5-MW FCR power reserve to the TSO.

Regarding the set values of BESS operating parameters in the previous sections, the optimal operating band of 10% and 90% SoC and the resulting minimum ratio of 1.6 between the installed capacity and maximum power are specified in Section 4.2.2. Permitted SoC Range, we should install storage equipment with an installed capacity of at least 8 MWh.

However, this paper's main idea was to use discarded batteries from electric cars during their second lives. These batteries have a lower initial capacity, greater propensity to accelerate degradation, and sufficient capacity reserve in the FCP service provision. Therefore, our proposed storage has an initial installed capacity of 10 MWh.

The proposed battery energy system storage consists of five separate container storages, and all of them are connected to the distribution system (DS) on the voltage level of 22 kV. These containers can be operated separately in the case of necessity or accident, so each has its technical components needed to connect to the network. This fact will provide us with a greater possibility of variability in the use of storage in the future if we are not successful in the tender in the market to provide FCR power backups. Should this case occur, we can use the BESS, for example, for energy arbitrage, trading in electricity, or charging it in times of negative prices in the electricity market. This technical design and separation of the total installed capacity will result in a reduction of the economic risk of the BESS operation during its estimated lifetime.

Table 5 summarizes all technical and operational parameters of the proposed BESS, which will be applied as basic inputs for the simulation. A more detailed overview of the components used will be given in Section 5.3.1. Initial investment, where their prices calculate the calculation of capital expenditures to acquire the proposed BESS.

Table 5. Basic technical parameters of the battery energy system storage (BESS).

BESS Parameter	Value
Installed capacity	10 MWh
Maximum FCR provided	5 MW
Round-trip efficiency	94%
Voltage DS level connected	22,000 V
Maximum BESS Voltage (DC)	1000 V
Nominal frequency	50 Hz

4.5. Annual Operation Model Results

This section applied the corrected historical data of frequency deviations on the BESS programmed model, already having selected the technical parameters, as listed in Table 5.

To illustrate and prove the proposed model's correct functionality, we examined the limit case that may occur during operation in the power system. The characteristics and waveforms of the individual BESS output operating parameters (SoC, power, and P-f) will be examined for the day of the maximum positive frequency deviation in the transmission system in 2018.

The day of the maximum positive frequency deviation represents when the BESS must extract lots of energy from the grid, so the presumption is that the SoC of the BESS could rise above the upper allowed SoC operation limit. The following Figures 10–15 show the effects of the proposed optimal SoC control strategies in contrast to the BESS operating without any control strategies allowed when the output power of the storage is linearly equal to the current Δf within this day.

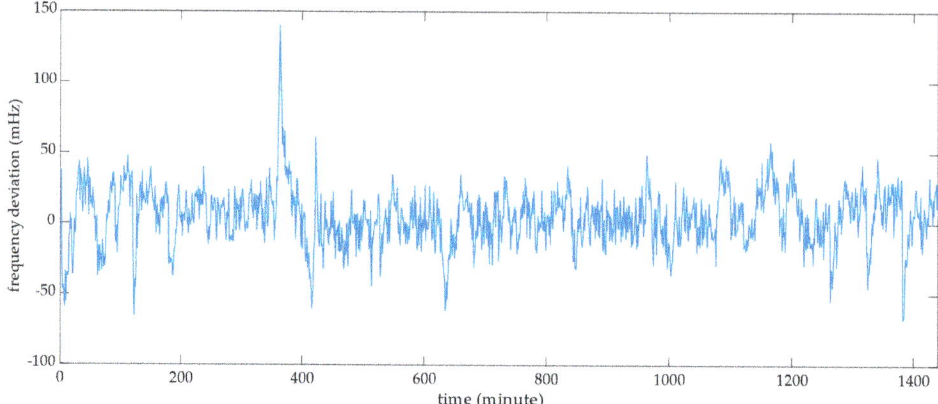

Figure 10. Frequency deviation values in the day of maximum positive deviation in the grid in 2018.

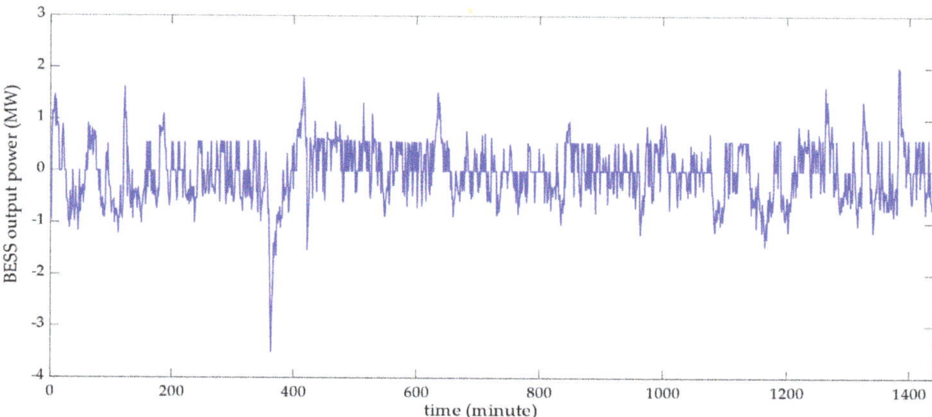

Figure 11. Battery storage power output using the proposed SoC DoF correction methodology.

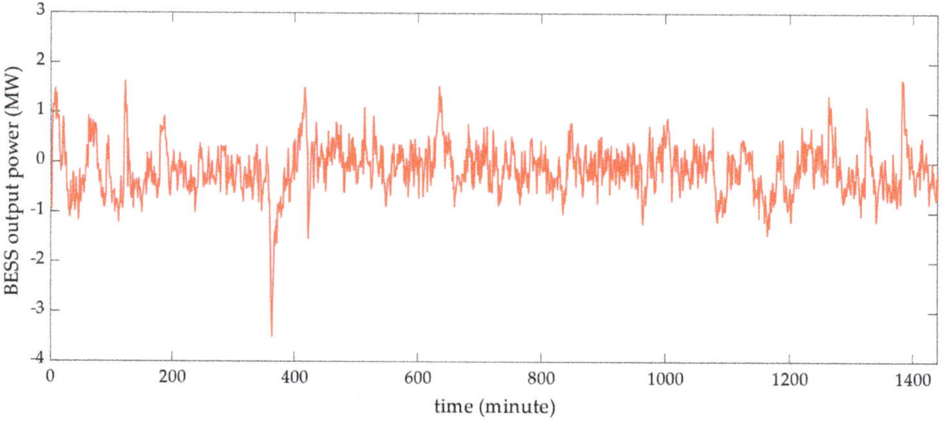

Figure 12. Battery storage power output without DoF strategies. P is given by Equation (2).

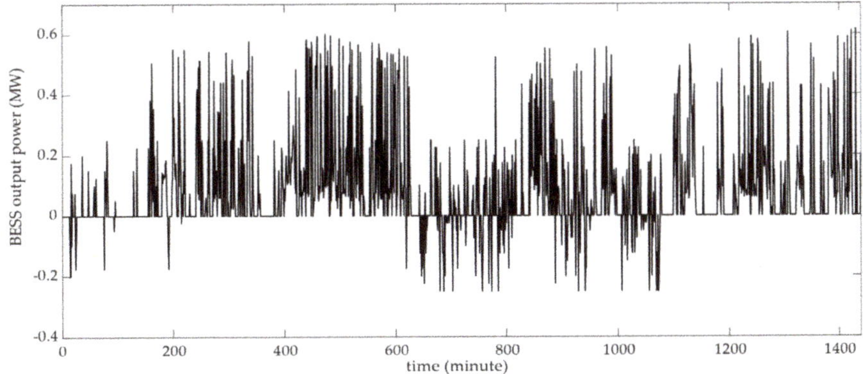

Figure 13. Differential power between the BESS operation with activated control strategies and without control strategies.

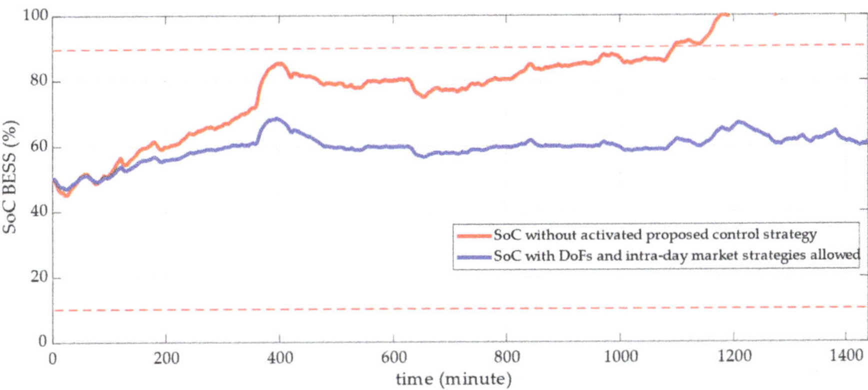

Figure 14. Impact of the use of the proposed SoC control strategy on the charge level of the BESS.

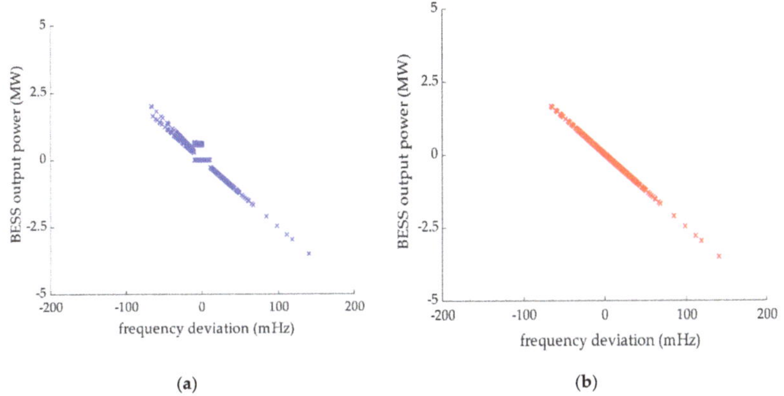

Figure 15. (**a**) Working power to frequency ratio (ΔP-Δf) characteristics of the BESS operation during the day of maximum positive frequency deviation (Δf) when all DoF strategies are in usage. (**b**) Working P-f characteristic of the BESS operation during the day of maximum positive Δf without any correction strategies activated.

Verification of the proposed control strategy's proper design and functionality for managing the optimal charge level of the large-capacity storage in operation was successful. The impact of using allowed strategies of degrees of freedom is more than evident in the BESS power characteristic shown in Figure 13. This characteristic represents times during the day of maximum positive frequency deviations when the BESS mostly increases its output power to discharge itself additively and to remain with the SoC level in the permitted working range.

The Table 6 summarizes the primary and crucial outputs from the annual simulations, working with frequency datasets from 2015 to 2018. Some of them are being used for the final lifespan estimation, and some of them are needed to calculate the economy of the BESS investment.

Table 6. Annual simulation outputs of BESS participating on providing FCR power reserves.

Simulation Output	2015	2016	2017	2018
Positive regulation FCP energy [MWh]	1661	1665	1770	1704
Negative regulation FCP energy [MWh]	1283	1288	1386	1350
Total amount of FCP energy [MWh]	2944	2953	3156	3054
Energy used within DoFs * [MWh]	677	676	646	728
Energy sold on intraday market [MWh]	125	104	99	178
Energy bought on intraday market [MWh]	20	31	23	38
System total year power losses [MWh]	242	247	251	249
Total number of the year EFC	453	455	475	473
Number of states SoC outside the allowed range	47	43	38	55
Number of emergency states, 30 min crit. required	19	16	12	22
Number of hours BESS not providing FCP	66	59	50	77
Yearly total availability of BESS for FCP needs [%]	99.237	99.318	99.422	99.109

* Energy obtained within DoFs strategies is free of charge, does not cause operating costs.

The number of times the BESS charge level falls outside the allowable operating range indicates that even the complex proposed SoC battery management strategy cannot fully and timely correct the increase or decrease in battery charge.

During the annual simulated BESS operations, this condition also occurred on other occasions than only during the warning state of the frequency recovery process within the transmission system according to the conditions analyzed in Section 3.1. Frequency Containment Process.

When the BESS reached the lower SoC limit, although not a single warning state condition was met, it occurred on 7.2.2018 and was represented in the following Figures 16 and 17.

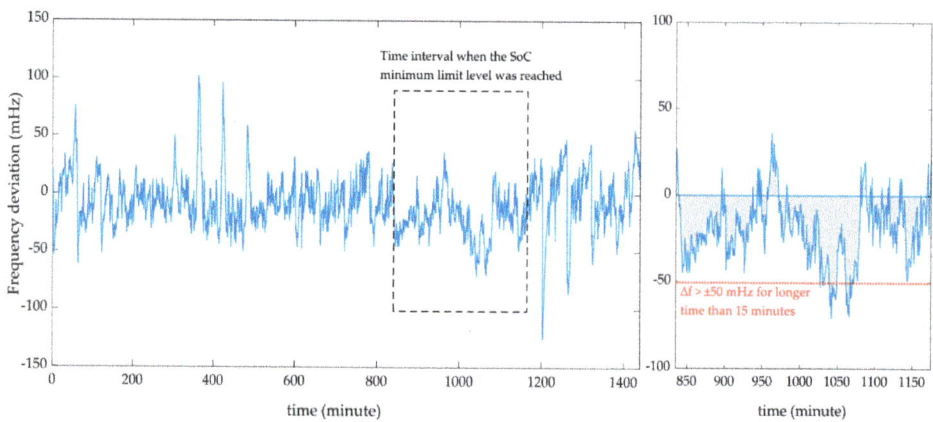

Figure 16. Example of the frequency deviation curve of the day the BESS reached the minimum allowed SoC level during the normal operation state of the grid.

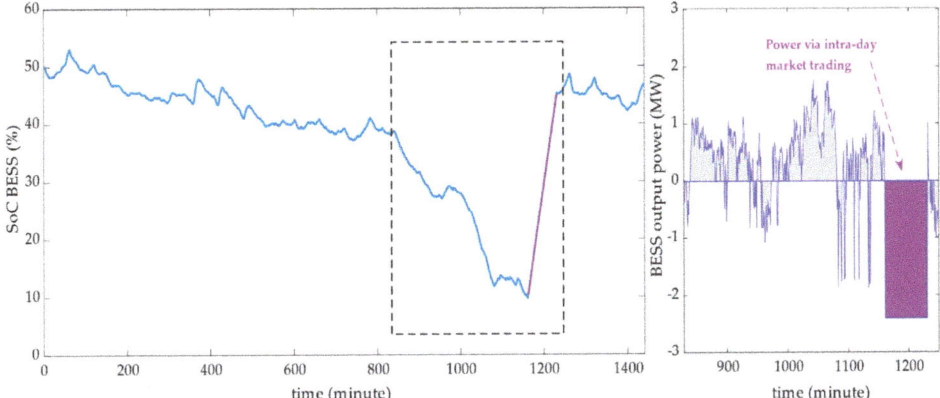

Figure 17. BESS SoC and output power curve during the exemplary day of reaching the minimum level of allowed SoC.

From the course of the frequency curve during the example of days in Figure 16, it may not be evident at first glance why the lower limit of the SoC was violated during the provision of the FCP service. We examined the values of frequency deviations in the network within the interval when the SoC drop occurred in more detail. The value of the network deviation lasted for almost five h between 0 mHz and −50 mHz. Although, the essence of declaring an alert condition in the network was not fulfilled within this time interval, the state of long-lasting low negative-frequency deviation did not allow the battery to use correction strategies according to the rules given in Table 4, and the energy storage system slowly discharged despite the relatively low output power.

The correction of this limit situation of the BESS operation through the purchase of additional corrective energy on the intraday market is described in the diagrams in Figure 17. The SoC level course shows how its level decreased almost continually within the analyzed five h.

Avoiding limit situations such as this could be achieved by combining the real-time market trading of electricity and the simultaneous provision of FCP without interruption. This advanced combined strategy would be reflected in the BESS output power by the power offset in the direction of the physical flow of energy currently traded on the market.

However, the question is whether this strategy will be allowed by the national regulator and TSO. There would be a shift in the sense of power output against the direction of the required application of the FCR power reserve. This situation could be counterproductive and cause a worsening of the balance situation in the network.

4.6. Second Life Lifespan Estimation

Determining the correct lifespan based on the simulation outputs is a critical part of this paper. The investment's economic efficiency is directly affected by the asset's final technical lifetime under review.

Many studies have already dealt with the degradation principle of battery cells during operation. We based the lifespan estimation at the battery cells aging principles in storage on the knowledge from studies [42,44,45]. A look at the aging of the battery cell, which most authors agreed on, is the possibility of dividing the aging of the battery into two independent principles according to the following equation:

$$\text{Cell aging}_{\text{total}} \approx \text{Calendar aging} + \text{Cycle aging} \tag{21}$$

Degradation of the battery due to time ("calendar") is entirely independent of the number of cycles performed and the depth of discharging or charging the battery. The rate of decrease in battery capacity is affected by the calendar storage conditions: SoC (%), temperature (°C), time (years), and cathode chemistry of the cells used as described by Equation (22).

$$\text{Degradation}_{\text{Calendar}} = f(\text{SoC}, t, T, \text{Chemistry}_{\text{Cell}}) \tag{22}$$

Rodrigo Martins et al., in their study, focused on the LSS for industrial applications; they linearized the course of calendar degradation over 10 years [46]. Their linearization for new NMC battery cells is shown in the following equation:

$$C_{\text{fade,cal,lin}}(\text{SoC})_i = 3.676 \times 10^{-7} \cdot \text{SoC} + 6.246 \times 10^{-6} \tag{23}$$

We modified this linearized model within our model based on using discarded batteries, which already have a different degradation process, possible different SoC states, and operating temperatures.

Unlike the calendar capacity fade, the decrease in the battery cell's capacity in operation is strongly dependent on the number of performed work cycles and their average depth of discharge. The operating parameters of the BESS affecting the rate of degradation of the batteries in operation are shown as a function in Equation (24).

$$\text{Degradation}_{\text{Cyclic}} = (\text{Number}_{\text{Cycle}}, \text{DoD}, \text{SoC}, T, \text{Chemistry}_{\text{Cell}}) \tag{24}$$

Concerning the function above, we decided to use the EFC discharge capacity retention methodology for NMC battery cells described by Yuliya Preger et al. [44] to determine the cyclic degradation's final contribution to the overall capacity fade.

Within our model, we introduced the assumption of maintaining a constant operating temperature of the BESS at 25 °C. This assumption can be considered valid, as the containers are equipped with a cooling system, air conditioning, and the C-rate coefficient value is set so that there is no additive warming of the battery.

We identified the average BESS charged level within the annual operation via simulated output SoC values and its probability distribution function shown in the following statistical analysis in Figure 18. We compared the most different years, 2018 and 2016, regarding the distribution of frequency deviations in the power grid. For the lifespan estimation, we calculated with an annual average SoC equal to 55%.

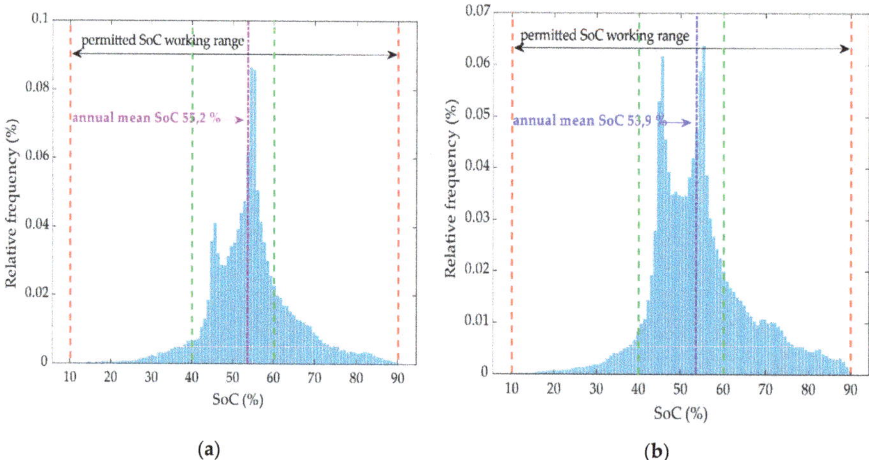

Figure 18. (**a**) Probability distribution function of the BESS SoC during 2016. (**b**) Probability distribution function of the BESS SoC during 2018.

We determined the DoD parameters to calculate the battery degradation based on Equation (25), where N = individual BESS FCR activations during the year, ΔP = power curve during one activation, t1 = start time of a given FCR activation request, and t2 = FCR activation request end time.

$$\text{DoD}_{\text{avg}}^{\text{annual}} = \frac{\frac{1}{C_{\text{installed}}^{\text{BESS}}} \cdot \sum_{n=1}^{N} \int_{t1}^{t2} \Delta P \, dt}{N} \tag{25}$$

The average discharge depth of the simulated battery storage ranged from 17% to 20% during the examined frequency historical datasets from the available years. We decided to calculate within the degradation model with an annual DoD_{avg} = 18%.

Figure 19 shows methodologies for determining the resulting BESS lifetime. According to the analyzed simulation outputs, the simulated repository was most often operated at the SoC level of 55%, considered the middle level. On the contrary, the average depth of the performed cycles did not exceed 20%. We consider this value to be a low cycling window. The EFC input value was selected as 460 cycles per year based on the data in Table 6.

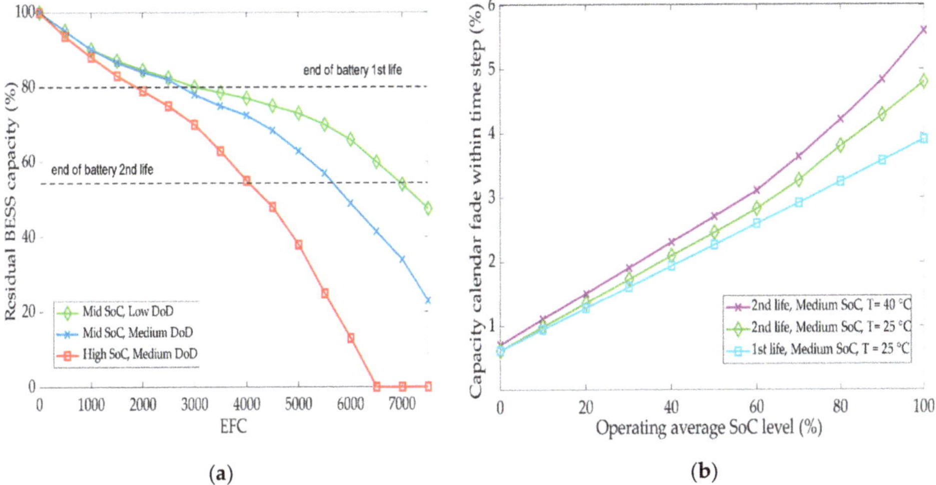

Figure 19. (a) The course of degradation of the battery capacity due to cycling dependent on the level of SoC and depth of discharge (DoD) and disposal of the battery when it reaches 55% of the original capacity. (b) Comparison of the calendar aging of a new battery and a secondary battery and the effects of the operating temperature within a time step of 10 years.

With an average of 460 EFC cycles/year, assuming an average operating temperature of 25 °C, an average operation SoC level 55%, and a DoD < 20%, the BESS lifetime is estimated at 8.7 years in terms of cyclic capacity degradation. The contribution of calendar degradation is, according to the assumed constant operating parameters for 10 years for the second life of the battery, according to Figure 19b, a 2.35% decrease of the initial residual capacity after decommissioning from EVs (80% of the original capacity). Our model estimates an effect of calendar degradation of 2% because of our time step <10 years, which, according to the capacity retention methodology, corresponds to 320 EFC. The lifetime of the modeled high-capacity secondary battery storage is determined as follows:

$$\text{Lifespan}_{\text{BESS}} = \frac{\text{Cycle life}_{\text{SoC 55 \%}}^{\text{BESS}} - \text{Calendar fade}_{\text{time step}}^{\text{EFC}}}{\text{EFC}_{\text{year avg}}} = \frac{4000 - 320}{460} = 8 \text{ years.} \tag{26}$$

To calculate the investment's economic efficiency, the life of battery cells in the repository of eight years is considered in the following sections.

We can consider this secondary battery cell lifespan under the established assumption that, in the future, there will be no massive changes in the behavior of the power system or a massive change in the distribution and frequency of frequency deviations in the grid.

5. Economic BESS Model

5.1. Electricity Market Prices Analysis

Electricity prices purchased or sold within the additional BESS charge level correction strategy were determined based on a historical analysis of power electricity prices on the Czech Republic's short-term power markets, operated by the Czech electricity and gas market operator (OTE a.s.). Figure 20 represents the historical development of the price on the internal market in the electricity market from 1 January 2017 to 31 March 2019.

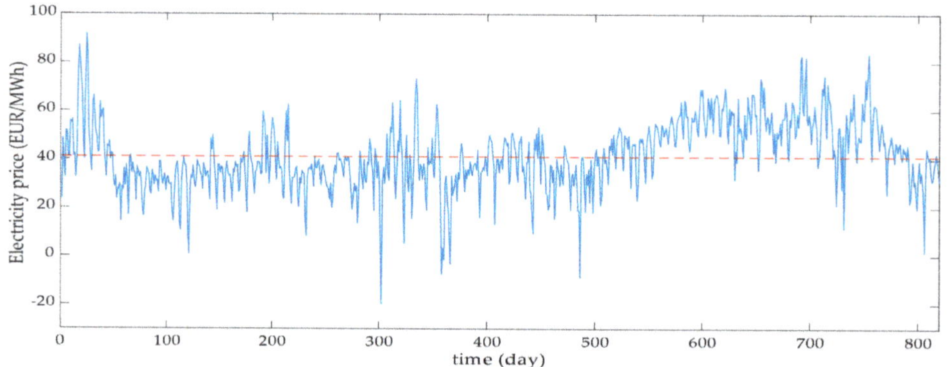

Figure 20. Historical day weighted electricity prices on the spot intraday power market.

For a correct economic evaluation, and to prevent an overestimation of the return on the investment, we decided, based on the economic precautionary principle, that the following prices of sold and purchased electricity listed in the Table 7 will be used in the economic efficiency calculations.

Table 7. Short-term market electricity prices used in the economic model.

Input Parameter	Value
Price of sold electricity (P_{sold})	35 EUR/MWh
Price of bought electricity (P_{bought})	50 EUR/MWh
Annual growth in electricity prices	2%

The primary source of the BESS annual revenues will be the payment for the provided power unit reserve within the FCP service. An analysis of the historical development of the average weekly hour FCR payments within the FCR interconnected market closely described in Section 3.4. Frequency Containment Reserves Market in the period 1.1.2017 to 1.7.2019 was performed.

The development of FCR prices in Figure 21 shows a long-term decline in payments. This situation was mainly due to the interconnection of individual FCR markets and increased competition between FCP service providers. No further decline is expected in FCR payments, and, on the contrary, an increase is expected due to the shutdown of conventional energy sources. For the relevant economic evaluation, we propose three different scenarios for the future development of payments for provided FCR. In all scenarios, the turning point is 2024, when the new emission limits for large combustion plants from the EU BAT BREF LCP document [47] come into force.

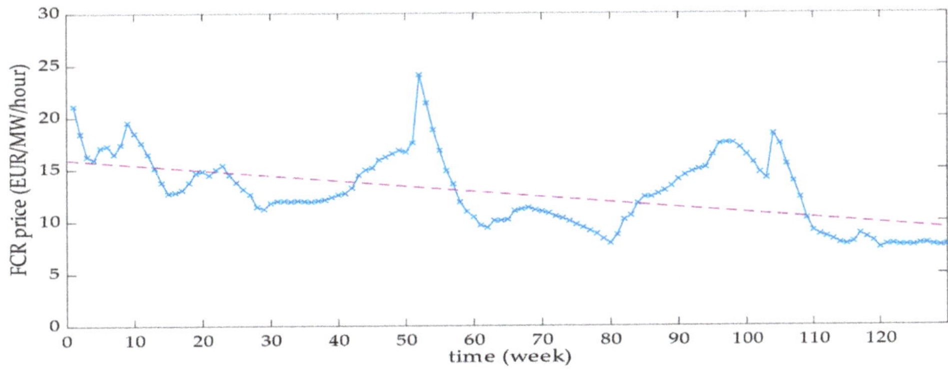

Figure 21. Power reserve payment on the FCR cooperation interconnected market.

- FCP 1 scenario: Optimal increase in the number of LSS in the Czech Republic, sufficient replacement of decommissioned combustion energy production capacities, and equalization of prices within the interconnected FCR market.
- FCP 2 scenario: Only a slight increase in the number of LSS in the Czech Republic, incomplete replacement of the combustion decommissioned energy production capacities, a slight shortage of FCR service providers after 2024, and FCR payment increasing slightly.
- FCP 3 scenario: Low increase in the number of LSS in the Czech Republic, insufficient replacement of decommissioned combustion energy production capacities, a significant shortage of FCR service providers after 2024, and FCR payment increases.

The values of payments for the provided reserves in the Czech Republic within the individual proposed scenarios are given in Appendix A.

5.2. Price Estimation of Second-Life Batteries

To determine the price of discarded battery cells, we estimated the battery cell's expected life cycle. We identified the main parameters influencing the secondary batteries' price in the emerging market with second-life batteries. We based the life cycle battery model shown in the Figure 22 on the principle of circular economy proclaimed by the European Commission.

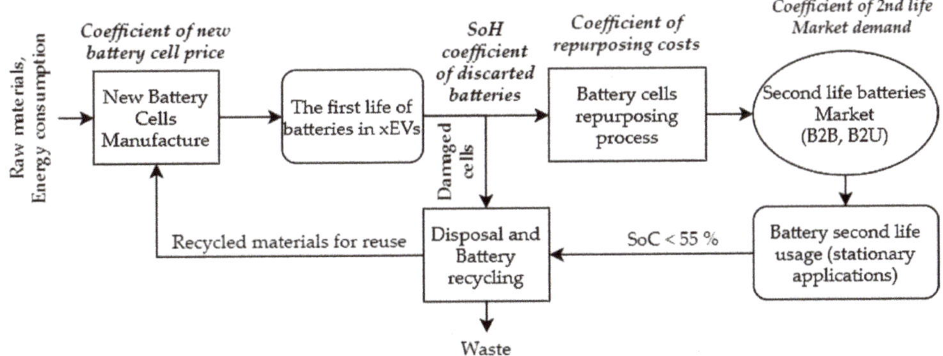

Figure 22. Battery cell circular life cycle. xEVs: electric vehicles.

Based on the model of the extended life cycle of the battery cell, we determined the price of discarded battery capacities according to the following equation:

$$\text{Price}_{2\text{nd life},t}^{\text{Battery}} = \text{Price}_{\text{new},t}^{\text{Battery}} \cdot C_{\text{SoH}} \cdot C_{\text{Repurposing},t} \cdot C_{\text{Market demand},t} \tag{27}$$

$C_{\text{Repurposing}}$ represents the repurposing process's costs, including material costs, wage costs, energy costs, transport costs, costs of production premises, and production equipment, etc. Since the idea of second-life batteries is at the very beginning, we assume the repurpose process's costs in full so that the $C_{\text{Repurposing}}$ coefficient will be equal to 1.

In the refurbishment factory, the measurements determine the "health" condition of discarded batteries represented by C_{SoH}. This coefficient can be described by the relationship given in Equation (28). For the discarded batteries used in our model, we decided on the values of residual resistance and capacity based on the study Lifetime Analyses of Li-Ion Batteries by P. Keil et al. [30].

$$SoH = \frac{Capacity_{residual}}{Capacity_{initial}} \cdot \frac{Resistance_{initial}}{Resistance_{increased}} = \frac{0.8}{1} \cdot \frac{1}{1.25} = 0.64 \tag{28}$$

The last coefficient identified in Equation (27) is the $C_{\text{Market demand}}$, which represents the demand for secondary battery cells. This coefficient is a prerequisite for future growth, as the popularity and portfolio of applications where second-life batteries can be used will grow. For our economic model, we set the value of this coefficient at 0.4.

We determined the price of new NMC battery cells according to general knowledge about battery cost breakdown and the study's results. The cost of modeling of lithium-ion battery cells for automotive applications [48], as in this paper, deal exclusively with batteries originating from electric vehicles.

The current price of NMC battery technology on the market is around 200 EUR/kWh. After applying our coefficients representing the complexity and cost of the process and battery life cycle, we considered the price of discarded battery capacities of 55 EUR/kWh for the economic model.

5.3. Economic Efficiency Evaluation

The economic evaluation of our proposed battery storage was performed through the indicators of economic efficiency net present currency (NPV), internal rate of return (IRR), and discounted payback period (DPP). In all these methods, the time value of money was respected.

The NPV indicator must have a value greater than or equal to 0 for economically profitable investments. If the investment's NPV value is equal to 0, the investment met our financial requirements precisely in the amount of the required rate of return (r).

$$NPV_{BESS} = -Investment + \sum_{t=1}^{T} \frac{CF_t}{(1+r)^t} \tag{29}$$

The IRR value should be equal to or higher than the ranked rate of return for economically profitable projects and investments. If the IRR is less than r, then the investment did not meet our economic expectations.

$$0 = -Investment + \sum_{t=1}^{T} \frac{CF_t}{(1+\textbf{IRR})^t} \tag{30}$$

The discounted payback period indicates the period during which the investment income will occur in the total amount of the initial investment, respecting the time value of money.

$$0 = -Investment + \sum_{t=1}^{DPP} \frac{CF_t}{(1+r)^t} \tag{31}$$

To calculate these indicators, it is necessary to calculate the cash flows (CF) resulting from the BESS operations for each year of life. The basic equation for calculating cash flows is given below:

$$CF_t = (Total\ Incomes - Total\ Expenditures)_t \tag{32}$$

5.3.1. Initial Investment

The amount of the initial investment of the energy storage system includes all expenses before commissioning. A more detailed overview of the individual items of the initial investment is given in Table 8.

Table 8. Assessment of the BESS initial investment (authors own calculations).

BESS Component	Unit Price	Total Amount	Investment Cost
2nd-life battery capacity	55 EUR/kWh	10 MWh	550,000 EUR
Main Storage components	335 ths. EUR/MW	5 MW	1,675,000 EUR
Connection of containers and DS 22 kV	77 ths. EUR	1	77,000 EUR
Energy licenses and fees	8000 EUR	1	8000 EUR
One-time fee for connecting installed power to the DS 22 kV	24.6 ths. EUR/MW	5 MW	123,000 EUR
Certification of power unit for ancillary service	7.5 ths. EUR	1	7500 EUR
State authorization of the power unit	3.25 ths. EUR	1	3250 EUR
Total BESS Initial Investment			**2,443,750 EUR**

The most expensive items of the initial investment are the main technical components of the container storage. The main components included in this monetary amount are used transformers of a voltage level 1 kV/22 kV; four-quadrant converters with a nominal power of 1 MVA; control systems (battery management system (BMS), EMS and thermal management system (TMS)); control software; high-voltage and low-voltage cabling; protections and switchboards; air conditioning; and safety systems.

5.3.2. Economic Lifetime Model

Based on the principles of degradation and operating conditions, the battery cell's technical service life was calculated to be eight years. Due to the technical lifetimes of the other BESS components (transformer and inverter have lifespans longer than 15 years), we consider at the end of the eighth year of operation a complete replacement of battery capacities and extension of the total economic lifespan to 16 years. The second-life batteries will again be used to keep reinvestment costs as low as possible.

The most fundamental point of the proposed economic model is the correct determination of the required rate of return (RoR) on the investment. At present, the RoR value of investments in modern decentralized energy and RES is around 7%. As the BESS are still considered more of pilot projects and their construction still carries certain increased risks (unclear distribution fees and incomplete legislative framework), we considered an r value of 10%.

The Table 9 represents the set values of the parameters entering the economic model. For simplicity, their values are constant throughout the evaluation of the economic efficiency. The values of the considered volumes of purchased and sold electricity per year for the SoC BESS correction are based on the simulated output average values from Table 6.

Table 9. Overall economic model parametrization.

Parameter	Value
Lifetime for economic evaluation	16 years
The required rate of return on investment	10%
Annual electricity sold on intraday market (IDM)	130 MWh
Annual electricity bought on intraday market (IDM)	30 MWh
Income tax rate	19%
BESS linear depreciation period	10 years
The annual value of tax depreciation	244,375 EUR
Replacement of the entire storage battery capacity	In the 8th year
Nominal annual price growth	2%
Average annual BESS readiness for FCP *	98%
BESS out of SoC, not providing contracted FCR	60 h/year
Penalties for non-delivery of FCR contracted	3 * FCR payment
Annual operating expenses (percentage of the investment)	2%
BESS annual insurance (percentage of investment)	0.5%
Replace defective batteries every three years	200 kWh
BESS decommissioning costs (percentage of investment)	3%

* Including disconnection of the BESS due to service and maintenance.

Based on all previous performed analyses, assumptions, and outputs of the annual MATLAB BESS simulations, we calculated the total annual revenues and expenditures for the maximum power FCR reserve of 5 MW based on the following proposed methodology described by Equations (33) and (34). These equations represent annual revenues and expenses resulting from the BESS operations and the provision of the network support service.

$$\text{Total Income}_t = \text{Payment}_{\text{FCR}} \cdot \text{FCR}_{\text{Reserve}}^{\text{BESS}} \cdot \text{Hours}_{\text{FCP,contracted}}^{\text{BESS}} + \text{Energy}_{\text{sold}}^{\text{annual}} \cdot \text{Price}_{\text{Sold}}^{\text{Electrictiy}} \tag{33}$$

$$\text{Total Expenditure}_t = \text{OPEX}_t + \text{Energy}_{\text{bought}}^{\text{annual}} \cdot \text{Price}_{\text{Bought}}^{\text{Electricity}} + \text{Penalties} \cdot \text{Hours}_{\text{FCP,unavailabe}}^{\text{BESS}} + \text{Tax}_t \tag{34}$$

5.3.3. Proposed Future Scenarios

Our study proposed three scenarios representing different future possible developments on the FCR interconnected EU market and on the short-term power electricity markets in the grid frequency stability and the differing success of the BESS in FCR tenders. Within these scenarios, we can investigate in more depth whether the BESS investment is economically rentable or not.

- Optimistic scenario: Annual average provided FCR in the full amount of 5 MW, 100% annual success in tenders on the FCR market, the complexity of frequency regulation in the network remains the same, constant growth of the power electricity prices 2% per year, and increase in traded volume of corrective electricity 0.5%.

- Realistic scenario: Since 2024, the average annual decline of the provided FCR by 1.5%, the annual success rate of tenders decreasing by 1% per year, a slight increase in the need to control frequency deviations in the network due to the integration of RES, an increase in traded volume of the corrective electricity 2% due to a higher requirement to maintain an optimal SoC, and a constant growth of power electricity prices 3% per year.

- Pessimistic scenario: Decrease in the success of won tenders 5% per year until 2024, and from the next year, a decrease of 2% per year until the end of the economic life, average annual decline of the provided FCR by 2.5%, high increase in the need to manage the network stability due to the integration of RES and an increase in the traded volume of corrective electricity of 3%, constant growth of the power electricity prices 3% per year until the year 2024, and since the next year, the increase will be 4% by the economic end of the project life.

Within each of these main scenarios, we evaluate the economic efficiency of the BESS operation with the proposed scenarios for the development of payments for the FCR reserve services defined in Section 5.1. Electricity market price analysis. With these combinations of the proposed scenarios, we get a 3 × 3 matrix of possible future developments for the high reliability of the obtained results.

6. Results and Discussion

The Table 10 summarizes the results for all evaluated combinations of the three main and three FCP scenarios. Based on these results, we can conclude that all possible combinations have a positive internal rate of return. This means that all are able to cover the initial investment and generate an additional return on the investment. With 10% of the required return on investment (discount rate), three out of nine scenarios showed negative results and would be rejected by potential investors, and the remaining six generated additional revenue above the required discount rate.

Table 10. Economic results of the proposed scenarios. NPV: net present currency, IRR: internal rate of return, and DPP: discounted payback period.

	Optimistic Main Scenario		
FCP Scenario	**NPV (EUR)**	**IRR (%)**	**DPP (year)**
FCP 1	280,000	12%	13
FCP 2	928,000	16%	9
FCP 3	2,416,000	25%	5
	Realistic Main Scenario		
FCP Scenario	**NPV (EUR)**	**IRR (%)**	**DPP (year)**
FCP 1	−219,000	8%	-
FCP 2	291,000	12%	11.5
FCP 3	1,584,000	21%	5.5
	Pessimistic Main Scenario		
FCP Scenario	**NPV (EUR)**	**IRR (%)**	**DPP (year)**
FCP 1	−618,000	2%	-
FCP 2	−276,000	7%	-
FCP 3	640,000	16%	6.5

However, the main problem in terms of setting a discount rate to 10% is the amount of risk connected with the bidding strategy and short-term contracts. Some investors may ask for significantly higher values, especially at the first phases of their entry to the new harmonized short-term ancillary market. Therefore, there is an urgent need for the further discussion of involved stakeholders how to reasonably minimize investors' risk with the help of legal framework so that there is no need for direct financial support caused by higher requested discount rates.

The usage of second-life batteries also significantly reduces the amount of initial investment costs compared to entirely new battery storage systems, and if we compare the technical parameters, storage systems based on discarded batteries deliver the complete same services comparable to new battery systems. Even from the point of view of operational safety, discarded battery systems do not pose any increased risk. After passing their initial measurement and inspection, they meet all safety requirements for the operation of battery systems.

The obtained results for the Czech case are fully transferable to all EU countries with interconnected electricity markets and ancillary services. The economic effectiveness of second-life batteries will be almost the same across all these countries. The reasons supporting this statement are:

(1)	The market for second-life batteries will be very likely a joined one within the whole EU, and therefore, the price for 1 MWh of a discarded battery will vary only to a minimal extent. However, there is also an open question as to what extent the manufacturers of batteries (or big car-producing companies) will control the market using specific and very likely closed softwares that allow cars to properly charge and control their batteries.

(2) The prices for the provision of ancillary services will be similar across all interconnected countries in the EU, copying the price development of electricity (power) prices. The reason for such predicted price unification lies in the new harmonized regulations for the ancillary services market, such as the European Balancing Guidelines (EBGL), System Operation Guideline SOGL, and Clean Energy Package CEB.

(3) The price of electricity, as one of the main operation expenses, is almost always similar thanks to interconnected and harmonized power markets in the EU. The only time when the electricity price may differ is the case of insufficient cross-border transmission capacities, which leads to a break-up of the one unified price area.

The only main factor that can significantly a change in the final economic efficiency is direct and indirect financial support of the BESS operation. This can then indirectly affect the whole market for cross-border shared ancillary services. For this reason, there should be an effort to avoid significantly different support across individual EU countries.

To find the limit values of the economic model's main input parameters to maintain the BESS investment's economic rentability, a set of two-parameter sensitivity analyses of the NPV are proposed. Assuming the uncertain development of the available power capacity and the number of future balancing service providers in the emerging interconnected market for ancillary network services, the following parameter combinations were examined: the amount of payment for the provided FCR reserve, average annual provided FCR, and number of won hour tenders.

To examine the effects of the changes in the values of these parameters on the resulting NPV, the other input parameters of the model were set to the basic values according to Table 9, the main realistic scenario.

In the Table 11, the green line indicates the limit combinations of the examined parameters for maintaining the economically profitable operation of the BESS in the provision of FCP service. These combinations are valid and relevant for the economic evaluation when considering the predicted FCP 2 price development scenario, the values of which are given in Appendix A. The coloring of Tables 11 and 12 highlights the BESS project's economic rentability regarding the input parameters combination. The green color represents the state when BESS's operation is economically efficient and the red color represents the state when BESS operation is non-profitable.

Table 11. Sensitivity analysis of the FCR power reserve and number of hours successful in the tender.

	ths. EUR	Average Annual Provided FCR within Won FCP Tenders (MW)						
		5	4.75	4.5	4.25	4	3.75	3.5
Annual hours of FCP service provision	8500	941	755	570	384	199	13	−172
	8000	725	550	375	200	26	−149	−324
	7500	508	344	181	17	−147	−311	−475
	7000	292	139	−14	−167	−320	−473	−626
	6500	76	−66	−209	−351	−493	−635	−778
	6000	−140	−272	−403	−535	−666	−798	−929
	5500	−357	−477	−598	−719	−839	−960	−1080

Table 12. Sensitivity analysis of the annual hour tenders won and average annual FCR payments.

		Annual Hours of FCP Service Provision (Tenders Won)						
	ths. EUR	8500	8000	7500	7000	6500	6000	5500
Average annual FCR payment (EUR/MW/hour)	20	2506	2189	1872	1555	1238	921	604
	19	2237	1935	1634	1333	1032	731	430
	18	1967	1682	1397	1111	826	541	256
	17	1698	1428	1159	890	620	351	81
	16	1428	1175	921	668	414	161	−93
	15	1159	921	683	446	208	−30	−267
	14	890	668	446	224	4	−220	−442
	13	620	414	208	2	−204	−410	−616
	12	351	161	−30	−220	−410	−600	−790
	11	81	−93	−267	−442	−616	−790	−965
	10	−188	−347	−505	−663	−822	−980	−1139
	9	−457	−600	−743	−885	−1028	−1171	−1313
	8	−727	−854	−980	−1107	−1234	−1361	−1488
	7	−996	−1107	−1218	−1329	−1440	−1551	−1662

The second two-parameter sensitivity analysis performed aimed to determine the amount of FCR payment with a variable volume of won hourly annual tenders is presented in Table 12. The provided power reserve is 5 MW and is constant throughout the economic evaluation.

Through the performed sensitivity analyses, it can be stated that maintaining the economy of the BESS operation with the required return of 10% is maintained even with relatively significant decreases in the examined key operating economic parameters. The predictability of future values of the examined parameters during real operations is very limited, and it is necessary to take into account very variable differences in prices for the provided power reserves for FCP during the individual tender periods (days to hours).

The BESS operators can defend themselves against sudden price changes in the FCP market by changing the bidding strategy or by allocating the installed capacity for different services on separate units in the event of a decrease in the amount of reserves. All these facts support the future motivation of investors to acquire a BESS, leading to their widespread increase in popularity.

Crucial to the future emergence and development of the market for discarded battery cells will be the position taken by electric vehicle manufacturers in this resulting battery value chain. The question is whether vehicle manufacturers will be interested in these discarded capacities or will the discarded capacities be collected by newly established companies that will directly specialize in the secondary use of batteries and the repurposing process. However, car manufactures are increasingly aware of the growing value of discarded batteries, which is influenced by the expected high participation of storage technologies in the electricity market.

There is, therefore, a high presumption that individual vehicle manufacturers will have a policy for the disposal of discarded batteries and will, therefore, use or redistribute them themselves after decommissioning from electric vehicles. A possible variant also remains that the battery modules will be owned by the car manufacturer throughout the life of the customer's electric vehicles, which will replace the full capacity for a lower fee at the end of life.

7. Conclusions

Battery cells discarded from the electromobility sector still offer sufficient technical and safety parameters for further usage. They can find their utilization in a wide range of applications, especially in stationary energy storage, where the ratio of power to weight is not a crucial parameter, as it is in the case of EVs. This fulfils the idea of a circular EU economy to save the primary energy resources needed in production and to achieve a careful use of natural resources and valuable raw materials.

Such utilization also contributes to the further development of renewable energy generation from intermittent sources, such as solar and wind. However, a further increase of such renewable sources will also create an increased demand for fast and reliable balancing services that are currently provided primarily by fossil fuel-based power plants and that can be effectively substituted by sufficient and economically available energy storage.

Battery systems can also significantly help with charging infrastructure development for the expected rapid increase in electromobility, especially in densely populated agglomerations. In these agglomerations, currently used power line transmission capacities may reach limit values. Thus, the BESS working in parallel with charging stations can relieve the high required power for charging purposes, delay the necessary investment in strengthening the line's transmission capacity, and minimize the risk of overload due to the impact of fast charging.

From an economic point of view, battery systems based on discarded batteries appear to be viable for the provision of FCP without direct financial support, even in scenarios reflecting relatively unfavorable price developments.

The proposed and presented procedure for the optimization and evaluation of battery systems can also be fully used (when resetting several input parameters) for the assessment of other types of stationary systems, such as charging hubs for e-mobility. Discarded batteries could thus significantly support the development of the charging infrastructure, with all the resulting benefits.

The proposed general methodology was verified on the Czech Republic case; however, the obtained results are fully transferable to all EU countries, thanks to interconnected electrical grids and, also, harmonized markets for power and ancillary services. The proposed methodology approach can also be easily adjusted to a vast number of similar tasks that evaluate the technical and economic feasibility of large battery systems. To conclude, the discarded batteries from the dynamically developing electromobility sector have high potential for their second lives. There are possible applications that can fully utilize discarded batteries with promising financial results. These utilizations also indirectly contribute to achieving the EU's climate goals towards carbon neutrality in 2050 and energy independence.

Author Contributions: Conceptualization, L.J., T.K., J.K.; Data curation, L.J.; Investigation, L.J.; Methodology, L.J.; Software, L.J.; Supervision, T.K. and J.K.; Visualization, L.J.; Writing—original draft, L.J.; Writing—review & editing, T.K. and J.K. All authors have read and agreed to the published version of the manuscript.

Funding: This research received no external funding.

Conflicts of Interest: The authors declare that they have no known competing financial interests or personal relationships that could have appeared to influence the work reported in this paper.

Abbreviations

BESS	Battery Energy System Storage
BEV	Battery Electric Vehicle
BMS	Battery Management System
BSP	Balancing Service Provider
CF	Cash Flow
CO_2	Carbon Dioxide
DB	Dead Band
DER	Decentralized Energy Resources
DoD	Depth of Discharge
DoF	Degrees of Freedom
DPP	Discounted Payback Period
EBEM	European Balancing Energy Market
EBGL	European Balancing Guideline
EFC	Equivalent Full Cycles
EMS	Energy Management System
ENTSOE	European Network of Transmission System Operators for Electricity

EU	European Union
EV	Electric Vehicle
FCP	Frequency Containment Process
FCR	Frequency Containment Reserve
FRR	Frequency Restoration Reserve
GHG	Greenhouse Gas
IEM	Intraday Energy Market
IRR	Internal Rate of Return
LCO	Lithium Cobalt Oxide
LFC	Load Frequency Control
LFP	Lithium Iron Phosphate
LSS	Large Storage System
Mtoe	Millions of Tonnes of Oil Equivalent
NCA	Nickel Cobalt Aluminum
NMC	Lithium Nickel Manganese Cobalt
NPV	Net Present Value
PHEV	Plug-in Hybrid Electric Vehicle
RE	Regulation Energy
RoR	Required Rate of Return
SLB	Second Life Batteries
SoC	State of Charge
SoH	State of Health
TMS	Thermal Management System
TSO	Transmission System Operator
UCTE	Union for the Coordination of Transmission of Electricity

Appendix A

Table A1. FCR payment within the proposed future scenarios (EUR/MW/hour).

Year	FCP 1	FCP 2	FCP 3
2020	11.5	11.5	11.5
2021	12.7	12.8	13.9
2022	9.1	9.3	15.1
2023	8.5	9.7	16.2
2024	8.7	10.2	17.2
2025	8.9	11.2	17.6
2026	9.1	12.1	17.9
2027	9.2	13	18.3
2028	9.4	13.9	18.7
2029	9.6	14.3	19
2030	9.8	14.6	19.4
2031	10	14.9	19.7
2032	10.2	15.2	20.1
2033	10.4	15.5	20.5
2034	10.6	15.8	20.8
2035	10.8	16.1	21.2

References

1. Miyamoto, M.; Takeuchi, K. Climate agreement and technology diffusion: Impact of the Kyoto Protocol on international patent applications for renewable energy technologies. *Energy Policy* **2019**, *129*, 1331–1338. [CrossRef]
2. Oberthür, S.; Kelly, C.R. EU leadership in international climate policy: Achievements and challenges? *Int. Spect.* **2008**, *43*, 35–50. [CrossRef]
3. Gouardères, F. Energy Policy. General Principles. 2018. Available online: https://www.europarl.europa.eu/factsheets/en/sheet/68/energy-policy-general-principles (accessed on 25 September 2019).

4. European Commission. A European Strategy for Low-Emission Mobility—Fact Sheet. 2016. Available online: https://eur-lex.europa.eu/resource.html?uri=cellar:e44d3c21-531e-11e6-89bd-01aa75ed71a1.0002.02/DOC_1&format=PDF (accessed on 20 July 2016).

5. European Environment Agency. GHG Emissions by Sector in the EU-28, 1990–2016. Available online: https://www.eea.europa.eu/data-and-maps/daviz/ghg-emissions-by-sector-in#tab-chart_1 (accessed on 31 May 2018).

6. European Environment Agency. *Greenhouse Gas Emissions from Transport in Europe*; European Environment Agency, 2019. Available online: https://www.eea.europa.eu/data-and-maps/indicators/transport-emissions-of-greenhouse-gases/transport-emissions-of-greenhouse-gases-12 (accessed on 15 February 2019).

7. Statharas, S.; Moysoglou, Y.; Siskos, P.; Zazias, G.; Capros, P. Factors Influencing Electric Vehicle Penetration in the EU by 2030: A model-based policy assessment. *Energies* **2019**, *12*, 2379. [CrossRef]

8. European Commission. Regulation (EU) 2019/631 of the European Parliament and of the Council of 17 April 2019 setting CO_2 emission performance standards for new passenger cars and for new light commercial vehicles. *Off. J. Eur. Union* **2020**, *8*, 2–7.

9. International Energy Agency (IEA). Global EV Outlook 2020: Entering the decade of electric drive? *Glob. EV Outlook* **2020**, *2*, 273.

10. Pietrzak, K.; Pietrzak, O. Environmental effects of electromobility in a sustainable urban public transport. *Sustainability* **2020**, *12*, 1052. [CrossRef]

11. Tucki, K.; Orynycz, O.; Świć, A.; Mitoraj-Wojtanek, M. The Development of Electromobility in Poland and EU States as a Tool for Management of CO_2 Emissions. *Energies* **2019**, *12*, 2942. [CrossRef]

12. Wang, L.; Wang, X.; Yang, W. Optimal design of electric vehicle battery recycling network—From the perspective of electric vehicle manufacturers. *Appl. Energy* **2020**, *275*, 115328. [CrossRef]

13. Donald, C.; Emma, E.; Shriram, S. Automotive Lithium-ion Cell Manufacturing: Regional Cost Structures and Supply Chain Considerations. *Joule* **2016**, *1*, 229–243. [CrossRef]

14. Martinez-Laserna, E.; Gandiaga, I.; Sarasketa-Zabala, E.; Badeda, J.; Stroe, D.-I.; Swierczynski, M.; Goikoetxea, A. Battery second life: Hype, hope or reality? A critical review of the state of the art. *Renew. Sustain. Energy Rev.* **2018**, *93*, 701–718. [CrossRef]

15. Richter, S.; Rehme, M.; Temmler, A. Second-Life Battery Applications-Market potentials and contribution to the cost effectiveness of electric vehicles. In Proceedings of the 5th Conference on Future Automotive Technology (CoFAT), München, Germany, 4 May 2016; pp. 2017–2018. [CrossRef]

16. Canals, L.C.; Amante García, B.; González Benítez, M.M. Amante García, B.; González Benítez, M.M. A Cost Analysis of Electric Vehicle Batteries Second Life Businesses. In *Project Management and Engineering Research*; Springer: Berlin/Heidelberg, Germany, 2016; pp. 129–141. [CrossRef]

17. Canals, L.C.; Amante García, B.; Cremades, L.V. Electric vehicle battery reuse: Preparing for a second life. *J. Ind. Eng. Manag.* **2017**, *10*, 266–285. [CrossRef]

18. Casals, L.C.; García, B.A.; Aguesse, F.; Iturrondobeitia, A. Second life of electric vehicle batteries: Relation between materials degradation and environmental impact. *Int. J. Life Cycle Assess.* **2017**, *22*, 82–93. [CrossRef]

19. Reid, G.; Julve, J. Second Life-Batteries as Flexible Storage for Renewables Energies. *Bundesverb. Erneuerbare Energ. E.V.* **2016**, *1*, 46.

20. Bobba, S.; Podias, A. Sustainability Assessment of Second Life Application of Automotive Batteries (SASLAB). *JRC Explor. Res. (2016–2017)* **2018**. [CrossRef]

21. Garozzo, D.; Tina, G.M. Evaluation of the Effective Active Power Reserve for Fast Frequency Response of PV with BESS Inverters Considering Reactive Power Control. *Energies* **2020**, *13*, 3437. [CrossRef]

22. Nordic Analysis Group. Review of the Frequency Containment Process (FCP)—FCR-D Design Executive Summary Nordic Analysis Group (NAG). 2019. Available online: https://www.statnett.no/globalassets/for-aktorer-i-kraftsystemet/utvikling-av-kraftsystemet/review-of-the-frequency-containment-process.pdf (accessed on 17 May 2019).

23. Figgener, J.; Stenzel, P.; Kairies, K.-P.; Linßen, J.; Haberschusz, D.; Wessels, O.; Angenendt, G.; Robinius, M.; Stolten, D.; Sauer, D.U. The development of stationary battery storage systems in Germany—A market review. *J. Energy Storage* **2020**, *29*, 101153. [CrossRef]

24. Mathews, I.; Xu, B.; He, W.; Barreto, V.; Buonassisi, T.; Peters, I.M. Technoeconomic model of second-life batteries for utility-scale solar considering calendar and cycle aging. *Appl. Energy* **2020**, *269*, 115127. [CrossRef]

25. Neubauer, J.S.; Wood, E.; Pesaran, A. A Second Life for Electric Vehicle Batteries: Answering Questions on Battery Degradation and Value. *SAE Int. J. Mater. Manuf.* **2015**, *8*, 21–23. [CrossRef]

26. Saez-de-ibarra, A.; Martinez-laserna, E.; Stroe, D.; Swierczynski, M.; Rodriguez, P. Sizing Study of Second Life Li-ion Batteries for enhancing renewable energy grid integration. *IEEE Trans. Ind. Appl.* **2016**, *52*, 4999–5008. [CrossRef]

27. Canals Casals, L.; Barbero, M.; Corchero, C. Reused second life batteries for aggregated demand response services. *J. Clean. Prod.* **2019**, *212*, 99–108. [CrossRef]

28. Murnane, M.; Ghazel, A. A Closer Look at State of Charge (SOC) and State of Health (SOH) Estimation Techniques for Batteries. *Analog Devices* **2017**, *2*, 426–436.

29. Werner, D.; Paarmann, S.; Wiebelt, A.; Wetzel, T. Inhomogeneous temperature distribution affecting cyclic aging of Li-ion cells. Part ii: Analysis and correlation. *Batteries* **2020**, *6*, 12. [CrossRef]

30. Keil, P.; Schuster, S.F.; von Lueders, C.; Hesse, H.; Arunachala, R.; Jossen, A. Lifetime Analyses of Lithium-Ion EV Batteries. In Proceedings of the 3rd Electromobility Challenging Issues Conference (ECI), Singapore, 1–4 December 2015; pp. 1–22.

31. Fournié, L.; Andrey, C.; Hentschel, J.; Wilkinson, G. Integration of electricity balancing markets and regional procurement of balancing reserves. *Energy* **2016**, *4*, 18–24. [CrossRef]

32. European Network of Transmission System Operators for Electricity (ENTSO-E). An Overview of the European Balancing Market in Europe. *Balancing* **2018**, *15*, 1–16. Available online: https://www.entsoe.eu/news/2018/12/12/electricity-balancing-in-europe-entso-e-releases-an-overvi ew-of-the-european-electricity-balancing-market-and-guideline/ (accessed on 23 May 2019).

33. Cauret, L.; Belhomme, R.; Raux-Defossez, P. Benchmark of markets and regulations for electricity, gas and heat and overview of flexibility services to the electricity grid D3.1—Benchmark of markets and regulations for electricity, gas and heat and overview of flexibility services to the elec. *Magnitude* **2020**, *46*, 53–150.

34. The European Commission. System Operations Guideline 2017. *Off. J. Eur. Union* **2017**, *220*, 2017.

35. Tenne, T. End Report FCR Pilot. 2018, pp. 1–35. Available online: https://www.tennet.eu/electricity-market/a ncillary-services/fcr-documents/ (accessed on 15 April 2019).

36. Rominger, J.; Ludwig, P.; Kern, F.; Loesch, M.; Schmeck, H. Utilization of local flexibility for charge management of a battery energy storage system providing frequency containment reserve. *Energy Procedia* **2018**, *155*, 443–453. [CrossRef]

37. Zeh, A.; Müller, M.; Naumann, M.; Hesse, H.C.; Jossen, A.; Witzmann, R. Fundamentals of using battery energy storage systems to provide primary control reserves in germany. *Batteries* **2016**, *2*, 29. [CrossRef]

38. ENTSO-E. Frequency Containment Reserves (FCR). Web Network Codes. 2018. Available online: https://www.entsoe.eu/network_codes/eb/fcr/ (accessed on 13 October 2019).

39. Hermiyanty, W.A.B.; Sinta, D. State of Lithium-ion battery. *J. Chem. Inf. Model.* **2017**, *8*, 1–58. [CrossRef]

40. Madani, S.S.; Schaltz, E.; Kær, S.K. Effect of current rate and prior cycling on the coulombic efficiency of a lithium-ion battery. *Batteries* **2019**, *5*, 57. [CrossRef]

41. Ovejas, V.J.; Cuadras, A. Effects of cycling on lithium-ion battery hysteresis and overvoltage. *Sci. Rep.* **2019**, *9*, 1–9. [CrossRef]

42. Li, T.; Yuan, X.-Z.; Zhang, L.; Song, D.; Shi, K.; Bock, C. Degradation Mechanisms and Mitigation Strategies of Nickel-Rich NMC-Based Lithium-Ion Batteries. *Electrochem. Energy Rev.* **2020**, *3*, 43–80. [CrossRef]

43. Deutschen, T.; Gasser, S.; Schaller, M.; Siehr, J. Modeling the self-discharge by voltage decay of a NMC/graphite lithium-ion cell. *J. Energy Storage* **2018**, *19*, 113–119. [CrossRef]

44. Preger, Y.; Barkholtz, H.M.; Fresquez, A.; Campbell, D.L.; Juba, D.W.; Romàn-Kustas, J.; Ferreira, S.R.; Chalamala, B. Degradation of Commercial Lithium-Ion Cells as a Function of Chemistry and Cycling Conditions. *J. Electrochem. Soc.* **2020**, *167*, 120532. [CrossRef]

45. Andrenacci, N.; Pede, G.; Chiodo, E.; Lauria, D.; Mottola, F. Tools for Life Cycle Estimation of Energy Storage System for Primary Frequency Reserve. In Proceedings of the 2018 International Symposium on Power Electronics, Electrical Drives, Automation and Motion (SPEEDAM), Amalfi, Italy, 20–22 June 2018; pp. 1008–1013. [CrossRef]

46. Martins, R.; Hesse, H.C.; Jungbauer, J.; Vorbuchner, T.; Musilek, P. Optimal component sizing for peak shaving in battery energy storage system for industrial applications. *Energies* **2018**, *11*, 2048. [CrossRef]

47. Lecomte, T.; Neuwahl, F.; Canova, M.; Pinasseau, A.; Jankov, I.; Brinkmann, T.; Roudier, S.; Sancho, L.D.; Neuwahl, J.F.F.d.l.F. *Best Available Techniques (BAT) Reference Document for Large Combustion Plants*; Publications Office of the European Union: Rue Mercier, Luxembourg, 2017. Available online: https://eippcb.jrc.ec.europa.eu/sites/default/files/2019-11/JRC_107769_LCPBref_2017.pdf (accessed on 3 September 2019).

48. Patry, G.; Romagny, A.; Martinet, S.; Froelich, D. Cost modeling of lithium-ion battery cells for automotive applications. *Energy Sci. Eng.* **2015**, *3*, 71–82. [CrossRef]

Publisher's Note: MDPI stays neutral with regard to jurisdictional claims in published maps and institutional affiliations.